Lecture Notes in Computer Science 9242

Commenced Publication in 1973
Founding and Former Series Editors:
Gerhard Goos, Juris Hartmanis, and Jan van Leeuwen

Editorial Board

More information about this series at http://www.springer.com/series/7412

Xiaofei He · Xinbo Gao
Yanning Zhang · Zhi-Hua Zhou
Zhi-Yong Liu · Baochuan Fu
Fuyuan Hu · Zhancheng Zhang (Eds.)

Intelligence Science and Big Data Engineering

Image and Video Data Engineering

5th International Conference, IScIDE 2015
Suzhou, China, June 14–16, 2015
Revised Selected Papers, Part I

 Springer

Editors
Xiaofei He
Zhejiang University
Hangzhou
China

Xinbo Gao
Xidian University
Xi'an
China

Yanning Zhang
Northwestern Polytechnical University
Xi'an
China

Zhi-Hua Zhou
Nanjing University
Nanjing
China

Zhi-Yong Liu
Chinese Academy of Sciences
Beijing
China

Baochuan Fu
Suzhou University of Science
    and Technology
Suzhou
China

Fuyuan Hu
Suzhou University of Science
    and Technology
Suzhou
China

Zhancheng Zhang
Suzhou University of Science
    and Technology
Suzhou
China

ISSN 0302-9743   ISSN 1611-3349   (electronic)
Lecture Notes in Computer Science
ISBN 978-3-319-23987-3   ISBN 978-3-319-23989-7   (eBook)
DOI 10.1007/978-3-319-23989-7

Library of Congress Control Number: 2015948171

LNCS Sublibrary: SL6 – Image Processing, Computer Vision, Pattern Recognition, and Graphics

Printed on acid-free paper

Springer International Publishing AG Switzerland is part of Springer Science+Business Media
(www.springer.com)

Preface

IScIDE 2015, the International Conference on Intelligence Science and Big Data Engineering, took place in Suzhou, China, June 14–16, 2015. As one of the annual events organized by the Chinese Golden Triangle ISIS (Information Science and Intelligence Science) Forum, this meeting was scheduled as the fifth in a series of annual meetings promoting the academic exchange of research on various areas of intelligence science and big data engineering in China and abroad. In response to the call for papers, a total of 416 papers were submitted from 14 countries and regions. Among them, 18 papers were selected for oral presentation, 32 for spotlight presentation, and 76 for poster presentation, yielding an acceptance rate of 30.3 % and an oral presentation rate of about 4.3 %. We would like to thank all the reviewers for spending their precious time on reviewing papers and for providing valuable comments that helped significantly in the paper selection process.

We would like to express special thanks to the Conference General Co-chairs, Yanning Zhang, Zhi-Hua Zhou, and Zhigang Chen, for their leadership, advice, and help on crucial matters concerning the conference. We would like to thank all Steering Committee members, Program Committee members, Invited Speakers' Committee members, Organizing Committee members, and Publication Committee members for their hard work. We would like to thank Prof. Lionel M. Ni, Prof. Deyi Li, and Prof. Lei Xu for delivering the keynote speeches, and Tony Jebara, Houqiang Li, Keqiu Li, Shutao Li, Bin Cui, and Xiaowu Chen for delivering the invited talks and sharing their insightful views on ISIS research issues. Finally, we would like to thank all the authors of the submitted papers, whether accepted or not, for their contribution to the high quality of this conference. We count on your continued support of the ISIS community in the future.

June 2015   Xiaofei He
                                                                                               Xinbo Gao

Organization

General Chairs

Yanning Zhang	Northwestern Polytechnical University, Xi'an, China
Zhi-Hua Zhou	Nanjing University, Nanjing, China
Zhigang Chen	Suzhou University of Science and Technology, Suzhou, China

Technical Program Committee Chairs

Xiaofei He	Zhejiang University, Hangzhou, China
Xinbo Gao	Xidian University, Xi'an, China

Local Arrangements Chair

Baochuan Fu	Suzhou University of Science and Technology, Suzhou, China

Local Arrangements Members

Fuyuan Hu	Suzhou University of Science and Technology, Suzhou, China
Xue-Feng Xi	Suzhou University of Science and Technology, Suzhou, China
Ze Li	Suzhou University of Science and Technology, Suzhou, China
Hongjie Wu	Suzhou University of Science and Technology, Suzhou, China

Publicity Chair

Zhi-Yong Liu	Chinese Academy of Sciences, Beijing, China

Program Committee Members

Deng Cai	Zhejiang University, Hangzhou, China
Fang Fang	Peking University, Beijing, China
Jufu Feng	Peking University, Beijing, China
Xinbo Gao	Xidian University, Xi'an, China
Xin Geng	Southeast University, Nanjing, China
Ziyu Guan	Northwest University, Xi'an, China
Xiaofei He	Zhejiang University, Hangzhou, China
Kalviainen Heikki	Lappeenranta University of Technology, Finland
Akira Hirose	The University of Tokyo, Japan
Dewen Hu	National University of Defense Technology, Changsha, China
Hiroyuki Iida	Japan Advanced Institute of Science and Technology, Japan
Zhong Jin	Nanjing University of Science and Technology, Nanjing

Contents – Part I

Contents – Part II

Fast and Accurate Text Detection in Natural Scene Images

Chengqiu Xiao$^{(\boxtimes)}$, Lixin Ji, Chao Gao, and Shaomei Li

The National Digital Switching System Engineering and Technological Research Center, Zheng Zhou 450001, China
xiao125c@gmail.com

Abstract. Repeating component filtering is problematic for the accuracy and speed of scene text detection. This paper proposes a fast and accurate method for detecting scene text. A novel MSER tree pruning algorithm is proposed to extract unique Maximally Stable Extremal Regions (MSERs) as character candidates. Two cues specially designed for capturing the intrinsic features of characters are integrated by a Bayesian classifier. Character candidates are grouped into text candidates by some characteristics between words and then they are verified with some efficient rules in a crossing line. Experimental results on the ICDAR 2011 Robust Reading Competition dataset demonstrate that the performance of our much simpler method is slightly lower than the state-of-the-art performance, however, the processing speed of this algorithm is at least four times faster.

Keywords: Maximally stable extremal region · Repeating components · Multi-cue integration

1 Introduction

Text is a class of informative feature which provides contextual cues to obtain and manage valuable information in natural scene images. Text detection is an important prerequisite for text information extraction. But it has been suffering from the factors, such as complex background, various fonts, sizes, colors, orientations and lightness, etc. All these factors make text within images a challenge task to be robustly located. Text detection has been considered in recent studies which can roughly be categorized into three groups: texture-based methods [1,2], connected component (CC)-based methods [3–5] and hybrid methods [6,7]. Texture-based methods consider text regions as a special texture which is different with non-text regions. First, features (including Histogram of Gradients (HOG), Local Binary Patterns (LBP), Gabor filters, wavelets and etc.) are extracted over a certain region and then machine learning methods or heuristics are used to train a classifier to separate text regions and non-text regions. Connected component (CC)-based methods first extract CCs from an image exploiting a specific method (e.g. MSER [4,5] or Stroke Width Transform (SWT) [3]) and then to

© Springer International Publishing Switzerland 2015
X. He et al. (Eds.): IScIDE 2015, Part I, LNCS 9242, pp. 1–10, 2015.
DOI: 10.1007/978-3-319-23989-7_1

estimate whether these CCs are characters or non-characters. Hybrid methods are a combination of two methods above. Usually, the initial step is to extract CCs which are potential characters. Then, non-character candidates and repeating components are pruned with classifiers or heuristic rules. This integration usually achieves better results than any single method.

MSER has been proved to be an effective detector to deal with scene-text detection. As the name implies, MSER is a special case of Extremal Region (ER) for its size almost maintaining stable over a range of thresholds. Although it has appeared several methods [8,9] abandoned this stability of ERs. But in contrast with traditional connected component features, MSER achieves detecting most characters in condition of low resolution, lightness and low contrast, etc. However, current MSER-based algorithms still suffer the pitfall: a large amount of repeating components (as shown in Fig. 1, which contains repeating characters and part of non-characters) and non-characters are left after MSER detection. To obtain better results, repeating components which consist of repeating characters and part of repeating non-characters should be firstly filtered. This problem has been studied by Chen et al. [10], Neumann and Matas [11] and Xucheng Yin [12], etc. Chen et al. [10] employed MSER as letter candidate and pruned out MSER pixels outside the boundary which were detected by canny edge detector [13]. Neumann and Matas [11] proposed a novel MSER++ as character candidate, which used rather complicated features and prunes it by exhaustive search. Xucheng Yin [12] proposed an MSER pruning algorithm and pruned most repeating components. But there is still much room to improve the pruning efficiency to acquire more accurate results.

Thus, in this paper, firstly, by exploring the hierarchical structure of MSERs, we adopt simple features such as spatial constraints, area and aspect ratios to design a novel MSER tree pruning algorithm to filter the repeating components. Repeating components are significantly reduced with a high accuracy. Secondly, we propose to use a Bayesian classifier to integrate multi-cues to estimate the posterior probabilities of character candidates and remove the vast majority of non-character candidates. Finally, single character candidates are grouped into text candidates by some characteristics between words and text candidates are verified with the properties and the number of character candidates on the crossing lines.

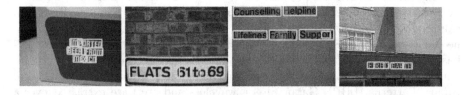

Fig. 1. The situation which contains detected repeating components

2 Scene Text Detection Algorithm

Figure 2 shows the flowchart of our algorithm. Firstly, The repeating components are excluded so that we can filter non-character candidates with a much higher accuracy. Then we filter non-character candidates to get pure characters. Finally, characters are grouped and verified into text lines.

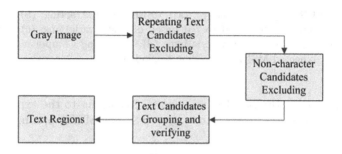

Fig. 2. The flowchart of our algorithm

2.1 Repeating Components Filtering Using MSER Tree Pruning

There are a large amount of repeating components and non-characters left after extracting MSERs from the original image and these repeating components are severe influencing factors for later candidate construction. Before excluding non-characters, we should first filter the repeating components which contain repeating characters and part of repeating components. Motivated by [12], we designed a MSER tree pruning method which have two steps: repeating components excluding and node accumulation. The proper nodes are chosen by a minimizing regularized variation strategy.

The variation of MSER is defined as follows. If C_i is an MSER, $S(C_i) = (C_i, C_{i+1}, \cdots, C_{i+\delta})$ (δ is the number of children nodes) represents the set of the branches of the root node C_i. Thus the variation of C_i is $v(C_i) = \frac{|C_{i+\delta} - C_i|}{|C_i|}$. Where C_i represents the number of pixels in C_i. When its variation is lower and more stable than its parent node C_{i-1} and child node C_{i+1}, then C_i is the MSER. When the MSER has lower variation and sharper boundary, it is more likely to be a character. Our excluding strategy is to choose the parent or child node which has the smallest variation. Let v and a represent variation and aspect ratio respectively. The aspect ratio of character candidate is between $[a_{min}, a_{max}]$, the regularization variation is defined as (1):

$$v = \begin{cases} v + \varepsilon_1(a - a_{max}) & if\, a > a_{max} \\ v + \varepsilon_2(a_{min} - a) & if\, a < a_{min} \\ \quad\quad v & else \end{cases} \tag{1}$$

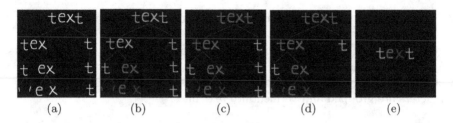

Fig. 3. The MSER pruning procedure. (a): MSER tree Extracting; (b): MSERs are colored from green to yellow and to red based on their variations; (c): MSERs are recolored according to regularized variations; (d): Repeating Components Excluding; (e): Extracting expected MSERs according to Node Choosing Algorithm

where ε_1 and ε_2 are the penalize arguments. According to the experiment, let $\varepsilon_1 = 0.05$, $\varepsilon_2 = 0.1$, $a_{\min} = 0.1$ and $a_{\max} = 0.8$. The MSER pruning procedure is shown in Fig. 3.

Repeating Components Excluding. When there is only one child of the MSER tree. This algorithm chooses the lower variation one from parent and child node and then excluding others. The process is described below: for a node n, the number of children node is examined using this algorithm. If no child, return to the given node directly. If there is one child, return to the root node r. If the variation of n is lower than r, the children of r are connected with n, and return to n, otherwise, return to r. If there are multiple children, the children nodes are processed with this algorithm and the processing results are connected to n before returning to n.

Node Choosing Algorithm. When the MSER has multiple children nodes, this algorithm is performed. The details of this algorithm is as follows: first, for a node n, the number of nodes is checked, if no child, return n. If there are multiple nodes, an empty set E is created and this algorithm is performed on each child node of n and then the result is put into E, if there is a variation of some node lower than n, return E, otherwise, exclude all the children nodes and return n.

2.2 Non-characters Filtering Using Bayesian Multi-cue Integration

The repeating components are filtered during the process above. But there are still a large amount of non-characters can be found among these MSERs. Then, we propose a Bayesian multi-cue integration method to separate characters and non-characters.

eHOG. eHOG is a gradient vector based on boundary of the region, namely histogram of gradients of edges. It is aimed at identifying that the edge pixels of characters are usually located on the end of the opposing gradient directions. Firstly, canny edge detector is exploited to detect the edge pixels of region L. Secondly, HOG features of these pixels are computed and the gradient orientations of them are calculated into four orientations, namely Orientation 1: $0 < \theta < \frac{\pi}{4}$ or $\frac{7\pi}{4} < \theta < 2\pi$, Orientation 2: $\frac{\pi}{4} < \theta < \frac{3\pi}{4}$, Orientation 3: $\frac{3\pi}{4} < \theta < \frac{5\pi}{4}$, Orientation 4: $\frac{5\pi}{4} < \theta < \frac{7\pi}{4}$ (Four different colors are used to describe different types of the edge pixels). Orientation 1 is expected to be near to Orientation 3 and the same to Orientation 2 and Orientation 4. eHOG on region L is defined as follows:

$$eHOG(L) = \frac{\sqrt{(\theta_1(L) - \theta_3(L))^2 + (\theta_2(L) - \theta_4(L))^2}}{\sum_{j=1}^{4} \theta_j(L)}. \qquad (2)$$

Where $\theta_j(L)$ represents the number of edges pixels of Orientation j in L. The denominator $\sum_{j=1}^{4} \theta_j(L)$ is the factor of scale invariance of eHOG.

Stroke Width (SW). Stroke width is one of the most popular cue to be exploited to identify characters. It is an intrinsic property of character. Epstein [3] proposed SWT to compute the distance between two boundary pixels detected by canny edge detector as a preprocessing step. Different from Epstein [3], we first apply a morphological operation on the binary image obtained above. Then, motivated by Chen [10], the Euclidean distance transform is employed to find the shortest path to the edge of corresponding MSER. Finally, we get the skeleton map and skeleton-distance map. Through comparing the variance on skeleton-map of each MSER which is used to measure the difference between character candidates and non-character candidates, we note that variances of character candidates are much smaller than non-character candidates. We develop stroke width computation method Li [14] with the extracted MSER m as:

$$SW(m) = \frac{Variance(w)}{[Mean(w)]^2}. \qquad (3)$$

Where $Mean(w)$ and $Variance(w)$ represent mean value of stroke width and its variance respectively.

Learning Probabilities of Cues. To obtain probability distribution of the cues above and make our algorithm extract character candidates more efficient, we employ the training set from the text segmentation task of ICDAR 2011 reading competition (challenge 2). This is a benchmark dataset with ground truth. Details of this dataset are discussed in Sect. 3.

To learn positive cues, we calculate two cues of character on the region with given ground truth. To learn negative cues, the regions are the negative samples which are left after ground truth character extracted, we also compute above two cues on them.

Bayesian Integration of Multiple Cues. The cues computed above separate characters and non-characters from different views. Stroke Width and eHOG are different in their internal architecture. They are complementary with each other. Through integration of cues above, we can generate a more excellent cue than any other single cue.

We suppose that each cue is independent of each other. Hence, the probabilities of character candidates are computed according to the Bayesian Theorem as 4:

$$p(x|U) = \frac{p(U|x)p(x)}{p(U)} = \frac{p(x) \prod_{cue \in U} p(cue|x)}{\sum_{u \in \{x,y\}} p(U) \prod_{cue \in U} p(cue|u)}. \qquad (4)$$

Where $U = \{eHOG, SW\}$, $p(x)$ and $p(y)$ represent posterior probabilities of characters and non-characters which is determined based on relative frequency. As $p(x|U)$ and $p(y|U)$ obtained above represent probabilities of characters and non-characters respectively. Thus we can figure out whether regions are characters or not by the probability distribution of cues on positive samples and negative samples which is learned above.

2.3   Character Candidates Grouping

Text Line Formulation. According to algorithms proposed in recent years, the main goal of this step is to group the identified character candidates in the previous steps into readable text lines. Clustering techniques are exploited most in this step, but the performance is not satisfactory. Through the analysis of the results, the main causes are that character candidate data is extremely small, the distinguishing degree of some of adjacent character candidates are too low and the speed of clustering algorithms are too slow.

The alignment of scene-text usually consists of a few words in a line. The distance between adjacent characters in a word is usually lower than 1/5 of the width of characters and the distance between two words are often bigger than 2/3. This distinguishable characteristic between words is the basis of our text line formulation method.

Text Candidates Verification. Experimental results show that the lengths of most text candidates merged by non-characters are shorter. So we verify text candidates based on the number of character candidates and an additional crossing line validation step. Text candidates containing more than three characters are determined as the true word candidates. Most characters have low proportion of the pixels in the convex hull; a crossing line should be rejected if characters on the crossing line have high proportion of pixels in the convex hull. Finally, characters on the crossing line are classified into character space and word space. They are labeled with the rectangles.

3   Experimental Results

In this section, we present the experimental results of the proposed method on the ICDAR 2011 competition dataset [18]. Then compare it with the leading algorithms in this field. All the experiments are conducted on a regular PC (CPU: Intel core i5 (Dual-core), 2.5 GHZ and Windows 64-bit). As mentioned in Sect. 2.2, the proposed algorithm is evaluated on the dataset from the text segmentation task of the ICDAR 2011 reading competition (challenge 2) which is a benchmark with ground truth. It consists of 255 images for testing and 229 natural scene images for training. We use the training set to learn cues in the Bayesian framework and the testing set to evaluate the performance of proposed method.

Note that the evaluation criterion of the ICDAR 2011 reading competition uses three criteria evaluate performance of algorithms: precision P, recall R and f measure. In this way, only one-to-one matches are taken into account.

To solve this drawback, the ICDAR 2011 competition evaluate the results through software DetEval [19] which supports one-to-one matches, one-to-many matches and many-to-one matches. Part of our results are shown below:

Fig. 4. Experimental results on ICDAR 2011 dataset

As is shown in Fig. 4, our approach not only can handle text variations, such as orientation, complex background, color and size, etc. But also can handle a wide range of challenging task, such as lightness, shadow, flexible surface and some characters in different scales and fonts, etc.

As is shown in Fig. 5, note that there are three situations that our method cannot handle, (i) characters with rather same color with the background, (ii) characters in seriously non-uniform illumination (with reflective surfaces), (iii) characters overlaid by objects on the foreground.

Fig. 5. False negatives of our method

The main factor of this algorithm is the elimination of repeating components and non-characters. As character candidates classifier has a strong connection with the elimination stage, higher performance the elimination stage achieved, more powerful the classifier is. To approve the efficient of repeating components excluding and non-character filtering, we conduct the proposed method on the ICDAR 2011 dataset above in following ways (as shown in Table 1):

Table 1. Performance (%) of the methods on ICDAR 2011 dataset in different ways

Experiments	R	P	f	S(s)
Overall	88.5	68.3	77.1	0.29
Experiment-I	80.4	61.2	69.5	0.25
Experiment-II	79.6	60.7	68.9	0.22

Experiment-I. The repeating components excluding stage is not performed. We use regular MSER extraction method and then directly perform the non-character filtering stage to integrate multi-cues to estimate posterior probabilities of character candidates. Finally, grouping character candidates in established method.

Experiment-II. We conduct the proposed method without the non-character filtering stage, character candidates are directly grouped into words after repeating components been excluded.

As can be seen in Table 1, the performance of our method has achieved Recall (88.5 %), Precision (68.3 %) and f (77.1 %). The average processing speed of our algorithm is 0.29 s per image. Compared with the overall performance, the recall and precision of Experiment-I and Experiment-II both decrease. Experiment-I could explained that repeating components have a significantly effect on the performance of this algorithm, and the non-characters filtering has a more serve effect on it. The combination of repeating components excluding and non-characters filtering could achieve a much better performance with a little loss of the speed.

Table 2 shows the performance of our method, Huang [15] (the state-of-the-art performance), Neumann and Matas [8], a very recent MSER-based method by Yin [12] who won the ICDAR 2013 Robust Competition and a top scoring method (Li [5]). The f measure is computed as $f = \frac{2R \cdot P}{R+P}$, where P and R are precision and recall rates. The comparison results are separated into two aspects: the performance of our approach which is evaluated by the criteria obtained from

Table 2. Performance (%) on ICDAR 2011 dataset

Algorithms	R	P	f
Huang [15]	88.0	71.0	78.0
Ours	88.5	68.3	77.1
Yin [12]	68.2	86.2	76.2
Shi [16]	63.1	83.3	71.8
Neumann and Matas [8]	64.7	73.1	68.7
Yi and Tian [17]	76.0	68.0	67.0
Li [5]	62.0	59.0	60.5

the evaluation software DetEval [19] and assess our approach by the mean time per an image. Though the performance of this algorithm is slightly lower than the state-of-the-art performance, the results in Table 3 prove that our method offers obvious speed advantage compared with all the listed methods over the same experimental environment.

Table 3. Evaluation on the speed of algorithms

Algorithms	Experimental environment (CPU)	Time
Huang [15]	Intel core i5 (Dual-core), 2.5 GHZ	3.20 s
Yin [12]	Intel core i5 (Dual-core), 2.5 GHZ	1.20 s
Neumann and Matas [8]	Intel core i5 (Dual-core), 2.5 GHZ	1.80 s
Ours	Intel core i5 (Dual-core), 2.5 GHZ	0.29 s
Li [14]	Intel core i5 (Dual-core), 2.5 GHZ	3.30 s

4 Conclusion

A fast and accurate scene-text detection method is proposed in this paper. The purpose of our method is to overcome the problem of repeating components affections on later grouping algorithm. To the best of our knowledge, we are the first to process both repeating components and non-character candidates. Compared with the state-of-the-art scene-text detection approaches, this approach is also competitive. Our method can process a wide range of challenging conditions, but there are still three conditions which may be false negatives: first, text and background are rather similar; second, characters in seriously non-uniform illumination (with reflective surfaces); third, characters overlaid by objects on the foreground. Our further study is focus on training more effective character classifiers, creating better integrated features which are not sensitive to illumination and enlarging the training datasets can achieve better performance.

References

1. Lee, J.-J., Lee, P.-H., Lee, S.-W., Yuille, A.L., Koch, C.: Adaboost for text detection in natural scene. In: Proceedings of the IEEE International Conference on Document Analysis and Recognition, pp. 429–434 (2011)
2. Chen, X., Yuille, A.L.: Detecting and reading text in natural scenes. In: Proceedings of IEEE Conference on Computer Vision and Pattern Recognition, vol. 2, pp. 366–373 (2004)
3. Epshtein, B., Ofek, E., Wexler, Y.: Detecting text in natural scenes with stroke width transform. In: CVPR, pp. 2963–2970 (2010)
4. Li, Y., Shen, C., Jia, W., van den Hengel, A.: Leveraging surrounding context for scene text detection. In: Proceedings of IEEE International Conference on Image Processing, pp. 2264–2268 (2013)

5. Li, Y., Lu, H.: Scene text detection via stroke width. In: Proceedings of IEEE International Conference on Pattern Recognition, pp. 681–684 (2012)
6. Pan, Y.-F., Hou, X., Liu, C.-L.: A hybrid approach to detect and localize texts in natural scene images. IEEE Trans. Image Process. **20**, 800–813 (2011)
7. Neumann, L., Matas, J.: A method for text localization and recognition in real-world images. In: Kimmel, R., Klette, R., Sugimoto, A. (eds.) ACCV 2010, Part III. LNCS, vol. 6494, pp. 770–783. Springer, Heidelberg (2011)
8. Neumann, L., Matas, J.: Real-time scene text localization and recognition. In: 2012 IEEE Conference on Computer Vision and Pattern Recognition (CVPR), pp. 3538–3545 (2012)
9. Matas, J.G., Zimmermann, K.: A new class of learnable detectors for categorisation. In: Kalviainen, H., Parkkinen, J., Kaarna, A. (eds.) SCIA 2005. LNCS, vol. 3540, pp. 541–550. Springer, Heidelberg (2005)
10. Chen, H., Tsai, S., Schroth, G., Chen, D., Grzeszczuk, R., Girod, B.: Robust text detection in natural images with edge-enhanced maximally stable extremal regions. In: Proceedings of IEEE International Conference on Image Processing, pp. 2609–2612 (2011)
11. Neumann, L., Matas, J.: Text localization in real-world images using efficiently pruned exhaustive search. In: ICDAR, pp. 687–691 (2011)
12. Yin, X.C., Yin, X.W., Huang, K.: Robust text detection in natural scene images. IEEE Trans. Pattern Anal. Mach. Intell. **36**, 970–983 (2014)
13. Zhang, J., Kasturi, R.: Text detection using edge gradient and graph spectrum. In: Proceedings of IEEE International Conference on Pattern Recognition, pp. 3979–3982 (2010)
14. Li, Y., Jia, W., Shen, C., van den Hengel, A.: Characterness: an indicator of text in the wild. In: Proceedings of IEEE International Conference on Image Processing, pp. 1666–1677 (2014)
15. Huang, W., Qiao, Y., Tang, X.: Robust scene text detection with convolution neural network induced MSER trees. In: Fleet, D., Pajdla, T., Schiele, B., Tuytelaars, T. (eds.) ECCV 2014, Part IV. LNCS, vol. 8692, pp. 497–511. Springer, Heidelberg (2014)
16. Shi, C., Wang, C., Xiao, B., Zhang, Y., Gao, S.: Scene text detection using graph model built upon maximally stable extremal regions. Pattern Recogn. Lett. **34**, 107–116 (2013)
17. Yi, C., Tian, Y.: Text extraction from scene images by character appearance and structure modeling. CVIU **117**(2), 182–194 (2013)
18. Shahab, A., Shafait, F., Dengel, A.: ICDAR 2011 robust reading competition challenge 2: reading text in scene images. In: Proceedings of International Conference on Document Analysis and Recognition, pp. 1491–1496 (2011)
19. DetEval. http://liris.cnrs.fr/christian.wolf/software/deteval/

Orthogonal Procrustes Problem Based Regression with Application to Face Recognition with Pose Variations

Ying Tai$^{(\boxtimes)}$, Jian Yang, Lei Luo, and Jianjun Qian

Nanjing University of Science and Technology,
Nanjing, People's Republic of China
tyshiwo@gmail.com, {csjyang, csjqian}@njust.edu.cn,
zzdxpyy3001@163.com

Abstract. Recently, sparse representation and collaborative representation based classifiers for face recognition have been proposed and achieved great attention. However, the two linear regression analysis based methods are sensitive to pose variations in the face images. In this paper, we combine the orthogonal Procrustes problem (OPP) with the regression model, and propose a novel method called orthogonal Procrustes problem based regression (OPPR) for face recognition with pose variations. An orthogonal matrix is introduced as an optimal linear transformation to correct the pose of test image to that of training images as far as possible. According to where the matrix is multiplied to deal with pose variations in vertical or horizontal direction, we propose the left or the right side OPP based regression, respectively. What's more, we further fuse the two models and propose a bilateral OPP based regression. The proposed model is solved via the efficient alternating iterative algorithm and experimental results on public face databases verify the effectiveness of our proposed models for handling pose variations.

Keywords: Face recognition · Linear regression analysis · Orthogonal Procrustes Problem (OPP) · Orthogonal matrix · Linear transformation

1 Introduction

Face recognition is a classical topic in computer vision and pattern recognition community for its great need in many areas. Although great progress has been made by many researchers, it is still a challenging problem because of the large variations existed in the face images.

Recently, linear regression analysis based methods have become a hot topic in face recognition field. Wright et al. [1] proposed a new face recognition framework called sparse representation based classification (SRC), which casts the recognition problem as seeking a sparse linear representation of the query image over the training set. Furthermore, Naseem et al. [2] developed a linear model representing a probe image as a linear combination of class-specific galleries and proposed the Linear Regression Classification (LRC) algorithm. Zhang et al. [3] proposed a new classification scheme, namely collaborative representation based classifier (CRC), which emphasizes the role of collaborative representation in classification task. More recently, Yang et al. [4]

© Springer International Publishing Switzerland 2015
X. He et al. (Eds.): IScIDE 2015, Part I, LNCS 9242, pp. 11–19, 2015.
DOI: 10.1007/978-3-319-23989-7_2

proposed a two-dimensional image matrix based model and employed nuclear norm constraint as a criterion to make full use of the low-rank structural information caused by some occlusion and illumination changes. These popular classifiers have achieved some interesting results. However, they are sensitive to pose variations in the face images.

To solve this problem, several regression based methods for face recognition with pose variations have been proposed. Chai et al. [5] performed linear regression on local patches for virtual frontal view synthesis, which directly convert non-frontal faces into frontal by keeping the regression coefficients. More recently, Sharma and Jacobs [6] used partial least square (PLS) regression to project samples from two poses to a common latent subspace, with one pose as regressor and another pose as response. Li et al. [7] also performed PLS to maximize the squares of the intra-individual correlations via training on the coupled face images of the same identities and across two different poses. What's more, Li [8] further improved the regressor-based cross-pose face representation by striking a coupled balance between bias and variance in regression for different poses. Competitive results have been achieved by these methods. However, these methods need to conduct a pose pair for each probe sample, which is a strong restriction and unrealistic for a real-world application.

In this paper, we combine the orthogonal Procrustes problem (OPP) with the traditional popular classifiers, and propose a novel method called orthogonal Procrustes problem based regression (OPPR) for face recognition with pose variations. In our method, we don't have the assumption that for each non-frontal test image, there should exist a corresponding frontal image in the gallery set. Since we extend the traditional SRC and CRC models directly, we can handle pose correction and face representation simultaneously. The main contribution of this paper is that we combine the orthogonal Procrustes problem with the popular classifiers to deal with face recognition with pose variations task for the first time. Inspired by the ability of OPP [9], which seeks an optimal transformation so as to make the transformed object best fits the other one in the sense of Frobenius norm, we introduce an orthogonal matrix Q to the traditional regression model for correcting the pose of the non-frontal test image to that of the training images. Figure 1 gives an example to exhibit our model intuitively. According to where the orthogonal matrix is multiplied, three different models are introduced, including the left side OPP based regression, the right side OPP based regression and the bilateral OPP based regression. Specifically, from the viewpoint of linear algebra, the left side version indicates performing elementary row operation on the image so as to handle pose variations in vertical direction; the right side version indicates performing elementary column operation on the image so as to handle pose variations in horizontal direction and the bilateral version handles variations in both directions.

2   Literature Review of Orthogonal Procrustes Problem

In this section, we briefly introduce the orthogonal Procrustes problem. OPP originates from factor analysis in psychometrics in the 1950s and 1960s [10, 11]. The task of OPP is to determine an orthogonal matrix, which rotates a factor (data) matrix A, to fit some hypothesis matrix B. In statistics, Procrustes analysis is a standard technique for correcting the rigid-body geometric transformations between two matrices, with

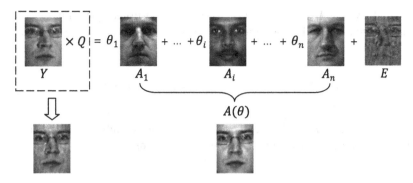

Fig. 1. The illustration of our proposed model.

applications in registration and shape analysis. In linear algebra [9], the orthogonal Procrustes problem can be seen as a matrix approximation problem, which seeks the optimal *rotation* or *reflection* for the transformation of an object with respect to another. Generally, letting Q be an orthogonal matrix and X, B be known real matrices of correct dimensions, the optimization problem

$$\min_{Q} \|QX - B\|_F^2, \ s.t. \quad Q^T Q = I \tag{1}$$

is called an *orthogonal Procrustes problem*, where I is an identity matrix and $\|\cdot\|_F$ denotes the Frobenius norm. The OPP has an analytical solution, that can be derived by using the singular value decomposition (SVD) of XB^T. Let $USV^T = XB^T$, the solution $\hat{Q}$ to Eq. (1) is given by $\hat{Q} = VU^T$ [9, 12].

Actually, the main applications of OPP are related to determination of rigid motion, factor analysis and multidimensional scaling [9, 12]. Meanwhile, there are still some works which applied Procrustes analysis to the face alignment or recognition tasks. For example, Bellino et al. [13] applied Procrustes analysis in the context of alignment of faces in images. Sujith and Ramanan [14] turned to face recognition task and proposed a Procrustes analysis based classifier. Yu et al. [15] used Procrustes analysis to achieve pose-free facial landmark initialization. What's more, Chen et al. [16] introduced a rotation transformation to promote the sparsity of a projection, which is used for dimensionality reduction. However, these works didn't use the Procrustes analysis to handle face recognition with pose variations task. Since OPP finds an optimal transformation so as to make the transformed object most approximate the other one in the sense of Frobenius norm, we combine it with the popular classifiers and propose a regression based model for face recognition with pose variations.

3 Orthogonal Procrustes Problem Based Regression

In this section, we apply OPP into the regression model and present the proposed models in detail. As mentioned in Sect. 1, three different models are proposed according to where the orthogonal matrix is multiplied. Given a set of frontal images

$A_1, A_2, \ldots, A_n$, where $A_i \in R^{p \times q}$, $i = 1, 2, \ldots, n$ and a non-frontal query image $Y \in R^{p \times q}$. Since the image is stacked in the form of matrix in our models, we firstly define a linear mapping as follows

$$A(\theta) = \sum_{i=1}^{n} \theta_i A_i = \theta_1 A_1 + \theta_2 A_2 + \ldots + \theta_n A_n, \qquad (2)$$

where $\theta_i, i = 1, 2, \ldots, n$ is the coefficient of the corresponding image. Then, we introduce the different models in the following subsections.

3.1 The Left Side OPP Based Regression

First, we present the left side OPP based regression (OPPR_L). Here, an orthogonal matrix is introduced to the original SRC or CRC models and the proposed model is defined as follows:

$$\min_{\theta, P} \|PY - A(\theta)\|_F^2 + \lambda R(\theta), \ s.t. \quad P^T P = I_p, \qquad (3)$$

where P is an orthogonal matrix, λ is a parameter, I_p is a $p \times p$ identity matrix and $R(\cdot)$ represents some kind of regularizer, which is used to avoid over-fitting. Two popular regularizers are introduced in our model: the L_2-norm based regularizer used in the CRC model and the L_1-norm based one used in the SRC model. We solve this problem by breaking it into two sub-problems, and alternatively solving the unknown variables. To be specific, the minimization in Eq. (3) can be divided into two parts: fix the orthogonal matrix P to optimize the coefficients θ and fix the coefficients θ to optimize the orthogonal matrix P. Each sub optimization problem is an independent problem and can be solved efficiently.

3.2 The Right Side OPP Based Regression

In this section, we introduce the right side OPP based regression (OPPR_R) in which the orthogonal matrix is multiplied in the right side of the query image Y and the model is formulated as

$$\min_{\theta, Q} \|YQ - A(\theta)\|_F^2 + \lambda R(\theta), \ s.t. \quad Q^T Q = I_q. \qquad (4)$$

where I_q is a $q \times q$ identity matrix. Just like Eq. (3), this problem can also be solved through alternatively optimizing the variables. However, the unilateral OPP based regression can only handle the variations in one direction. Actually, the left side version deals with vertical direction and the points to be rotated in the images are ordered column wise, while the right side version deals with horizontal direction and the points are ordered row wise. To overcome this drawback, we integrate the two versions and propose the bilateral OPP based Regression (OPPR_B).

3.3   The Bilateral OPP Based Regression

Two different orthogonal matrices are multiplied in both sides of the query image Y. We formulate the model as follows:

$$\min_{\theta,P,Q}\|PYQ - A(\theta)\|_F^2 + \lambda R(\theta), \; s.t. \quad P^TP = I_p, Q^TQ = I_q. \tag{5}$$

Different from the unilateral versions, there are three unknown variables. However, we can still adopt the alternative optimization strategy to solve this problem.

4   Algorithm

Since Eq. (5) is the integration of Eqs. (3) and (4), it contains the optimization of the matrix P in Eq. (3), the optimization of the matrix Q in Eq. (4) and the optimization of the coefficients θ in both versions. For convenience, we just take Eq. (5) as the example and present our algorithm. The minimization of OPPR_B in Eq. (5) can be divided into three sub-problems: the traditional SRC or CRC model by fixing P and Q; the left side OPP by fixing θ and Q; and the right side OPP by fixing θ and P. Next, we spell out how to solve these problems in detail.

Update of θ: Fix P, Q and set $Y_1 = PYQ$. We have

$$\min_{\theta}\|Y_1 - A(\theta)\|_F^2 + \lambda R(\theta). \tag{6}$$

According to different regularizers mentioned in Sect. 3.1, Eq. (6) can be solved by the solutions mentioned in [1, 3] efficiently.

Update of P: Fix $\hat{\theta}, Q$ and set $Y_2 = YQ, H = A(\hat{\theta})$. We have

$$\min_{P}\|PY_2 - H\|_F^2, \; s.t. \quad P^TP = I_p. \tag{7}$$

This problem has an analytical solution, which has been discussed in Sect. 2. Here, by using SVD: $U_P S_P V_P^T = Y_2 H^T$, the solution $\hat{P}$ to Eq. (7) is given by $\hat{P} = V_P U_P^T$.

Update of Q: Fix $\hat{\theta}, \hat{P}$ and set $Y_3 = \hat{P}Y$. We have

$$\min_{Q}\|Y_3 Q - H\|_F^2, \; s.t. \quad Q^TQ = I_q. \tag{8}$$

Similar as Eq. (7), this problem also has an analytical solution [12]. By using SVD: $U_Q S_Q V_Q^T = H^T Y_3$, the solution $\hat{Q}$ to Eq. (8) is given by $\hat{Q} = V_Q U_Q^T$.

The entire algorithm for OPPR_B is summarized in Algorithm 1. We stop the algorithm until the values of the objective function in adjacent iterations are close enough. Since in each round of alternative minimization, the objective function of OPPR_B will decrease, the proposed algorithm will converge. Figure 2 plots the

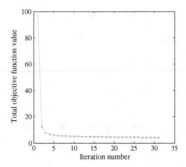

Fig. 2. The convergence curve of OPPR_B objective function on the CMU Multi-PIE database.

empirical convergence curve of OPPR_B objective function on the CMU Multi-PIE database [17], from which we can see that the algorithm converges quickly.

Algorithm 1 The bilateral OPP based Regression (OPPR_B)

Input: A set of image matrices $A_1, \cdots, A_n \in R^{p \times q}$ and an image matrix $Y \in R^{p \times q}$, the model parameters λ, the termination condition parameter ε.

1: **Initialization:** $k = 0, P_k, Q_k, \theta_k$;
2: **While** not converged ($k = 0,1,\ldots$) **do**
3: Update θ_{k+1} with fixed P_k, Q_k by solving Eq. (6);
4: Update P_{k+1} with fixed θ_{k+1}, Q_k by solving Eq. (7);
5: Update Q_{k+1} with fixed θ_{k+1}, P_{k+1} by solving Eq. (8);
6. **End while**

Output: Solution θ^*, P^*, Q^* to Eq. (5)

---

5 Experiments

In this section, we perform face recognition with pose variations on two popular face databases, including CMU PIE database [18] and CMU Multi-PIE database [17]. We compare the proposed models with the traditional SRC and CRC models. It should be noted that for fair comparison, we perform the same solutions used in the SRC [1] and CRC [3] models for solving Eq. (6) in our algorithm. Default parameters mentioned in [1, 3] are used for SRC and CRC. As for our models, we fix the parameter $\lambda = 0.001$ for the L_2-norm based regularizer and $\lambda = 0.01$ for the L_1-norm based regularizer.

5.1 Experiments on the CMU PIE Database

We first test the robustness of our models on the CMU PIE database. In this database, there are 68 persons whose face images are captured in 13 poses with yaw- and pitch-angle differences. We choose five different poses, including poses {C05, C07, C27, C09, C29}. The example images of one person are shown in Fig. 3 and the yaw-angle difference of the neighbor pose is about 22.5°. Here, 10 frontal images with extreme illuminations are used as the training set and 4 images with moderate illuminations are for testing. The image is with the resolution of 64×64 and the results of different methods are shown in Table 1. We know that poses C05 and C29 vary in horizontal direction, and poses C07 and C09 vary in vertical direction. As a result, OPPR_R performs well in C05 and C29 but contributes little in C07 and C09, and OPPR_L is just the opposite. As an integration of these two versions, OPPR_B gives competitive results in both directions. We can see that our method improves the performance of the classifiers when handling pose variations significantly. The reason why the versions with L_1-norm based regularizer slightly outperform the ones with L_2-norm based regularizer may be the fact that a sufficient dictionary is conducted here.

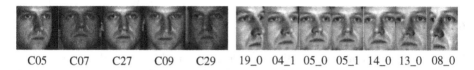

C05   C07   C27   C09   C29   19_0   04_1   05_0   05_1   14_0   13_0   08_0

Fig. 3. The example images of one person. **Left:** examples from the CMU PIE database. **Right:** examples from the CMU Multi-PIE database.

Table 1. Recognition rates (%) vs. pose variations on the CMU PIE database

Recognition (%)	C05	C07	C27	C09	C29
CRC	49.7	62.9	100	79.8	49.6
OPPR_R_L_2	**82.8**	60.7	100	84.2	**77.2**
OPPR_L_L_2	49.9	**86.5**	100	**92.4**	54.8
OPPR_B_L_2	76.5	83.3	100	91.8	75.1
SRC	64.4	78.6	100	88.9	63.2
OPPR_R_L_1	**85.9**	75.6	100	86.1	**78.3**
OPPR_L_L_1	60.5	**87.5**	100	**92.9**	61.1
OPPR_B_L_1	82.7	83.9	100	92.6	76.3

5.2   Experiments on the CMU Multi-PIE Database

We then conduct the experiments on the CMU Multi-PIE database. Images with the neutral expression from the first 100 subjects in Session 1 are used and seven different poses are chosen, including poses {19_0, 04_1, 05_0, 05_1, 14_0, 13_0, 08_0}. The example images of one person are also shown in Fig. 3. The yaw-angle difference of the neighbor pose is about 15°. Since the pose varies only in the horizontal direction in these poses, we just present the right side version for convenience. Here, seven frontal images with extreme illuminations {0, 1, 7, 13, 14, 16, 18} are used as the training set and the images with illumination {6} from different poses are used for testing. The image is with the resolution of 80×60 and the results of different methods are shown in Fig. 4. As we can see, our models outperform the traditional classifiers in all cases. Also, the version with L_1-norm based regularizer performs a litter bit better than the one with L_2-norm based regularizer.

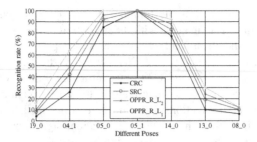

Fig. 4. Recognition rates (%) of different methods vs. pose variations on the CMU Multi-PIE database

6   Conclusion

In this paper, we combine the orthogonal Procrustes problem with the popular classifiers, and propose a novel method named orthogonal Procrustes problem based regression (OPPR) to handle face images with moderate pose variations. Experimental results verify the effectiveness of our method for improving the performance of those traditional classifiers when dealing with face recognition with pose variations.

References

1. Wright, J., Yang, A., Ganesh, A., Sastry, S., Ma, Y.: Robust face recognition via sparse representation. IEEE Trans. PAMI **31**, 210–227 (2009)
2. Naseem, I., Togneri, R., Bennamoun, M.: Linear regression for face recognition. IEEE Trans. PAMI **32**(11), 2106–2112 (2010)
3. Zhang, L., Yang, M., Feng, X.: Sparse representation or collaborative representation which helps face recognition?. In: ICCV (2011)
4. Yang, J., Qian, J., Luo, L., Zhang, F., Gao, Y.: Nuclear norm based matrix regression with applications to face recognition with occlusion and illumination changes (2014). arXiv:1207.0023
5. Chai, X., Shan, S., Chen, X., Gao, W.: Locally linear regression for pose-invariant face recognition. IEEE Trans. Image Process. **16**(7), 1716–1725 (2007)
6. Sharma, A., Jacobs, D.W.: Bypassing synthesis: Pls for face recognition with pose, low-resolution and sketch. In: CVPR (2011)
7. Li, A., Shan, S., Chen, X., Gao, W.: Cross-pose face recognition based on partial least squares. Pattern Recogn. Lett. **32**(15), 1948–1955 (2011)
8. Li, A., Shan, S., Gao, W.: Coupled bias-variance tradeoff for cross-pose face recognition. IEEE Trans. Image Process. **21**(1), 305–315 (2012)
9. Gower, J., Dijksterhuis, G.: Procrustes Problems. Oxford University Press, Oxford (2004)
10. Green, B.F.: The orthogonal approximation of an oblique simple structure in factor analysis. Psychometrika **17**, 429–440 (1952)
11. Hurley, J.R., Cattell, R.B.: The procrustes program: producing direct rotation to test a hypothesized factor structure. Behav. Sci. **6**, 258–262 (1962)

12. Viklands, T.: Algorithms for the weighted orthogonal procrustes problem and other least squares problems (2006)
13. Bellino, K.: Computational algorithms for face alignment and recognition. M.S thesis, Virginia Polytechnic Institute and State University (2002)
14. Sujith, K.R., Ramanan, G.V.: Procrustes analysis and moore-penrose inverse based classifiers for face recognition. In: Li, S.Z., Sun, Z., Tan, T., Pankanti, S., Chollet, G., Zhang, D. (eds.) IWBRS 2005. LNCS, vol. 3781, pp. 59–66. Springer, Heidelberg (2005)
15. Yu, X., Huang, J., Zhang, S., Yan, W., Metaxas, D.N.: Pose-free facial landmark fitting via optimized part mixtures and cascaded deformable shape model. In: ICCV (2013)
16. Chen, D., Cao, X., Wen, F., Sun, J.: Blessing of dimensionality: high dimensional feature and its efficient compression for face verification. In: CVPR (2013)
17. Gross, R., Matthews, I., Cohn, J., Kanade, T., Baker, S.: Multi-pie. Image Vis. Comput. **28** (5), 807–813 (2010)
18. Sim, T., Baker, S., Bsat, M.: The cmu pose, illumination, and expression database. IEEE Trans. PAMI **25**(12), 1615–1618 (2003)

A New Fisher Discriminative K-SVD Algorithm for Face Recognition

Hao Zheng[1,2,3,4(✉)]

[1] Key Laboratory of Trusted Cloud Computing and Big Data Analysis,
Nanjing XiaoZhuang University, Nanjing, China
[2] State Key Laboratory for Novel Software Technology, Nanjing University,
Nanjing, China
[3] Jiangsu Key Laboratory of Image and Video Understanding for Social Safety,
Nanjing University of Science and Technology, Nanjing, China
[4] Key Laboratory of Computer Network and Information Integration
(Southeast University), Ministry of Education, Nanjing, China
zhh710@163.com

Abstract. In a sparse-representation-based face recognition scheme, dictionary learning has attracted growing attention for its good performance. Discriminative K-SVD (D-KSVD) is one of conventional dictionary learning algorithm, which can effectively solve the face recognition problem. However, D-KSVD doesn't consider the discrimination of the sparse coding coefficients. To address this issue, a new algorithm named Fisher Discriminative K-SVD (FD-KSVD) is proposed. In the new algorithm, the Fisher discrimination criterion is imposed on the sparse coding coefficients to make them discriminative through small within-class scatter and big between-class scatter. The optimization is employed by the Iterative Projective Method and K-SVD method alternatively. The experimental results of face databases indicated recognition performance of the new algorithm is superior to other state-of-the-art algorithms.

Keywords: Discriminative K-SVD · Fisher discrimination criterion · Face recognition

1 Introduction

Recently sparse coding has been successfully applied to face recognition. The sparse representation classification (SRC) [1] algorithm firstly codes a testing sample as a sparse linear combination of all the training samples, and then classifies the test sample by evaluating which class leads to the minimum representation error. SRC is much more effective than state-of-art methods in dealing with face occlusion, corruption, lighting and expression changes, etc. But if the number of training samples is very large, a big training dictionary will be employed, which makes the sparse coding process very computationally expensive and make the real-time processing very difficult [2,3]. Therefore, small-size dictionary and sparse coding is essential for large scale face recognition. K-SVD

X. He et al. (Eds.): IScIDE 2015, Part I, LNCS 9242, pp. 20–28, 2015.
DOI: 10.1007/978-3-319-23989-7_3

[4] is a conventional method which can represent the test sample by the suitable training dictionary. However, K-SVD lacks of discrimination capability [5–7], so Discriminative K-SVD(D-KSVD) [8–11] has been proposed. D-KSVD is based on extending the K-SVD algorithm by incorporating the classification error into the objective function, so that retains the representational power while making the dictionary discriminative. Although D-KSVD has improved the discrimination performance, it does not consider the discriminative information of the sparse coding coefficients.

In this paper, we propose a new algorithm named Fisher Discriminative K-SVD (FD-KSVD) which employs the fisher discrimination criterion to make the sparse coding coefficients discriminative. In the new algorithm, we make the sparse coding coefficients have small within-class scatter but big between-class scatter. In addition, for including the classification error, a linear predictive classifier is introduced. So not only the reconstruction error associated with each class can be used, but also the discriminative error and the classification error have been represented. To demonstrate the effectiveness and advantage of the new algorithm for face recognition, extensive experiments have been carried out on the face image database: the extended YaleB database [12] and the AR database [13]. The experimental results show that the proposed algorithm is superior to other state-of-the-art methods. Furthermore, we analyze and compare the performance of the classifiers.

The rest of paper is organized as follow. In Sect. 2, we firstly present the proposed FD-KSVD algorithm. In Sect. 3, the experimental results and analysis are reported. Finally, conclusion and discussion is presented in Sect. 4.

2   Fisher Discrimination K-SVD Algorithm (FD-KSVD)

We aim to learn a reconstructive and discriminative dictionary, the discriminative spare coding coefficients is important for the dictionary learning. Thus we add a coefficients discrimination term into objective function for learning a dictionary with more discriminative power.

2.1   Fisher Discrimination K-SVD Model

Suppose Y is the matrix of all training sample, X is the sparse representation coefficients, $L = [l_1 \ldots l_N]$ is the labels of Y, W is the linear predictive classifier. Consider the reconstruction error and the classification error, an objective function for learning a dictionary D can be defined as follow:

$$< D, X, W >= \underset{D,X,W}{\arg\min} ||Y - DX||_2^2 + \lambda_1 ||L - WX||_2^2 \ \ s.t. \forall i, ||x_i||_0 \leq T, \quad (1)$$

where the term $||Y - DX||_2^2$ represents the reconstruction error, the term $||L - WX||_2^2$ represents the classification error, T is a parameter to impose the sparsity prior, λ_1 is the scalar.

To make the sparse representation coefficient more discriminative, we can employ the fisher discrimination criterion to make the sparse coding coefficients have small within-class scatter but big between-class scatter. Suppose the within-class scatter of X denoted by $Sw(X)$, between-class scatters of X denoted by $S_B(X)$ are defined:

$$S_w(X) = \sum_{i=1}^{c} \sum_{j=1}^{N_i} (x_j^i - \mu_i)(x_j^i - \mu_i)^T \qquad (2)$$

$$S_B(X) = \sum_{i=1}^{c} N_i(\mu_i - \mu)(\mu_i - \mu)^T, \qquad (3)$$

where x_j^i is the j-th sample of class i, μ_i is the mean of class i, c is the number of classes, and N_i is the number of samples of class i.

Then we define $g(x) = tr(S_w(X)) - tr(S_B(X))$. Because $g(x)$ is non-convex, an elastic term $\|X\|_F^2$ need be added into $g(x)$. So $g(x)$ is defined as

$$g(x) = tr(S_w(X)) - tr(S_B(X)) + \eta \|X\|_F^2, \qquad (4)$$

where η is a parameter.

Finally we join the Eqs. (1) and (4), and an objective function for learning a dictionary D with reconstructive and discriminative power can be define as follow:

$$< D, X, W > = \underset{D,X,W}{\arg\min} \|Y - DX\|_2^2 + \lambda_1 \|L - WX\|_2^2 + \lambda_2(tr(S_w(X)) - tr(S_B(X)))$$

$$+ \mu \|X\|_F^2) s.t. \forall i, \|x_i\|_0 \leq T, \qquad (5)$$

where λ_2 is a parameter. When D and X is fixed alternatively, optimization of D and X can be conducted. In the following section, we describe the optimization procedure for FD-KSVD.

2.2 Optimization

In the optimization, we can optimize D and X alternatively by updating D and W by fixing X, and updating X by fixing D and W.

First, suppose X is fixed, the objective function (5) is reduced to:

$$< D, W > = \underset{D,W}{\arg\min} \|Y - DX\|_2^2 + \lambda_1 \|L - WX\|_2^2 \ s.t. \forall i, \|x_i\|_0 \leq T. \qquad (6)$$

We can use the K-SVD algorithm to find the optimal solution for D and W. Equation (6) can be rewritten as

$$\langle D, W \rangle = \underset{D,W}{\arg\min} \left\| \left(\frac{Y}{\sqrt{\lambda_1 L}} \right) - \left(\frac{D}{\sqrt{\lambda_2 W}} \right) X \right\|_2^2. \qquad (7)$$

Let $Y_{new} = (Y^t, \sqrt{\lambda_1} L^t)^t$, $D_{new} = (D^t, \sqrt{\lambda_2} W^t)^t$. The optimization of Eq. (7) is equivalent to solving the following problem:

$$< D_{new}, W > = \underset{D_{new}, W}{\arg\min} \|Y_{new} - D_{new} X\|_2^2. \tag{8}$$

Now, the problem of Eq. (8) can be efficiently solved by updating the dictionary atom by atom with the following method. Each atom d_k and the corresponding coefficient x_k can be computed by

$$< d_k, x_k > = \underset{d_k, x_k}{\arg\min} \|E_k - d_k x_k\|_F, \tag{9}$$

where $E_k = Y - \sum_{i \neq k} d_k x_k$. This can be solved by original K-SVD, so the solution of Eq. (9) is giving by

$$\begin{aligned} U * \Sigma * V &= SVD(E_k) \\ d_k &= U(:,1) \\ \widetilde{x}_k &= \Sigma(1,1) * V(1,:). \end{aligned} \tag{10}$$

Finally the nonzero values in x_k are replaced by $\widetilde{x}_k$, and the desired dictionary D and W are computed as follow:

$$D = \left\{ \frac{d_1}{\|d_1\|_2} \cdots \frac{d_k}{\|d_k\|_2} \right\}, W = \left\{ \frac{w_1}{\|d_1\|_2} \cdots \frac{w_k}{\|d_k\|_2} \right\} \tag{11}$$

Let we discuss X when D and W is fixed, the objective function in Eq. (5) is equivalent to:

$$< X > = \underset{X}{\arg\min} \sum_{i=1}^{c} (\|Y_i - D_i X_i\|_2^2 + \lambda_1 \|L_i - W X_i\|_2^2) + \lambda_2 (\|X_i - \mu_i\|_F^2$$
$$- \sum_{k=1}^{c} \|\mu_i - \mu\|_F^2 + \eta \|X_i\|_F^2) \ s.t. \forall i, \|x_i\|_0 \leq T \tag{12}$$

where μ_i and μ are the mean matrices of class k and all classes, respectively. Equation (12) can be solved by many optimization methods, such as Iterative Projection Method [14]. Suppose

$$J(X_i) = \|Y_i - D_i X_i\|_2^2 + \lambda_1 \|L_i - W X_i\|_2^2 + \lambda_2 (\|X_i - \mu_i\|_F^2 - \sum_{k=1}^{c} \|\mu_i - \mu\|_F^2) \tag{13}$$

and X_i^t is the t-th iteration of X_i, X_i^1 is initialized 0. when the iteration number is t, X_i^t is computed by

$$X_i^t = \phi(X_i^{t-1} - (1/\sigma)\nabla J(X_i^{t-1})) \tag{14}$$

where $\nabla J(X_i^{t-1})$ is the derivation of $J(X_i^t)$ with regard to X_i^{t-1}, and ϕ is soft thresholding function, further details can be referred [14]. Finally when the optimization is convergence, the optimal value of X_i^t is obtained. Thus the proposed FD-KSVD algorithm is summarized in Algorithm 1.

Algorithm 1. Fisher Discrimination K-SVD algorithm (FD-KSVD)

---

1. Input: $\{Y_i, L_i\}$, $i = 1, 2, \cdots, N$, where $Y_i \in \Re^{n \times d}$ is the training samples, L_i is the corresponding label vector.
2. Initialize D and X with K-SVD.
3. Calculate D and W when X is fixed by Eq. (6).
4. Calculate X when D and W is fixed by Eq. (12).
5. Go back to step 3 until the condition of convergence is met.

2.3   Classification

Though we obtain the learned dictionary D and the corresponding classifier W, the dictionary D does not readily support a sparse coding based representation of a test face image. With the normalized D, we can find the sparse coefficients for a test face image y by solving

$$< \hat{x} > = \arg\min_{\hat{x}} (||y - Dx||_2^2 + \lambda ||x||_1), \tag{15}$$

which can be solved by $l1$ optimization method, such as basic pursuit [15] and so on. Finally we obtain the label of the test image:

$$l = \hat{W} * \hat{x}. \tag{16}$$

The label of the test image y is the index corresponding to the largest element of l.

Fig. 1. Sample images of the extended YaleB database

3   Experiments

In this section, we perform experiments on face databases to demonstrate the efficiency of FD-KSVD. To evaluate more comprehensively the performance of

FD-KSVD, in Sect. 3.1 we test face recognition on the Extended YaleB database, and then in Sect. 3.2 we test face recognition on the AR database. We compare our algorithm with SRC [1], K-SVD [11], D-KSVD [9]. In both experiments, we adopted cross-validation strategy for recognition.

3.1   Extended YaleB Database

The Extended Yale B dataset consists of 2,414 frontal face images of 38 subjects. They are captured under various lighting conditions and cropped and normalized to 192×168 pixels. The face images were captured under various illumination conditions. We randomly split the database into two halves. One half (about 32 images per person) was used for training, and the other half for testing. Here, for computational convenience, the size of image is cropped to 32×32. Figure 1 shows some samples of the Extended YaleB database. The learned dictionary contains 380 atoms, which corresponds to roughly 10 atoms for each person.

We tested four methods: SRC, K-SVD, DK-SVD, and the new algorithm FD-KSVD. For a fair comparison, all the parameters about dictionary learning used in the methods were set to be the same in our experiments. The sparsity prior T, λ_1, λ_2 is set to 20, 3 and 4 respectively. Table 1 illustrates the face recognition rates under different methods. We can see that the proposed algorithm achieves the best in all the methods. FD-KSVD has at least 3% improvement than other methods. In addition, the accuracy of SRC is only 83.5%, this is because the performance of SRC degrade with the number of the dictionary is small.

To evaluate the performance of FD-KSVD, we also compared their speed performance by classifying one test sample. The computation time is obtained by an average over all the test samples. Table 2 shows computation time on Extended YaleB database. We ran 10 times and calculated the average result, as shows on Table 2. The second row shows the computation time when the dictionary contains 380 atoms, while the third is the result when the dictionary contains 304 atoms. In Table 2, the computation time of FD-KSVD is only 0.63, and 20 times faster than that of SRC. Moreover we can see that a smaller dictionary may need less computation time, which is expected in the fact.

Table 1. Accuracy on the extended YaleB database

Method	SRC(%)	K-SVD(%)	D-KSVD(%)	FD-KSVD(%)
Accuracy	83.5	93.4	95.7	**98.8**

Table 2. Computation time on the extended YaleB database

Method	SRC	K-SVD	D-KSVD	FD-KSVD
Accuracy Time(ms)	13.22	0.48	0.52	0.63
Accuracy Time(ms)	13.18	0.45	0.49	0.57

3.2 AR Database

The AR dataset consists of over 3,000 frontal images of 126 individuals. There are 26 images of each individual, taken at two different occasions. The faces in AR contain variations such as illumination change, expressions and facial disguises. We selected 100 subjects (50 male and 50 female) for out experiments. For each subject, we randomly take the 20 images for training and the other 6 for testing. Some samples of AR database are shown as Fig. 2. Here, for computational convenience, the size of image is cropped to 33×24. The learned dictionary contains 500 atoms, i.e., about 5 atoms for each person. In the experiments, the sparisty prior T, λ_1, λ_2 is set to 15, 2 and 4 respectively.

Similar to the above experiments, we also tested SRC, K-SVD, DK-SVD and FD-KSVD. In the experiments, all four methods used the same parameters. The results from all the methods are listed in Table 3. It can be seen that the proposed algorithm outperforms all the competing method. Compared in the extended YaleB database, the performance of SRC increases in the AR database. This is partially because that the size of the training dictionary has increased, therefore the number of the atoms per person affects the accuracy of the recognition. In further experiments, we found that the performance of SRC degraded dramatically when the size of the learned dictionary is small, while our proposed algorithm is still stable.

In addition, we also compared the speed performance of FD-KSVD and other methods for classifying one test sample. The result is shown in Table 4. We can see that the computation time of FD-KSVD is much faster than that of SRC. Tables 3 and 4 illustrate that FD-KSVD not only promotes the accuracy of recognition, but also saves the computation time. It is the reason that FD-KSVD adds the discriminative coefficients, while avoids large learning dictionary.

Fig. 2. Sample images of AR database

Table 3. Accuracy on the AR database

Method	SRC(%)	K-SVD(%)	D-KSVD(%)	FD-KSVD(%)
Accuracy	90.4	89.4	91.3	**96.8**

Table 4. Computation time on the AR database

Method	SRC	K-SVD	D-KSVD	FD-KSVD
Accuracy Time(ms)	20.06	0.583	0.641	0.694

4 Conclusion

This paper proposed a Fisher Discriminative K-SVD (FD-KSVD) based fisher criterion. On the high-dimensional data such as face images, D-KSVD could get better performance than K-SVD, but D-KSVD does not consider the sparse coding coefficients discriminative. FD-KSVD can solve this problem by combining small within-class scatter and big between-class scatter. On face data sets containing varying illumination change, expressions, FD-KSVD achieves the best performance. In the future, we will extend the new algorithm to other visual pattern recognition.

Acknowledgement. This work is partially supported by the Project funded by China Postdoctoral Science Foundation Under grant No. 2014M5615556. And, it is partially supported by grants KFKT2014B18 from the State Key Laboratory for Novel Software Technology from Nanjing University, Supported by grants 30920140122007 from Project supported by Jiangsu Key Laboratory of Image and Video Understanding for Social Safety. And, this research was partly supported by the National Science Foundation of China (61273300, 61232007) and the Jiangsu Science Foundation (BK20140022). Finally, the authors would like to thank the anonymous reviewers for their constructive advice.

References

1. Wright, J., Yang, A.Y., Ganesh, A., Sastry, S.S., Ma, Y.: Robust face recognition via sparse representation. TPAMI **31**, 210–227 (2009)
2. Zhang, L.: Sparse representation or collaborative representation which helps face. In: International Conference on Computer Vision (2011)
3. Shi, Q., Eriksson, A., Hengel, A.: Is face recognition really a compressive sensing problem? In: Computer Vision and Pattern Recognition, pp. 553–560 (2011)
4. Aharon, M., Elad, M., Bruckstein, A.: K-SVD: an algorithm for designing over-complete dictionaries for sparse representation. IEEE Trans. Sig. Process. **54**(1), 4311–4322 (2006)
5. Guan, N., Tao, D., Luo, Z., Yuan, B.: Online nonnegative matrix factorization with robust stochastic approximation. IEEE Trans. neural Netw. Learn. Syst. **23**(7), 1087–1099 (2012)
6. Guan, N., Tao, D., Luo, Z., Yuan, B.: Non-negative patch alignment framework. IEEE Trans. Neural Netw. **22**(8), 1218–1230 (2011)
7. Guan, N., Tao, D., Luo, Z., Yuan, B.: NeNMF: an optimal gradient method for non-negative matrix factorization. IEEE Trans. Sig. Process. **60**(6), 2882–2898 (2012)
8. Lian, X., Li, B., Zhang, L.: Max-margin dictionary learning for multiclass image categorization. In: Proceedings of European Conference on Computer Vision (2010)
9. Pham, D.-S., Venkatesh, S.: Joint learning and dictionary construction for pattern recognition. In: Proceedings of Computer Vision and Pattern Recognition (2008)
10. Zhang, Q., Li, B.: Discriminative k-svd for dictionary learning in face recognition. In: Proceedings of Computer Vision and Pattern Recogntion (2010)
11. Mairal, J., Bach, F.: Discriminative learned dictionaries for local image analysis. In: Proceedings of Computer Vision and Pattern Recognition (2008)

12. Lee, J., Ho, J., Kriegman, D.: Acquiring linear subspaces for face recognition under variable lighting. TPAMI **27**, 684–698 (2005)
13. Martinez, A., Benavente, R.: The AR face database. University of Purdue, CVC Technical report, 24 (1998)
14. Rosasco, L., Verri, M., Santoro, A., Villa, S.: Iterative projection methods for structured sparsity regularization. MIT Technical reports MIT-CSAIL-TR-2009-050 CBCL-282 (2009)
15. Chen, S., Donoho, D., Saunders, M.: Atomic decomposition by basis pursuit. SIAM J. Sci. Comput. **20**, 33–61 (1998)

Learning Sparse Features in Convolutional Neural Networks for Image Classification

Wei Luo$^{(\boxtimes)}$, Jun Li, Wei Xu, and Jian Yang

School of Computer Science and Engineering, NJUST,
Nanjing 210094, People's Republic of China
{cswluo,junli.njust,csweix}@gmail.com, csjyang@njust.edu.cn

Abstract. The Neural Network (NN) with Rectified Linear Units (ReLU), has achieved a big success for image classification with large number of labelled training samples. The performance however is unclear when the number of labelled training samples is limited and the size of samples is large. Usually, the Convolutional Neural Network (CNN) is used to process the large-size images, but the unsupervised pre-training method for deep CNN is still progressing slowly. Therefore, in this paper, we first explore the ability of denoising auto-encoder with ReLU for pre-training CNN layer-by-layer, and then investigate the performance of CNN with weight initialized by the pre-trained features for image classification tasks, where the number of training samples is limited and the size of samples is large. Experiments on Caltech-101 benchmark demonstrate the effectiveness of our method.

Keywords: Rectified linear units · Convolutional neural networks · Image classification

1 Introduction

Image classification is still a challenging task in computer vision. Bag-of-Features (BoF) [10] and deep learning based models [2,5] are the two mainstream methods to solve this problem. The deep learning methods directly learn representations from images if the image size is small, e.g. $\leq 32 \times 32$, by using Restricted Boltzmann Machines (RBMs) [5] or regularized Auto-Encoders (AEs) [2], e.g. Sparse Auto-Encoder (SAE) [17,18], Denoising Auto-Encoder (DAE) [21], Contractive Auto-Encoder (CAE) [20], etc. With the image size increasing, it's impractical to learn representations directly from the original images, since the large-size will cause the weight matrix in RBMs or AEs too large to learn efficiently and effectively. Therefore, the mainstream to learn new representations is reconstructing the image patches with some regularizations [7,19]. And then the image is processed patch-by-patch to obtain its global representations by concatenating each patch's representation.

The deep learning method for image classification has achieved high performances with the labelled training data increasing in recent years [4,9].

© Springer International Publishing Switzerland 2015
X. He et al. (Eds.): IScIDE 2015, Part I, LNCS 9242, pp. 29–38, 2015.
DOI: 10.1007/978-3-319-23989-7_4

The most surprising result is the classification performance obtained by purely supervised training surpassed the traditional unsupervised learning with supervised fine tuning [4] on some benchmarks when using rectified linear units (ReLU) in Deep Neural Networks (DNN). ReLU has been empirically verified to achieve the highest performance in various applications, but the size of the training samples constrained its applications. Until recently, Krizhevsky et al. [9] incorporated ReLU into the Convolutional Neural Networks (CNN) to achieve the best performance on ImageNet classification task [3] by purely supervised learning, where large labelled training data is available. But what's the performance of ReLU will be when the number of labelled training data is limited and the size of training samples is large? As far as we known, it's still unclear in machine learning and computer vision field. This is the main investigation in this study.

The basic models, e.g. RBMs and AEs, for building DNN are seldom be used to train the weight matrix for CNN, except their sparse variants: Sparse RBMs [12] and SAE [18]. Because studies reveal that the neurons in the early area of mammalian's primal visual cortex preserve population sparsity and life-time sparsity [15]. While DAE and CAE argued that they can find the low-dimensional manifold of data distributions and preserve robustness to noise, they indeed do not introduce the sparse responses of hidden units because of their contractive properties [1].

In this study, we first explore the possibility of pre-training CNN by using DAE. Specifically, we instead employ ReLU as the activation unit in DAE, while not utilizing its original logistic sigmoid unit. Because around 50 % of the ReLU hidden units will continuously output zero values after uniform initializing the weights. This simple operation can introduce the sparse responses of hidden units, which is exactly the reasons why SAE and sparse RBMs can be used to initialize CNN. Therefore, we may expect higher sparsity of the latent representations (code or activations of the hidden units) when coding the input by DAE with ReLU. Furthermore, the robustness property of DAE with higher sparsity signifies it can find the main factors of variations embedded in the training data. We then empirically investigate the classification performance of CNN with weight initialized by the pre-trained DAE features. As we stated in the above, we mainly focus on the classification tasks where the number of labelled training data is limited and the data itself is large-size.

2   Related Work

With the success of deep learning methods for image classification [4,9], many variants of the basic models are proposed. While the variant for RBMs is progressing slowly, the variant for AEs are developed fast with different regularization strategy, e.g. SAE, DAE and CAE. These models are mainly dealing with small-size image classification tasks, e.g. MNIST[1], NORB[2], CIFAR-10[3].

---

[1] http://yann.lecun.com/exdb/mnist/.
[2] http://www.cs.nyu.edu/~ylclab/data/norb-v1.0/.
[3] http://www.cs.toronto.edu/~kriz/cifar.html.

Because the small number of classes and low-dimensional structure (after vectorized) classification task can be used to fully explore the model's properties. But with the size of training samples increasing, vectorizing them to train these models becomes impractical.

These models' properties can be controlled by the activation units in the hidden layer. While the widely used saturated activation units, e.g. logistic or hyperbolic tangent units, have nice mathematical nature and simplicity, it is biologically implausible and hurts gradient-based optimization [4]. Jarrett et al. [7] explored various rectified nonlinearities (including the $\max(0, x)$ units, which they refer to positive party) in the context of convolutional networks and found them have stronger discriminative ability. Nair and Hinton [14] found the discriminative performance of RBMs can be steadily improved when using noisy ReLU to replace stochastic hidden binary units. In [4], Glorot et al. investigated the biological property of ReLU, and reported the purely supervised trained model can achieve better performance than the model with unsupervised per-training on some benchmarks. But all these works were worked in NN except [7]. Our work is closely related to [7], but the unsupervised pre-training is totally different. While [7] pre-trains their CNN by using Predictive Sparse Decomposition (PSD) [8] method, we pre-train our CNN by taking advantage of DAE with ReLU.

3 The Model

In this section, we first introduce the basic CNN used in our experiments, and then investigate the learnable ability of DAE with ReLU on different datasets. Especially, we focus on the robustness and sparseness of the learned representations by DAE with ReLU.

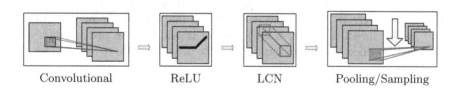

Convolutional   ReLU   LCN   Pooling/Sampling

Fig. 1. CNN flowchart

3.1 The Convolutional Neural Networks (CNN)

CNN [11] works as the early area of mammalian primary visual cortex [6], it models the simple cell and complex cell by convolutional (**C**) layer and pooling (**P**) layer, respectively. Given a 2D input x, the **C** layer outputs a 3D feature map y by:

$$y^{(k)} = f(W_k * x) \tag{1}$$

where W_k is the convolutional kernel for each 2D feature map $y^{(k)}$, $f(x)$ is a component-wise nonlinear activation function. Then the feature maps pass through the **P** layer to pool neighboring responses to increase feature's robustness. Usually, average pooling or max pooling is used:

$$\text{Ave pooling}: p_{i,j}^{(k)} = \frac{1}{N^2} \sum_{p,q=-N/2}^{N/2} y_{i+p,j+q}^{(k)} \tag{2}$$

$$\text{Max pooling}: p_{i,j}^{(k)} = \max\{y_{p,q}^{(k)}\}, \ y_{p,q}^{(k)} \in \mathcal{N}y_{i,j}^{(k)} \tag{3}$$

Finally, the pooling layer is down-sampled to decrease the features' dimension. Modern CNN usually adds one Local Contrast Normalization (**LCN**) layer after the rectified operation to enhance the most active features and suppress the others [7]. In our implementation, we do not include the **LCN** layer, since the main investigation of this work is to verify the effectiveness of a new unsupervised pre-training method for CNN. This architecture can be extended to deep models by stacking one layer after another alternatively. Figure 1 is a flowchart of the modern CNN [7].

3.2   DAE with ReLU

DAE [21] proposes to reconstruct a clean "repaired" input from a corrupted, partially destroyed one. This is done by first corrupting the initial input x to get a partially destroyed version $\tilde{x}$ by means of a stochastic mapping $\tilde{x} \sim qD(\tilde{x}|x)$, and then minimizing the loss function between the reconstructed output from $\tilde{x}$ and the clean pattern x. Specifically, given $\tilde{x} \in R^d$, it first maps $\tilde{x}$ to hidden representation $y \in [0,1]^{d'}$ by a deterministic function $y = f_\theta(\tilde{x}) = s(W\tilde{x}+b)$, parameterized by $\theta = \{W, b\}$. W is the $d' \times d$ weight matrix and b is the bias vector. The resulting latent representation y is then mapped back to "reconstructed" vector $z \in R^d$ in input space $z = g_{\theta'}(y) = s(W'y + b')$ with $\theta' = \{W', b'\}$. The weight matrix W' of the reverse mapping may optionally be constrained by $W^T = W'$, in which case the auto-encoder is said to have tied weights. The parameters are then learned by minimizing the loss function:

$$\theta, \theta' = arg\min_{\theta,\theta'} \frac{1}{N} \sum_{i=1}^{N} L(x^{(i)}, z^{(i)})$$

$$= arg\min_{\theta,\theta'} \frac{1}{N} \sum_{i=1}^{N} L(x^{(i)}, g_{\theta'}(f_\theta(\tilde{x}^{(i)}))) \tag{4}$$

where L is a loss function, e.g. squared loss $L(x, z) = \|x - z\|_2^2$. The deterministic function is usually a saturated nonlinear unit, e.g. logistic function $f(x) = \frac{1}{1+e^{-x}}$ or hyperbolic tangent function $f(x) = \frac{e^x - e^{-x}}{e^x + e^{-x}}$.

In this study, we instead of the original saturated nonlinear unit in DAE by a piecewise linear unit $f(x) = \max(0, x)$, which is called Rectified Linear Units (ReLU) [4,14], to learn the DAE's parameters. To verify our statements in Sect. 1 that the higher sparsity of the latent representations learned by DAE

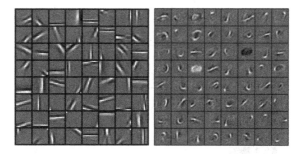

Fig. 2. *Left image*: features on BSD500. *Right image*: features on MNIST.

Table 1. Degree of sparsity with different activation units on different datasets

Activation units	BSD500	MNIST	Average
Logistic sigmoid	0.686	0.481	0.584
ReLU	0.889	0.644	0.767

The activation below 0.01 is considered to be sparse.

with ReLU can preserve the main factors of variations embedded in the data, we experiment on Berkeley Segmentation Dataset (BSD500) [13] and MNIST. The weight matrix of DAE is tied, so that $W^T = W'$ and the number of hidden units is set to 64. We make use of $L1$ penalty on the activations of hidden units with weight $\alpha = 1$ and 0.1 for BSD500 and MNIST datasets, respectively. We random sample 100,000 patches of size 12×12 on BSD500 and 50,000 samples from MNIST training set to learn the weight matrix (features) and compare the sparsity of the mapped latent representations. The learned features with ReLU are displayed in Fig. 2. As we can see the features learned on BSD500 are looks like the Gabor filters, features on MNIST are mostly related to the digit strokes. Table 1 compares the sparsity of the learned latent representations by DAE with ReLU to DAE with logistic units. The sparse penalty ($L1$ norm) for the logistic unit is set to 2 and 0.5 for BSD500 and MNIST, respectively. In all these experiments, we use linear output units and mean square loss function.

We further compare the distance and Coincidence Degree (C.D.) of the latent representations between instances to study the learnability of DAE with ReLU and demonstrate a specific example in Table 2. Two instances from the same class are sampled and the other one is selected to have some degree of the same structure. From Table 2, we can find that the distance of latent representations computed from DAE with ReLU is more distinguishable than that of distance computed from DAE with logistic sigmoid units, and that the category distance can be preserved even though the shared structures between different classes. C.D. computes the number of the same states of hidden units for different inputs. The results in Table 2 reflect the C.D. between instances that come from the same

Table 2. The distance and C.D. of the latent representations between instances come from the same class and different classes

	ReLU		logistic sigmoids	
	dist.[a]	C.D.[a]	dist.	C.D.
	6.739	0.672	2.214	-[b]
	8.420	0.610	2.794	-
	7.972	0.625	2.816	-

[a] dist. and C.D. mean Euclidean distance and coincidence degree, respectively.

[b] The '-' means unavailable.

class is higher than that come from different classes, which can be considered as the ability to find different factors embedded in the data. In contrast to [22], the number of activations of hidden units is not constrained by hand, it is determined automatically from the data. Therefore, we say DAE with ReLU has the ability to find main factors of variations embedded in the data. Note that C.D. is unavailable for logistic sigmoid units, because it doesn't explicitly output 0.

4   Experiments and Results

We evaluate the performance of CNN with different unsupervised pre-training methods for large-size image classification with limited number of labelled training samples on Caltech-101 dataset. Especially, we compare our proposal with random initialization and PSD pre-training methods in [7]. Although our evaluation is focused on two-hierarchy model, it can be extended to more deeper architecture naturally.

4.1   Implementations

We follow [7] to build a two-hierarchy CNN. The number of feature maps in the first and second **C** layers is set to 64 and 256, respectively. The second **P** layer is vectorized to feed to a multi-nominal logistic regression classifier that produces a 102-dimensional output vector representing a posterior distribution over class labels. Each feature map in the second **C** layer is randomly connected to 16 feature maps in the previous **P** layer by 9×9 kernels. So the total number of kernels in the second **C** layer is 4096. We do not consider pooling over different feature maps, but only over the spatial dimensions. Therefore, the numbers of input and output feature maps are identical in the **P** layer, while the spatial resolution is decreased. The average pooling is employed in our experiments.

The pooling modular use a 10×10 boxcar filter with a 5×5 down-sampling and a 6×6 boxcar filter with a 4×4 down-sampling in the first and second $\mathbf{P}$ layers, respectively.

The training images are pre-processed as follows: (1) Each image is converted into gray-scale and down-sampled so that the longest side is 151 pixels, (2) the mean is subtracted and each pixel is divided by the image standard deviation, (3) the image is locally normalized by subtracting the weighted local mean from each pixel and dividing it by the weighted norm if this norm is larger than 1. The weights are forming a 9×9 Gaussian window centered on each pixel, and (4) the image is 0-padded to 143×143 pixels.

To train a model (DAE with ReLU) which is compatible with the first $\mathbf{C}$ layer of CNN, we randomly sample 100,000 patches of size 9×9 from the pre-processed images and whiten these samples. The number of hidden units is set to 64. We train 50 epochs of stochastic gradient descent (SGD) with 100 samples per batch. The learning rate is $\epsilon = 0.001$ and decayed by $\eta = 0.95$ after each epoch. The initial momentum is $\kappa = 0.5$ and linear increased to 0.99 after 5 epochs. The weights of DAE are tied and $L1$ norm is utilized to force the sparse responses of hidden units with penalty coefficient $\rho = 0.5$. We also add $L2$ weights decay on the weight matrix with coefficient $\beta = 0.0001$. The DAE is destructed with $\nu = 0.25$, which means 25 % pixels of each input are forced to 0, while the others are left untouched. When training is done, the weight matrix is used to initialize the kernels in the first $\mathbf{C}$ layer.

We train the second layer of the model of by randomly sampling 100,000 patches of size $9 \times 9 \times 64$ from the first $\mathbf{P}$ layer. 64 DAEs with ReLU are trained separately by using data only comes from the same channel. Thus, the size of training data for each DAE is $81 \times 100,000$ after vectorizing each 9×9 patch. The number of hidden units of each DAE is determined by the connections between the first $\mathbf{P}$ layer and the second $\mathbf{C}$ layer in CNN. We train 30 epochs SGD with 100 samples per batch, and $L1$ norm is used to force the sparse responses of hidden units with penalty coefficient $\rho = 0.2$. The other training parameters are the same with the first layer DAE model. After training is done, the weight matrix of each DAE is used to initialize the kernels connected with the corresponding feature maps.

4.2   Results

The Caltech-101 dataset contains 101 classes and one background class. The number of images per category varies from 31 to 800. All images are pre-processed as specified in the previous section. We follow the common experiment setup, training on 15 and 30 images per category and testing on no more than 50 images per category. To adjust hyper-parameters in the stage of fine tuning, a validation set of 5 samples per category was taken out of the training set. The hyper-parameters were selected to maximize the performance on the validation set. The hyper-parameters in our system involve learning velocity and training epoch. The learning velocity $\epsilon = \{0.005, 0.01, 0.1\}$ were chosen in our experiment and tested on the validation set. The optimal learning epoch was determined by

Table 3. Classification performance on Caltech-101 with 15 and 30 training samples from each category for different initialization methods. The number in parentheses is the optimal training epoch determined by GL criterion.

Training images	15(epochs)	30(epochs)
Random	50.2(27)	59.5(24)
PSD [7]	-[a]	60.5
Ours	50.1(**12**)	60.0(**14**)

[a]'-' means unavailable.
The variances are all below 0.05 in all setups.

GL criterion [16] with maximal 30 epochs. Then, the system was trained over the entire training set with the learning rate which has the lowest test error on the validation set. We report the average correct classification rate over 5 random trials.

Table 3 compares the correct classification rate of our method with random and PSD initialization methods. Due to our CNN do not include the **LCN** layer, we compare our results with [7] in the case of not containing the **N** stage in their report. When trained on 15 samples per category, the result for random initialization and pre-trained by DAE with ReLU is very close. *But the random initialization needs **more than twice training epochs** to achieve the same level of classification rate as pre-trained initialization method does* (see Table 3). While we not investigate whether other pre-training method can bring this effect, the pre-training with DAE with ReLU indeed accelerates the optimization for CNN. This phenomenon is also observed when the training data is enlarged to 30 samples per category. The classification rate under 30 training samples per category is also very similar for different initialization methods, which is consistent with [7]. From Table 3 and reports in [7], we find the pre-training mainly brings the advantages of fast optimizing CNN when compared to random initialization, and the mechanic behinds the success of these pre-training methods is the induced sparse activations of hidden units.

5   Conclusions

In this paper, we explored the property of DAE with ReLU for finding the main factors of variations embedded in the data by inducing higher sparse activations of its hidden units. Based on this property, we proposed to initialize CNN by using DAE with ReLU. Experiments on Caltech-101, where the number of labelled training samples is limited and the size of samples is large, verified the effectiveness of our method and validated the importance of the sparse responses of hidden units in CNN. Future work includes evaluating the performance under max-pooling method and extending to contain the **LCN** layer in our CNN, also testing on more datasets to verify the robustness of our method.

Acknowledgments. The authors would like to thank the editor and the anonymous reviewers for their critical and constructive comments and suggestions. This work was partially supported by the National Science Fund for Distinguished Young Scholars under Grant Nos. 61125305, 91420201, 61472187, 61233011 and 61373063, the Key Project of Chinese Ministry of Education under Grant No. 313030, the 973 Program No. 2014CB349303, Fundamental Research Funds for the Central Universities No. 30920140121005, and Program for Changjiang Scholars and Innovative Research Team in University No. IRT13072.

References

1. Alain, G., Bengio, Y.: What regularized auto-encoders learn from the data generating distribution. In: ICLR (2013)
2. Bengio, Y., Lamblin, P., Popovici, D., Larochelle, H.: Greedy layer-wise training of deep networks. In: NIPS (2006)
3. Deng, J., Dong, W., Socher, R., Li, L.J., Li, K., Fei-Fei, L.: Imagenet: a large-scale hierarchical image database. In: CVPR (2009)
4. Glorot, X., Bordes, A., Bengio, Y.: Deep sparse rectifier neural networks. In: International Conference on Artificial Intelligence and Statistics (2011)
5. Hinton, G.E., Salakhutdinov, R.R.: Reducing the dimensionality of data with neural networks. Science **313**(5786), 504–507 (2006)
6. Hubel, D., Wiesel, T.: Receptive fields and functional architecture in two nonstriate visual areas (18 and 19) of the cat. J. Neurophys. **28**, 229–289 (1965)
7. Jarrett, K., Kavukcuoglu, K., Ranzato, M., LeCun, Y.: What is the best multi-stage architecture for object recognition? In: ICCV (2009)
8. Kavukcuoglu, K., Ranzato, M., LeCun, Y.: Fast inference in sparse coding algorithms with applications to object recognition. Technical report, New York University (2008)
9. Krizhevsky, A., Sutskever, I., Hinton, G.E.: Imagenet classification with deep convolutional neural networks. In: NIPS (2012)
10. Lazebnik, S., Schmid, C., Ponce, J.: Beyond bags of features: spatial pyramid matching for recognizing natural scene categories. In: CVPR (2006)
11. LeCun, Y., Bottou, L., Bengio, Y., Haffner, P.: Gradient based learning applied to document recognition. Proc. IEEE **86**(11), 2278–2324 (1998)
12. Lee, H., Ekanadham, C., Ng, A.Y.: Sparse deep belief net model for visual area v2. In: NIPS (2008)
13. Martin, D., Fowlkes, C., Tal, D., Malik, J.: A database of human segmented natural images and its application to evaluating segmentation algorithms and measuring ecological statistics. In: ICCV (2001)
14. Nair, V., Hinton, G.E.: Rectified linear units improve restricted boltzmann machines. In: ICML (2010)
15. Ngiam, J., Koh, P.W., Chen, Z., Bhaskar, S., Ng, A.Y.: Sparse filtering. In: NIPS (2011)
16. Prechelt, L.: Early stopping — but when? In: Montavon, G., Orr, G.B., Müller, K.-R. (eds.) Neural Networks: Tricks of the Trade, 2nd edn. LNCS, vol. 7700, pp. 53–67. Springer, Heidelberg (2012)
17. Ranzato, M., Poultney, C., Chopra, S., LeCun, Y.: Efficient learning of sparse representations with an energy-based model. In: NIPS (2006)
18. Ranzato, M., Boureau, Y.L., LeCun, Y.: Sparse feature learning for deep belief networks. In: NIPS (2007)

19. Ranzato, M., Huang, F.J., Boureau, Y.L., LeCun, Y.: Unsupervised learning of invariant feature hierarchies with applications to object recognition. In: CVPR (2007)
20. Rifai, S., Vincent, P., Muller, X., Glorot, X., Bengio, Y.: Contractive auto-encoders: explicit invariance during feature extraction. In: ICML (2011)
21. Vincent, P., Larochelle, H., Bengio, Y., Manzagol, P.A.: Extracting and composing robust features with denoising autoencoders. In: ICML (2008)
22. Wang, J., Yang, J., Yu, K., Lv, F., Huang, T., Gong, Y.: Locality-constrained linear coding for image classification. In: CVPR (2010)

Semi Random Patches Sampling Based on Spatio-temporal Information for Facial Expression Recognition

Haiying Xia[1,2,3](✉)

[1] College of Electronic Engineering, Guangxi Normal University,
Guilin 541004, China
xhyhust@gmail.com
[2] Guangxi Key Laboratory of Automatic Detecting Technology and Instruments,
Guangxi 541004, China
[3] Key Laboratory for the Chemistry and Molecular Engineering of Medicinal
Resources (Guangxi Normal University), Ministry of Education of China,
Guangxi 541004, China

Abstract. This paper presents a novel method for facial expression recognition based on semi random patches sampling. Different from most of the facial expression methods that use spatio expression descriptor, temporal expression descriptor or both, we extract spatio-temporal expression information by a technology of semi random patches sampling. In the facial feature extraction, expression salient features are first determined; face images are second normalized and warped to a standard face; then temporal expression information is obtained by the differences between the warped expression and its corresponding reference expression model; thirdly, a small set of patches is extracted from both spatio expression information and temporal expression information around each expression salient features. The semi random patches are used to perform facial expression recognition. The proposed semi random patches extraction is simple, yet by leveraging the sparse nature of expression images. Experimental results demonstrate that our approach is able to capture subtle texture variations caused by expression, even under occlusion and corruption.

Keywords: Facial expression recognition · Semi random patches · Sparse representation · Spatio-temporal · Expression salient features

1 Introduction

The successfully recognition of one's expressions in images or videos has many applications such as human-computer interaction, virtual reality, video conferencing and synthetic face animation [1]. A commonly used facial expression recognition approach is to classify each given facial image into the basic expression types, e.g., happy, sad, disgust, surprise, fear and angry, proposed by Ekman and Friesen [2]. Numerous expression detectors that operate on an image or a video have been developed in the past few years. However, the issue of recognizing facial expression in complicated environment, such as illumination, occlusion, disguise and corruption, needs deeper study for its' practical utilization [3].

© Springer International Publishing Switzerland 2015
X. He et al. (Eds.): IScIDE 2015, Part I, LNCS 9242, pp. 39–49, 2015.
DOI: 10.1007/978-3-319-23989-7_5

The design of a facial expression recognition system generally includes three major steps: face detection, feature extraction and classification. Most research in facial expression recognition focuses on the feature extraction step. Early approaches include Gabor filters [4], extended Haar-like features [5], Local Phase Quantization from Three Orthogonal Planes [6], Local Phase Quantization, local binary pattern from Three Orthogonal Planes [7], among many others. A quantum jump occurred in the early 21st century when face can be reliably detected from complex background in real time. With further advances in texture description and face tracking, there emerge plenty of feature extraction methods, such as local binary pattern (LBP) [8], advanced local binary pattern [9], and many others. All of those feature extraction methods can be classified in three ways: spatio information based, temporal information based and spatio-temporal based. However, no single approach does perform best or very close to the best for face expression images. Those above mentioned methods encounter some drawbacks in application.

(1) Accurate feature location. Most of temporal information extraction methods need accurately locate the expression features. The expression features will constantly change their positions when subjects are expressing emotions. In result, the same feature in different images generally has different positions. So we can to track the expression features to explore the variations in positions, or shapes (e.g., shape vectors [10], facial animation parameters [11], distance and angular [12], and trajectories [13]). Although they achieved promising results, these approaches often require accurate location and tracking of facial points, which remains problematic [14].

(2) Extraction complexity. Spatio information mainly involves appearance difference between different expressions in static images (e.g. Gabor filters), or between holistic facial regions in consequent frames (e.g., optical flow [15], and differential-AAM [16], surface deformation [17], motion units [18], spatio-temporal descriptors FV [19], animation units [20], and pixel differences [21]). No matter static images or dynamic videos, most of those feature extraction methods rely on the results of face alignment and some of them are time consuming. It is still an open question as to how to balance the extraction complexity and the recognition rate.

(3) Sensitive to occlusion. Occlusion poses a significant obstacle to robust real-world facial expression recognition. When face has occlusion, it is difficult to predict the correct positions of expression features and it will cause some loss of expression information. In most facial expression systems, this will decrease the system performance to some extent.

(4) Noise sensitivity. Noise is more or less inevitable in face images. It will randomly corrupt pixels and influence the accuracy of feature extraction. Some methods which require accurate feature location or depend on pixel difference, cannot obtain high recognition performance as predicted.

Recently, sparse representation has shown great potential in various applications such as face recognition, super resolution reconstruction and object tracking. Inspired by this methodology, we exploit the discriminative nature of sparse representation to perform facial expression recognition. The success of our sparse approach largely depends on the type of feature extraction based on semi random patches sampling. Instead of using the facial features discussed above, we represent the test sample in an overcomplete dictionary based on semi random patches sampling whose base elements are the training patches themselves. If sufficient training patches are available from each expression, it will be possible to represent the test samples as a linear combination of just those training patches from the same class. This representation is naturally sparse, involving only a small fraction of the overall training database. We argue that face images contain a lot of information, such as face identification information, age information, gender information and expression information. Expression information is just a small part of the overall face information. We not only consider the sparsest linear representation of the test face image, but also pay more attention to improve the recognition performance. Semi random patches sampling is a way to sample face images around facial expression features. Different patches extracted from different expression features are connected to form an expression vector. A large number of expression vectors are utilized to train a desired dictionary. After the dictionaries are obtained, we can realize facial expression recognition via sparse representation over learned dictionaries.

2 Expression Classification Using Semi Random Patches Sampling

2.1 Face Preprocessing

For expression recognition, we first use the detection algorithm of the face salient features to locate the face salient features. Here, we take active shape model (ASM) [22], which are constrained by the PDM (point distribution model) to vary only in ways seen in a training set of labeled examples, to find the face salient features. Note that other face feature location algorithms are also appropriate choices in our situation. Secondly, delaunay triangulation, which is a technique for connecting points in a space into triangular groups such that the minimum angle of all angles in the triangulation is a maximum, is used to warp the face to a reference face. An example of triangulated face is demonstrated in Fig. 1. Finally, face texture is warped to a standard frame by the affine transformation. The transformation formula is shown in (1).

Fig. 1. The triangulated faces

$$\begin{bmatrix} x' \\ y' \end{bmatrix} = \begin{bmatrix} a_1 & a_2 & m_x \\ b_1 & b_2 & m_y \end{bmatrix} \times \begin{bmatrix} x \\ y \\ 1 \end{bmatrix} \tag{1}$$

where (x', y') is the coordinate position after warping, and (x, y) is the coordinate position before warping. a_1, a_2, b_1, b_2 determine the size and rotation of the face image, and m_x, m_y control the translation of the face image. During the warping procedure, we use bilinear interpolation. Compared with bicubic interpolation, the speed of the warping of bilinear interpolation is faster, and the results are satisfactory. Figure 2 gives one example face and the face after preprocessing.

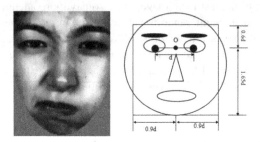

Fig. 2. (a) Preprocessed face; (b) the face reference model

2.2   Feature Extraction

2.2.1   Spatio-temporal Expression Information

To improve the performance of the expression recognition system, we not only consider spatio information, which can be described by a static appearance descriptors, but also take appearance changes into consideration, that is differential texture (DT). The extraction of DT can be divided into the following steps. First, the reference face model is established and face salient features are located by ASM in a given expression image. Then, by utilizing the relative location information we have obtained, we can warp both neutral expression and non-neutral expression to the reference face model. After the above process, the difference between neutral images and non-neutral images can be formed as DT images. Figure 3 lists some examples of DT images. In Fig. 4, the first row is the expression images after warping, the second is the corresponding neutral images and the DT images are the difference between the expression images and its corresponding neutral images.

2.2.2   Semi Random Patches Sampling

In fact, random blocks are small image patches extracted randomly from the face. To minimize the interruption from those facial features that contribute significantly to face recognition problems, our random blocks are sampled near expression salient features, instead of completely random on the whole face. The size of the block is relevant to the size of the face. This kind of random blocks has two advantages. One is that these blocks encode efficient features for facial expression recognition, which discards much unwanted information. The other is that the block sampling has a low requirement for the position precision of facial salient features. As for the location of facial salient

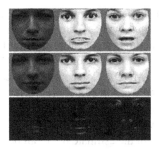

Fig. 3. The extraction of DT images

features, they are located by ASM to guide the block extraction. In this way, these blocks are easy to extract and more suitable for the problem of recognizing facial expressions.

We can find evidence in previous studies that the important facial salient features for facial expression recognition are distributed mainly on areas, such as eye, eyebrow and mouth. Here, we use ASM to locate the 26 facial expression salient features along the contours of eyebrows, eyes and mouth. The expression patches are extracted from two ways: one is to sample the warped expression images; the other is to sample the DT images. Figure 4(a) shows the distribution of those salient features and Fig. 4(b) indicates the way to sample blocks within the areas of salient features, where the square in solid line is the area of salient features and the square in dotted line is the sample block.

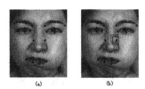

Fig. 4. (a) Facial salient features. (b) Random patch sampling

2.2.3 Feature Vectors

How to organize so many patches is vital to the success of our FER system. A good expression feature vector should contain both the global expression structure and the local expression variation. To obtain a good feature vector, we organize the semi random patches in such a way that one feature vector contains 26 patches, which come from 26 different expression salient features.

2.3 Facial Expression Recognition

Sparse representation aims at describing a high-dimensional signal as a linear combination of a few generating elements. Expression dictionary training can extract the important information, which reduces the dimension of the input samples, but can describe that expression more effectively. There are several ways to learn a dictionary that that yields sparse representations for a set of random blocks, such as K-SVD, the generalized lloyd algorithm (GLA), maximum likelihood methods (MLM) and the method of optimal directions (MOD). Compared to other dictionary learn methods, K-SVD is more flexible and can work in conjunction with any pursuit algorithm. Here, we choose the K-SVD to train an expression dictionary. The detail of the K-SVD algorithm is described in the work of Michal's [23].

Ideally, the nonzero entries in the estimate b will all be associated with the columns of D from a single expression class, and we can easily assign the test sample y to that class. However, modeling error may lead to small nonzero entries associated with multiple expression classes. Based on the sparse representation, we can design many possible ways to classify this expression this. For example, we can simply assign test expression y to the class with the single largest entry in b. However, such approach does not harness the subspace structure associated with facial expression images. To better exploit such linear structure, we instead classify y based on the distribution of the coefficients among all training samples.

For each class i, let δ_i be the nonzero entries in b associated with the ith class. Using all the nonzero entries in b, we can get the distribution of the coefficients among all expression classes by summing the values of the nonzero entries in each expression class. We then classify test expression y based on distribution analysis by assigning it to the expression class that maximizes the sum of nonzero entries in each expression class.

$$\hat{i} = \arg\max_i (sum(\delta_i)) \qquad (2)$$

3 Experiments

Our experiments are conducted on the JAFFE and CK+ facial expression database to recognize six basic expressions (sadness, anger, surprise, happy and disgust). The JAFFE, which are widely used, contains 213 facial expressions of 10 female expressers. The expressers posed 3 or 4 examples for each of seven expressions (happy, anger, sadness, surprise, fear, disgust and neutral). The image size is 256×256 pixels. The CK+ database contains a large number of sequences and subjects. CK+ databases have been widely used for evaluating facial activity recognition system. One advantage is that the results on the CK+ and JAFFE database can be used to compare with other published methods. The other advantage of using CK+ and JAFFE database is that it contains a large number of subjects that display facial expressions with six basic expression and neutral expression.

3.1 Experimental Setup

We divide the face images in JAFFE into two disjoint sets: training set and test set. Among 10 subjects, we randomly took 7 subjects for train set and the rest for test set.

90 subjects that contain the basic six expressions and neutral expression are selected from the CK+ database. Among 90 subjects, we randomly took 60 subjects for train set and the rest for test set. In all, we have 67 subjects for train and 33 subjects for test. All the face images are cropped and warped to a standard frame with size 100 × 100.

Next, semi random patches are extracted from the training subjects to form the training samples. One feature vector of the training subject contains 26 patches, which come from 26 expression salient features. One feature vector of the training sample has 1664 elements, which come from 26 patches and each patch is normalized to 8 × 8. Here, we take ten random patches around each expression salient features. From one expression face, we take 200 feature vectors through variously combining patches in each expression salient feature. So those feature vectors are sent to the K-SVD algorithm to learn an expression dictionary. Assume the learned expression dictionary has 100 column vectors.

3.2 Experimental Results

From the Table 1, we can see our method achieves a good performance on both JAFFE and Cohn-Kanade database. Happy and Surprise are all correctly recognized and Fear also obtains a relative higher recognition. The average recognition rate reaches to 98.8 %.

Table 1. The recognition rate of our method on JAFFE and Cohn-Kanade database

	Anger	Disgust	Fear	Happy	Sad	Surprise
Anger	97.9 %	0	2 %	0	2.08 %	0
Disgust	1.04 %	98.94 %	0	0		0
Fear	0	0	98 %	0	1.04 %	0
Happy	0	0	0	100 %		0
Sad	1.04 %	1.06 %	0	0	97.9 %	0
Surprise	0	0	0	0	1.04 %	100 %

We also test our SRC algorithm using three conventional features, namely single-level LBP (Local Binary Pattern) features, multi-level LBP features and Gabor features, and compare their performance with our random patches. The single-level, multi-level LBP and Gabor features are all extracted from the cropped face images, which are normalized to 100 × 100 with the minimization of the variations in the different face images. So we get 59 single-level LBP features, 4956 multi-level LBP features and 400,000 Gabor features from one cropped face images.

Table 2 shows the recognition performance for the various features, in conjunction with three different classifiers: SRC, SVM, and NN. From the distribution of recognition rates, we can observe that Happy and Surprise can be recognized with high accuracy, but Fear is easily confused with others. Multi-level LBP features have a much better performance on expression recognition than single-level LBP features. Our SRC algorithm using random blocks provides better overall performance than single-level LBP features or multi-level LBP features. And, the performance of our

SRC algorithm also exceeds the performances of the two classical methods, SVM and NN. Otherwise, the performances of the other two classifiers depend strongly on a good choice of "optimal" features. Under the same conditions, the features with feature selection by AdaBoost receive much better performance than without feature selection. But compared to Boosting-based methods, the recognition rate of our SRC algorithm is still a bit higher.

Table 2. Recognition performance of Gabor based NN or SVM, LBP based SVM or SRC vs Random patches based SRC

Method	Happy	Sadness	Fear	Disgust	Surprise	Anger	Average
Gabor + NN	83.1 %	72.9 %	60 %	70.3 %	82.9 %	70 %	73.2 %
Multi-level LBP + SVM	96.7 %	93.3 %	86.7 %	96.7 %	96.7 %	93.3 %	92.8 %
Gabor + SVM	98.9 %	98 %	94.5 %	98.9 %	100 %	97.9 %	98 %
Single-level LBP + SRC	74.3 %	68.6 %	51.4 %	74.3 %	77.1 %	65.7 %	68.5 %
Multi-level LBP + SRC	98 %	96.4 %	90.6 %	94.2 %	96.8 %	97.9 %	95.6 %
Random patches + SRC	100 %	97.9 %	98 %	98.9 %	100 %	97.9 %	98.9 %

We analyze the time and memory costs of feature extraction process for LBP features and Gabor-wavelet features in Table 2, where the Gabor-filter convolutions were calculated in spatial domain. It is observed that LBP features bring significant speed benefit, and, compared to the high dimensionality of the Gabor-wavelet features, LBP features lie in a much lower dimensional space (Table 3).

Table 3. Time and memory costs for extracting LBP features and Gabor-filter features

	Single-level LBP	Multi-level LBP	Gabor	Random patches
Memory	59	4956	400,000	650
Time	0.3 ms	60 ms	30 s	80 ns

3.3   Recognition with Noise and Occlusion

Next, we test our SRC algorithm on corrupted or occluded test faces. The corrupted faces are obtained by randomly chosen pixels from each face image, replacing their values with independent and identically distributed samples from a Gassian distribution. The corrupted pixels are randomly chosen for each test image, and the locations are unknown to the algorithm. Figure 5(a) shows the examples of the corrupted faces with the Gaussian variance 0.01, 0.03, 0.05, 0.1, 0.2, 0.5.

We simulate the occluded test faces by occluding each test image with a square range from 5×5 to 50×50, where the test image is 256×256. The square is randomly positioned on the test face such that it lies within the face. An example of this type of occlusion is shown in Fig. 5(b) where the occlusion is a black square of size 10, 20, 30, 40, 50, 60 pixels.

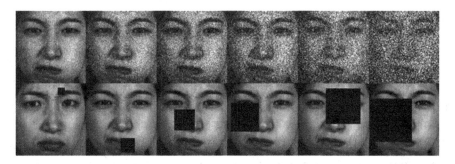

Fig. 5. (a) the corrupted face; (b) the occluded face

From Fig. 6(a), we can see when the variance is less than 0.1, noise has little effect on the facial expression classification; and when the variance is 0.5, the expression information is almost covered by noise. The recognition rate of our method can still reach 78 %, while the method of Gabor plus SVM only obtains 48 %. Also, from Fig. 6 (b), we can be see our method for occlusion shows better robustness, significantly higher than that of Gabor combined with SVM classifier.

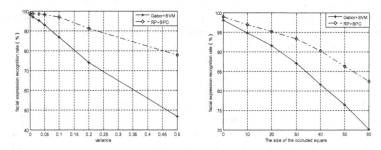

Fig. 6. (a) Noise effect on two kinds of FER methods; (b) occlusion effect on two kinds of FER methods.

4 Conclusions and Future Work

In this paper, we proposed a novel algorithm using semi random patches sampling via sparse representation, which takes advantage of both spatio and temporal expression information. The proposed semi random patches extraction is simple, yet by leveraging the sparse nature of expression images. Extensive experiments showed that our semi random patches based on SRC were highly suited for the problem of FER, outperforming the classic works that reported on this task. Even, under corrupted or occluded, our method still can get a good performance.

Given the discriminative power of the dictionary in terms of classifying different expressions, it seemed natural to extend our work to further attach much attention to discriminability of the trained dictionary, instead of only emphasizing the data representation.

References

1. Zeng, Z., Pantic, M., Roisman, G.I., Huang, T.S.: A survey of affect recognition methods: audio, visual, and spontaneous expressions. IEEE Trans. PAMI **31**(1), 39–58 (2009)
2. Ekman, P., Friesen, W.V.: Pictures of facial affect. In: Human Interaction Laboratory. University of California Medical Center, San Francisco (1976)
3. Fasel, B., Luettin, J.: Automatic facial expression analysis: a survey. Pattern Recogn. **36**(1), 259–275 (2003)
4. Tong, Y., Liao, W., Ji, Q.: Facial action unit recognition by exploiting their dynamic and semantic relationships. IEEE Trans. Pattern Anal. Mach. Intell. **29**(10), 1683–1699 (2007)
5. Kim, D.H., Jung, S.U., et al.: Extension of cascaded simple feature based face detection to facial expression recognition. Pattern Recogn. Lett. **29**(11), 1621–1631 (2008)
6. Zhao, G., Pietikainen, M.: Dynamic texture recognition using local binary pattern with an application to facial expressions. IEEE Trans. Pattern Anal. Mach. Intell. **2**(6), 915–928 (2007)
7. Jiang, B., Valstar, M., Martinez, B., Pantic, M.: A dynamic appearance descriptor approach to facial actions temporal modeling. IEEE Trans. Cybern. **44**(2), 161–174 (2014)
8. Feng, X., Hadid, A., Pietikainen, M.: Facial expression recognition with local binary patterns and linear programming. Pattern Recogn. Image Anal. **15**(2), 546–554 (2005)
9. Shan, C., Gong, S., McOwan, P.W.: Facial expression recognition based on local binary patterns: a comprehensive study. Image Vis. Comput. **27**, 803–816 (2009)
10. Yan, T., Jixu, C., Qiang, J.: A unified probabilistic framework for spontaneous facial action modeling and understanding. IEEE Trans. Pattern Anal. Mach. Intell. **32**(2), 258–273 (2010)
11. Aleksic, P.S., Katsaggelos, A.K.: Automatic facial expression recognition using facial animation parameters and multistream HMMs. IEEE Trans. Inf. Forensics Secur. **1**(1), 3–11 (2006)
12. Hamdi, D., Roberto, V., Ali, S.A., Theo, G.: Eyes do not lie: spontaneous versus posed smiles. In: Proceedings of the international conference on Multimedia, pp. 703–706 (2010)
13. Wang, T.-H., Lien, J.-J.J.: Facial expression recognition system based on rigid and non-rigid motion separation and 3D pose estimation. J. Pattern Recogn. **42**(5), 962–977 (2009)
14. Peng, L., Prince, S.J.D.: Joint and implicit registration for face recognition. In: Proceedings of IEEE Conference on Computer Vision and Pattern Recognition, pp. 1510–1517 (2009)
15. Yeasin, M., Bullot, B., Sharma, R.: Recognition of facial expressions and measurement of levels of interest from video. IEEE Trans. Multimedia **8**(3), 500–508 (2006)
16. Cheon, Y., Kim, D.: Natural facial expression recognition using differential-AAM and manifold learning. Pattern Recogn. **42**, 1340–1350 (2009)
17. Tsalakanidou, F., Malassiotis, S.: Real-time 2D+3D facial action and expression recognition. Pattern Recogn. **43**, 1763–1775 (2010)
18. Cohen, I., Sebe, N., Garg, A., Chen, L.S., Huang, T.S.: Facial expression recognition from video sequences: temporal and static modeling. Comput. Vis. Image Underst. **91**, 160–187 (2003)
19. Zhao, G., Pietikainen, M.: Boosted multi-resolution spatio-temporal descriptors for facial expression recognition. Pattern Recogn. Lett. **30**, 1117–1127 (2009)
20. Qiao, L., Chen, S., Tan, X.: Sparsity preserving projections with applications to face recognition. Pattern Recogn. **43**(1), 331–341 (2010)
21. Lai, Z., Jin, Z., Yang, J.: Global sparse representation projections for feature extraction and classification. In: Proceedings of Chinese Conference on Pattern Recognition, pp. 1–5, November 2009

22. Shan, C., Gong, S., McOwan, P.W.: Facial expression recognition based on local binary patterns: a comprehensive study. Image Vis. Comput. **27**, 803–816 (2009)
23. Aharon, M., Elad, M.: K-SVD: an algorithm for designing overcomplete dictionaries for sparse representation. IEEE Trans. Sig. Process. **54**(11), 4311–4323 (2006)

Band Selection of Hyperspectral Imagery Using a Weighted Fast Density Peak-Based Clustering Approach

Sen Jia[1,2(✉)], Guihua Tang[1,2], and Jie Hu[1,2]

[1] College of Computer Science and Software Engineering,
Shenzhen University, Shenzhen, China
senjia@szu.edu.cn, tangguihua@email.szu.edu.cn,
hujie0513@163.com
[2] Shenzhen Key Laboratory of Spatial Information Smarting Sensing
and Services, Shenzhen University, Shenzhen, China

Abstract. Based on the search strategy of representative bands in Hyperspectral Imagery, various existing unsupervised band selection approaches are mainly classified into two parts: ranking-based and clustering-based ones. Recently, a fast density peak-based clustering (abbreviated as FDPC) algorithm has been proposed. The product of two factors (the computation of local density and intra-cluster distance) is sorted in decreasing order and cluster centers are recognized as points with anomalously large values, hence the FDPC algorithm can be considered as a ranking-based clustering method. In this paper, the FDPC algorithm has been modified to make it suitable for hyperspectral band selection by weighting the normalized local density and intra-cluster distance. It is called a weighted fast density peak-based clustering (W-FDPC) method. Experimental results demonstrate that the bands selected by W-FDPC approach can achieve higher overall classification accuracies than FDPC and other state-of-the-art band selection techniques.

Keywords: Hyperspectral imagery · Band selection · Density-based clustering

1 Introduction

Hyperspectral image data is collected by the hyperspectral sensors (its wavelength range is typically 400–2500 nm) in a continuous spectral bands at different wavelengths. These images form a image cube, which can be used for surface material classification and identification according to the continuous spectrum of each pixel [1–4]. However, compared with the large amount of spectral bands, it is difficult and

This work was jointly supported by grants from National Natural Science Foundation of China (61271022 and 61272050), Guangdong Foundation of Outstanding Young Teachers in Higher Education Institutions (Yq2013143), and Shenzhen Scientific Research and Development Funding Program (JCYJ20140418095735628).

X. He et al. (Eds.): IScIDE 2015, Part I, LNCS 9242, pp. 50–59, 2015.
DOI: 10.1007/978-3-319-23989-7_6

laborious to obtain sufficient training samples in practice, which often leads to ill-conditioned problems, such as the Hughes phenomenon [5].

Moreover, due to the high correlation among some bands, which makes the task of data processing and analyzing time-consuming. Therefore, dimensionality reduction (DR) should be applied as a preprocessing step to discard redundant information [6, 7]. Roughly speaking, the DR methods can be divided into two main categories: feature extraction [8, 9] and band (or feature) selection [10, 11]. Generally, the features obtained by feature extraction methods can not be related to the original wavelength. Alternatively, band selection methods can preserve the relevant original information of the spectral bands, which could simplify the image acquisition process and save the data storage. As the task of hyperspectral data analysis is always unsupervised, there exist various unsupervised band selection approaches (UBS), which can be further classified into two main parts: ranking-based [12–14] and clustering-based ones [7, 15, 16].

Recently, a fast density peak-based clustering (abbreviated as FDPC) algorithm has been proposed, which identifies the cluster centers through investigating the local density and intra-cluster distance of each point [17]. Based on a reasonable assumption that cluster centers are surrounded by neighbors with lower local density and that they are at a relatively large distance from any points with a higher local density. In this paper, through investigating the characteristics of the hyperspectral UBS problem, the FDPC method has been used for the task of band selection by the ranking score (the product of the local density and the intra-cluster distance as equally important) of each band. Specifically, after the bands with extremely large scores have been picked out, in order to choose more informative and non-redundant bands from the rest ones, the intra-cluster distance should be considered more favorable than the local density. As a result, the FDPC algorithm is modified by normalizing the two factors and increasing the weight of intra-cluster distance, we called it a weighted fast density peaks clustering (W-FDPC) approach.

The remainder of this paper is organized as follows. In Sect. 2, we present the recently proposed FDPC clustering algorithm and Sect. 3 will introduce its enhanced version (W-FDPC). Experimental results on real hyperspectral remote sensing data are showed in Sect. 4. Finally, conclusions are given in Sect. 5.

2 The Fast Density Peak-Based Clustering Algorithm (FDPC)

By defining the cluster centers as local maxima in the density of data points [18], the FDPC algorithm has presented a simple and effective criterion to rank the data points and find the cluster centers automatically. That is, for the image of each band i; $1 \leq i \leq L$, which can be considered as a data point, its representativeness is decided by two factors, its local density ρ_i and its distance δ_i from points of higher density. Both these quantities depend only on the similarity matrix S, which has been scaled down by the following transformation:

$$D_{ij} = \sqrt{S_{ij}}/L; \tag{1}$$

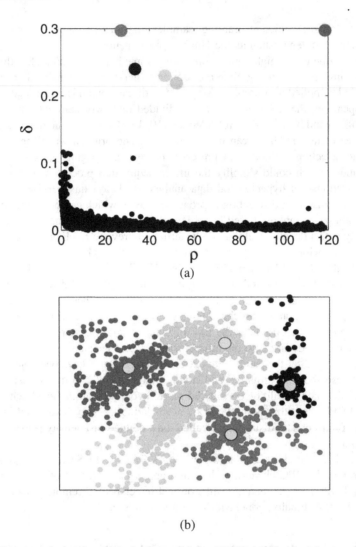

Fig. 1. Cluster analysis of a synthetic data point for FDPC. (a) The decision graph with the centers colored by cluster. (b) The distribution in a two-dimensional space.

$$\rho_i = \sum_{j=1,j\neq i}^{L} \exp(-(D_{ij}/d_c)^2) \tag{2}$$

where dc is a cutoff distance used to keep a region for each data point. It is worth to point out that the gaussian kernel function is adopted to estimate the local density ρ_i for each band image to decrease the negative impact of statistical errors caused by the small number of bands. Likewise, δ_i is measured by computing the minimum distance between the point i and any other point with higher density:

$$\delta_i = \min_{j:\rho_j > \rho_i}(D_{ij}) \tag{3}$$

To illustrate the idea of method, Fig. 1(a) shows the plot of δ_i as a function of ρ_i for each point in a synthetic data distributions, which is called "decision graph". Clearly, the only points of high δ and relatively high ρ are the cluster centers (here five representative points are picked out). Figure 1(b) displays the points' distribution in a two-dimensional space, points are colored according to the cluster to which they are assigned, and the big yellow circles represent the five chosen cluster centers. It can be easily observed from the figure that the cluster centers identified via FDPC are able to reveal the non-spherical structure of this set. Further, to make the clustering process more convenient, the two factors can be multiplied together to obtain a score for each point as follows:

$$\gamma_i = \rho_i \times \delta_i \tag{4}$$

3 The Weighted Fast Density Peak-Based Clustering Algorithm (W-FDPC)

Weight Based Ranking Score Scheme for Each Band. Although the FDPC's idea of combining the local density ρ and the distance δ from points of higher density is far more reliable than using either of the two factors alone, how to effectively integrate the two factors for hyperspectral band selection is still an unsolved issue. Actually, the weight information is important to promote the performance of the algorithm, and the ranking score of each band should be computed through weighting the two factors. Based on this idea, instead of taking δ and ρ into account equally, the proposed W-FDPC algorithm increases the weight of δ through a normalization and square product process. After δ has been computed by Eq. (3), it is normalized to the scale of [0,1], which can be done by

$$\delta = (\delta - \delta_{min})./(\delta_{max} - \delta_{min}) \tag{5}$$

where ./ represents the element-wise division operator, and ρ is normalized in the same way. Because the range of ρ is much larger than that of δ, the weight of δ has been significantly increased after the normalization. Contrarily, in order to compensate ρ for the loss of weight during the normalization procedure, the ranking score γ_i for any band i is finally obtained by

$$\gamma_i = \rho_i \times \delta_i^2 \tag{6}$$

In the respect of band selection, the appropriate number of selected bands is typically much larger than that of clusters in the synthetic data which is showed above. Concerning the FDPC clustering algorithm, when the number of selected bands k is small,

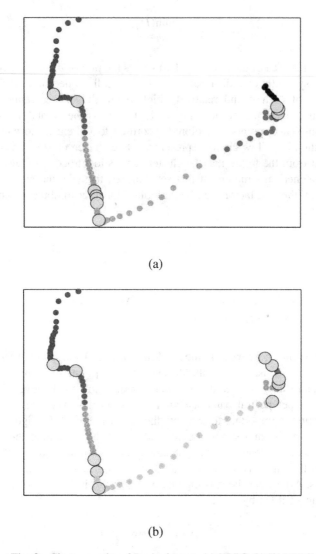

(a)

(b)

Fig. 2. Cluster results of Pavia dataset: (a) FDPC (b) W-FDPC

the representative bands are obviously the cluster centers identified. As the increase of k, FDPC may choose the bands with large ρ and relatively small δ, i.e., the points around the cluster centers, which are highly correlated to the bands already selected.

To illustrate, Fig. 2(a) shows the band select by FDPC of the Pavia hyperspectral data which will be described in detail later. It can be clearly seen the chosen points are strongly dominated by local density ρ and concentrated in a small area (here k is equal to 10), while from Fig. 2(b), it can be observed that the clustering results obtained by W-FDPC are scattered in a relative uniform manner, which is in line with the goal of band selection for choosing independent and representative bands from the original

Fig. 3. Ground-truth map: (a) Indian Pines dataset (b) Pavia data set

hyperspectral data. Therefore, the W-FDPC method can be more advantageous than FDPC.

4 Experimental Results

Having presented our method in the previous section, we now turn our attention to demonstrate its utility for the purpose of two real-world hyperspectral remote sensing data sets. Here, Indian Pines data set and Pavia data set with different spatial resolution have been employed.

The first real-world data set to be used is the commonly-used Indian Pine data set acquired by the AVIRIS instrument over the agricultural area of North-western Indiana in 1992, which has spatial dimension of 145×145 and 224 spectral bands. The spatial resolution of the data is 20 m per pixel. After discarding four zero bands and 35 lower signal-to-noise ratio (SNR) bands affected by atmospheric absorption, 185 channels are preserved. The data set contains 10366 labeled pixels and 16 ground-truth classes, most of which are different types of crops. Figure 3(a) shows the ground truth map containing 16 mutually exclusive land-cover classes.

Another data set was acquired by the ROSIS-03 sensor over the center of Pavia, Italy with 115 spectral bands. The Pavia center image was originally 1096×1096

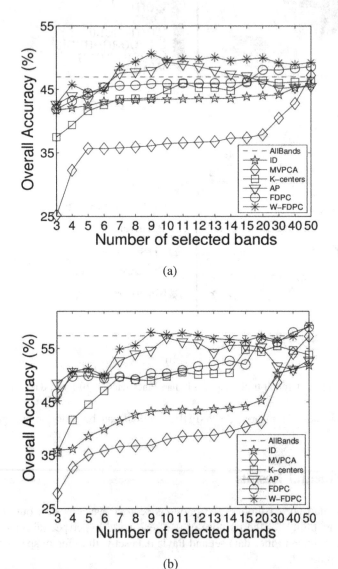

(a)

(b)

Fig. 4. Classification accuracy of Indian Pines dataset: (a) KNN (b) SVM

pixels. A 381-pixel-wide black stripe in the left part of image was removed, resulting in an image with 1096 × 715 pixels. After removing the 13 noisy bands, the remaining 102 channels are processed. There are 148152 labeled samples in total, and nine classes of interest are considered. Figure 3(b) shows the Pavia data's ground truth map containing 9 mutually exclusive land-cover classes. The experiments are conducted by comparing the W-FDPC algorithm with four popular band selection techniques, including two ranking-based ones (ID and MVPCA) and two clustering-based ones (K-centers and AP). Meanwhile, the FDPC algorithm is also considered. With respect to the classification algorithms, K-Nearest Neighborhood (KNN) and support vector machine (SVM) have demonstrated their status as the "state of the art" in classification

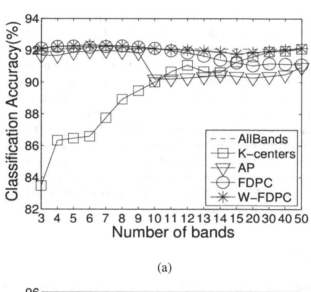

(a)

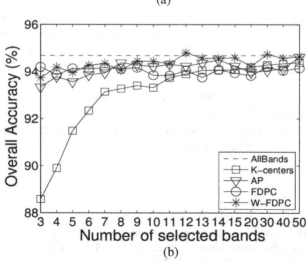

(b)

Fig. 5. Classification accuracy of Pavia dataset: (a) KNN (b) SVM

problem. The parameter of the number of neighbors in KNN was set to be 3, and linear kernel and one-against-all scheme in SVM was used for multi-class classification.

Besides, the C parameter of SVM is estimated by 10-fold cross validation. To evaluate the performance of the compared band selection methods, the available labeled samples of each class are divided into train set and test set. Each experiment is repeated 10 times with different training sets to reduce the influence of random effects, and the average results are reported. The overall accuracy (OA) measure defined by (7) is used to assess the classification results.

$$OA = \frac{number\ of\ correctly\ classified\ samples}{number\ of\ test\ samples} \tag{7}$$

Figure 4 demonstrates the classification accuracy results of Indian Pines data set using KNN and SVM, the bands are obtained by the former four clustering-based methods and two ranking-based methods (ID and MVPCA). Both datasets randomly pick out ten samples from each class to form the training set. To investigate the impact of different number of selected bands on the classification accuracy, the number ranges from 3 to 50, as represented by the x-axis. The results with all bands is also given for comparison, with the legend "AllBands". From Fig. 4, it can be seen that the results of the four clustering-based methods (K-centers, AP, FDPC and W-FDPC) are better than those of the two ranking-based methods in most cases, which is mainly due to the high correlation of the bands selected by the ranking-based methods. Figure 5 illustrates the classification accuracy results obtained by four clustering-based methods of Pavia data set using KNN and SVM, the effectiveness of the proposed W-FDPC approach for hyperspectral band selection is also verified.

5   Conclusion

In this paper, we present an unsupervised band selection approach for hyperspectral data. The proposed W-FDPC method is a ranking-based clustering band selection approach which is different from the conventional clustering-based methods (like K-centers and AP) and ranking-based methods (like ID and MVPCA). The experimental results consistently show that W-FDPC exhibits better performance. We would like to emphasize that the W-FDPC method presented here is quite general in nature and can be readily applied to other features (such as wavelet, Gabor features) in which the ranking-based clustering combination is a valuable computational tool for unsupervised band selection of hyperspectral imagery.

References

1. Manolakis, D., Mardon, D., Shaw, G.A.: Hyperspectral image processing for automatic target detection applications. Lincoln Lab. J. **14**(1), 79–115 (2003)
2. Chang, C.-I.: Hyperspectral Imaging: Techniques for Spectral Detection and Classification. Kluwer Academic/Plenum Publishers, New York (2003)

3. Plaza, A., Benediktsson, J.A., Boardman, J., Brazile, J., Bruzzone, L., Camps-Vails, G., Chanussot, J., Fauvel, M., Gamba, P., Gualtieri, A., Marconcinie, M., Tilton, J.C., Trianni, G.: Recent advances in techniques for hyperspectral image processing. Remote Sens. Environ. **113**, 110–122 (2009)
4. Bioucas-Dias, J.M., Plaza, A., Camps-Valls, G., Scheunders, P., Nasrabadi, N.M., Chanussot, J.: Hyperspectral remote sensing data analysis and future challenges. IEEE Geosci. Remote Sens. Mag. **1**(2), 6–36 (2013)
5. Hughes, G.F.: On the mean accuracy of statistical pattern recognizers. IEEE Trans. Inf. Theor. **14**(1), 55–63 (1968)
6. Khodr, J., Younes, R.: Dimensionality reduction on hyperspectral images: a comparative review based on artificial datas. In: Proceedings of 4th International Conference on Image and Signal Processing (CISP) Congress, vol. 4, pp. 1875–1883 (2011)
7. Martinez-uso, A., Pla, F., Sotoca, J.M., Garcia-sevilla, P.: Clustering-based hyperspectral band selection using information measures. IEEE Trans. Geosci. Remote Sens. **45**(12), 4158–4171 (2007)
8. Kumar, S., Ghosh, J., Crawford, M.M.: Best-bases feature extraction algorithms for classification of hyperspectral data. IEEE Trans. Geosci. Remote Sens. **39**(7), 1368–1379 (2001)
9. Jimenez-Rodrguez, L.O., Arzuaga-Cruz, E., Velez-Reyes, M.: Unsupervised linear feature-extraction methods and their effects in the classification of high-dimensional data. IEEE Trans. Geosci. Remote Sens. **45**(2), 469–483 (2007)
10. Serpico, B.S., Bruzzone, L.: A new search algorithm for feature selection in hyperspectral remote sensing images. IEEE Trans. Geosci. Remote Sens. **39**(7), 1360–1367 (2001)
11. Guyon, I., Elisseeff, A.: An introduction to variable and feature selection. JMLR **3**, 1157–1182 (2003)
12. Chang, C.-I., Du, Q., Sun, T.L., Althouse, M.L.G.: A joint band prioritization and band-decorrelation approach to band selection for hyperspectral image classification. IEEE Trans. Geosci. Remote Sens. **37**(6), 2631–2641 (1999)
13. Chang, C.-I., Wang, S.: Constrained band selection for hyperspectral imagery. IEEE Trans. Geosci. Remote Sens. **44**(6), 1575–1585 (2006)
14. Du, Q., Yang, H.: Similarity-based unsupervised band selection for hyperspectral image analysis. IEEE Geosci. Remote Sens. Lett. **5**(4), 564–568 (2008)
15. Zhao, Y.-Q., Zhang, L., Kong, S.G.: Band-subset-based clustering and fusion for hyperspectral imagery classification. IEEE Trans. Geosci. Remote Sens. **49**, 747–756 (2011)
16. Jia, S., Ji, Z., Qian, Y., Shen, L.: Unsupervised band selection for hyperspectral imagery classification without manual band removal. IEEE J. Sel. Top. Appl. Earth Observ. Remote Sens. **5**(2), 531–543 (2012)
17. Rodriguez, A., Laio, A.: Clustering by fast search and find of density peaks. Science **344** (6191), 1492–1496 (2014)
18. Cheng, Y.: Mean shift, mode seeking, and clustering. IEEE Trans. Pattern Anal. Mach. Intell. **17**(8), 790–799 (1995)

Modified Supervised Kernel PCA for Gender Classification

Yishi Wang[1]([⊠]), Cuixian Chen[1], Valerie Watkins[1], and Karl Ricanek[2]

[1] Mathematics & Statistics, UNCW, Wilmington, USA
{wangy,chenc,vjw1275}@uncw.edu
[2] Computer Science, UNCW, Wilmington, USA
ricanekk@uncw.edu

Abstract. In this work we investigate the problem of gender recognition and develop a novel approach based on Supervised Kernel Principal Components Analysis that demonstrates a significant advantage over more traditional approaches of Linear Discriminant Analysis (LDA), Mixture Discriminant Analysis (MDA), and Support Vector Machines (SVM with RBF-kernel) through 5-fold cross validation. To evaluate the effectiveness of the proposed approach for gender recognition, we use FG-NET Aging database, since it contains faces of very young children as well as senior adults. These two subsets of human faces, young children and senior adults, have been shown by prior researchers to be challenging for gender classification. Both simulation and experiment on FG-NET database suggest that the modified supervised manifold learning approach deconvolves high dimensional features into linearly separable projections that can be easily separated with standard techniques.

1 Introduction

Gender classification has many useful commercial applications, in particular it would be useful in the interaction between human users and computers. It is also useful for very large scale face recognition systems, e.g. millions of enrolled, where binning the search space based on sex will greatly reduce the number of matches. Human variability makes this two-class classification problem extremely difficult. What makes a face female or male has been studied by anthropologist, psychologists and clinicians alike, however there does not exist a unified theory on feminine or masculine faces. As an example, work by [22] demonstrated the difficulty that humans and machines have at determining gender of young child faces. This work hypothesized that amorphous nature of toddler and baby faces does not lend itself to easy human classification: In fact, the human observers were often fooled by non-face attributes, i.e. hair length or style, color of clothing, and even the type of clothing, whereas in this work the machine dramatically out performed the human.

It has been observed that senior faces may also present a problem in gender classification as these faces tend to lose their characteristic maleness or femaleness. In support of this observation, Guo *et al.* [14] found that gender

X. He et al. (Eds.): IScIDE 2015, Part I, LNCS 9242, pp. 60–71, 2015.
DOI: 10.1007/978-3-319-23989-7_7

Fig. 1. FG-NET sample facial images. Can you guess the gender of each facial image? Answers from left to right and from top to bottom: 1-M 2-F 3-F 4-M 5-F 6-F 7-M 8-F.

classification accuracy on young and senior faces can be much lower than the one on adult faces.

1.1   Prior Work

The pioneering work on face-based gender classification is found in [12]. Similar with other face image analysis problems, there are three major aspects in gender classification: feature extraction, manifold learning/feature selection, and classification. Major feature extraction methods include using geometric features derived from fiducial (landmark) points or geometry and regional texture derived features as in [6,7,22], local texture techniques as in local binary pattern and its variates [8,14,19,20,25], Gabor filter [23], and biologically-inspired features [14].

Our interests lie with understanding and developing manifold learning and classification algorithms that can effectively process face features and can differentiate the gender regardless of age or ethnicity. Traditional approaches of manifolding learning for image data include principal components analysis (PCA) [8] and locality preserving projection (LPP) [22], however the drawback of these unsupervised manifold learning algorithms is the selection of features after linear/non-linear transformations. There have been efforts to marry feature selection such as memetic algorithm [18] and Random Forest [5] with varying success. Work by [6,7,22] demonstrate the effectiveness variable selection methods for gender recognition.

Support Vector Machines (SVM) [21] has been considered as one of the most effective classifiers among publications involving face processing such as gender [7,14,17,19,23,25]. Other applications include Boosting Algorithm [3,11,20,25], Linear Discriminant Analysis (LDA) [23,25], etc.

Questions that need to be addressed are the effectiveness of those aforementioned manifold learning algorithms in terms of contribution to the response variable, the compatibility of third party variable selection schemes with classifiers, and the non-intuitiveness of the non-linear classifiers.

1.2   Contribution of Work

We propose an innovative modified Supervised Kernel PCA (SKPCA) following [4] with a new link function as our new manifold learning algorithm. This supervised approach maintains strong correlations between the response and the transformed features, and hence variable selection algorithms are unnecessary. The modification avoids the overall loss function becoming trivial and makes the transformation contains good local structure, which eventually helps improving the recognition rate.

According to the preliminary study of simulation examples, the transformed features are linearly separable. This may be due to the effect of kernel functions. Similar patterns are also observed in the FG-NET database used in this study.

1.3   Overview

The organization of this paper is laid out as follows: Sect. 2 presents the techniques of SKPCA and its main numerical computational technique. Numerical simulation and experiment results are evaluated systematically in Sect. 3; and conclusions are drawn in the final section of this paper.

2   Supervised Kernel PCA

The benefit of supervised manifold learning is that it incorporates both the original feature and response information, and transfers them into new features. Specifically, let $\mathcal{Z} := \{(X_1, y_1), \cdots, (X_n, y_n)\} \subseteq \mathcal{X} \times \mathcal{Y}$ be a series of n independent observations drawn from joint probability measure p_{xy}. Here the X_i's are the feature vectors in $\mathbb{R}^p$ and y_i's are the labels. The idea is to find a transformation $\tau_{\mathcal{Z}}(\cdot) := \tau(\cdot | \{(X_1, y_1), \cdots, (X_n, y_n)\})$, such that $\tau_{\mathcal{Z}}(X_1), \cdots, \tau_{\mathcal{Z}}(X_n)$ contain less correlations and more relevance with the labels. Methods along this approach include Fisher's LDA [9], Mixture Discriminant Analysis (MDA) [15], Metric Learning [24], sufficient dimension reduction [16], supervised principal component analysis algorithms such as in [2] and [4]. In this work, we follow the framework from [4], which utilize the Hilbert Schmidt independent criterion (HSIC) [13], stated as follows:

$$HSIC(p_{xy}, \mathcal{F}, \mathcal{G}) := E_{x,x',y,y'}[k(x,x')l(y,y')] \tag{1}$$
$$+ E_{x,x'}[k(x,x')]E_{y,y'}[l(y,y')]$$
$$- 2E_{x,y}[E_{x'}[k(x,x')]E_{y'}[l(y,y')]],$$

where (x, y) and (x', y') are i.i.d copies from distribution p_{xy}, and $\mathcal{F}$ and $\mathcal{G}$ are reproducing kernel Hilbert spaces of functions on $\mathcal{X}$ and $\mathcal{Y}$ respectively. Let $\phi(\cdot)$ be the feature map of kernel function $k(\cdot, \cdot)$, such that $k(x, x') = <\phi(x), \phi(x')>$, and similarly let $\psi(\cdot)$ be the feature map of kernel function $l(\cdot, \cdot)$. From statistics perspective, HSIC measures the strength of association between $\phi(x)$ and $\psi(y)$. Maximizing HSIC is thus a good strategy in order to create better features as the new features are more correlated with the response y.

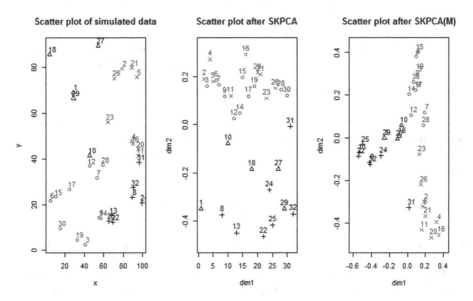

Fig. 2. A simulation study toward the effectiveness of SKPCA

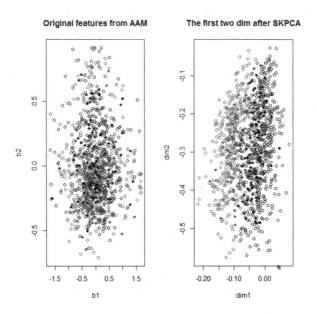

Fig. 3. The before and after scatter plots of observations from FG-NET, by using SKPCA

Due to the high dimensionality and non-linearity of face features, kernel functions are a reasonable tool for analysis. If selected appropriately, the kernel transformation will result in a linearly separable space such that well studied linear machine learning approaches can be used to categorize the original data into labels. In this work, we use Gaussian kernel for the features, which is defined as

$$k(x, x') := \exp(-\delta||x - x'||_2), \tag{2}$$

where δ is the tuning parameter, and $||\cdot||_2$ is the L_2 norm. Discussion about the strengths of Gaussian kernels can be found in [10].

In terms of application, according to [13], the empirical estimator of HSIC from (1) is

$$HSIC(\mathcal{Z}, \mathcal{F}, \mathcal{G}) := \frac{1}{(n-1)^2} \operatorname{tr}(KHLH), \tag{3}$$

where $H, K, L \in \mathbb{R}^{n \times n}$, with $K_{ij} := k(X_i, X_j)$, $L_{ij} := l(y_i, y_j)$ and $H_{ij} := [1(y_i = y_j) - \frac{1}{n}]$.

When $L_{ij} = 1$ and $k(x, x') =< x, x' >$ the linear kernel, it can be verified that $HSIC(\mathcal{Z}, \mathcal{F}, \mathcal{G}) = \operatorname{tr}(\Sigma)$, where Σ is the sample covariance matrix from X_i's. This indicates that when the link matrix L is made of a positive constant, HSIC is reduced to traditional PCA. This makes sense, since when L is a positive constant, it basically allows HSIC ignore the diversity of $y'i$'s and only focus on the X_i's, which is exactly the way PCA works.

Based on the discussion in [4], maximization of Eq. (3) is equivalent with

$$\max_{\beta}\{\operatorname{tr}(\beta KHLHK\beta^T)\}, \tag{4}$$

where $\beta \in \mathbb{R}^{m \times n}$, $m \leq n$ and $\beta K \beta^T = I$, the m dimensional identity matrix.

In the following discussion, we will verify that the solution to (4) is equivalent to generalized eigenvalue problem $Av = \lambda Bv$, where both A and B are symmetric matrices, λ is the eigenvalue, and v is the eigenvector. Let $A = KHLHK$ and $m = 1$, then (4)= $\text{Max}_{\beta_1}(\beta_1^T A \beta_1)$ with $\beta_1 \in \mathbb{R}^n$, $\beta_1^T K \beta_1 = 1$. Let $r(v) = \frac{v^T Av}{v^T Kv}$ where v is the candidate for β_1 so the denominator is 1. In order to maximize $r(v)$ we will take the derivative with respect to v and set it equal to zero, and we obtain

$$0 = \nabla r(v) = \frac{\partial(r(v))}{\partial(v)}$$

$$= \frac{\frac{\partial(v^T Av)}{\partial(v)}(v^T Kv) - (v^T Av)\frac{\partial(v^T Kv)}{\partial(v)}}{(v^T Kv)^2}$$

$$= \frac{(2Av)(v^T Kv) - (v^T Av)(2Kv)}{(v^T Kv)^2} \tag{5}$$

Setting the numerator to be zero, we have

$$Av = \frac{v^T Av}{v^T Kv}(Kv). \tag{6}$$

Therefore, by letting $\lambda = \dfrac{v^T Av}{v^T Kv}$, and $B = K$, (4) is proved to be a generalized eigenvalue problem.

As to the numerical solution of (6), there are two cases:

1. When K is a non-singular matrix the generalized eigenvalue problem is equivalent with the regular eigenvalue problem because $Av = \lambda Kv \Leftrightarrow K^{-1}Av = \lambda v$.
2. When K is a singular matrix we use simultaneous denationalization. Assume that the $\min(\text{rank}(A), \text{rank}(K)) = m$ with $m < n$ and both A and K are semi-positive definite matrices.

Since K is semi-positive definite with $\text{rank}(K) \leq m$, there exist eigenvalues

$$\lambda_{K1} \geq \cdots \geq \lambda_{Km} \geq 0, \tag{7}$$

and an $n \times m$ orthogonal matrix U_K such that $U_K^T K U_K = \text{diag}(\lambda_{K1}, \cdots, \lambda_{Km})$.
Let

$$V_K := U_K \left[\text{diag}\left(\frac{1}{\sqrt{\lambda_{K1}}}, \cdots, \frac{1}{\sqrt{\lambda_{Km}}} \right) \right],$$

and define $B := V_K{}^T A V_K$. Then B is a positive definite matrix with rank m.
Let

$$\lambda_1, \cdots, \lambda_m \text{ be the eigenvalues of } B \tag{8}$$

and corresponding eigenvectors form orthogonal matrix U_B. Let

$$V = V_K U_B. \tag{9}$$

Then it can be verified that V diagonalizes A, and $V^T K V = I$.

In summary, for generalized eigenvalue problem $Av = \lambda Kv$, the eigenvalues are $\lambda_1, \cdots, \lambda_m$ from Eq. (8) with eigenvectors in matrix V defined in (9).

2.1   Modified Link Matrix L

In [4], the link function is defined as follows:

$$L_{ij} := 1(y_i = y_j) \tag{10}$$

However, in the two-class gender classification problem, the direct consequence is that $\text{rank}(L) = 2$ and $\text{rank}(KHLKH) \leq 2$. Therefore no more than two positive eigenvalues may be obtained. Fewer dimension may directly cause fewer information kept from the original features. Also the link function L lacks regulation over local structures in the original features.

Hence, in this work, we propose an innovative modified link function to obtain modified features that are related with the labels y, while maintaining local information from the original features. The proposed link function is given by:

$$L_{ij}^X := 1(y_i = y_j) \times k(X_i, X_j). \tag{11}$$

Thus L^X is the Hadamard product of the original link matrix L with kernel matrix K. The merits of this new link function are:

- rank($KHLKH$) is not longer a concern, and the total number of non-trivial eigenvalues can be controlled by choosing appropriate tuning parameters.
- Such construction preserves locality structure of the original features $X_i's$ among the transformed new features $\tau_{\mathcal{Z}}(X_i)'s$.

The detailed algorithm for SKPCA is given in the following paragraph.

---

input : Kernel matrix of training data, K as after (3), link function L^X as defined in (11), unlabeled feature $X_{new} \in \mathbb{R}^p$.
output: Transformed training and testing features, and predicting model.

1 $A := \beta KHLHK\beta^T$;
2 Find eigenvectors $\beta_1, \cdots, \beta_m$ of generalized eigenvalue problem $Av = \lambda Kv$;
3 Set $\beta^T = (\beta_1, \cdots, \beta_m)$;
4 Transformation of training set $(\tau_{\mathcal{Z}}(X_1), \cdots, \tau_{\mathcal{Z}}(X_n)) = K\beta^T$;
5 Transformation of testing/unlabeled data X_{new} is
   $\tau_{\mathcal{Z}}(X_{new}) = (k(X_{new}, X_1), \cdots, k(X_{new}, X_n)) \beta^T$;
6 Build a SVM model M_{skpca} by using $(\tau_{\mathcal{Z}}(X_1), y_1), \cdots, (\tau_{\mathcal{Z}}(X_n), y_n)$;
7 Prediction of the testing/unlabeled data by applying M_{skpca} on $\tau_{\mathcal{Z}}(X_{new})$.

Algorithm 1. SKPCA Algorithm

---

3 Experiment

In this section we shall systematically evaluate the effectiveness of the SKPCA approach through a simulation study as well as an application to gender recognition against the FG-NET dataset.

3.1 Preliminary Simulation Study

In this study, the simulated data are uniformly distributed in $(0, 100) \times (0, 100)$, with the diagonal section belonging to a group (black) and the off-diagonal ones belonging to the second group (red), as shown in the left plot of Fig. 2. To evaluate the performance of the manifold learning algorithm in terms of keeping local structure, each section of the data is marked with a different symbol. The black sections have $\triangle$ and $+$, while the red sections have $\circ$ and $\times$.

One important property of a good manifold learning algorithm is to maintain the local structures of the data. This behavior is highly prized in the context of supervised learning. Hence, the SKPCA should exhibit this property in the simulated study as well as in the real world test of gender recognition. The desired behavior is observed when the simulated data is modeled by the modified SKPCA as demonstrated in the Fig. 2.

We apply the algorithm from above in the following manner. First, we apply the supervised linear PCA from [4] with $k(x, x') = < x, x' >$. The resulting features are simple rotations and translations of the original data (features), which makes no significant difference with the original scatter plot. Second, we analyze the simulated data with SKPCA by using group link function L as in (10) and kernel function as in (2). As seen in the middle plot of Fig. 2, the two groups become linearly separable. However, the $\circ$ and $\times$ subgroups within the red group are mixed, as well as the $\triangle$ and $+$ subgroups. This is due to the lack of detailed supervision of the link function in (10), with which the data are not 'required' to stay within their own subgroups. And, finally, the simulated data is transformed by using SKPCA with the link function in (11) and kernel function as in (2). The resulting transformed data is displayed in the right plot of Fig. 2. The resulting transformation separates the groups while maintaining the subgroup structures.

This promising simulation study encourages us to apply the last approach of SKPCA with proposed modified link function (11) and kernel function (2) for the gender classification problem.

3.2   Face Aging Database

The FG-NET Aging dataset [1] is a publicly available image dataset. It contains 1002 color or grey scale face images from 82 subjects, with age ranging from 0 to 69 years. Each image is annotated with 68 landmarks which characterize both the shape features and regional texture. Illumination variation, pose variation and expression variation are all captured in the generated parameters of the Active Appearance Model (AAM), with dimension size 109. Hereafter, the original imagines are represented by the set of new features. Pie chart for the gender distribution of 82 subjects on FG-NET dataset is shown in Fig. 4. It shows that among 82 subjects, 47 of them are males, while 35 are female.

The FG-NET dataset is a popular dataset for face age estimation and age progression, but not on gender classification. This may be due to the fact that over 70 % of the subjects were 20 or younger, which makes it challenging for gender classification. (See Fig. 1 for sample images.) Table 1. shows the age distribution of the subjects in the dataset.

3.3   Experiment

We compare the performance of the proposed SKPCA approach against the traditional SVM with Gaussian kernel, LDA and MDA. The SKPCA algorithm

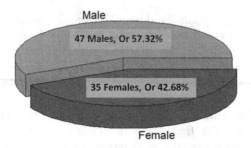

Fig. 4. Pie Chart for gender distribution of 82 subjects in FG-NET dataset.

Table 1. Composition of male and female at different age groups on FG-NET dataset.

	Male	Female	Both
Young (0–18)	389	298	**687**
Adult (19–55)	161	145	306
Senior (56–69)	5	4	9
All ages (0–69)	555	447	1002
Children (0–10)	251	160	411

used in this work is completely developed in R with the simultaneous diagonalization technique from Sect. 2. The other approaches used in this work come from the R package 'MASS' for LDA, contributed R package 'MDA' for MDA, and contributed R package 'e1071' for SVM. For a general comparison, we use 5-fold cross validation.

3.4   Results

For the purpose of accessing the linearity of the transformed features $\tau_{\mathcal{Z}}(X_i)$'s, we consider the features in each of the 5-fold cross validation. Figure 3 shows the before and after scatter plots on the first two dimensions for the modified SKPCA for one of the 5-folds. (Note: All other folds exhibited similar patterns). In the figure, the red and black colors denote features from both genders and circles represent subjects from the training set, while the solid dots represent subjects from the testing sets. The left plot suggests that the original features of the AAM are not linearly separable for gender, i.e. the AAM a multi-factored PCA does not separate out gender. Hence, only a nonlinear classification algorithm like the SVM used in this work will be capable of detangling the data. After SKPCA, as illustrated in the right plot showing the first two dimension of $\tau_{\mathcal{Z}}(X_i)$'s, it is clear that the red and black are becoming linearly separable, and the solid points, testing set, are reasonably projected along the border. This picture suggests using linear SVM on the transformed features $\tau_{\mathcal{Z}}(X_i)$'s.

Table 2. Overall gender recognition on the FG-NET dataset.

Gender recognition Accuracy on 5-fold CV (%)			
Range	SKPCA	LDA	MDA	SVM
All Ages(0–69)	**85.43**	80.14	76.45	83.93
Standard error	1.87	2.04	3.24	1.11

Next, we evaluate the performances of the four approaches, SKPCA, LDA, MDA, and SVM with Gaussian kernel under the general framework of gender classification. The tuning parameters are selected by using cross validations.

Table 2 reports the overall gender recognition rate on the FG-NET dataset. The recognition rate of using SKPCA is the best under 5-fold cross validation. Since in 5-fold cross validation, similar images from the same subject (at different ages) are very likely in the training set already, the training model already 'learn' more about how to classify the subject. SKPCA benefits the most from such extra information because of the HSIC criteria and the way the link function L^X is defined. It clearly suggests the strengths of using SKPCA over other algorithms.

What interests us the most are the results from Table 3, where recognition rates for age groups are reported. In this table, the significant differences are observed only in the 'Young' and 'Children' groups. As discussed previously these groups are expected to be more difficult to be recognized than the adult group, minus seniors. With SKPCA, the gender recognition rates for Young and Childern group are much higher than the ones from the other approaches.

LDA and MDA do not outperform SVM. This is due to the lack of preservation of local structures, which can be extended from the simulation study.

While comparing the results from the top section to the bottom section of Table 3, the local structure does help SKPCA significantly improve the gender recognition rate. Together with the simulation results, both suggest the effectiveness of our proposed methodology.

Table 3. Gender recognition over different age ranges on the FG-NET dataset.

Gender recognition accuracy on 5-fold CV (%)			
Range	SKPCA	LDA	MDA	SVM
Young(0–18)	82.24	74.82	70.31	79.91
Adult(19–55)	92.48	91.83	89.54	92.81
Senior(56–69)	88.89	88.89	100	88.89
Children(0–10)	79.08	72.63	71.27	76.16

4   Conclusion

In this paper, we proposed a modified version of SKPCA and applied the algorithm in both simulations and FG-NET database for gender classification. The simulation results suggest that SKPCA can transform overall non-linear data structure into linearly separable case, while at the same time it maintains the local structures from the original input features. Experiment results on gender classification suggest that when enough information is presented, SKPCA outperforms other supervised and unsupervised algorithms.

Some interesting directions on this work include: (1) more efficient data-driven tuning parameter selection methods for kernel functions; (2) online learning and fast implementation of SKPCA for large dataset; (3) theoretical justification on keeping the local neighborhood structure after SKPCA transformation; and (4) extension of SKPCA toward regression problems.

References

1. FG-NET aging database. http://www.fgnet.rsunit.com
2. Bair, E., Hastie, T., Paul, D., Tibshirani, R.: Prediction by supervised principal components. J. Am. Stat. Assoc. **101**(473), 119–137 (2006)
3. Baluja, S., Rowley, H.A.: Boosting sex identification performance. Int. J. Comput. Vis. **71**(1), 111–119 (2007)
4. Barshan, E., Ghodsi, A., Azimifar, Z., Jahromi, M.Z.: Supervised principal component analysis: visualization, classification and regression on subspaces and submanifolds. Pattern Recogn. **44**(7), 1357–1371 (2011)
5. Breiman, L.: Random forests. Mach. Learn. **45**(1), 5–32 (2001)
6. Chang, Y., Wang, Y., Chen, C., Ricanek, K.: Improved image-based automatic gender classification by feature selection. J. Artif. Intell. Soft Comput. Res. **1**, 241–253 (2011)
7. Chang, Y., Wang, Y., Ricanek, K., Chen, C.: Feature selection for improved automatic gender classification. In: 2011 IEEE Workshop on Computational Intelligence in Biometrics and Identity Management (CIBIM), pp. 29–35. IEEE (2011)
8. Fang, Y., Wang, Z.: Improving LBP features for gender classification. In: International Conference on Wavelet Analysis and Pattern Recognition, ICWAPR 2008, **1**, pp. 373–377. IEEE (2008)
9. Fisher, R.A.: The use of multiple measurements in taxonomic problems. Ann. Eugenics **7**(2), 179–188 (1936)
10. Francois, D., Wertz, V., Verleysen, M., et al.: About the locality of kernels in high-dimensional spaces. In: International Symposium on Applied Stochastic Models and Data Analysis, pp. 238–245 (2005)
11. Gao, W., Ai, H.: Face gender classification on consumer images in a multiethnic environment. In: Tistarelli, M., Nixon, M.S. (eds.) ICB 2009. LNCS, vol. 5558, pp. 169–178. Springer, Heidelberg (2009)
12. Golomb, B.A., Lawrence, D.T., Sejnowski, T.J.: Sexnet: a neural network identifies sex from human faces. In: NIPS, pp. 572–579 (1990)
13. Gretton, A., Bousquet, O., Smola, A.J., Schölkopf, B.: Measuring statistical dependence with hilbert-schmidt norms. In: Jain, S., Simon, H.U., Tomita, E. (eds.) ALT 2005. LNCS (LNAI), vol. 3734, pp. 63–77. Springer, Heidelberg (2005)

14. Guo, G., Dyer, C.R., Fu, Y., Huang, T.S.: Is gender recognition affected by age? In: 2009 IEEE 12th International Conference on Computer Vision Workshops (ICCV Workshops), pp. 2032–2039. IEEE (2009)
15. Hastie, T., Tibshirani, R.: Discriminant analysis by gaussian mixtures. Journal of the Royal Statistical Society, Series B (Methodological), pp. 155–176 (1996)
16. Li, K.-C.: Sliced inverse regression for dimension reduction. J. Am. Stat. Assoc. **86**(414), 316–327 (1991)
17. Moghaddam, B., Yang, M.-H.: Learning gender with support faces. IEEE Trans. Pattern Anal. Mach. Intell. **24**(5), 707–711 (2002)
18. Moscato, P., et al.: On evolution, search, optimization, genetic algorithms and martial arts: Towards memetic algorithms. Caltech concurrent computation program, C3P. Report 826, 1989 (1989)
19. Shan, C.: Learning local binary patterns for gender classification on real-world face images. Pattern Recogn. Lett. **33**(4), 431–437 (2012)
20. Sun, N., Zheng, W., Sun, C., Zou, C., Zhao, L.: Gender classification based on boosting local binary pattern. In: Wang, J., Yi, Z., Żurada, J.M., Lu, B.-L., Yin, H. (eds.) ISNN 2006. LNCS, vol. 3972, pp. 194–201. Springer, Heidelberg (2006)
21. Vapnik, V.: Statistical Learning Theory. Wiley-Interscience, New York (1998)
22. Wang, Y., Ricanek, K., Chen, C., Chang, Y.: Gender classification from infants to seniors. In: 2010 Fourth IEEE International Conference on Biometrics: Theory Applications and Systems (BTAS), pp. 1–6. IEEE (2010)
23. Xia, B., Sun, H., Lu, B.-L.: Multi-view gender classification based on local gabor binary mapping pattern and support vector machines. In: IEEE International Joint Conference on Neural Networks, 2008, IJCNN 2008, (IEEE World Congress on Computational Intelligence), pp. 3388–3395. IEEE (2008)
24. Xing, E.P., Ng, A.Y., Jordan, M.I., Russell, S.: Distance metric learning, with application to clustering with side-information. In: Advances In Neural Information Processing Systems 15, pp. 505–512. MIT Press (2003)
25. Yang, Z., Li, M., Ai, H.: An experimental study on automatic face gender classification. In: 18th International Conference on Pattern Recognition, 2006, ICPR 2006, vol. **3**, pp. 1099–1102. IEEE (2006)

Fast Film Genres Classification Combining Poster and Synopsis

Zhikang Fu[1], Bing Li[2], Jun Li[3], and Shuhua Wei[1(✉)]

[1] College of Electronic Information Engineering,
North China University of Technology, Beijing, China
fzklove@126.com, jslwsh@hotmail.com
[2] Chinese Academy of Sciences, Institute of Automation, Beijing, China
bli@nlpr.ia.ac.cn
[3] School of Automation, Southeast University, Nanjing, China
Lijun_automation@seu.edu.cn

Abstract. In this paper, we present an efficient approach to fast classify film genre by making use of film posters and synopsis simultaneously. Compared with traditional video content-based classification methods, the proposed method is much faster and more accurate. In the proposed method, a film poster is represented as multiple features including color, edge, texture, and the number of faces. On the other hand, we employ Vector Space Model (VSM) to characterize the texts in the synopsis. Then, we train a poster classifier and a text classifier using the Support Vector Machine (SVM). Finally, a test film is classified based on the 'OR' operation on the outputs of the two classifiers. We verify our scheme on our collected film poster and synopsis dataset. The experimental results demonstrate the promise of our method which achieves the desirable performance by combining posters with synopsis.

Keywords: Film genre · Film poster · VSM · Synopsis · SVM · OR

1 Introduction

More and more films come into our life along with the rapid development of the Internet. Recent years witness the extensive research conducted in the film genre classification. However, limited progress has been made due to the challenge of big data and the ambiguity in the definition of film genres. In this paper, we classify the film into four categories as illustrated in Fig. 1, and present a generic framework for film genre classification.

1.1 Related Work

Rasheed et al. [1] extracted low-level visual features from movies manually and classified them into four genres, namely: drama, action, comedy, horror. Zhou et al. [2] simultaneously adopted three kinds of features, i.e. GIST, CENTRIST and W-CENTRIST scene features, to describe a collection of temporally-ordered static key frames for the sake of representation. Genre classification and test on 1239 movies

© Springer International Publishing Switzerland 2015
X. He et al. (Eds.): IScIDE 2015, Part I, LNCS 9242, pp. 72–81, 2015.
DOI: 10.1007/978-3-319-23989-7_8

trailers were based on visual vocabulary structured by these features. Huang et al. [3] employed the same features used in [1] to categorize movies into three genres which are action, drama, and thriller. Ivasic-Kos et al. [4] utilized film posters to achieve effective film genre classification. Specifically, they proposed to use a set of low-level features for multi-label poster classification. The multi-label poster classification refers to the scenario in which the film poster simultaneously contains two informative labels from the label set of action, animation, comedy, drama, horror and war, which poses more challenges in contrast to the conventional genre classification problem in which only a single label is taken into account. But, its accuracy is very low for inadequate features. Subashini et al. [11] proposed a method for combing audio and video for classifying the genre of a movie. His results are better, but vast audio data and video data are used for his experiments. Paris et al. [12] made use of a thematic intensity extracted from synopsis and movie content to detect animated movies. However, his method can not detect other genres of films.

Fig. 1. The illustrative four films genres (from left to right are horror, love, comedy and action)

1.2 Our Work

On account of different definitions of film genres currently, it is required to determine the task-specific film genres beforehand. Specifically, the classical genres available on the popular film websites are given in Table 1. Without loss of generality, we refer to massive relevant literatures [1, 3, 4] and divide films into four groups: the horror films, comedies, love stories and action movies.

In this paper, we take advantage of film posters and synopsis to classify films into four genres. We train image set and text set by SVM separately. So, we get two respective predictions. If any prediction is right, we choose the right prediction as the last prediction of a film. Otherwise the last prediction is decided by the prediction based on poster.

The rest of the paper is organized as follows: Sect. 2 gives the whole framework of our proposed method. The following Sect. 3 introduces our elaborated devised features extracted from images and texts. We provide our experimental results in Sect. 4. Section 5 concludes this paper.

Table 1. The illustrative film genres on major domestic and foreign film website

	Genres
So Hu	Comedy Love Action Thriller War Science Fiction Disaster
Teng Xun	Action Adventure Comedy Love War Crime Thriller Science Fiction
You Ku	Comedy Horror Love Action Science Fiction War Crime
iqiyi	Love Comedy Action Horror Ethic Science Fiction Crime
YouTube	Adventure Animation Comedy Drama horror Love Action
The Movie DataBase	Thriller Adventure Science Fiction Romance Action Crime Horror Drama

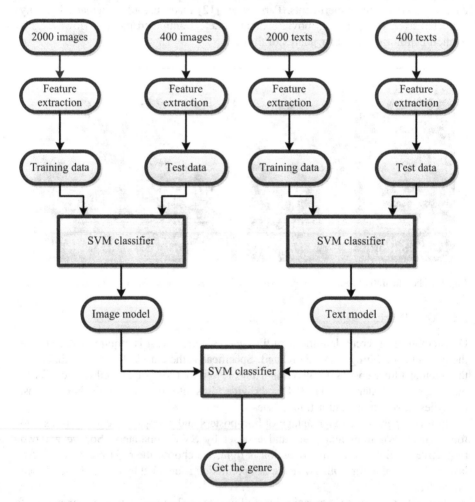

Fig. 2. The frame work of our method

2 Proposed Method

Figure 2 illustrates the processing pipeline of our method. First, we obtain high-resolution film posters and relevant synopsis from several popular foreign film websites. We operate on six feature modalities for the fine description of posters: color emotion, color harmony, edge feature, texture, color variance and the number of faces. Besides, the film synopsis is represented by VSM. We put them into SVM classifier separately and derive respective detectors from dual modalities, namely image and text model. Then, we get the first prediction Y1based on image and the second predict-ionY2 based on text. If any prediction is true, we choose the right prediction as the last prediction Y of a film. Otherwise Y is decided by theY1.

We simultaneously employ film posters and synopsis to detect film genres for many advantages.

- "Fast." It is faster to get the detector of a film, comparing with using the video content.
- "Accuracy." We get a high accuracy with combing posters and synopsis. The last result is up to 88.5 %.
- "Convenience." We can get the genre of a film with its poster and synopsis at the situation of non-existent video content.

Last, we classify films by posters or synopsis singly, comparing with our method.

3 Feature Extraction

Under our framework, image features and text features are simultaneously extracted. Specifically, the features of film posters are obtained by utilizing six low-level attri-butes: color emotion, color harmony, edge feature, texture, color variance and the number of faces. Additionally, the texts in the film synopsis are described by making use of VSM. The feature generations are detailed in the following sections.

3.1 Image Feature

Color Emotion. In real world, color is a chromatic cue which significantly influences our emotion and feelings. We respond to different colors in very different moods. For example, we are likely to feel excited, nervous while in an environment full of red objects. Conversely, lush scenery can make us feel light-hearted and comfortable. Likewise, blue enables bringing us the feeling of warmness and serenity.

In order to better delineate the color and correlate it with human emotion in mathematical formulation, Ou et al. [5, 6] proposed that human emotion are closely related with three factors relevant to color cues: Activity, Weight, and heat:

$$activity = -2.1 + 0.06\left[(a^* - 3)^2 + (L^* - 50)^2 + (\frac{b^* - 17}{1.4})^2\right]^{1/2}$$
$$weight = -1.8 + 0.45\cos(h - 10^o) + 0.04(100 - L^*)$$
$$heat = -0.5 + 0.02(C^*)^{1.07}\cos(h - 50^o)$$

(1)

Where $(L*, C*, h)$ and $(L*, a^*, b^*)$ are the color values in CIELCH and CIELAB color spaces respectively.

We define each pixel's color emotion *EI* as:

$$EI(x, y) = \sqrt{activity^2 + weight^2 + heat^2}.$$

(2)

Color Harmony. Color harmony of two-color combinations has been investigated in several empirical experiments. Ou et al. [7] proposed a model based on a psychophysical experiment of two-color combinations for predicting color harmony of two-color combinations. The model includes H_H (hue effect), H_L lightness effect and H_C (chromatic effect.)

$$H_C = 0.04 + 0.53\tanh(0.8 - 0.045\Delta C)$$
$$\Delta C = \left[(\Delta H_{ab}^*)^2 + (\frac{\Delta C_{ab}^*}{1.46})^2\right]^{\frac{1}{2}}$$
$$H_L = H_{Lsum} + H_{\Delta L}$$
$$H_{Lsum} = 0.28 + 0.54\tanh(-3.88 + 0.029\Delta L_{sum})$$
$$L_{sum} = L_1^* + L_2^*$$
$$H_{\Delta L} = 0.14 + 0.15\tanh(-2 + 0.2\Delta L).$$
$$\Delta L = |L_1^* - L*_2|.$$
$$H_H = H_{SY1} + H_{SY2}$$
$$H_{SY} = E_C(H_S + E_Y)$$
$$E_C = 0.5 + 0.5\tanh(-2 + 0.5C_{ab}^*)$$
$$H_s = 0.08 - 0.14\sin(h_{ab} + 50^o) - 0.07\sin(2h_{ab} + 90^o)$$
$$E_Y = \frac{0.22L^* - 12.8}{10}\exp\{\frac{90^o - h_{ab}}{10} - \exp\{\frac{90^o - h_{ab}}{10}\}\}$$

(3)

Where h_{ab} = CIELAB hue angle, C_{ab}^* = CIELAB chroma, ΔC_{ab}^* and ΔH_{ab}^* are the difference of two-color in CIELAB color space, L_1^* and L_2^* are the lightness of two different colors in CIELAB color space. Color harmony (CH) is defined as:

$$CH = H_H + H_C + H_L.$$

(4)

Edge Feature. Given an image, we begin with its transform from RGB into HSV color space. The derived value (V) channel is blurred by the 3×3 Gaussian filter. Next, the result is convolved by the Sobel edge detector. Finally, the outlier pixels are filtered by using the predefined threshold which is empirically set to be 0.5 in our experiment.

Texture Feature. Geusebroek et al. [8] proposed a six-stimulus basis to express stochastic texture perception. The texture statistics of an image is assumed to drawn from Weibull-distribution.

$$wb(y) = \frac{\gamma}{\beta} (\frac{x}{\beta})^{\gamma-1} e^{-\frac{1}{\gamma}(\frac{x}{\beta})^{\gamma}} Z \tag{5}$$

The parameters of the distribution enable the fine description of the spatial structure of the texture. The wild size is given by β which represents the contrast of an image while the gain size γ denotes the peakedness of the distribution.

Color Variance. To detect the color variability exhibited in the film poster, we employ the CIELuv color space, since it is designed to match with human perception. The three-order covariance matrix ρ is defined as:

$$\rho = \begin{pmatrix} \sigma_L^2 & \sigma_{Lu}^2 & \sigma_{Lv}^2 \\ \sigma_{Lu}^2 & \sigma_u^2 & \sigma_{uv}^2 \\ \sigma_{Lv}^2 & \sigma_{uv}^2 & \sigma_v^2 \end{pmatrix}. \tag{6}$$

Color variance is thus represented by the determinant Δ_F:

$$\Delta_F = \det(\rho). \tag{7}$$

The Number of Faces. Our observation implies the absence of normal human faces in the horror film posters and frequent occurrences of frontal faces and profiles in the comedy posters. Thus, we consider the number of faces in the film poster as an independent feature and detect human faces in the poster. In implementation, the detection of front faces is achieved by employing OpenCV containing a haarcascade_frontalface_alt model. The illustrative result is demonstrated in Fig. 3.

3.2 Text Feature

The English film synopsis is crawled from the Movie Data Base (TMDB) [10] website. We adopt the BOWs framework for obtaining text feature. The synopsis of every film is taken as a text document, removing the stop word of every text document, getting the stem of every word in the text document with porter's [9] algorithm, selecting feature word with information gain, structuring the Bag-of-words based feature word and representing every text document in term of VSM.

Reduction of English Stem. There are many forms in the same English word, such as adjective tense, past tense, progressive tense and so on. So, we must get the stem of every word for reducing the dimension of features. It has been demonstrated that we can get a better result compare with others, using porter's algorithm for reducing English stem.

Fig. 3. The results of face detection

Structure of Bag-of-Words. We should have typical feature word which can represent the content of every document and the genres of films. Information Gain (IG) is used to choose the feature word in this paper. IG formula is following:

$$IG(T) = H(C) - H(C|T)$$

$$H(C) = -\sum_{i=1}^{n} P(C_i) \log_2 P(C_i).$$

$$H(C|T) = - P(t)\sum_{i=1}^{n} P(C_i|t) \log_2 P(C_i|t) - P(\bar{t})\sum_{i=1}^{n} P(C_i|\bar{t}) \log_2 P(C_i|\bar{t})$$

(8)

Where $P(t)$ is the document frequency of the feature word T. It counts the number of documents in S where T appears. |S| is the total number of documents in the corpus. $P(C_i|t)$ is the document frequency of the feature word T under the situation of C_i category. It counts the number of documents in D where T appears. |D| is the total number of documents in the C_i category.

Last Bag-of-Words is been constructed by feature word. At the same time every document can be described by VSM based on Bag-of-Words.

3.3   Classification

We construct two irrelevantly training set which are image training and text training. Then they are passed into SVM classifier. We get image model and text model. Subsequently, we get result Y1 by using image model to predict image test set. We get result Y2 by using text model to predict text test set. Last, if any prediction is true, we choose the right prediction as the last prediction Y of a film. Otherwise Y is decided by the Y1.

4   Experiments

4.1   Dataset

For performance measure of our proposed method, the experiments are carried out on the collection of websites including the English text and images. We collect 2400 film posters and 2400 text documents obtained from TMDB and select 4 genres (Horror, Comedy, Romance, Action) each of which has 600 training examples. We employ 2000 posters and 2000 text document for training and the rest are used for test. Our dataset is balanced. Each genre has 500 samples in training set. Meantime, each genre has 100 samples in test set.

4.2   The Result of Experiments

To demonstrate our proposed method, we have performed three experiments: film genres classification using posters, film genres classification using synopsis, and film genres classification combining posters and synopsis.

First, we extract the poster features, using SVM classifier which is double Radial Basis Function (RBF) kernel to predict the genre of a film. The result is shown in Table 2. We can see that there is low accuracy for classifying films by posters singly, especially the accuracy of action.

Table 2. The result of predicting the posters

	Horror	Comedy	Love	Action	All test set
Accuracy	67 %	61 %	64 %	51 %	60.75 %

Then, we extract the synopsis features. SVM classifier with RBF kernel is used for predict the genre of a film. The result is shown in Table 3. We can see that text features can perform better than image features. The accuracy of each genre has been improved obviously, especially the accuracy of action which is up to 89 %. Meantime, the computing time in second experiment is shorter than the previous experiment. However, the accuracy of comedy is very low according to Tables 2 and 3.

Table 3. The result of predicting the texts

	Horror	Comedy	Love	Action	All test set
Accuracy	70 %	62 %	72 %	89 %	73.25 %

Last, we feed image model and text model which are got from previous two experiments into the same SVM as before. We fuse the prediction further. The best result is shown in Table 4. We can see that the accuracy has exceeded 90 % in horror, love and action. The lowest is comedy which is 81 %. The accuracy of test set is up to 88.5 %. However, the computing time in third experiment is longer than the previous

Table 4. The result of predicting the fusion

	Horror	Comedy	Love	Action	All test set
Accuracy	91 %	81 %	90 %	92 %	88.5 %

two experiments. We come to the conclusion that classifying films combing posters and synopsis can get high accuracy.

5   Conclusions

In this paper, the genres of films are detected by combining posters and synopsis. The posters are detected with color emotion, color harmony, edge feature, texture, color variance and the number of faces. At the same time, the synopsis is represented in VSM. We employ image model to predict image test set and text model to predict text test set separately. The last fusion is based on the OR operation of two detectors. Experimental results show that the proposed method is fast, high accuracy and convenient in film classification.

Acknowledgements. This work is partly supported by the 973 basic research program of China (Grant No. 2014CB349303), the Natural Science Foundation of China (Grant No. 61472421), the National 863 High-Tech R&D Program of China (Grant No. 2012AA012504), and the Project Supported by Guangdong Natural Science Foundation (Grant No. S2012020011081), and the Scientic Research Project of Beijing Educational Committee (No. KM201410009005).

References

1. Rasheed, Z., Sheikh, Y., Shah, M.: On the use of computable features for film classification. IEEE Trans. Circ. Syst. Video Technol. **15**(1), 52–64 (2005)
2. Zhou, H., Hermans, T., Karandikar, A.V., Rehg, J.M.: Movie genre classification via scene categorization. In: International Conference on Multimedia, pp. 747–750 (2010)
3. Huang, H.-Y., Shih, W.-S., Hsu, W.-H.: A film classifier based on low-level visual features. In: International Workshop on Multimedia Signal Processing, pp. 465–468 (2007)
4. Ivasic-Kos, M., Pobar, M., Mikec, L.: Movie Posters Classification into Genres Based on Low-level Features. In: International Convention on Information and Communication Technology, pp. 1198–1203 (2014)
5. Ou, L.C., Luo, M.R., Woodcock, A., Wright, A.: A study of colour emotion and colour preference. part I: colour emotions for single colours. Color Res. Appl. **29**(3), 232–240 (2004)
6. Ou, L.C., Luo, M.R., Woodcock, A., Wright, A.: A study of colour emotion and colour preference. part III: colour preference modeling. Color Res. Appl. **29**(5), 381–389 (2004)
7. Ou, L.C., Luo, M.R.: A colour harmony model for two-colour combinations. Color Res. Appl. **31**(3), 191–204 (2006)
8. Geusebroek, J., Smeulders, A.: A six –stimulus theory for stochastic texture. IJCV **62**, 7–16 (2005)
9. Porter, M.F.: An algorithm for suffix stripping. Program **40**(3), 211–218 (2006)

10. The movie database. http://www.themviedb.org/
11. Subashin, K. Palanivel, S., Ramaligam, V.: Audio-Video based segmentation and classification using SVM. In: International Conference on Computing, Communication and Networking Technologies (2012)
12. Paris, G., Lambert, P., Beauchene, D., Deloule, F., Ionescu, B.: Animated Movie genre detection using symbolic fusion of text and image descriptors. In: International Workshop on Content-Based Multimedia Indexing, pp. 37–42 (2012)

A Heterogeneous Image Transformation Based Synthesis Framework for Face Sketch Aging

Shengchuan Zhang[1]([✉]), Nannan Wang[2], Jie Li[1], and Xinbo Gao[1]

[1] VIPS Lab, School of Electronic Engineering, Xidian University,
Xi'an 710071, China
zsc_2007@163.com, {leejie,xbgao}@mail.xidian.edu.cn
[2] State Key Laboratory of Integrated Services Networks,
School of Telecommunications Engineering, Xidian University, Xi'an 710071, China
nannanwang.xidian@gamil.com

Abstract. Face sketch aging (FSA) simulation is a challenging task with many real applications. Although researches on face aging have achieved great progress, most of researchers focus on face photos. The main reason is that it is time consuming and expensive to build databases of face sketch aging. In order to escape the process of collecting sketch aging sequences and make use of existing face aging methods, a novel heterogeneous image transformation (HIT) based synthesis framework for face sketch aging is proposed. In the proposed framework, face sketches to be aged are first transferred to pseudo-photos by existing HIT methods. Then existing face aging methods are employed to obtain corresponding aged pseudo-photos. Finally, aged face sketches can be synthesized from obtained aged pseudo-photos via age-related HIT methods. Experimental results demonstrate that the proposed framework achieves exciting performances.

Keywords: Face sketch aging · Heterogeneous image transformation · Face aging

1 Introduction

Face sketch aging is an interesting computer vision task with many applications, such as law enforcement and digital entertainment. For example, it could assist with the police looking for lost children and wanted fugitives. What is more, it could be designed to digital entertainment. In the face sketch aging process, we need to keep the identity and aging effect simultaneously for the input face sketch to be aged. Due to the high cost of human face sketch aging sequences acquisition, it is a challenging task for face sketch aging via example based methods. Since existing face aging methods have achieved great progress, it is feasible to make use of the knowledge learned from these methods to guide the face sketch aging problem. In order to bridge the relationship between sketch aging and photo aging, heterogeneous image transformation (HIT) methods are introduced to assist face sketch aging.

© Springer International Publishing Switzerland 2015
X. He et al. (Eds.): IScIDE 2015, Part I, LNCS 9242, pp. 82–89, 2015.
DOI: 10.1007/978-3-319-23989-7_9

Researchers have proposed many algorithms about face aging modeling and simulation. Previous works on face aging could be categorized into two groups: physical model based [1] and example based [2–4]. With the aid of hierarchical and-or graph, Suo *et al.* [2] presented a compositional and dynamic model for face aging. Due to lack of long run dense aging sequences, Suo *et al.* [3] proposed a new face aging model learning long run face aging patterns from partially dense aging dataset. With the help of prototype, Kemelmacher-Shlizerman *et al.* [4] displayed the first compelling results for the whole life aging.

The aim of HIT methods is to transform heterogeneous images into homogeneous ones. Face sketch-photo synthesis is one of the most famous of HIT methods. Researches on face sketch-photo synthesis could be mainly grouped into three categories: subspace learning based [5], sparse representation based [6] and Bayesian inference based methods [7,8]. Recently, great progress to face sketch synthesis has been made with the aid of Bayesian inference (BI) [9].

In this paper, we apply [2] to conduct the aging process of pseudo-photos corresponding to sketches to be aged. The pseudo-photos are first synthesized from sketches to be aged by performing existing HIT methods. After obtaining the aged pseudo-photos by employing [2], aged sketches can be acquired by performing HIT methods which is designed considering the age factor. By incorporating the face aging methods and HIT methods, the proposed face sketch aging framework can ignore the database of the human face sketch aging sequences.

The contributions of this paper can be summarized as follows: (1) we propose a framework to simulate the aging process of the human face sketch; (2) the presented framework skips the time consuming process of collecting sketch aging sequences.

2   The Proposed Framework

In this paper, we propose a HIT based synthesis framework for face sketch aging. The framework includes three parts: sketch based photo synthesis, face photo aging and photo based sketch synthesis, as shown in Fig. 1. The proposed framework can both escape the time consuming process for collecting sketch aging sequences which is also expensive and make full use of existing state-of-the-art face aging methods. The determinant step to successfully implement the proposed framework is the HIT algorithms. We will describe each part of the framework in details bellow.

2.1   Sketch Based Photo Synthesis

To obtain pseudo-photos from sketches to be aged, we can perform HIT method proposed in [8]. Sketch based photo synthesis methods are not limited to [8].

Figure 2 shows a diagram of the Bayesian inference framework for face photo synthesis. Given a training dataset including M photo-sketch pairs, we divide each photo into N overlapping patches. Likewise, we divide each sketch into N overlapping patches. Now given a test sketch to be aged, we divide it into N

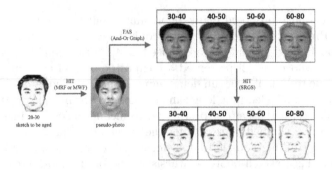

Fig. 1. Overall framework for face sketch aging.

overlapping patches. For each test patch, we find K candidate sketch patches from the training dataset. By applying a Markov Weight Fields model [8], a list of weights W for the K candidate sketch patches to reconstruct test patch are obtained. Finally, corresponding target photo patch is represented by a linear combination of K candidate photo patches with weights W. Fuse these N synthesized photo patches into a pseudo-photo with overlapping areas averaged.

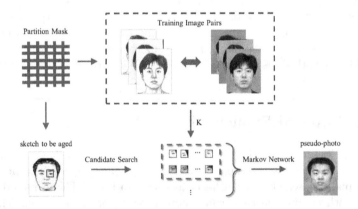

Fig. 2. Bayesian inference framework for sketch based photo synthesis.

2.2 Face Photo Aging

For a sketch to be aged, we first apply HIT algorithm described in Subsect. 2.1 to synthesize its corresponding pseudo-photo. Then existing face aging approaches can be used to age the pseudo-photo. Without loss of generality we employ the compositional and dynamic model which considers the human hair proposed by Suo *et al.* [2] in this subsection.

The compositional and dynamic model represents the face in an And-Or graph which considers the global appearance changes, the difference of facial components, and wrinkles and white hairs emergence. Given a pseudo-photo, we apply the And-Or graph to represent it, then a dynamic Markov process on the And-Or graph is adopted to simulate face aging. 50,000 face photo images are used as training set to train the generative model and the dynamics. Figure 3 is an illustration of dynamic model for face aging.

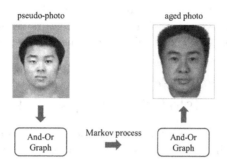

Fig. 3. Modeling the aging process as a Markov chain on And-Or graphs.

2.3 Photo Based Sketch Synthesis

After obtaining the aged pseudo-photo through Subsect. 2.2, aged sketches can be acquired by performing HIT methods. However, subjects in the dataset used for HIT methods are young people and existing HIT methods cannot synthesize the high quality sketch of the aged pseudo-photo based on young photo-sketch pair training set. There are two solutions to overcome the above drawback: (1) propose an improved face sketch synthesis method; (2) collect old photo-sketch pairs for training. Since the motivation of the proposed framework is to ignore the time consuming collection process for sketch aging sequences, we choose the first solution to handle the drawback. The improved face sketch synthesis method is performed as follow:

Given the training dataset, we divide it into two sets: set I for dictionary learning, set II for sketch synthesis.

In the dictionary learning stage, we divide photos in set I into patches and each patch overlaps. We perform dictionary learning on the obtained photo patch set $\mathbf{Y}$ to learn photo patch feature dictionary $\mathbf{D}_p \in \mathbb{R}^{l \times m}$. Dictionary learning is equal to solve the following objective function:

$$\min_{\{\mathbf{D}_p, \mathbf{C}\}} \|\mathbf{Y} - \mathbf{D}_p\mathbf{C}\|_2^2 + \lambda\|\mathbf{C}\|_1$$
$$s.t. \ \textstyle\sum_i \mathbf{D}_{pj}^2 \leq t, \ \forall j = 1, \ldots, m \tag{1}$$

where λ is experimentally set to 0.15 in our experiments.

We divide photos in set II into overlapping patch set denoted as $\mathbf{S}_p$. Then dictionary $\mathbf{D}_p$ is applied to obtain the sparse feature vector set $\mathbf{C}_p$ corresponding to $\mathbf{S}_p$. The sparse coding function is defined as follow:

$$\min_{\mathbf{C}_p} \|\mathbf{S}_p - \mathbf{D}_p\mathbf{C}_p\|_2^2 + \lambda\|\mathbf{C}_p\|_1 \tag{2}$$

Given the aged pseudo-photo, we divide it into overlapping patches. For each patch x_t, we can employ the same way we do for training patch set S_p to obtain corresponding sparse feature vector c_t. Then a greedy search strategy is applied to find K candidate sketch patches for each aged pseudo-photo patch over the whole training patch set S_p.

For greedy search strategy, we iterate two standards to measure the similarity between sparse feature vector c_t and the sparse feature vector set C_p. The two criteria include sparse coefficient value and dimension selection order of each sparse feature vector:

$$\{c\}_t = \left\{ c \left| p_o^j(c) = p_o^j(c_t), c \in C_p \right. \right\}$$
$$\{c\}_t = \left\{ c \left| \left\| p_v^j(c) - p_v^j(c_t) \right\|_2^2 < \varepsilon, c \in C_p \right. \right\} \tag{3}$$

where $p_o^j(c)$ and $p_v^j(c)$ represent the dimension selection order and the sparse coefficient value of j th atom in sparse feature vector c, respectively. The similar work can be seen in [10].

Through the foregoing operation, we can select K candidate sketch patches for each aged pseudo-photo patch. Then a Markov Network described in [7] is applied to synthesize the final aged sketch corresponding to the input aged pseudo-photo.

Since improved face sketch synthesis method extends the candidate search region from local area to the whole training set, it can synthesize the high quality sketch of aged pseudo-photo based on young photo-sketch pair training set. Due to the sparsity of the sparse feature vector, performing candidate search in the whole training set is without loss of much time and memory.

3   Experimental Results

To verify the effectiveness of the proposed framework for face sketch aging, experiments on database including sketches and corresponding photos sampled from [2] are executed. Some examples are shown in Fig. 4. Given a sketch, our framework for face sketch aging is implemented in three steps. The first step is sketch based photo synthesis. Since the database includes photos, we can skip the first step. The second step is to perform face aging using the corresponding photos of sketches to be aged. A detailed description of face aging used in this paper can be found in [2]. Photo based sketch synthesis is implemented in the last step. We accomplish photo based sketch synthesis in the CUHK student database [7]. The CUHK student database consists of 188 sketch-photo pairs. We choose 30 sketch-photo pairs for dictionary training and another 30 sketch-photo pairs for sketch synthesis. In our experiments, parameter settings are as follows: the size of each patch is 10×10, the size of overlap is 5×5 and the number of nearest candidates is 10. Figures 5 and 6 show some of the aging results synthesized by our framework.

Commonly, there are two criteria to evaluate the performance of the face sketch aging system: (1) whether the aged sketch looks like the expected age and (2) whether the aged sketch can be identified as the same person.

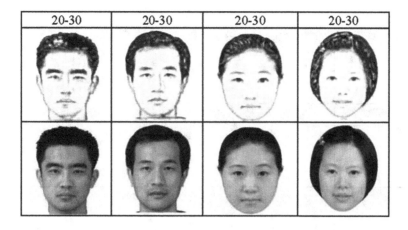

Fig. 4. Sketches and corresponding photos sampled from [2].

In order to validate the goodness of the age-related HIT method we proposed, we compare it with the state-of-the-art HIT method [11]. we choose 134 photo-sketch pairs from the CUHK student database for training and some old photos grabbed from Internet are applied for test. Some synthesized results can be found in Fig. 7. From Fig. 7, we can see that the age-related HIT method achieves better results for the old photos.

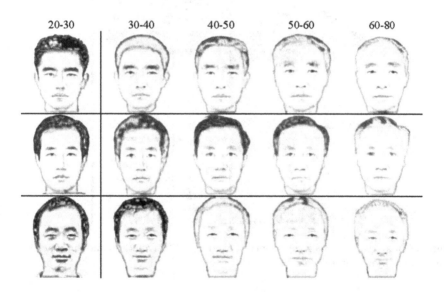

Fig. 5. Some sketch aging simulation results for male subjects

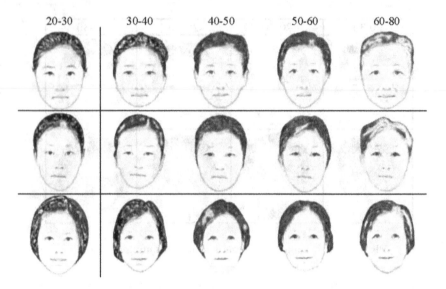

Fig. 6. Some sketch aging simulation results for female subjects.

Fig. 7. Comparison between the proposed age-related HIT method and approach introduced in [11] for synthesized sketches.

4   Conclusion

In this paper, we present a HIT based synthesis framework for face sketch aging. The framework consists of three parts: sketch based photo synthesis, face photo aging and photo based sketch synthesis. Our framework obtains compelling results due to two factors: the age-related HIT algorithm and the effective face aging method. Experimental results demonstrate the feasibility and effectiveness for combining HIT methods and face aging methods to simulate face sketch aging. The work further proves that the collection of sketch aging sequences, which is time consuming and expensive, can be ignored.

We will devote ourselves to develop novel face aging methods and advanced age-related HIT algorithms in the future.

Acknowledgement. This research was supported partially by the National Natural Science Foundation of China (Grant Nos. 61201294, 61472304, 61125204, 61432014, and 61172146), the Program for Changjiang Scholars and Innovative Research Team in University of China (No. IRT13088) and the Shaanxi Innovative Research Team for Key Science and Technology (No. 2012KCT-02).

References

1. Ramanathan, N., Chellappa, R.: Modeling shape and textural variations in aging faces. In: 8th IEEE International Conference on Automatic Face Gesture Recognition, pp. 1–8 (2008)
2. Suo, J., Zhu, S.C., Shan, S., Chen, X.: A compositional and dynamic model for face aging. IEEE Trans. Pattern Anal. Mach. Intell. **32**, 385–401 (2010)
3. Suo, J., Chen, X., Shan, S., Gao, W.: Learning long term face aging patterns from partially dense aging databases. In: 12th IEEE International Conference on Computer Vision, pp. 622–629 (2009)
4. Kemelmacher-Shlizerman, I., Suwajanakorn, S., Seitz, S.M.: Illumination-aware age progression. In: IEEE Conference on Computer Vision and Pattern Recognition, pp. 3334–3341 (2014)
5. Liu, Q., Tang, X., Jin, H., Lu, H., Ma, S.: A nonlinear approach for face sketch synthesis and recognition. In: IEEE Conference on Computer Vision and Pattern Recognition, pp. 1005–1010 (2005)
6. Chang, L., Zhou, M., Han, Y., Deng, X.: Face sketch synthesis via sparse representation. In: 20th International Conference on Pattern Recognition, pp. 2146–2149 (2010)
7. Wang, X., Tang, X.: Face photo-sketch synthesis and recognition. IEEE Trans. Pattern Anal. Mach. Intell. **31**, 1955–1967 (2009)
8. Zhou, H., Kuang, Z., Wong, K.Y.: Markov weight fields for face sketch synthesis. In: IEEE Conference on Computer Vision and Pattern Recognition, pp. 1091–1097 (2012)
9. Wang, N., Tao, D., Gao, X., Li, X., Li, J.: A Comprehensive Survey to Face Hallucination. Int. J. Comput. Vis. **106**, 9–30 (2014)
10. Jia, K., Wang, X., Tang, X.: Image transformation based on learning dictionaries across image spaces. IEEE Trans. Pattern Anal. Mach. Intell. **35**, 367–380 (2013)
11. Song, Y., Bao, L., Yang, Q., Yang, M.-H.: Real-time exemplar-based face sketch synthesis. In: Fleet, D., Pajdla, T., Schiele, B., Tuytelaars, T. (eds.) ECCV 2014, Part VI. LNCS, vol. 8694, pp. 800–813. Springer, Heidelberg (2014)

A Novel Image Segmentation Algorithm Based on Improved Active Contour Model

Jiasheng Song[✉], Leyang Dai, Yongjian Wang, and Di Sun

Marine Engineering Institute, Jimei University,
Shigu Road 176, Xiamen 361021, China
shengzisong@163.com

Abstract. During executing the image segmentation algorithm based on classical geometric active contour model, to obtain accurate segmentation results always involves a redundantly iterative process. And what's more, sometimes this tedious iteration does not make the algorithm converge on the desired edge and even brings out some overshoot. To improve the segmentation efficiency and accuracy, a novel image segmentation algorithm was presented. First, the gradient image is calculated out based on the vector-valued image and then an adaptive edge indicator is proposed. Second, the revised active contour evolution model using variational level set method is put forward. The experiments demonstrate that the model has significantly increased the convergence rate and accuracy. And the proposed segmentation algorithm has also greatly improved the flexibility of the control of active contour evolution by means of its adaptive parameters adjustment.

Keywords: Geodesic active contour · Level set method · Image segmentation

1 Introduction

Image segmentation is an important process for image analysis tasks. By segmentation methods, an image is partitioned into separate regions which ideally correspond to different real-world objects. Many approaches to image segmentation have been proposed over these years. Of these various methods, the variational method is one of the most versatile [1–5]. Its essence is to translate the problem of image segmentation into the functional extreme value problem, whose solutions can be found by solving the corresponding Euler equations. Usually, the functional model of a curve for segmentation is composed of two components. One is the internal energy to maintain curves' continuation and smoothness, and the other is the external energy to pull the curves toward features such as lines and edges. The minimization of the functional is supposed to be the segmentation process. For example, Kass, Witkins, and Terzopoulos introduced the dynamic curves based on sort of energy functional to segment objects in images [1]. This sort of curve models is often called active contour models (ACM). These active contour models can be broadly classified into parametric active contour models [1–3] and geometric active contour models according to their representation and implementation. In particular, the parametric active contours are represented explicitly as parameterized curves in a Lagrangian framework, while the geometric

© Springer International Publishing Switzerland 2015
X. He et al. (Eds.): IScIDE 2015, Part I, LNCS 9242, pp. 90–99, 2015.
DOI: 10.1007/978-3-319-23989-7_10

active contours are represented implicitly as level sets of a two-dimensional function that evolves in an Eulerian framework. Compared to the parametric one, the geometric have two outstanding advantages [6]. It can adapt to the changing of topologic structures, such as split-and-merge. Moreover, the level set function is confined to the image lattice space, which facilitates the model's numerical analysis and calculation optimization.

The early geometric ACM is contour model, which is based on curve evolution and level set method, adopts a level set function to represent curves and executes variational method to get the curve evolution model. Furthermore, the segmentation problem comes down to solving partial differential equations (PDEs) [5]. The PDEs for the level set function can be directly gotten by minimizing the energy functional of level set functions. This method is called variational level set method [7]. Its most important advantage is to make it flexible to add some other features to the functional model. For example, the CV model [8] encompasses the region feature information. To maintain the stability and convergence in the curve evolution, the level set function must approximate to the signed distance function (SDF) [9]. Usually, we can reinitialize the level set function at regular intervals. But the reinitialization could result in many negative effects, such as greatly increasing of the calculation time. Moreover the selection of the interval and reinitialization method can make influence in the revolution results [10]. To address the reinitialization, Li Chunming introduced a distance regularization item into the curve functional in his proposed DRLSE model. This item is a potential function and the diffusing effect caused by minimizing it keep the function's shape attributes approximate to SDF, especially in the narrow band about at zero level set. Thus, this scheme dovetails the maintenance of level set function into the curve evolution model and cuts off the periodical reinitialization.

However, the DRLSE model also had some problems in our experiments. It could result in some overshoots on the weak edge. Moreover, the iteration times is too many in order to get a fine segmentation. To solve these problems, our work is comprised of the following three sections. The edge feature extraction method is introduced in the second section, which uses three channels of color information. Then we represent the improvement of the DRLSE model as a new active contour evolution model in the third section. Finally, the segmentation algorithm is proposed and the tests are presented and discussed in the fourth section.

2   Edge Feature Extraction and Edge Indicator

The edge indicator (EI) is important for curve evolving because the curve evolution is based on this information and EI is the function of edge information. In the algorithm proposed by Li, it is calculated according to the Eq. (1), where r denotes the gradient module of the gray image in the region of interest (ROI) and p and K are constants.

$$g(r) = [1 + (r/K)^p]^{-1} \tag{1}$$

As the equation shows, its calculation is based on the gray or monochromatic image. When the image segmented is vector-valued, this method probably would lead to miss some true edges. To overcome this problem, a novel edge indicator function is proposed, which is defined as follows. First, the gradient of a ROI is calculated based on vector-valued images. Let $I(x,y)\mathbb{R}^2 \to \mathbb{R}^3$ be a vector-valued image whose 3 channel images are denoted as $I^{(i)}(x,y) : \mathbb{R}^2 \to \mathbb{R}, i = 1, 2, 3$. Thus, the value at a given pixel (x, y) in the image, I, is a vector in $\mathbb{R}^3$ and the partial derivatives at the pixel are denoted as

$$I_x = \left[\frac{\partial I^{(1)}}{\partial x}, \frac{\partial I^{(2)}}{\partial x}, \frac{\partial I^{(3)}}{\partial x}\right]$$

$$I_y = \left[\frac{\partial I^{(1)}}{\partial y}, \frac{\partial I^{(2)}}{\partial y}, \frac{\partial I^{(3)}}{\partial y}\right]$$

According to Riemannian geometry, $I(x,y)$ is regarded as a parametric surface in 3-dimensional Euclidean space $\mathbb{R}^3$ with x and y parameters. And the arc element at a given pixel is defined as

$$dI = I_x dx + I_y dy$$

Its squared norm is represented as the following equation

$$|dI|^2 = \sum_{i=1}^{3}\left(\frac{\partial I^{(i)}}{\partial x}dx, \frac{\partial I^{(i)}}{\partial y}dy\right)^2 = \begin{bmatrix} dx \\ dy \end{bmatrix}^T A \begin{bmatrix} dx \\ dy \end{bmatrix}$$

where,

$$A = \begin{bmatrix} \langle I_x, I_x \rangle & \langle I_x, I_y \rangle \\ \langle I_x, I_y \rangle & \langle I_y, I_y \rangle \end{bmatrix}$$

For a unit vector $(\cos\theta, \sin\theta)$, $|dI|^2$ is a measure of the rate of change of the image in the θ direction. The eigenvectors of A provide the direction of maximal and minimal changes at a given point in the image and the eigenvalues are the corresponding rates of change [12], which is calculated according the Eq. (2).

$$\lambda_{\pm} = \frac{1}{2}\left(\langle I_x, I_x \rangle + \langle I_y, I_y \rangle \pm \sqrt{(\langle I_x, I_x \rangle - \langle I_y, I_y \rangle)^2 + 4\langle I_x, I_y \rangle^2}\right) \qquad (2)$$

Where, λ_+ is the maximal change rate of the $|dI|$ and λ_- is its minimal change rate. Considering the difference of λ_+ and λ_- shows the maximum of gradient change in the two different directions, we defined it as the gradient module of vector-valued images:

$$|\nabla I| \triangleq \sqrt{\lambda_+ - \lambda_-} \tag{3}$$

Finally, the distribution threshold of the gradient module in the ROI is chosen adaptively by Otsu method [13]. And the edge indicator function is defined as the Eq. (4), where r is the gradient module derived from Eq. (3) and σ is the band parameter.

$$g(r) = e^{-\left(\frac{r-\mu}{\sigma}\right)^2} \tag{4}$$

3 Improvement of the Controlling of Active Contour Evolution

Caselles [5] proposed a geometric active contour model called geodesic active contour (GAC) model, whose energy functional is defined as a circuit integral along the curve arc. To get its numerical solution, variational level set methods have been widely used. It is indispensible for these methods to periodically reinitialize the level set function because of the convergence requirement. Because periodical reinitialization is rather time-consuming, it is deserved to improve and pursue a level set method that does not require reinitialization. Li introduced a distance regularization term in his variational level set formulation [11]. This regularization term is defined with a potential function and forces the gradient magnitude of the level set function to its minimum, which particularly results in a signed distance profile near its zero level set. Therefore, the need for reinitialization is eliminates. The functional with the regularization term is defined as the Eq. (5).

$$E(u) = \mu \iint p(|u|)\,\mathrm{d}x\mathrm{d}y + \lambda \iint \delta(u)g|\nabla u|\,\mathrm{d}x\mathrm{d}y + \alpha \iint gH(-u)\,\mathrm{d}x\mathrm{d}y \tag{5}$$

In this equation, $H(x)$ is the Heaviside function and $\delta(x)$ is the Dirac delta function. For regularization, both of them are approximated by the following smooth functions as in many level set methods, defined by

$$H_\varepsilon(x) = \begin{cases} 1, x > \varepsilon \\ 0, x < -\varepsilon \\ \frac{1}{2}(1 + \frac{x}{\varepsilon} + \frac{1}{\pi}\sin\frac{\pi x}{\varepsilon}), |x| \le \varepsilon \end{cases} \tag{6}$$

$$\delta_\varepsilon(x) = \frac{\mathrm{d}}{\mathrm{d}x}H_\varepsilon(x) = \begin{cases} \frac{1}{2\varepsilon}(1 + \cos\frac{\pi x}{\varepsilon}), |x| \le \varepsilon \\ 0, x > \varepsilon \end{cases} \tag{7}$$

The potential function, $p(x)$, is a double-well potential for distance regularization [11]. The first term on the right hand side in (5) is associated with the distance regularization energy, while the second and third terms are associated with the edge energy term and area energy term, respectively. According to the variational theory, the gradient descent flow of the level set function shows as Eq. (8).

$$\frac{\partial u}{\partial t} = \mu \text{div}\big(d_p(|\nabla u|)\nabla u\big) + \left[\alpha g + \lambda \text{div}\left(g\frac{\nabla u}{|\nabla u|}\right)\right]\delta(u) \tag{8}$$

According to the definition of divergence, the Eq. (8) can be expanded as the Eq. (9) shows.

$$\frac{\partial u}{\partial t} = \mu \text{div}\big(d_p(|\nabla u|)\nabla u\big) + \left[\alpha g + \beta \text{div}\left(\frac{\nabla u}{|\nabla u|}\right)g + \gamma\left(\nabla g \cdot \frac{\nabla u}{|\nabla u|}\right)\right]\delta(u) \tag{9}$$

Compared to (8), the expanded model in (9) separates out one more term. Except for the distance regularization term $\text{div}\big(d_p(|\nabla u|)\nabla u\big)$ and area term $g\delta(u)$, the expanded model includes curvature term $\text{div}\left(\frac{\nabla u}{|\nabla u|}\right)g\delta(u)$ and new edge term $\left(\nabla g \cdot \frac{\nabla u}{|\nabla u|}\right)\delta(u)$ with the parameters α, β and γ respectively.

As we previously represent, the edge indicator is gotten by means of the previously mentioned edge feature extracting module. And the edge indicator image is comprised of two discriminating regions. The first region with $g \approx 1$ and $\nabla g \approx 0$ approximately corresponds to the little changed edge features, while the second region with $g \approx 0$ corresponds to the severely changed edge features. According to (9), the dynamic curve in the first region, defined as zero level set function, will evolve to the direction of both of the curvature and area decreasing because the gradient of EI nearly equal to zero. The curve in the second region will be forced to verge to the desired edge location because the EI is equal to zero. The corresponding evolution speeds of these different motions can be adjusted by the parameters α, β and γ. Therefore, the revised model (9) has a fine defined physical meaning and better flexibility to adjust the parameters to the controlling of the curve evolution. In [11], evolution model corresponding to (8) is called DRLSE. We refer to the expanded model as GAC with improved evolution. For simplicity, it is called a IEGAC model.

4 Image Segmentation Algorithm Based on the IEGAC Model

The image segmentation algorithm is proposed as the Fig. 1 shows, which is based on the previous edge extraction method and the IEGAC model.

First, a vector-valued image is input and the edge feature is extracted as an edge indicator image according to the Eq. (4).

Second, the level set function is initialized with its zero level set function being a rectangle a little small than the image.

Third, the curve evolves based on the IEGAC model by iterating the Eq. (9). To increase the evolution speed, we select the area parameter α to be a bigger value and curvature parameter β to be zero when the curve is far away from the desired edge.

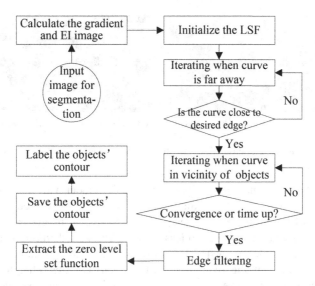

Fig. 1. Flow chart of the proposed image segmentation algorithm

To avoid the overshoot across the weak edge, we select the area parameter α to be zero and the edge parameter γ to be a bigger value.

Finally, the curve corresponding to the zero level set is filtered and extracted and labeled in the image.

The algorithm is applied to three different images to perform the segmentations of cells, wear debris and the pedestrian's gait.

4.1 Segmentation of Cell Image

We do some tests to compare the DRLSE model and IEGAC model. The first test is the segmentation of two cells image. The parameters' values in the test are listed in the Table 1. Figure 2 is the segmentation results of the DRLSE model with $\lambda = 8$. As it shows, the zero level set function doesn't separate the two cells, until the evolution model has been iterated 110 times. Nevertheless, the cell on the right hand side isn't segmented out because the dynamic curves don't converge on the desired edge. And what's worse, the dynamic curve brings out an overshoot on the weak edge.

Table 1. Parameters in the evolution models

Models	Δt	μ	α	λ	β	γ
DRLSE	2	0.1	1	8	–	–
	2	0.1	1	20	–	–
IEGAC	2	0.1	1	–	8	20

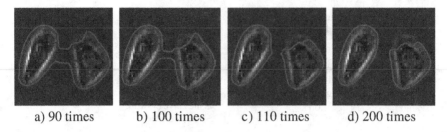

a) 90 times   b) 100 times   c) 110 times   d) 200 times

Fig. 2. Iteration results of the DRLSE model with $\lambda = 8$

Figure 3 is the segmentation results of the DRLSE model with $\lambda = 20$. Although the zero level set function can eventually converge on the desired edge, it needs at least 266 times of iterations. Several tests show the DRLSE model can segment the two cells correctly when lambda is no less than 13, but it needs more than 250 times iterations. The segmentation tests also show smaller lambda can lead to more quickly convergence but the overshoot will happen. On the other hand, greater lambda can result in more correctly segmentation but it means no less than 266 times iterations.

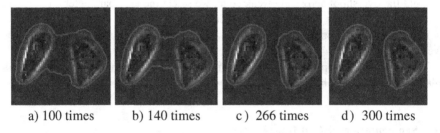

a) 100 times   b) 140 times   c) 266 times   d) 300 times

Fig. 3. Iteration results of the DRLSE model with $\lambda = 20$

Figure 4 is the segmentation results of the IEGAC model with $\beta = 8$, $\gamma = 20$. By only 107 times iteration, the two cells in the image are segmented correctly. And more times iteration do not result in overshoot, which demonstration the model can converge on the desired edge steadily. Further tests show we can always segment the two cells by the model when lambda is greater than 2β with about 110 times iterations.

These tests indicate that there is a contradiction between the segmentation accuracy and iteration times. This contradiction can't be resolved by controlling the single parameter lambda in the DRLSE model. On the contrary, there are two different kinds of parameters in the IEGAC model and the appropriate selection of the parameters will always result in accurate and stable segmentation results with lesser iterations. Therefore, the IEGAC model has greatly improved in terms of accuracy and efficiency.

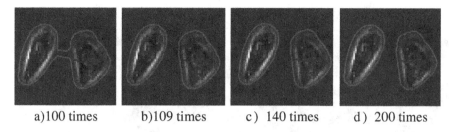

a)100 times   b)109 times   c) 140 times   d) 200 times

Fig. 4. Iteration results of the IEGAC model with $\beta = 8$, $\gamma = 20$

4.2 Segmentation of Ferrographic Image

In Fig. 5(a), there are three wear particles, and the segmentation algorithm with the IEGAC model can separate them and converge on the desired edge without overshoot at the weak edge location. In (b) and (c), there is only one piece of wear debris resulted from cutting wear and fatigue wear respectively. Both of them are segmented out clearly.

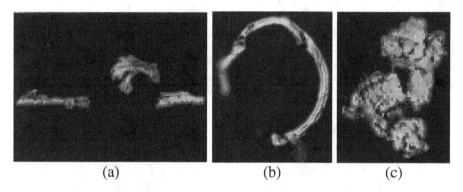

(a)   (b)   (c)

Fig. 5. Segmentation results applying the IEGAC to wear debris image

4.3 Segmentation of the Pedestrian's Gait

The Fig. 6 is the segmentation results of a pedestrian applying to a pedestrian video. The frame range is from 147 to 257 and one per 10 frames is shown in the figure. The detection and tracking algorithm is based on the IEGAC model in the framework of the unscented Kalman filtering. As the figure shows, the IEGAC model can adjust its zero level set function adaptively and converge on the desired edge according to the pedestrian's gait.

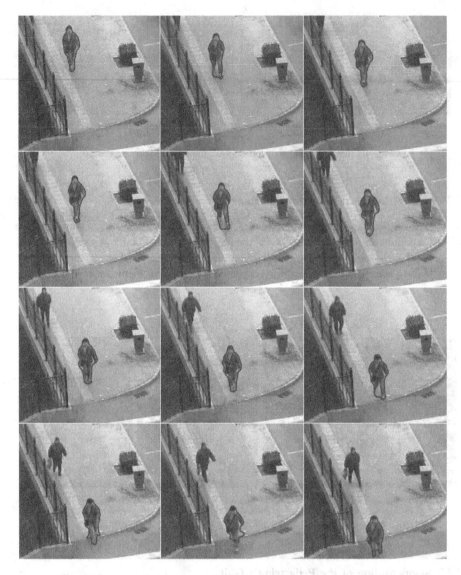

Fig. 6. Segmentation results applying the IEGAC to the pedestrian's gait

5 Conclusion

We have presented a new image segmentation algorithm based on active contour model. It is improved in two ways comparing to the conventional algorithm based on active contour model. First, the images for segmentation are viewed as vector-valued images and its edge indicator images can be calculated adaptively. The edge image can indicate the desired edge more clearly. Second, the active contour evolution model is improved by adding more control parameters, which make it easier to control the evolution process. As an application example, we have applied the algorithm to three different image segmentation problems. The test results shows the algorithm have more efficiency and control flexibility and the segmentation results be more accurate with little overshoot, especially on the weak edge.

Acknowledgment. This material is based upon the work which is supported by the Foundation of Jimei University Li Shangda Subjects and Scientific Research Project of Educational Commission of Fujian Province in China (No. JA14175).

References

1. Kass, M., Witkin, A., Terzopoulos, D.: Snakes: active contour model. Int. J. CV **1**(4), 321–331 (1988)
2. Xu, C., Prince, J.L.: Snakes, shapes, and gradient vector flow. IEEE Trans. Image Process. **7**(3), 359–369 (1998)
3. Li, B., Acton, S.T.: Active contour external force using vector field convolution for image segmentation. IEEE Trans. Image Process. **16**(8), 2096–2106 (2007)
4. Caselles, V., Catté, F., Coll, T., et al.: A geometric model for active contours in image processing. Numer. Math. **66**(1), 1–31 (1993)
5. Caselles, V., Kimmel, R., Sapiro, G.: Geodesic active contours. Int. J. Comput. Vis. **22**(1), 61–79 (1997)
6. Sethian, J.A.: Level set methods and fast marching methods. J. Comput. Inf. Technol. **11**(1), 1–2 (2003)
7. Zhao, H.-K., Chan, T., Merriman, B., et al.: A variational level set approach to multiphase motion. J. Comput. Phys. **127**(1), 179–195 (1996)
8. Chan, T.F., Vese, L.A.: Active contours without edges. IEEE Trans. Image Process. **10**(2), 266–277 (2001)
9. Peng, D., Merriman, B., Osher, S., et al.: A PDE-based fast local level set method. J. Comput. Phys. **155**(2), 410–438 (1999)
10. Gomes, J., Faugeras, O.: Reconciling distance functions and level sets. In: Proceedings of the 5th IEEE EMBS International Summer School on Biomedical Imaging, pp. 15–20. IEEE press, New York (2002)
11. Li, C., Xu, C., Gui, C., et al.: Distance regularized level set evolution and its application to image segmentation. IEEE Trans. Image Process. **19**(12), 3243–3254 (2010)
12. Chung, D.H., Sapiro, G.: On the level lines and geometry of vector-valued images. IEEE Signal Process. Lett. **7**(9), 241–243 (2000)
13. Otsu, N.A.: Threshold selection method from gray-level histograms. Automatica **11**(285), 23–27 (1975)

Supervised Spectral Embedding
for Human Pose Estimation

Yukun Guo[1], Zhonggui Chen[1], and Jun Yu[2(✉)]

[1] Xiamen University, Xiamen, China
[2] Hangzhou Dianzi University, Hangzhou, China
guoyukun@live.com, chenzhonggui@xmu.edu.cn, yujun@hdu.edu.cn

Abstract. In exemplar-based approaches for human pose estimation, it is common to extract multiple features to better describe the visual input data. However, simply concatenating multiview features into a long vector has two shortcomings: (1) it suffers from "curse of dimensionality"; (2) it is not physically meaningful and may be incapable of fully exploiting the complementary properties of multi-view features. To address such problems, in this paper we present a dimension reduction method based on supervised spectral embedding, followed by an ensemble of nearest neighbor regressions in multi-view feature space, to infer 3D human poses from monocular videos. The experiments on HumanEva dataset show the effectiveness of the proposed method.

Keywords: Human pose estimation · Spectral embedding · k-NN regression

1 Introduction

We consider the problem of estimating 3D human body configurations from monocular 2D video frames. Among discriminative methods, exemplar-based approaches recover poses from the training examples in a straightforward way. Built on top of a dense database of *visual feature – pose* pairs, when a novel visual input is captured, exemplar-based methods find the most visually similar instance in the database by nearest neighbor search, and return the corresponding pose as the estimation.

Feature design is one of the major concerns of such algorithms. To well characterize the images, researchers often use multiple features. However, directly concatenating theses features together to form a long vector seems improper. Firstly, it worsens the "curse of dimensionality" problem which degrades the performance of the estimation system; secondly, it lacks physical meanings and may fail to fully exploit the discriminative power of multi-view features.

We address such problems from two perspectives accordingly. First we propose a dimension reduction algorithm based on supervised spectral embedding. Spectral embedding based dimension reduction techniques have shown excellent results for clustering. However for exemplar-based pose estimation, unsupervised

© Springer International Publishing Switzerland 2015
X. He et al. (Eds.): IScIDE 2015, Part I, LNCS 9242, pp. 100–109, 2015.
DOI: 10.1007/978-3-319-23989-7_11

ones can provide very limited improvement. This is due to the nature of the exemplar-based approach: it has implicitly modeled the manifold of data points, so even if the spectral embedding successfully takes care of the non-linearities, we will hardly observe any significant performance boost. After all, many manifold learning algorithms themselves rely on k-NN to build local patches. In this paper, we resort to a supervised method to bridge the semantic gap between visual feature space and pose space.

On the other hand, the NN regressor from each view can work in a standalone fashion. Our goal is combining them to achieve better accuracy. Besides naive vector concatenation, a method one can easily come up with is to put different weights on different views. We can further consider such weights to be varying depending on the location in the multi-view feature spaces. It is reasonable to assume that the weight is roughly a function of concatenated long vector, i.e., $\mathbf{w} = \mathcal{W}(\mathbf{x})$, which can be approximated in the training stage. The ensemble of NN regressions in the multi-view feature spaces can be done by first determining the weights and then finding the nearest neighbor under the weighted distance measure.

2   Related Work

2.1   Vision-Based Pose Estimation

After decades of research, there have been a plethora of publications on human pose estimation. Detailed comparison of them is beyond the scope of this paper, so we refer the readers to [15] for a survey of this subject. Using silhouette-based visual features is a very popular choice among published work, including some most significant ones in this field [1,9], as silhouettes can preserve sufficient information for pose estimation while being invariant to clothing and illumination. Silhouette-based algorithms assume foreground can be reliably extracted, which is true for most indoor scenarios.

The newly developed part-based algorithms, e.g. pictorial structures [2], have shown very promising results. Part-based methods search each body part in an independent way with the help of rich feature descriptors such as HOG (Histogram of Oriented Gradients), hence they are capable of recovering highly versatile human poses "in the wild" by annotating 2D joints/limbs on the pictures. It may seem that those methods have rendered silhouette-based ones outdated, however they still have some disadvantages: (1) They are computational intensive and commonly cannot operate in real-time; (2) It is nontrivial to lift 2D poses estimated by them to 3D configurations, while silhouette-based approaches can recover the 3D pose directly in a holistic way.

2.2   Subspace Learning

Manifold learning algorithms Laplacian Eigenmaps [4], Locally Linear Embedding [16] and IsoMap [18], as well as their linearized counterparts Locality Preserving Projection [12], Neighborhood Preserving Embedding [11] and Isometric

Projection [7] etc., have drawn great attention of researchers recently. Specifically, subspace learning techniques have been incorporated to solve pose estimation problems. Closely related to our work, Elgammal et al. [9] explicitly learn activity manifolds along with the mapping functions between visual input space and the 3D body pose space. Chen et al. [8] use trace-ratio criterion to select effective visual feature components. BenAbdelkader applies supervised manifold learning to estimate head pose [5].

2.3 Multi-View Feature Combination

Multi-view spectral embedding algorithms further extend single view methods to make embeddings in different views agree with each other [13, 19]. The effectiveness of multi-view learning has been confirmed by its applications in lots of real world problems [20, 21].

Another way to combine multi-view features is aggregating individual nearest neighbor regressor over all views. Unfortunately, most done researches [3, 10] in this line are designed with the classification tasks in mind and are not suitable for the regression scenarios. Hence we propose an ad hoc method that better fits our specific problem domain.

3   Proposed Algorithms

3.1 Preliminary

The exemplar database consists of the multi-view features of images $\{X^{(v)}\}_{v=1}^{V}$ and the associated ground truth 3D poses $Y = [\mathbf{y}_1, \mathbf{y}_2, \cdots, \mathbf{y}_n]$, where $X^{(v)} = [\mathbf{x}_1^{(v)}, \mathbf{x}_2^{(v)}, \cdots, \mathbf{x}_n^{(v)}] \in \mathbb{R}^{d_v \times n}$ is the image features from the vth view, and $\mathbf{y}_k = [\cdots, \mathbf{y}_{k,j}^{\top}, \cdots]^{\top} \in \mathbb{R}^{3 \cdot M}$ is the XYZ coordinates of all M joints concatenated together.

Denoting the concatenated features as

$$X = \begin{bmatrix} X^{(1)} \\ X^{(2)} \\ \cdots \\ X^{(v)} \end{bmatrix} \in \mathbb{R}^{d \times n}, \quad d = \sum_{v=1}^{V} d_v,$$

the goal of supervised spectral embedding is to find a low-dimensional representation $\tilde{X} = [\tilde{\mathbf{x}}_1, \tilde{\mathbf{x}}_2, \ldots, \tilde{\mathbf{x}}_n] \in \mathbb{R}^{d' \times n}$, $d' < d$ in the hope that the Euclidean distances of low-dimensional features can faithfully reflect the (dis)similarities of the poses.

3.2 Visual Feature Extraction

Visual feature is extracted from the silhouette for robustness. Following [8], we choose 5 feature descriptors in total[1], which have proved to be efficient in the computer vision literature.

---

[1] We do not use *Poisson features* because it is very time-consuming to compute.

1. **Occupancy map**: Crop the silhouette according to its bounding box, and divide it by 12×8 grids. Calculate the percentage of foreground pixels inside each grid, and concatenate those values together columnwise. This generates a 96D vector, each element of which ranges from 0 to 1.
2. **Contour signature**: Start from the topmost point, uniformly sample 64 points along the contour of the silhouette. Under three variations of measures: coordinates, distances to the centroid, and tangent angles respectively, we get a $128 + 64 + 128 = 320$ dimensional feature.
3. **Fourier descriptor**: We treat the output feature of contour signature above as discrete signal, and apply DFT (Discrete Fourier Transform) to it. Eliminating the DC component results in a feature of $62 + 32 + 62 = 156$D.
4. **Hu moments**: Compute 7 image moments in Hu's invariant set, by considering the shape as either contour or area, a $7 \times 2 = 14$ dimensional feature is obtained.
5. **Shape contexts**: We use shape contexts containing 12 angular bins and 5 radial bins. Construct a codebook with the size of 100 by running k-means clustering on the generated 60D features, and quantize them to form the 100D histogram of shape contexts.

The feature vectors are normalized to prevent the dimensions with large ranges overpowering others.

3.3 Pose Distance Function

Here we describe the distance measure of poses. For evaluating retrieval results, we use *mean per joint position error*:

$$d_{\text{pose}}(\mathbf{y}, \hat{\mathbf{y}}) = \frac{1}{M} \sum_{j=1}^{M} \|m_j(\mathbf{y}) - m_j(\hat{\mathbf{y}})\| \tag{1}$$

where $m_j(\mathbf{y}) \in \mathbb{R}^3$ is the 3D coordinate of the jth joint centered at the root (pelvis joint) of the body, and $M = 14$ since we use a 15-joint skeleton model to represent human body and exclude the root joint, the coordinate of which is always $(0, 0, 0)$. This function is also employed to calculate the pairwise relationship in our supervised framework.

3.4 Supervised Spectral Embedding

Spectral Embedding: In this subsection we briefly recapitulate the spectral embedding algorithm [4]. First of all we construct the weighted undirected graph $\mathcal{G}$ to encode the structure of the manifold. This procedure is done in 2 steps:

1. *Construct the adjacency graph*: we use k-NN graph, i.e., nodes i and j are connected if either one is among the k nearest neighbors of the other. The resulting graph is sparse so it has the potential to scale well for larger data set.

2. *Set the weights*: The weight of each edge of $\mathcal{G}$ is set based on the similarity of the corresponding pair of data. The heat kernel is used here, i.e., for adjacent pair (i, j),

$$W_{ij} = \exp(-\frac{\|\mathbf{x}_i - \mathbf{x}_j\|^2}{t}). \tag{2}$$

Let D be a diagonal matrix where $D_{ii} = \sum_j W_{ji}$ and $L = D^{-\frac{1}{2}} W D^{-\frac{1}{2}}$ be the normalized graph Laplacian, the spectral embedding algorithm solves the following optimization problem [14]:

$$U = \arg\max_{U} \ \text{tr}\left(U^\top L U\right), \quad \text{s.t. } U^\top U = I. \tag{3}$$

Note that each row of U is the low-dimensional representation to be kept in the exemplar database.

Supervised Spectral Embedding: SE relies on the distribution of the feature points ("manifold hypothesis") to work, in an unsupervised way. The challenge here is that data points being close in visual feature space are not necessarily close in pose space. we can improve it by taking into consideration the information from the pose space.

Firstly, we build the pose similarity graph following almost the same procedures above, only differing in the weighting scheme:

$$W_{ij}^* = \exp(-d_{\text{pose}}^2(\mathbf{y}_i, \mathbf{y}_j)/\sigma^2) \tag{4}$$

where σ is a tuning parameter. The spectral embedding $U^* \in \mathbb{R}^{n \times d^*}$ in the pose space itself can be computed by (3), which encodes the pose proximities:

$$U^* = \arg\max_{U^*} \ \text{tr}\left(U^{*\top} L^* U^*\right), \quad \text{s.t. } U^{*\top} U^* = I. \tag{5}$$

In order to encourage U to be compatible with U^*, we introduce the cost function below to measure the disagreement of the two low-dimensional representations, as proposed in [13]:

$$D(U, U^*) = \left\| \frac{K_U}{\|K_U\|_F^2} - \frac{K_{U^*}}{\|K_{U^*}\|_F^2} \right\|_F^2 \tag{6}$$

where K is the similarity matrix, and $\| \cdot \|_F$ denotes the Frobenius norm. To make it easier to optimize, we choose $k(\mathbf{u}_i, \mathbf{u}_j) = \mathbf{u}_i^\top \mathbf{u}_j$ so that $K_U = UU^\top$. Provided $U^\top U$ and $U^{*\top} U^*$ being imposed to be I, (6) can be reformulated as

$$D(U, U^*) = -\text{tr}\left(UU^\top U^* U^{*\top}\right). \tag{7}$$

The optimal embedding should preserve locality in the feature space as well as respecting the pairwise relationship in the pose space (or equivalently, has lower value of D). Combining both unsupervised and supervised parts together, trading-off two terms by λ, we get

$$U = \arg\max_{U} \; \mathrm{tr}\left(U^{\top}LU\right) + \lambda \,\mathrm{tr}\left(UU^{\top}U^{*}U^{*\top}\right) \tag{8}$$

$$\text{s.t.} \quad U^{\top}U = I.$$

Linearization: The original SE obtains a nonlinear mapping to the low-dimensional space without explicitly finding the projection. There exist some out-of-sample extensions, e.g. [6] to deal with unseen data. Unfortunately, we cannot apply them directly because novel visual input with unknown poses cannot fit in our supervised framework. For the task of pose estimation, we have to seek the projection matrix by linearizing the embedding [12].

Therefore (8) is changed to

$$U = \arg\max_{U} \; \mathrm{tr}\left(U^{\top}XLX^{\top}U\right) + \lambda \,\mathrm{tr}\left(X^{\top}UU^{\top}XU^{*}U^{*\top}\right)$$

$$= \arg\max_{U} \; \mathrm{tr}\left(U^{\top}X\left(L + \lambda\,U^{*}U^{*\top}\right)X^{\top}U\right) \tag{9}$$

$$\text{s.t.} \quad U^{\top}XX^{\top}U = I$$

where U is an orthogonal projection matrix for the mapping $\tilde{X} = U^{\top}X$ and it can be solved by generalized eigen decomposition. We apply the algorithm to all views to compute the corresponding projections $U^{(v)} \in \mathbb{R}^{d^{v} \times d'}$.

3.5   Ensemble of Nearest Neighbor Regressions

The linearization discussed above has found the globally optimal projection. Now let us focus on the locally weighted distance function.

The motivation to use multi-view features is to take advantage of their complementary properties. "Complementary" basically means that in particular region, some views can truthly reflect the underlying pose similarities while the others are just somewhat misleading. Intuitively, if on some view, the Euclidean distances between feature vectors are small while the distances between their corresponding poses are large, we can suppose the Euclidean distance at the region of this view is unreliable.

The following two-phase algorithm is proposed. For convenience, we write the low-level feature of the ith example on the vth view as

$$\tilde{\mathbf{x}}_{i}^{(v)} = U^{(v)^{\top}}\mathbf{x}_{i}^{(v)}$$

and the concatenated low-level feature as

$$\tilde{\mathbf{x}}_{i} = [\tilde{\mathbf{x}}_{i}^{(1)}; \tilde{\mathbf{x}}_{i}^{(2)}; \cdots ; \tilde{\mathbf{x}}_{i}^{(V)}] \in \mathbb{R}^{d'}.$$

Training. In the training phase, the whole exemplar database is preprocessed to calculate the weight around every instance:

Step 1: For each training data pair $(\tilde{\mathbf{x}}_i, \mathbf{y}_i)$, find its k-nearest neighbors in concatenated low-dimensional feature space, denoted by $\mathcal{N}(\tilde{\mathbf{x}}_i)$;

Step 2: Compute the average *pose distance* to *feature vector distance* ratio of the neighborhood

$$\alpha_i^{(v)} = \frac{1}{k} \sum_{\tilde{\mathbf{x}}_{ij}^{(v)}: \; \tilde{\mathbf{x}}_{ij} \in \mathcal{N}(\tilde{\mathbf{x}}_i)} \frac{d_{\text{pose}}(\mathbf{y}_i, \mathbf{y}_{ij})}{\|\tilde{\mathbf{x}}_i^{(v)} - \tilde{\mathbf{x}}_{ij}^{(v)}\|}. \tag{10}$$

A smaller value of α indicates less variation in this direction, so the regression should bias toward this view.

Step 3: Set the weights accordingly

$$\mathbf{w}_i = [1/\alpha_i^{(1)}, 1/\alpha_i^{(2)}, \cdots, 1/\alpha_i^{(V)}]^\top. \tag{11}$$

Optionally one can normalize $\mathbf{w}_i$ by $\hat{\mathbf{w}}_i = \mathbf{w}_i/|\mathbf{w}_i|_1$. We process all training data sequentially to set the weight matrix with the space complexity of $O(V \cdot n)$.

Estimating. For novel data, the pose retrieval phase is done by the algorithm outlined below. Note that the weight is averaged among the query data's k^--nearest neighbors and $k^+ > k^-$ determines the number of candidates.

Algorithm. Recover pose from novel data

Input: The low-dimensional feature $\tilde{\mathbf{x}}_q = [\tilde{\mathbf{x}}_q^{(1)}; \tilde{\mathbf{x}}_q^{(2)}; \cdots; \tilde{\mathbf{x}}_q^{(V)}]$;
   the weight matrix $W = [\mathbf{w}_1, \mathbf{w}_2, \cdots, \mathbf{w}_n]$;
   neighborhood sizes k^- and k^+;
Output: Estimated pose $\mathbf{y}_q$;
 1: Find the k^--nearest neighbors of $\tilde{\mathbf{x}}_q$ in the concatenated low-dimensional space;
 2: Compute the mean weight

$$\bar{\mathbf{w}} = \frac{1}{k^-} \sum_{j: \; \tilde{\mathbf{x}}_j \in \mathcal{N}^-(\tilde{\mathbf{x}}_q)} \mathbf{w}_j \; ;$$

 3: Find the k^+-nearest neighbors $\mathcal{N}^+(\tilde{\mathbf{x}}_q)$ of $\tilde{\mathbf{x}}_q$ as candidates;
 4: Search $\mathcal{N}^+(\tilde{\mathbf{x}}_q)$ for the best match using the distance function weighted by $\bar{\mathbf{w}}$:

$$p = \arg\min_p \sum_{v=1}^{V} (\bar{\mathbf{w}})_v \|\tilde{\mathbf{x}}_p^{(v)} - \tilde{\mathbf{x}}_q^{(v)}\|, \quad \text{where } \tilde{\mathbf{x}}_p \in \mathcal{N}^+(\tilde{\mathbf{x}}_q) ,$$

   return the estimated pose $\mathbf{y}_q = \mathbf{y}_p$.

---

4 Experiments

We evaluate the performance of the proposed algorithm on synthetic data by retargetting MoCap data from HumanEva dataset [17] to the Humanoid model in MotionBuilder. All action types ('Walking', 'Jog', 'ThrowCatch', 'Gesture' and 'Box') are used in the experiments.

Due to the repetitive nature of MoCap data, we first select 300 most representative poses using k-medoids, and then rotate them around vertical axis with a 15 degrees interval, which leads to $300 \times 24 = 7200$ poses. Finally 40 % of the data is taken for training while the remaining 60 % for testing. For the testing part, we deliberately add noise to the generated silhouette to verify the robustness of the feature extraction.

To gain a basic idea of the prepared database, we report that the error of a random retrieval algorithm (average pairwise pose distance) is 275.3 mm while directly returning the smallest pose distance (lower bound of all possible estimators) has an error of 46.8 mm.

Since it is hard to determine the optimal dimensions of the low-dimensional embeddings, we simply keep half of dimensions ($d'/d = 0.5$). We decide to use $k = 15, d'/d = 0.5, \lambda = 3.0, d^* = 25, k^- = 10, k^+ = 60$ as the basic setting.

Table 1 shows the overall improvements of our proposed algorithms SSE (supervised spectral embedding) and NNE (nearest neighbor regression ensemble) against the baseline methods. Clearly SSE is superior than using concatenated features. It also demonstrates the effectiveness of NNE. Pose estimations in both the original feature space and the embedded low-dimensional space benefit from NNE. As a comparison, using weighted k-NN regression actually compromises the accuracy, perhaps due to the distribution of our exemplar database.

Table 1. Evaluate the proposed algorithms on synthetic data ($k_{\text{WKNN}} = 5$)

Method	Concat+NN	Concat+WKNN	Concat+NNE	SSE	SSE+NNE
Error (mm)	90.8	93.6	85.3	76.5	72.5

And we further investigate the effect of different parameter settings. Table 2 suggests that the performance is insensitive to k within the commonly used range of k.

Table 2. The influence of neighborhood size k (SSE followed by NN or NNE)

k	5	15	25
NN	76.0	76.5	76.7
NNE	72.7	72.5	72.2

Table 3. The influence of parameter λ (SSE + NN)

λ	0.0	0.5	1.0	3.0	10.0
Error (mm)	87.0	81.9	76.9	76.5	78.8

Table 3 explores the effect of trade-off parameter λ. When $\lambda = 0$, the supervised algorithm degenerates to an unsupervised one, which is significantly worse than the basic setting. The optimal value is around 3.0 since it well balances the two parts.

5   Conclusions

In this paper, we introduce the supervised spectral embedding and multi-view nearest neighbor regression combination based algorithm for human pose estimation. The experimental results demostrate that the proposed algorithm outperforms baselines. We would like to carry out detailed experiments on real data in our further work.

Acknowledgments. Zhonggui Chen was partially supported by the Fundamental Research Funds for the Central Universities (No. 20720140520).

References

1. Agarwal, A., Triggs, B.: Recovering 3D human pose from monocular images. IEEE Trans. Pattern Anal. Mach. Intell. **28**(1), 44–58 (2006)
2. Andriluka, M., Roth, S., Schiele, B.: Pictorial structures revisited: people detection and articulated pose estimation. In: IEEE Conference on Computer Vision and Pattern Recognition, 2009. CVPR 2009, pp. 1014–1021. IEEE (2009)
3. Bao, Y., Ishii, N., Du, X.-Y.: Combining multiple k-nearest neighbor classifiers using different distance functions. In: Yang, Z.R., Yin, H., Everson, R.M. (eds.) IDEAL 2004. LNCS, vol. 3177, pp. 634–641. Springer, Heidelberg (2004)
4. Belkin, M., Niyogi, P.: Laplacian eigenmaps and spectral techniques for embedding and clustering. NIPS **14**, 585–591 (2001)
5. BenAbdelkader, C.: Robust head pose estimation using supervised manifold learning. In: Daniilidis, K., Maragos, P., Paragios, N. (eds.) ECCV 2010, Part VI. LNCS, vol. 6316, pp. 518–531. Springer, Heidelberg (2010)
6. Bengio, Y., Paiement, J.-F., Vincent, P., Delalleau, O., Le Roux, N., Ouimet, M.: Out-of-sample extensions for lle, isomap, mds, eigenmaps, and spectral clustering. Adv. Neural Inf. Process. Syst. **16**, 177–184 (2004)
7. Cai, D., He, X., Han, J.: Isometric projection. In: Proceedings of the National Conference on Artificial Intelligence, vol. 22, p. 528. AAAI Press, MIT Press, Menlo Park, Cambridge (1999, 2007)
8. Chen, C., Yang, Y., Nie, F., Odobez, J.-M.: 3D human pose recovery from image by efficient visual feature selection. Comput. Vis. Image Underst. **115**(3), 290–299 (2011)

9. Elgammal, A., Lee, C.-S.: Inferring 3D body pose from silhouettes using activity manifold learning. In: Proceedings of the 2004 IEEE Computer Society Conference on Computer Vision and Pattern Recognition, CVPR 2004, vol. 2, pp. 681–688. IEEE (2004)
10. García-Pedrajas, N., Ortiz-Boyer, D.: Boosting k-nearest neighbor classifier by means of input space projection. Expert Syst. Appl. **36**, 10570–10582 (2009)
11. He, X., Cai, D., Yan, S., Zhang, H.-J.: Neighborhood preserving embedding. In: Tenth IEEE International Conference on Computer Vision, ICCV 2005, vol. 2, pp. 1208–1213. IEEE (2005)
12. He, X., Niyogi, P.: Locality preserving projections. In: Advances in Neural Information Processing Systems, pp. 153–160 (2004)
13. Kumar, A., Rai, P., Daume, H.: Co-regularized multi-view spectral clustering. In: Advances in Neural Information Processing Systems, pp. 1413–1421 (2011)
14. Ng, A.Y., Jordan, M.I., Weiss, Y., et al.: On spectral clustering: analysis and an algorithm. Adv. Neural Inf. Process. Syst. **2**, 849–856 (2002)
15. Poppe, R.: Vision-based human motion analysis: an overview. Comput. Vis. Image Underst. **108**(1), 4–18 (2007)
16. Roweis, S.T., Saul, L.K.: Nonlinear dimensionality reduction by locally linear embedding. Science **290**(5500), 2323–2326 (2000)
17. Sigal, L., Black, M.J.: Humaneva: synchronized video and motion capture dataset for evaluation of articulated human motion. Brown Univertsity TR, 120 (2006)
18. Tenenbaum, J.B., De Silva, V., Langford, J.C.: A global geometric framework for nonlinear dimensionality reduction. Science **290**(5500), 2319–2323 (2000)
19. Xia, T., Tao, D., Mei, T., Zhang, Y.: Multiview spectral embedding. IEEE Trans. Syst. Man Cybern. Part B Cybern. **40**(6), 1438–1446 (2010)
20. Jun, Y., Wang, M., Tao, D.: Semisupervised multiview distance metric learning for cartoon synthesis. IEEE Trans. Image Process. **21**(11), 4636–4648 (2012)
21. Yu, K., Wang, Z., Hagenbuchner, M., Feng, D.D.: Spectral embedding based facial expression recognition with multiple features. Neurocomputing **129**, 136–145 (2014)

Matrix Based Regression with Local Position-Patch and Nonlocal Similarity for Face Hallucination

Guangwei Gao[1($\boxtimes$)] and Jian Yang[2]

[1] Institute of Advanced Technology,
Nanjing University of Posts and Telecommunications,
Nanjing, People's Republic of China
csggao@gmail.com
[2] School of Computer Science and Engineering,
Nanjing University of Science and Technology,
Nanjing, People's Republic of China
csjyang@njust.edu.cn

Abstract. Learning based face hallucination methods have received much attention and progress in past decades. As opposed to the existing methods, where the input image patch matrix is first stacked into vectors before combination coefficients calculation, this paper directly uses the matrix based regression model for combination coefficients computation to preserve the essential structural information of the input matrix. For each low-resolution local patch matrix, its combination coefficients over the same position image patch matrices in training images can be computed. Then the desired high-resolution patch matrix can be obtained by replacing the low-resolution training samples with corresponding high-resolution counterparts. The nonlocal self-similarities are finally utilized to further improve the hallucination performance. Experiments conducted on the FERET face dataset indicate that our method could outperform other state-of-the-art algorithms in terms of both vision and quantity.

Keywords: Face hallucination · Position-patch · Matrix based regression · Nonlocal self-similarity

1 Introduction

Face hallucination, which is also referred as face super-resolution (SR), is a technology to obtain high-resolution (HR) face images from low-resolution (LR) inputs. The HR facial images are very useful for video surveillance, 3D facial modeling, automated face recognition, and so on [1].

In past few decades, learning based face hallucination methods have received a lot of attentions since they can utilize additional and relevant information from training samples, leading to better super-resolved results than that of reconstruction based ones [2–12]. How to utilize the training set is crucial in learning based methods. For example, Freeman et al. [4] utilized a Markov network to model the relationship between LR and HR patches to perform SR. Chang et al. [5] utilized the locally linear

© Springer International Publishing Switzerland 2015
X. He et al. (Eds.): IScIDE 2015, Part I, LNCS 9242, pp. 110–117, 2015.
DOI: 10.1007/978-3-319-23989-7_12

embedding (LLE) approach and proposed the neighbor embedding (NE) method. NE assumes that the two manifolds constructed by the LR and HR patches have similar structures. Li et al. [6] proposed to project the LR and HR patch pairs from the original manifolds into a common manifold with a manifold regularization procedure.

Wang and Tang [8] suggested the usage of principal component analysis (PCA) [9] to represent the structural similarity of face images. PCA is used to represent an input LR image as a linear combination of the LR training samples. The target HR image is obtained by using the same combination coefficients, but with the LR training samples replaced by the corresponding HR ones. However, the HR images usually contain ghost artifacts as a result of using global linear constraints on HR images and the PCA-based holistic appearance model. Recently, a novel face hallucination method based on position-patch has been proposed by replacing the global linear constraints with multiple local constraints learned from training patches [10]. Accordingly, the characteristics of the input image can be preserved since the input patch is approximated by position patches. Yang et al. [11] used a coupled dictionary learned from all training patches using sparse representation [13] for image SR. They assumed that there exist coupled dictionaries of LR and HR images, which have the same sparse representation for each pair of LR and HR patches. After learning the coupled dictionary pair, the LR image patch is first coded over the LR dictionary by sparse representation. Then the HR patch is reconstructed on HR dictionary with these sparse coefficients.

The previous methods, such as [5, 8, 10, 11], are all vector-based method. That is to say, before computing the combination coefficients of the input LR image (patch) matrix over the training samples, we have to convert the images (patches) matrix into vectors in advance (as shown in Fig. 1). However, in the converting step, some structural information (e.g. the rank of matrix) might be lost. The resulting HR images may contain some noisy artifacts along contours.

In this paper, we apply the matrix based regression model, which can compute the combination coefficients straightforward (without the matrix-to-vector conversion). We will make full use of the information of image matrices and use the minimal rank of representation residual matrix as a criterion to determine the combination coefficients. Inspired by the work in [10], we also use the position-patch instead of neighbor-patch in representation and reconstruction. First, each input LR image patch matrix is directly represented as a linear combination of the same position image patch matrices in each training image using nuclear norm based matrix regression. Then, the HR image patch matrix is reconstructed based on the corresponding HR training patch matrix with the same combination coefficients. The whole HR image can be synthesized by integrating the reconstructed patches. On the other hand, considering the fact that there are often many repetitive image structures in an image, the nonlocal (NL) self-similarity can be used to further enhance the result. The performance of the proposed method is examined and compared with other state-of-the-art methods.

The remainder of this paper is organized as follows. Section 2 formulates the nuclear norm based matrix regression. Section 3 presents our face hallucination algorithm. Section 4 reports the experimental results and Sect. 5 concludes the paper.

2 Nuclear Norm Based Coding

In this section, we introduce the nuclear norm based matrix regression and then use the alternating direction method of multipliers to solve the problem.

2.1 Formulation

Suppose that we are given a set of n matrices $A_1, \ldots, A_n \in R^{p \times q}$ and a query data matrix $D \in R^{p \times q}$. D can be represented linearly using $A_1, \ldots, A_n$:

$$D = A(x) + E, \tag{1}$$

where $A(x) = x_1 A_1 + x_2 A_2 + \cdots + x_n A_n$, $x_1, \ldots, x_n$ is representation coefficients and E is the representation residual. The nuclear norm based coding can be formulated as follows [14]:

$$\min_{x,E} \|x\|_1 + \lambda \|E\|_*, \quad s.t. \quad A(x) - D = E, \tag{2}$$

where $|\cdot|_1$ denotes the l_1-norm (the sum of the absolute values of each entry) of a vector, $\|\cdot\|_*$ denotes the nuclear norm (the sum of the singular values) of a matrix.

2.2 Algorithm

For the convenience of expression, problem (2) can be converted into the following equivalent problem:

$$\min_{x,z,E} \|z\|_1 + \lambda \|E\|_*, \quad s.t. \quad A(x) - D = E \ \text{ and } \ x = z. \tag{3}$$

The alternating direction method of multipliers (ADMM) or the augmented Lagrange multipliers (ALM) method has been applied to solve the nuclear norm optimization problems [15–17]. Here, we give the process of using ADMM to solve the regularized matrix regression problem:

The augmented Lagrangian function L of Eq. (3) is

$$L_\mu(x, z, E) = \|z\|_1 + \lambda \|E\|_* + tr\left(y_1^T (x - z)\right) + tr\left(Y_2^T (A(x) - D - E)\right) \\ + \frac{\mu}{2}\left(\|x - z\|_F^2 + \|A(x) - D - E\|_F^2\right), \tag{4}$$

where $\mu > 0$ is a penalty parameter, y_1 and Y_2 are the Lagrange multipliers, $tr(\cdot)$ is the trace operator. Following some simple algebraic steps, Eq. (4) can be rewritten as

$$L_\mu(x, z, E) = \|z\|_1 + \lambda \|E\|_* + \frac{\mu}{2}\left(\left\|x - z + \frac{1}{\mu}y_1\right\|_F^2 + \left\|A(x) - D - E + \frac{1}{\mu}Y_2\right\|_F^2\right) \\ - \frac{1}{2\mu}\|y_1\|_2^2 - \frac{1}{2\mu}\|Y_2\|_F^2. \tag{5}$$

The function L can be minimized by updating each of variables one at a time. The detailed algorithm via ADMM is summarized in Algorithm 1 [14]. For the step 1, the soft-thresholding operator [15] can be applied. For the step 3, the singular value thresholding algorithm [18] can be applied.

Algorithm 1[14]. Nuclear norm based coding via ADMM

Input: Training matrices $A_1,\ldots,A_n \in R^{p\times q}$ and a matrix $B \in R^{p\times q}$, parameters λ and μ, the termination condition parameter ε.

1: Fix the others and update z^{k+1} by

$$z^{k+1} = \arg\min_z \frac{1}{\mu}\|z\|_1 + \frac{1}{2}\left\|z - \left(x^k + \frac{1}{\mu}y_1^k\right)\right\|_F^2 ;$$

2: Fix the others and update x^{k+1} by

$$x^{k+1} = \left(M^T M + I\right)\backslash M^T\left[Vec\left(B + E^k - \frac{1}{\mu}Y_2^k\right) + z^{k+1} - \frac{1}{\mu}y_1^k\right];$$

$M = [Vec(A_1), \ldots, Vec(A_n)]$, $Vec(\cdot)$ is the vectorization operator.

3: Fix the others and update E^{k+1} by

$$E^{k+1} = \arg\min_E \frac{\lambda}{\mu}\|E\|_* + \frac{1}{2}\left\|E - \left(A(x^{k+1}) - B + \frac{1}{\mu}Y_2^k\right)\right\|_F^2 ;$$

4: Update the multiplies

$$y_1^{k+1} = y_1^k + \mu\left(x^{k+1} - z^{k+1}\right), \; Y_2^{k+1} = Y_2^k + \mu\left(A(x^{k+1}) - B - E^{k+1}\right);$$

5. If the termination condition is satisfied, go to 6; otherwise go to 1.

6. **Output:** Optimal regression coefficient vector x^{k+1}

---

3 Face Hallucination Based on Matrix Regression

In this section, we describe the proposed face hallucination algorithm based on local position-patch and nonlocal similarities.

3.1 Local Position-Patch Based Matrix Regression

Denote the patches of LR testing image, patches of the mth LR training image, patches of the mth HR training image located at position (i, j) as $\{B_L^p(i,j)\}_{p=1}^N$, $\{A_L^{mp}(i,j)\}_{p=1}^N$ and $\{A_H^{mp}(i,j)\}_{p=1}^N$ respectively. N is the number of patches. Each patch is represented in the form of matrix.

Let us define the following linear mapping:

$$A_L^p(i,j)(x) = \sum_{m=1}^M x_m(i,j)A_L^{mp}(i,j), \qquad (6)$$

where M is the number of training images.

For each patch matrix $B_L^p(i,j)$ in the LR testing image, its reconstruction weights can be obtained by

$$\hat{x}(i,j) = \arg\min_{x(i,j)} \|x(i,j)\|_1 + \lambda \|A_L^p(i,j)(x) - B_L^p(i,j)\|_* \qquad (7)$$

Equation (7) can be solved by Algorithm 1.

Replacing each LR image patch matrix $A_L^{mp}(i,j)$ by its corresponding HR sample $A_H^{mp}(i,j)$, the desired HR patch is denoted as

$$B_H^p(i,j) = \sum_{m=1}^{M} \hat{x}_m(i,j) A_H^{mp}(i,j) \qquad (8)$$

The whole HR image can be synthesized by integrating the reconstructed patches according to the original position. Pixels in the overlapping regions are obtained by averaging the pixels value in the overlapping regions between two adjacent patches reconstructed.

3.2   Exploiting Nonlocal Self-similarities

Recently many works have shown that the nonlocal (NL) redundancies are very useful for image restoration [19]. For each local patch b_i in B_H, we search for its similar patches in the whole image, and then predict this patch as: $\hat{b}_i = \sum_{i=1}^{L} w_i^l b_i^l$, where b_i^l is the lth most similar patch to b_i and w_i^l is the nonlocal weight as defined in [19]. The nonlocal based synthesis can be performed by

$$b_i = \arg\min_{b_i} \sum_{m=1}^{M} \hat{x}_m(i,j) A_H^{mp}(i,j) + \delta \left\| b_i - \sum_{i=1}^{L} w_i^l b_i^l \right\|_2^2 \qquad (9)$$

The proposed face hallucination algorithm is summarized in Algorithm 2.

Algorithm 2. Proposed face hallucination algorithm

Input: LR training image matrices $A_L^1, \ldots, A_L^M$, corresponding HR training image matrices $A_H^1, \ldots, A_H^M$, LR input image matrices B_L.

1: Divide the LR input image and training image into overlapped patch matrices;
2: For each patch $B_L^p(i,j)$:

    (a) Compute the reconstruct weights $x(i,j)$ by Eq. (7);
    (b) Synthesize the HR patch $B_H^p(i,j)$ by Eq. (8).
    (c) Update $B_H^p(i,j)$ by the nonlocal self-similarities in Eq. (9).
3:   Integrate the HR patches to form target HR image B_H.

---

4. **Output:** The Synthesized HR image B_H

4   Experimental Results

To verify the superiority of our method, experiments were conducted on the FERET database [20]. We selected 1196 face images of different people in our experiments. These face images were aligned manually using the locations of three points: centers of left and right eyeballs and center of the mouth. All faces are standardized to the size of 120 × 102.

These HR images were blurred and down-samples to generate LR images with size 40 × 34. We compared our approach with bicubic interpolation, Wang's method [8], Chang's method [5], Ma's method [10] and Yang's method [11]. In order to get the optimal results in Wang's method, we used image pairs of 1176 people for training and let variance contribution rate of PCA be 0.90. The number of neighbor-patches in Chang's method is 380. Image pairs of 380 people were used for training in Chang's method, Ma's method, Yang's method and our method. The other images of 20 people were used as test images. The LR image patch size was 3 × 3 and the patch overlapped with its adjacent patch by 3 pixels (i.e., one column). The parameters λ and μ in our method are set to be 0.01 and 1.

Figure 1 shows some hallucinated results visually, from which we can see that the results of Wang's method can hardly maintain global smoothness around the face contour and margin of the mouth. The results of Chang's method are somewhat

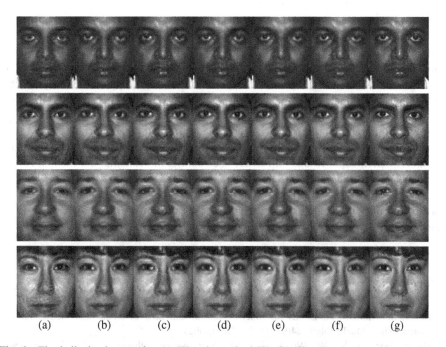

(a)   (b)   (c)   (d)   (e)   (f)   (g)

Fig. 1. The hallucination results. (a) Wang's method [8]. (b) Chang's method [5]. (c) Yang's method [11]. (d) Ma's method [10]. (e) Our method without NL. (f) Our method. (g) The original 120 × 102 HR faces.

Table 1. Average PSNR, SSIM and FSIM for each method.

Quality measure	PSNR	SSIM	FSIM
Wang's method [8]	27.3729	0.7521	0.8700
Chang's method [5]	29.5950	0.8795	0.9152
Yang's method [11]	29.7871	0.8704	0.9163
Ma's method [10]	30.2363	0.8883	0.9236
Our method without NL	**30.7514**	**0.8935**	**0.9280**
Our method	**30.8992**	**0.8944**	**0.9280**

blurred. There are jaggy artifacts along the eyes in Yang's method. While Ma's method and our method produce more nature looking images than them. To quantify the results from the ground truth data, Table 1 tabulates the average PSNR, SSIM [21] and FSIM [22] of the reconstructed facial images using respective methods. From the table, we can see that our method with nonlocal self-similarities achieves the best hallucination performance. Simultaneously, our method without nonlocal self-similarities also outperforms other methods. In our opinions, the proposed method without nonlocal self-similarities directly utilize image patch matrix based regression for combination coefficients calculation, which can capture the underlying structure information in the local patch matrix to some extent. Furthermore, the nonlocal self-similarities can provide complementary information for the final hallucinated results.

5 Conclusion

In this paper, we applied the matrix based regression with local position-patch and nonlocal self-similarity for efficient face hallucination. For each low-resolution local patch matrix, the nuclear norm based matrix regression was applied to compute the combination coefficients without matrix-to-vector conversion. Then the intermediate high-resolution patch matrix can be obtained based on the corresponding high-resolution training patch matrix with the same combination coefficients. The nonlocal self-similarities are finally utilized to further improve the hallucination performance. Experimental results on FERET face dataset indicate that our method could yield better reconstruction results than other state-of-the-art methods.

In the future work, we will still focus on the matrix regression based face hallucination with uncentralized face images and test the robustness for various variations in images.

References

1. Baker, S., Kanade, T.: Limits on super-resolution and how to break them. IEEE Trans. Pattern Anal. Mach. Intell. **24**(9), 1167–1183 (2002)
2. Dai, S., Han, M., Xu. W., Wu, Y., Gong, Y.: Soft edge smoothness prior for alpha channel super resolution. In: IEEE Computer Society Conference on Computer Vision and Pattern Recognition (CVPR), pp. 1–8 (2007)

3. Dong, W., Zhang, L., Shi, G., Wu, X.: Nonlocal back-projection for adaptive image enlargement. In: IEEE International Conference on Image Processing (ICIP), pp. 349–352 (2009)
4. Freeman, W., Jones, T.R., Pasztor, E.C.: Example-based super-resolution. IEEE Comput. Graph. Appl. **22**(2), 56–65 (2002)
5. Chang, H., Yeung, D.Y., Xiong, Y.M.: Super-resolution through neighbor embedding. In: IEEE Computer Society Conference on Computer Vision and Pattern Recognition (CVPR), pp. 1275–1282 (2004)
6. Li, B., Chang, H., Shan, S., Chen, X.: Aligning coupled manifolds for face hallucination. IEEE Signal Process. Lett. **16**(11), 957–960 (2009)
7. Par, J.S., Lee, S.W.: An example-based face hallucination method for single-frame, low-resolution facial images. IEEE Trans. Image Process. **17**(20), 1806–1816 (2008)
8. Wang, X., Tang, X.: Hallucinating face by eigentransformation. IEEE Trans. Syst. Man Cybern. Part C Appl. Rev. **35**(3), 425–434 (2005)
9. Jain, A.K., Duin, R.P.W., Mao, J.: Statistical pattern recognition: a review. IEEE Trans. Pattern Anal. Mach. Intell. **22**(1), 4–37 (2000)
10. Ma, X., Zhang, J., Qi, C.: Hallucinating face by position-patch. Pattern Recogn. **43**(6), 2224–2236 (2010)
11. Yang, J., Wright, J., Tang, H., Ma, Y.: Image super-resolution via sparse representation. IEEE Trans. Image Process. **19**(11), 2861–2873 (2010)
12. Zhang, W., Cham, W.K.: Learning-based face hallucination in DCT domain. In: IEEE Computer Society Conference on Computer Vision and Pattern Recognition (CVPR), pp. 1–8 (2008)
13. Wright, J., Yang, A., Ganesh, A., Sastry, S., Ma, Y.: Robust face recognition via sparse representation. IEEE Trans. Pattern Anal. Mach. Intell. **31**(2), 210–227 (2009)
14. Luo, L., Yang, J., Qian, J., Yang, J.: Nuclear norm regularized sparse coding. In: IEEE International Conference on Pattern Recognition (ICPR), pp. 1834–1839 (2014)
15. Lin, Z., Chen, M., Ma, Y.: The augmented lagrange multiplier method for exact recovery of corrupted low-rank matrices. UIUC Tech. Rep. UIUC-ENG-09-2215 (2009)
16. Yang, J., Qian, J., Luo, L., Zhang, F., Gao, Y.: Nuclear norm based matrix regression with applications to face recognition with occlusion and illumination changes. arXiv:1405.1207 (pdf)
17. Chen, J., Yang, J., Luo, L., Qian, J.: Matrix variate distribution induced sparse representation for robust image classification. In: IEEE Transactions on Neural Networks and Learning Systems, to appear
18. Cai, J.F., Candès, E.J., Shen, Z.: A singular value thresholding algorithm for matrix completion. SIAM J. Optim. **20**(4), 1956–1982 (2010)
19. Buades, A., Coll, B., Morel, J.. A non-local algorithm for image denoising. In: IEEE Computer Society Conference on Computer Vision and Pattern Recognition (CVPR), pp. 60–65 (2005)
20. Philips, P., Moon, H., Pauss, P., Rivzvi, S.: The FERET evaluation methodology for face recognition algorithms. In: IEEE Computer Society Conference on Computer Vision and Pattern Recognition (CVPR), pp. 137–143 (1997)
21. Wang, Z., Bovik, A.C.: Mean squared error: love it or leave it? A new look at signal fidelity measures. IEEE Signal Process. Mag. **26**(1), 98–117 (2009)
22. Zhang, L., Zhang, L., Mou, X., Zhang, D.: FSIM: a feature similarity index for image quality assessment. IEEE Trans. Image Process. **20**(8), 2378–2386 (2011)

Facial Occlusion Detection via Structural Error Metrics and Clustering

Xiao-Xin Li[1]([⊠]), Ronghua Liang[2], Jiaquan Gao[1], and Haixia Wang[2]

[1] College of Computer Science and Technology, Zhejiang University of Technology,
Hangzhou 310023, China
{mordekai,gaojq}@zjut.edu.cn
[2] College of Information Engineering, Zhejiang University of Technology,
Hangzhou 310023, China
{rhliang,hxwang}@zjut.edu.cn

Abstract. Facial occlusions pose significant obstacles for robust face recognition in real-world applications. To eliminate the effect incurred by occlusions, most of the popular methods concentrate on dealing with the error between the occluded image and its recovery. Inspired by the working mechanism of human visual systems in facial occlusion detection, we suggest that it should be the error metric and clustering rather than exact recovery that play important roles for occlusion detection. By considering the structural differences between faces and occlusions, such as colors and textures, we construct five structural error metrics. By considering the common structures shared by all occlusions, such as localization and contiguity, we construct a structured clustering operator. Furthermore, we select the optimal error metric via the minimum occlusion boundary regularity criterion. Integrating the above techniques, we propose the Structural Error Metrics and Clustering (SEMC) algorithm for facial occlusion detection. Experimental results demonstrate that, even just using the mean face of the training images as the recovery image, SEMC still achieves more accurate and robust performance compared to the related state-of-the-art methods.

Keywords: Unconstrained face recognition · Facial occlusion detection · Structural error metric · Structural clustering

1 Introduction

Recently, recognizing human faces with occlusions has received a lot of attention in computer vision and pattern recognition [4,7,11,13,15,16]. Facial occlusions, including accessories, shadows or other objects in front of a face, pose significant obstacles for robust face recognition in real-world applications [3]. To eliminate the effect incurred by occlusion, researchers have studied the solution schemes from different views. Most of these schemes concentrate on the recovery error $\hat{e} \in \mathbb{R}^m$ between the occluded image $y \in \mathbb{R}^m$ and its recovery $\hat{y} \in \mathbb{R}^m$ with respect to (w.r.t.) the training dictionary $A \in \mathbb{R}^{m \times n}$, with the assumption that the

© Springer International Publishing Switzerland 2015
X. He et al. (Eds.): IScIDE 2015, Part I, LNCS 9242, pp. 118–127, 2015.
DOI: 10.1007/978-3-319-23989-7_13

larger the entry values of the error $\hat{e}$, the higher the probability with which the corresponding pixels are occluded. By assuming that the error $\hat{e}$ can be sparsely coded w.r.t. some error coding dictionary $E \in \mathbb{R}^{m \times d}$, researchers proposed the following Error Coding Model (ECM) from different views [2,5,8,13,14]

$$\min_{x,c} \|x\|_1 + \|c\|_1 \quad s.t. \quad [A\ E] \begin{bmatrix} x \\ c \end{bmatrix} = \hat{y} + \hat{e} = y, \tag{1}$$

where $x \in \mathbb{R}^n$ and $c \in \mathbb{R}^d$ are the coding coefficients of the recovery image $\hat{y}$ and the error $\hat{e}$, respectively.

Another view on dealing with the error is the Error Weighting Model (EWM). According to the works of [4,7,15,16], the EWM can be summarized as

$$\min_{x,w} \|x\|_{\ell^a} + \mu \|w \odot \hat{e}\|_{\ell^b} + \lambda \phi(w, \hat{e}) \ s.t. \ \hat{e} = y - Ax, \tag{2}$$

where ℓ^a and ℓ^b are the norm indexes, μ and λ are the regularization parameters, $w \in \mathbb{R}^m$ is the error weight, $\phi(w, \hat{e})$ is the cost function of w w.r.t. the error $\hat{e}$, and $\odot$ is the Hadamard product. With $\mu = 1, \lambda = 0, w = 1, \ell^a = \ell^b = 1, E = I$, the EWM is equivalent to the ECM. The entries of the weight w indicate the occlusion support or the probabilities with which the corresponding pixels are occluded.

The occlusion location is mainly indicated by the reconstruction error $\hat{e} = y - \hat{y} = y - Ax$ in both ECM and EWM. It seems that the quality of the recovery image $\hat{y}$ determines the accuracy of the corresponding occlusion detection result. However, we know that an exact recovery does not be necessary for human visual system (HVS), that is, HVS could recognize the occluded region of a face without having to see this face before, i.e., without having to compare it with its ground truth. HVS captures the structural differences by comparing the occlusion with a *fuzzy* face model, which is learned from the faces emerging in everyday life ever before. We call that the learned face model is fuzzy as its identity is unclear and HVS only keeps its typical structure composed of eyes, nose, mouth and etc. Inspired by this observation, we explore the automatic facial occlusion detection technology via structural comparison, that is, structural error metrics and clustering. For simplicity, we choose the mean face $\bar{y} = \frac{1}{n}(A \times 1)$ $(1 \in \mathbb{R}^n)$ as the fuzzy face model in the subsequent work.

The rest of this paper is organized as follows. Section 2 presents several structural error metrics and a structural error clustering operator. Section 3 gives an optimal error metric selection criterion, which integrates the proposed structural error metrics and clustering operator together. Section 4 performs the experiments. Section 5 concludes the paper.

2   Structural Error Metrics and Clustering

In order to highlight the error incurred by occlusion, we measure the error $\tilde{e}$ between the test image y and the mean face $\bar{y}$ based on the potential differential structures between faces and occlusions; in order to cluster the occluded pixels, we consider the error $\tilde{e}$ and its clustering operator based on the common structures shared by all occlusions.

2.1 Structural Differences Induced Error Metrics

By observation, we note that the structural differences between occlusions and the fuzzy face model (the mean face) is mainly reflected in colors, textures and shapes. We now consider how to use these structural differences to construct error metrics. Suppose we could project an image into a structured subspace, which keeps or strengthens the preconceived structure and wipes off or weakens the unwanted structures. Then the error between two images can be measured in this new subspace

$$\tilde{e}_f = \mathcal{E}_f (y, \bar{y}) = |f (y) - f (\bar{y})|, \qquad (3)$$

where the function $f : \mathbb{R}^m \to \mathbb{R}^m$ is the structural difference descriptor embedded in the structured subspace. We call the error metrics based on (3) the structural error metrics (SEMs). Hence, what is critical for SEMs is to design the structural difference descriptors.

Color Difference Metrics. The color difference can be directly measured by the absolute error metric in the original image domain $\mathcal{E}_I (y, \bar{y}) = |I (y) - I (\bar{y})| = |y - \bar{y}|$. However, $\mathcal{E}_I$ does not consider the relative error between image pixels. For example, if there exists $y_j = y_i + a$ and $\bar{y}_j = \bar{y}_i + a$ for some $a \gg 1$, we have $\mathcal{E}_I (y_i, \bar{y}_i) = \mathcal{E}_I (y_j, \bar{y}_j)$. While it seems reasonable in mathematics, this error measurement result clashes with *Weber's law* [12] in psychophysics, which says that the relationship between the stimulus S and the perception p is logarithmic. That is, the lower the initial stimulus is, the more easily it could be perceived. By assuming that the occlusion is a stimulus with low intensities, we project images into the logarithmic domain and have the log-based error metric $\mathcal{E}_{\log} (y, \bar{y}) = |\log y - \log \bar{y}|$, which enhances the error caused by the low-value occlusions while suppresses the error incurred by high-value occlusions.

We then consider occlusions with high intensities. This problem might be well solved, if we can map the occluded image into a feature subspace, where most of the pixels with high values are transformed to the ones with low values. By supposing the pixel values of the occlusion in the local area change slowly and smoothly, this feature subspace can be described by the gradient $\nabla I = \sqrt{\left(\frac{\partial I}{\partial x}\right)^2 + \left(\frac{\partial I}{\partial y}\right)^2}$, where $\frac{\partial I}{\partial x}$ and $\frac{\partial I}{\partial y}$ are the gradients along the vertical and horizontal directions, respectively. We now have the log-gradient-based error metric $\mathcal{E}_\nabla (y, \bar{y}) = |\log_\nabla (y) - \log_\nabla (\bar{y})|$, where $\log_\nabla (y) = \log (\nabla y)$.

Texture Difference Metrics. While the log-based error metric $\mathcal{E}_{\log}$ and the log-gradient-based error metric $\mathcal{E}_\nabla$ are sensitive to occlusions with low intensities and high intensities changing uniformly, they are insensitive to occlusions with intensities changing rapidly or randomly. In this scenario, it is the texture differences rather than the color differences that dominate the structural differences between faces and occlusions. However, texture is easy to see but difficult to define [10], as its definition might be different for different applications.

In this work, we describe the texture by the density of edges per unit area, which can be computed over an image area by the Laplacian filtering $\nabla^2 I = \frac{\partial^2 I}{\partial x^2} + \frac{\partial^2 I}{\partial y^2}$. We then have a new error metric in the Laplacian filtered domain $\mathcal{E}_{\nabla^2}(y, \bar{y}) = |\nabla^2 y - \nabla^2 \bar{y}|$, which we call the Laplace-based error metric.

Color-Texture-Combined Difference Metrics. We now consider the scenario when both the color and texture differences are prominent. Here, we introduce the differential excitation (DE) of an image proposed by Chen *et al.* [1] $DE(I) = \arctan \frac{\nabla^2 I}{I}$. The DE operator simultaneously keeps the texture and color features of the original image, as the Laplacian filtered image $\nabla^2 I$ in the numerator calculates the texture feature and I in the denominator keeps the color feature. Specifically, the ratio $\frac{\nabla^2 I}{I}$ actually amounts to the *Weber fraction*, and the arctangent function limits the output of DE in $\left[-\frac{\pi}{2}, \frac{\pi}{2}\right]$ and is a logarithm-like function, which is also sensitive to the change incurred by low intensities. We now have the DE-based error metric $\mathcal{E}_{DE}(y, \bar{y}) = |DE(y) - DE(\bar{y})|$.

2.2   Sharing Structures Induced Error Clustering

Clearly, it is critical to seek the common structures shared by all occlusions for clustering occluded pixels. The common structures of occlusions explored in existing literature mainly includes locality [4] and contiguity [16].

Local Structure for Error Enhancement and Normalization. The correntropy induced metric (CIM) [4] measures the error between each pixel pair y_i and $\bar{y}_i$ as follows $CIM(y_i, \bar{y}_i) = 1 - g(y_i - \bar{y}_i)$, where $g(x) = \exp\left(-\frac{x^2}{2\sigma^2}\right)$ is the Gaussian kernel. CIM is a statistical local metric due to the utilized Gaussian kernel and its locality can be adjusted with the kernel size σ. In order to utilize the local structure of the error calculated by various error metrics, we extend the CIM to the following form $CIM_f(y_i, \bar{y}_i) = 1 - g(\mathcal{E}_f(y_i, \bar{y}_i))$. Note that the error metric LD proposed in [7] is just an instance of the error metric CIM_f.

Contiguous Structure for Error Clustering. The contiguous structure of occlusion is usually depicted by the adjacent relationships of the occlusion spatial support. However, the error support, instead of the occlusion support, is commonly used in literature, since the occlusion support is usually unknown. The contiguous filtering combined with the error clustering is the common way to obtain the error support. In [16], the authors explored the contiguous structure via Markov random field, and the error support is estimated by solving a binary GraphCut problem. In this work, for simplicity, we just use the morphological filtering [16] to obtain the contiguous structure. The key idea is to first cluster the errors by K-means (or just threshold the errors) and then to apply open and close operations to the binary error support, which can be formulated as $\mathcal{K}_s(\hat{e}) = f_\bullet(f_o(\mathcal{K}(\hat{e})))$, where $f_o(\cdot)$ and $f_\bullet(\cdot)$ are open and close operations, respectively.

Algorithm 1. Structural Error Metrics and Clustering (SEMC) for Facial Occlusion Detection

Input: data matrix $A \in \mathbb{R}^{m \times n}$, test sample $y \in \mathbb{R}^m$.
Output: detected occlusion support $\tilde{s}$.

1. Calculate the mean face $\bar{y} = \frac{1}{n} (A \times 1)$;
2. Set the structure difference operator ensemble: $F = \{I, \log, \nabla, \nabla^2, DE\}$;
3. **For** each $f \in F$
4.   Calculate the recovered error $\tilde{e}_f$: $(\tilde{e}_f)_i = CIM_f (y_i, \bar{y}_i)$;
5.   Cluster the recovered error: $\tilde{s}_f = \mathcal{K}_s (\tilde{e}_f)$;
6. **End For**
7. Select the optimal occlusion support: $\tilde{s} = \tilde{s}_{f^*}$, where $f^* = \arg\min_{f \in F} \mathcal{B} (\tilde{s}_f)$.

3   Optimal Error Metric Selection

We now have 5 structural error metrics, $\mathcal{E}_I$, $\mathcal{E}_{\log}$, $\mathcal{E}_{\log_\nabla}$, $\mathcal{E}_{\nabla^2}$ and $\mathcal{E}_{DE}$, for facial occlusion detection, which are designed for different occlusions with different structures, respectively. As the structure of a special occlusion is usually priori unknown, it seems difficult to automatically choose the optimal error metric. However, we find that the minimum occlusion boundary regularity criterion proposed in [7] can be used here to help selecting the optimal error metric. The idea is inspired by the observation that all natural occlusions usually have smooth and regular boundaries. We therefore deduce that if the shape of the detected occlusion based on an error metric is coarse and irregular, the corresponding utilized error metric might not be the optimal one. According to the morphological boundary detection algorithm presented in [7], the minimum boundary regularity criterion can be formulated as $\arg\min_f \mathcal{B} (s_f) = \|s_f - (s_f \boxminus T)\|_1$, where s_f is the detected error support, $\boxminus$ is the erosion operator, and $T = [1\,1\,1; 1\,1\,1; 1\,1\,1]$ is the structuring element.

Incorporating the 5 structural error metrics, $\mathcal{E}_I$, $\mathcal{E}_{\log}$, $\mathcal{E}_{\log_\nabla}$, $\mathcal{E}_{\nabla^2}$ and $\mathcal{E}_{DE}$ with the local error metric CIM, and using the structured clustering operator $\mathcal{K}_s$ and the minimum occlusion boundary regularity criterion, Algorithm 1, dubbed the Structural Error Metrics and Clustering (SEMC), summarizes the whole procedure of our method used to make facial occlusion detection.

4   Simulations and Experiments

To evaluate the proposed SEMC algorithm, we compare it with the state-of-the-art methods on two publicly available databases, namely, the Extended Yale B [6] database and the AR [9] database. Since the ECM (1) does not contain the occlusion detection mechanism, we just pay attention to the state-of-the-art EWM-based methods: the correntropy-based sparse representation (CESR) [4], the robust sparse coding (RSC) [15], and the structured sparse error coding (SSEC) [7]. Note that both CESR and RSC just calculate the probability w with which the pixels are occluded. We therefore cluster w to estimate the occlusion support: for CESR, we estimate the occlusion support by K-means clustering $s_{CESR} = \mathcal{K}(1 - w)$; for RSC, according to its open Matlab code, we estimate the occlusion support by threshold clustering $s_{RSC} = \left(\frac{w}{\max w} < 10^{-3}\right)$.

4.1   Synthetic Facial Occlusion Detection

In this section, we use the Extended Yale B database [6] to investigate the performance of SEMC under fixed feature dimension for various synthetic occlusions with varying levels and boundaries. We choose Subsets I and II (717 images, normal-to-moderate lighting conditions) for training and Subset III (453 images, more extreme lighting conditions) for testing. Synthetic occlusions with various boundaries and occlusion levels are imposed on the test samples. The images are resized to 96×84 pixels.

Detection With Various Occlusion Levels. To test the accuracy and stability of SEMC in occlusion detection, we first simulate various levels of occlusions from 10 % to 90 % by replacing a random located block of each test image with a mandrill image. Figure 1a gives nine occluded faces with various occlusion levels and their detailed detection results using the four compared methods. For each method against each occlusion level, the average true positive rates (TPRs) and false positive rates (FPRs) of the occlusion detection results over the 453 test images are also shown in Fig. 1c. For all the cases, the TPRs and FPRs of SEMC are almost always suboptimal compared to the optimal ones of the other methods, whereas the differences between TPRs and FPRs of SEMC are always the largest. This implies that SEMC achieves the optimal balance between TPRs and FPRs. Figure 1c also shows that SEMC achieves its optimal performance at the 50 %~60 % occlusion levels but not at the lowest ones. The reason is that when the occlusion levels are very low, the dominant differential structures between the mean face (without occlusions) and the occluded face are mainly determined by the differences between faces, which does not be considered by SEMC. The similar problem also exists for the other 3 methods, especially for SSEC.

Different from the other compared methods, SEMC not only detects occlusions but also *understands* occlusion structures. To illustrate this, Fig. 2a counts the number of the structural error metrics selected by SEMC at various mandrill occlusion levels. The main structure of mandrill occlusion actually changes with occlusion levels: when the occlusion levels are low, the texture feature is significant as the edge density is intensive; with the occlusion level increasing, both the texture and color features become more and more significant. This means that for low occlusion levels, SEMC should choose the Laplace-based error metric $\mathcal{E}_{\nabla}$, while for high occlusion levels, SEMC should choose the DE-based error metric $\mathcal{E}_{DE}$. Figure 2a shows that SEMC does perform as expected.

Detection With Various Occlusions. To test the adaptability of SEMC for various occlusions, we simulate 60 % occlusion levels with 7 different objects successively: mandrill, camera, dog, apple, sunflower, random block and white block, i.e., we have 453×7 occluded test images. Figure 1b and d gives the

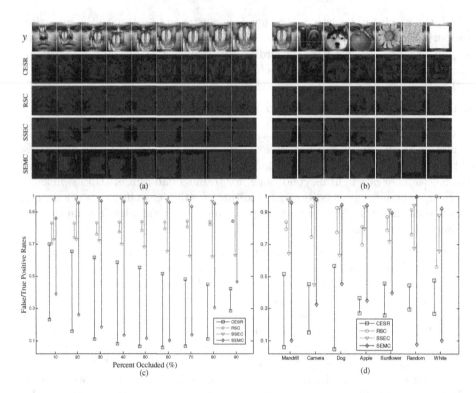

Fig. 1. (a)-(b) The occlusion detection results of various algorithms against various occlusions with various levels on the Extended Yale B database: (a) 10 %~90 % mandrill occlusions and (b) 60 % various occlusions. (c)-(d) The corresponding average TPRs and FPRs of the 453 occlusion detection results of various algorithms against various occlusions.

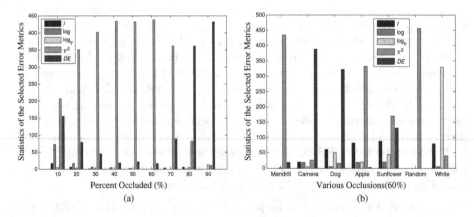

Fig. 2. The statistics of the structural error metrics selected by SEMC for 10 %~90 % mandrill occlusions (a) and for various occlusions (b).

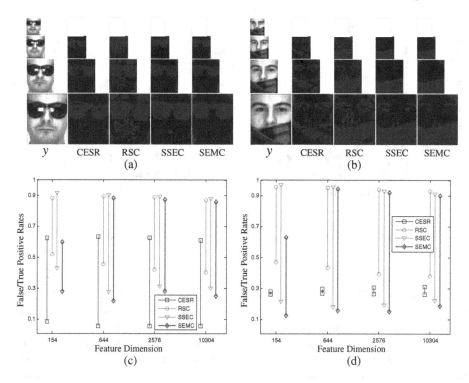

Fig. 3. (a)-(b): The occlusion detection results of various algorithms against various feature dimensions on the AR database. (c)-(d): The average TPRs and FPRs of the occlusion detection results of various algorithms against various feature dimensions on the AR database.

experimental results. Clearly, for all cases, SEMC achieves the optimal balance between TPRs and FPRs. Figure 2b states the occlusion structures that SEMC sees during its detection procedure.

4.2   Real-World Facial Occlusion Detection

We test the performance of SEMC in dealing with real disguises with the AR face database [9]. The grayscale images were resized to 112×92. We select a subset of the database that consists of 119 subjects (65 males and 54 females). For training, we choose 2 unoccluded frontal view images with neutral expressions for each subject from two sessions. For testing, we consider two separate test sets of the 119 subjects. The first/second test set contains 119×2 images of the subjects wearing sunglasses/scarves with neutral expressions from two sessions.

To test the accuracy and robustness of SEMC in occlusion detection for different feature dimensions, we use 4 different downsampled images of dimensions 154, 644, 2576, and 1,0304, which correspond to downsampling ratios of 1/8, 1/4, 1/2, and 1, respectively. The detailed occlusion detection results of the first

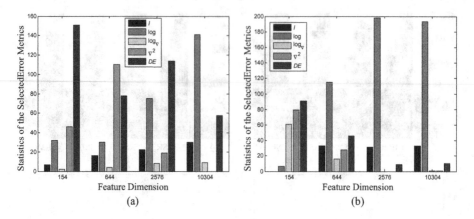

Fig. 4. The statistics of the structural error metrics selected by SEMC for various feature dimensions.

subject in the 4 different dimensions are shown in Fig. 3. Figure 3c and d compare TPRs and FPRs of the detection results of the competing methods. Clearly, SEMC achieves the optimal performance for the two types of disguises except for the lowest feature dimension. The statistical results in Fig. 4 show that with the feature dimension increasing, more and more log-based error metrics are selected by SEMC, since the dark color features of sunglasses/scarves become more and more significant. Figures 3 and 4 demonstrate that occlusion levels affect the performance of SEMC.

5   Conclusions

Most of the state-of-the-art methods in dealing with facial occlusion are based on the alternative iteration of image recovery and occlusion detection. In order to detect facial occlusions efficiently and accurately, we propose a novel method based on the structural error metrics and clustering (SEMC) without image recovery. Experiments show that, even just using the mean face of the training images as the recovery image, SEMC still achieves more accurate and robust performance compared to the related state-of-the-art methods. However, the experiments also show that the minimum occlusion boundary regularity criterion used by SEMC to select the optimal error metric limits its efficacy on occluded images with very low dimension features or with very low occlusion levels. This issue encourages us to further explore new criterion for the optimal error metric selection.

Acknowledgment. This work is partially supported by National Science Foundation of China (61402411, 61379017), Zhejiang Provincial Natural Science Foundation (LY14F020015, LY14F020014), and Program for New Century Excellent Talents in University of China (NCET-12-1087).

References

1. Chen, J., Shan, S., He, C., Zhao, G., Pietikainen, M., Chen, X., Gao, W.: Wld: a robust local image descriptor. IEEE Trans. Pattern Anal. Mach. Intell. **32**(9), 1705–1720 (2010)
2. Deng, W., Hu, J., Guo, J.: Extended src: undersampled face recognition via intraclass variant dictionary. IEEE Trans. Pattern Anal. Mach. Intell. **34**(9), 1864–1870 (2012)
3. Ekenel, H.K., Stiefelhagen, R.: Why is facial occlusion a challenging problem? In: Tistarelli, M., Nixon, M.S. (eds.) ICB 2009. LNCS, vol. 5558, pp. 299–308. Springer, Heidelberg (2009)
4. He, R., Zheng, W., Hu, B.: Maximum correntropy criterion for robust face recognition. IEEE Trans. Pattern Anal. Mach. Intell. **33**(8), 1561–1576 (2011)
5. Jia, K., Chan, T.-H., Ma, Y.: Robust and practical face recognition via structured sparsity. In: Fitzgibbon, A., Lazebnik, S., Perona, P., Sato, Y., Schmid, C. (eds.) ECCV 2012, Part IV. LNCS, vol. 7575, pp. 331–344. Springer, Heidelberg (2012)
6. Lee, K., Ho, J., Kriegman, D.: Acquiring linear subspaces for face recognition under variable lighting. IEEE Trans. Pattern Anal. Mach. Intell. **27**(5), 684–698 (2005)
7. Li, X.X., Dai, D.Q., Zhang, X.F., Ren, C.X.: Structured sparse error coding for face recognition with occlusion. IEEE Trans. Image Process. **22**(5), 1889–1900 (2013)
8. Luan, X., Fang, B., Liu, L., Yang, W., Qian, J.: Extracting sparse error of robust pca for face recognition in the presence of varying illumination and occlusion. Pattern Recogn. **47**(2), 495–508 (2014)
9. Martínez, A.: The ar face database. Technical report, Computer Vision Center (1998)
10. Tuceryan, M., Jain, A.K.: Texture analysis. Handb. Pattern Recogn. Comput. Vis. **276**, 235–276 (1993)
11. Tzimiropoulos, G., Zafeiriou, S., Pantic, M.: Subspace learning from image gradient orientations. IEEE Trans. Pattern Anal. Mach. Intell. **34**(12), 2454–2466 (2012)
12. Wandell, B.A.: Foundations of Vision. Sinauer Associates, Sunderland (1995)
13. Wright, J., Yang, A., Ganesh, A., Sastry, S., Ma, Y.: Robust face recognition via sparse representation. IEEE Trans. Pattern Anal. Mach. Intell. **31**(2), 210–227 (2009)
14. Yang, M., Zhang, L.: Gabor feature based sparse representation for face recognition with gabor occlusion dictionary. In: Daniilidis, K., Maragos, P., Paragios, N. (eds.) ECCV 2010, Part VI. LNCS, vol. 6316, pp. 448–461. Springer, Heidelberg (2010)
15. Yang, M., Zhang, L., Yang, J., Zhang, D.: Robust sparse coding for face recognition. In: Proceedings of the IEEE International Conference on Computer Vision and Pattern Recognition, pp. 625–632 (2011)
16. Zhou, Z., Wagner, A., Mobahi, H., Wright, J., Ma, Y.: Face recognition with contiguous occlusion using markov random fields. In: Proceedings of the IEEE International Conference on Computer Vision, pp. 1050–1057 (2009)

Graph Regularized Structured Sparse Subspace Clustering

Cong-Zhe You and Xiao-Jun Wu$^{(\boxtimes)}$

School of IOT Engineering, Jiangnan University, Wuxi, China
youcongzhe@gmail.com, xiaojun_wu_jnu@163.com

Abstract. High-dimensional data presents a big challenge for the clustering problem, however, the high-dimensional data often lie in low-dimensional subspaces. So, subspace clustering has been widely researched. Sparse subspace clustering (SSC) is considered as the state-of-the-art method for subspace clustering, it has received an increasing amount of interest in recent years. In this paper, we propose a novel sparse subspace clustering method named graph regularized structured sparse subspace clustering (GS3C) to jointly analyze the data under a single clustering framework and with shared underlying sparse representations. Two convex regularizations are combined and used in the model to enable sparsity as well as to facilitate multi-task learning. We also introduced the graph regularization to improve stability and consistency. The effectiveness of the proposed algorithm is demonstrated through experiments on motion segmentation and face clustering.

Keywords: Subspace clustering · Sparse representation · Graph regularization · Motion segmentation · Face clustering

1 Introduction

The past few decades witnessed the data explosion, we have entered the era of big data, and an overwhelming amount of data is generated and collected every day. This poses a great challenge to process such large scale datasets, especially the datasets are usually very high-dimensional, even though the computer processing speed becomes faster and faster. The high-dimensionality of data not only increases the computing time, but also decreases the performance due to the noise and insufficient samples in ambient space [1]. However, the intrinsic dimension of these data is often much smaller that the dimension of the ambient space. This has motivated a variety of techniques to find the low-dimensional representations of high-dimensional data, such as low-rank approximation (e.g. PCA [2], Matrix Completion [3]) and sparse representation [4].

In fact, in many problems, the data in the high-dimensional ambient space can be well represented by a low-dimensional subspaces. Subspace clustering is the problem of clustering data according their potential low-dimensional subspaces [5]. It has been widely used in many fields, such as motion segmentation and face clustering in computer vision, hybrid system identification in control, community clustering in social networks.

© Springer International Publishing Switzerland 2015
X. He et al. (Eds.): IScIDE 2015, Part I, LNCS 9242, pp. 128–136, 2015.
DOI: 10.1007/978-3-319-23989-7_14

Recently, the research of subspace clustering using the sparse representation or low-rank representation has received widespread attention, and a series of new subspace clustering algorithms relating to this have been proposed, such as sparse subspace clustering (SSC) [6, 7], low-rank representation (LRR) [5, 8] and its noisy variant LRSC [9]. The essence of this approach is that each data point can be represented as a sparse linear combination of other data via a convex formulation with L1 regularization. In [6, 7], sparse subspace clustering is first introduced by Elhamifar and Vida to solve the problem of motion segmentation and face clustering. SSC can automatically determine the number and dimension of subspace by analyzing the affinity graph constructed by the sparse representation of all the data points.

Despite SSC has achieved great success, there are still some problems unsolved. The solution of SSC is sometimes too sparse that cannot guarantee the connectivity of the affinity, which has a great influence on the later stage of spectral clustering. On the other hand, SSC is still a single-task learning framework, it lacks optimality as a result of the existence of the inner joint structure between data points [10]. However, we expect to estimate predictive clustering model for several related tasks together, not an individual one. The multi-task learning used L2,1-norm regularization to couple learning tasks using a strict assumption- all tasks share a common underlying representation. However, in many cases, the common pattern is shared by many tasks, but not all.

In order to solve the above problem, we propose a novel extension of SSC that named graph regularized structured sparse subspace clustering algorithm (GS3C). It aims at jointly analyzing the data under a single clustering framework and sharing underlying sparse representations. Two convex regularizations (L1 norm and L2,1 norm) are combined and used in the model to enable sparsity as well as to facilitate multi-task learning. We also introduced the graph regularization to improve stability and consistency. To demonstrate the effectiveness of the proposed GS3C algorithm, we apply it to the motion segmentation and face lustering problem on HOPKINS 155 dataset [11] and extended Yale B datasets [12]. The experimental results show that the proposed GS3C algorithm outperforms the main existing sparse subspace clustering algorithms.

2 Sparse Subspace Clustering

The work of Elhamifar and Vidal [6, 7] shows that, in the case of uncorrupted data, an affinity matrix for solving the subspace clustering problem can be constructed by expressing each data point as linear combination of all other data points. That is, we wish to find a matrix C such that $X = XC$ and $diag(C) = 0$. In principle, this leads to an ill-posed problem with many possible solutions. To resolve this issue, the principle of sparsity is invoked. Specifically, every point is written as a sparse linear combination of all other data points by minimizing the number of nonzero coefficients. That is

$$\min_{C} \sum_{i} \|C_i\|_0 \tag{1}$$

s.t. $X = XC$ and $diag(C) = 0$

where C_i is the i-th column of C. Since this problem is combinatorial, a simple L1 optimization is solved

$$\min \|C\|_1 \tag{2}$$

s.t. $X = XC$ and $diag(C) = 0$

It is shown in Elhamifar and Vidal that, under some conditions on the subspaces and the data, the solutions to the optimization problems in (1) and (2) are such that $C_{ij} = 0$ when points i and j are in different subspaces. In other words, the nonzero coefficients of the i-th column of C correspond to points in the same subspace as point i. Therefore, one can use C to define an affinity matrix as $|C| + |C^T|$. The segmentation of the data is then obtained by applying spectral clustering to this affinity.

In the case of data contaminated by noise G, the SSC algorithm assumes that each data point can be written as a linear combination of other points up to an error G, i.e., $X = XC + G$. We solve the following convex problem

$$\min \|C\|_1 + \frac{\alpha}{2} \|G\|_F^2 \tag{3}$$

s.t. $X = XC + G$ and $diag(C) = 0$

3 Graph Regularized Structured Sparse Subspace Clustering

We further consider that some important feature variables are only correlated to a subset of tasks. The l2,1-norm regularization cannot handle them properly. While L1-norm has the ability to impose the sparsity among all elements in C. Thus we combined L1-norm and L2,1 norm in the model to enable sparsity as well as to facilitate multi-task learning. The structured regularization term is

$$R_{structrued}(C) = \|C\|_1 + \lambda \|C\|_{2,1} \tag{4}$$

where λ is the trade-off parameter.

In addition, L1-norm and L2,1 norm can effectively seek sparsity and promote density within cluster. Here, we introduce the powerful graph regularization to effectively regularize C. the weights for inter-cluster coefficients should be as large as possible while for intra-cluster should be as small as possible. The graph regularization term is defined as

$$R_{graph} = \frac{\mu_g}{2} tr(C^T L C) \tag{5}$$

where L is the Laplacian matrix of the graph.

Thus, by combining the regularization terms, we proposed the new clustering method, named graph regularized structured sparse subspace clustering (GS3C). The objective function is defined as follows.

$$\min \ \|C\|_1 + \lambda \|C\|_{2,1} + \frac{\mu_n}{2} \|G\|_F^2 + \frac{\mu_g}{2} tr(CLC^T) \tag{6}$$

where G is the noise, each data point can be written as a linear combination of other points up to an error G, i.e., $X = XC + G$.

Next, we derive a computationally efficient algorithm to solve the proposed objective. Under the ADMM framework [13], we add two auxiliary terms, $C = C_1 = C_2$ to separate the two norms, and J to ensure each step has closed-form solution.

$$\min \ \|C_1\|_1 + \lambda \|C_2\|_{2,1} + \frac{\mu_n}{2} \|X - XJ\|_F^2 + \frac{\mu_g}{2} tr(JLJ^T) \tag{7}$$

s.t. $J = C_1 - diag(C_1), \ J = C_2 - diag(C_2)$
Its Augmented Lagrangian is

$$
\begin{aligned}
L = \ & \|C_1\|_1 + \lambda \|C_2\|_{2,1} + \frac{\mu_n}{2} \|X - XJ\|_F^2 + \frac{\mu_g}{2} tr(JLJ^T) \\
& + \frac{\mu_1}{2} \|J - C_1 + diag(C_1)\|_F^2 + \frac{\mu_2}{2} \|J - C_2 + diag(C_2)\|_F^2 \\
& + tr(\Lambda_1^T (J - C_1 + diag(C_1))) + tr(\Lambda_2^T (J - C_2 + diag(C_2)))
\end{aligned}
\tag{8}
$$

where μ_n, μ_g, μ_1 and μ_2 are numerical parameters to be tuned. By setting the partial gradient/subgradient of J, C1 and C2 iteratively and updating dual variable Λ_1, Λ_2 in every iterations, we obtain the update steps of ADMM

$$J = [\mu_n X^T X + (\mu_1 + \mu_2)I + \mu L]^{-1} (\mu_n X^T X + \mu_1 C_1 + \mu_2 C_2 - \Lambda_1 - \Lambda_2) \tag{9}$$

The soft-thresholding operators are defined as $T_\eta(v) = (|v| - \eta)_+ \mathrm{sgn}(v)$ and $\Upsilon_\eta(v) = v \cdot \max(1 - \eta/norm(v))$. The updating steps for C1 and C2 are given by

$$C_1 = T_{\frac{1}{\mu_1}}(J + \frac{\Lambda_1}{\mu_1}), \ C_1 = C_1 - diag(C_1) \tag{10}$$

$$C_2 = \Upsilon_{\frac{\lambda}{\mu_2}}(J + \frac{\Lambda_2}{\mu_2}), \ C_2 = C_2 - diag(C_2) \tag{11}$$

The dual variables are updated using gradient descending method as

$$\Lambda_1 = \Lambda_1 + \mu_1(J - C_1), \ \Lambda_2 = \Lambda_2 + \mu_2(J - C_2) \tag{12}$$

After solving the above optimization problem, we obtain the sparse coefficient matrix C. the next step is to do the final clustering using the sparse coefficient. We build a weighted graph $G = (v, \varepsilon, W)$, where $W = |C| + |C|^T$. Then we apply spectral clustering to the affinity graph to get the final clustering result.

Algorithm 1. Graph regularized structured sparse subspace clustering (GS3C)

Input: Data matrix $X \in R^{D \times N}$, tradeoff parameter λ, numerical parameters $\mu_n^{(0)}$, $\mu_g^{(0)}$, $\mu_1^{(0)}$, $\mu_2^{(0)}$ and (optional ρ_0, μ_{max}, η, ε)

Output: $ID_K = K$ Cluster Index.

Initialize: $C_1 = 0$, $C_2 = 0$, $J = 0$, $\Lambda_1 = 0$ and $\Lambda_2 = 0$

$\qquad$ Pre-compute $X^T X$, $H = [\mu_n X^T X + (\mu_1 + \mu_2)I + \mu L]^{-1}$ for later use.

1. Sparse Recovery

While not converged do

Update J by Eq. 9

Update C_1, C_2 by Eq. 10, Eq. 11

Update Λ_1, Λ_2 by Eq. 12

(optional) update parameter $(\mu_n, \mu_g, \mu_1, \mu_2) = \rho(\mu_n, \mu_g, \mu_1, \mu_2)$ and the pre-computed $H = H / \rho$ where

$$\rho = \begin{cases} \min(\mu_{max} / \mu_1), & \textit{if } \max(\sqrt{\eta} \|C_1 - C_1^{prev}\|_F) / \|X\|_F \leq \varepsilon \\ 1, & \textit{otherwise} \end{cases}$$

End while

2. Spectral Clustering

Compute ID_K using spectral clustering on graph $G = (v, \varepsilon, W)$, where

$$W = |C| + |C|^T$$

---

4 Experiments

In this section we evaluate the performance of GS3C on two computer vision tasks: motion segmentation and face clustering. For the motion segmentation problem, we consider the Hopkins 155 dataset, which consist of 155 video sequences of 2 or 3 motions corresponding to 2 or 3 low-dimensional subspaces in each video. For the face clustering problem, we consider the Extended Yale B dataset, which consists of face images of 38 human subjects, where images of each subject lie in a low-dimensional subspace.

In the experiments, we use the subspace clustering error,

$$\text{Subspace clustering error} = \frac{\#\text{ of misclassified points}}{\text{total}\#\text{of points}}$$

as a measure of performance, we compare GS3C to state-of-art subspace clustering algorithms based on spectral clustering, such as LRR, LRSC and SSC. We choose these methods as a baseline, because they have been shown to perform very well on the above tasks.

4.1   Experiments on Motion Segmentation

Motion segmentation refers to the problem of clustering a set of 2D point trajectories extracted from a video sequence into groups corresponding to different rigid-body motions. Here, the data matrix D is of dimension 2F × N, where N is the number of 2D trajectories and F is the number of frames in the video. Under the affine projection model, the 2D trajectories associated with a single rigid-body motion live in an affine subspace of R^{2F} of dimensional d = 1, 2 or 3. Therefore, the trajectories associated with n different moving objects lie in a union of n affine subspaces in R^{2F}, and the motion segmentation problem reduces to clustering a collection of point trajectories according to multiple affine subspaces.

We use the Hopkins 155 motion segmentation database to evaluate the performance of GS3C against that of other algorithms. The database consists of 155 sequences of two and three motions, where 120 of the videos have two motions and 35 of the videos have three motions. On average, in the dataset each sequence of two motions have N = 256 feature trajectories and F = 30 frames, while each sequence of three motions has N = 398 feature trajectories and F = 29 frames. For each sequence, the 2D trajectories are extracted automatically with a tracker and outlier are manually removed. Figure 1 shows some sample images with the feature points superimposed.

The results of applying subspace clustering algorithm to the data set when we sue the original 2F-dimensional feature trajectories and when we project the data into a 4n-dimensional subspace (n is the number of subspaces) using PCA are shown in Tables 1 and 2, respectively.

From Tables 1 and 2, we can found that, in both cases, GS3C obtains a small clustering error, outperforming the other algorithms. This suggests that the separation of different motion subspaces in terms of their principal angles and the distribution of the feature trajectories in each motion subspace are sufficient for the success of the graph regularized structured sparse optimization program, hence clustering. On the other hand, the clustering performance of different algorithms when using the 2F-dimensional feature trajectories or the 4n-dimensional PCA projections are close. This comes from the fact that the feature trajectories of n motion in a video almost perfectly lie in a 4n-dimensional linear subspace of the 2F-dimensional ambient sub-space preserves the structure of the subspaces and the data; hence, for each algorithm, the clustering error in Table 1 is close to the error in Table 2.

Fig. 1. Motion segmentation: given feature points on multiple rigidly moving objects tracked in multiple frames of a video (top), the goal is to separate the feature trajectories according to the moving objects (bottom).

Table 1. Clustering Error (%) of different algorithms on the Hopkins 155 dataset with the 2F-dimensional data points

Algorithm		LRR	LRSC	SSC	GS3C
2 motions	mean	4.10	2.57	2.07	1.73
	median	0.22	0.00	0.00	0.00
3 motions	mean	9.89	6.64	5.27	5.19
	median	6.22	1.76	0.40	0.43
all	mean	5.41	3.47	2.79	2.51
	median	0.53	0.09	0.00	0.00

Table 2. Clustering error (%) of different algorithms on the Hopkins 155 dataset with the 4n-dimensional data points obtained by applying PCA

Algorithm		LRR	LRSC	SSC	GS3C
2 motions	mean	4.83	2.57	2.14	1.74
	median	0.26	0.00	0.00	0.00
3 motions	mean	9.89	6.62	5.29	5.21
	median	6.22	1.76	0.40	0.43
all	mean	5.98	3.47	2.85	2.52
	median	0.59	0.00	0.00	0.00

4.2   Experiments on Face Clustering

Given face images of multiple subjects acquired with a fixed pose and varying illumination, we consider the problem of clustering images according to their subjects (see Fig. 2 as an example). It has been shown that, under the Lambertian assumption,

Fig. 2. Face clustering: given face images of multiple subjects (top), the goal is to find images that belong to the same subject (bottom).

images of a subject with a fixed pose and varying illumination lie close to a linear subspace of dimension 9. Thus, the collection of face images of multiple subjects lies close to a union of 9D subspaces.

In this section, we evaluate the clustering performance as well as the state-of-art methods on the Extended Yale B dataset. The dataset consists of 192×168 pixel cropped face images of $n = 38$ individuals, where there are $N_i = 64$ frontal face images for each subject acquired under various lighting conditions. To reduce the computational cost and the memory requirements of all algorithms, we downsample the images to 48×42 pixels and treat each 2016D vectorized image as a data point; hence $D = 2016$.

To study the effect of the number of subjects in the clustering performance of different algorithms, we devise the following experimental setting: we derive the 38 subjects into four groups, where the first three group correspond to subjects 1 to 10, 11 to 20, 21 to 30, and the fourth group corresponds to subjects 31 to 38. For each of the first three groups we consider all choices of $n \in \{2, 3, 5, 8, 10\}$ subjects and for the last group we consider all choices of $n \in \{2, 3, 5, 8\}$. Finally, we apply clustering algorithms for each trial, i.e., each set of n subjects.

Finally, we apply the clustering algorithms to the data points. The results are shown in Table 3.

We can see from the table that the proposed GS3C algorithm obtains a low clustering error for all numbers of subjects. The GS3C always outperforms others.

Table 3. Clustering error (%) of different algorithms on the extended Yale B dataset

Algorithm		LRR	LRSC	SSC	GS3C
2 subjects	mean	9.52	5.32	1.86	0.38
	median	5.47	4.69	0.00	0.00
3 subjects	mean	19.52	8.47	3.10	0.83
	median	14.58	7.81	1.04	0.52
5 subjects	mean	34.16	12.24	4.31	2.32
	median	35.00	11.25	2.50	1.25
8 subjects	mean	41.19	23.72	5.85	4.86
	median	43.75	28.03	4.49	4.88
10 subjects	mean	38.85	30.36	10.94	8.91
	median	41.09	28.75	5.63	4.84

5 Conclusion

We have proposed a novel method for sparse subspace clustering. The method is formulated in a sparse subspace representation and extends previous sparse subspace clustering method considerably in that we combined two convex regularizations to enable sparsity as well as to facilitate multi-task learning. We also introduce the graph regularization to improve stability and consistency. Experiments on real data such as face images and motion in videos showed the effectiveness of our algorithm and its superiority over the state of art methods.

Acknowledgements. This work is supported by the Specialized Research Fund for the Doctoral Program of Higher Education of China (Grant No. 20130093110009), the National Natural Science Foundation of China (Grant No. 61373055) and the Research Project on Surveying and Mapping of Jiangsu Province (Grant No. JSCHKY201109).

References

1. Bellman, R.E.: Dynamic Programming. Princeton University Press, Princeton (1957)
2. Jolliffe, I.T.: Principal Component Analysis. Springer Series in Statistics, vol. 487. Springer, New York (1986)
3. Lauer, F., Schnorr, C.: Spectral clustering of linear subspaces for motion segmentation. In: International Conference on Computer Vision (ICCV 2009), pp. 678–685. IEEE (2009)
4. Elad, M.: Sparse and Redundant Representations. Springer, New York (2010)
5. Liu, G., Lin, Z., Yu, Y.: Robust subspace segmentation by low-rank representation. In: Proceedings of ICML (2010)
6. Elhamifar, E., Vidal, R.: Sparse subspace clustering. In: Computer Vision and Pattern Recognition (CVPR 2009), pp. 2790–2797. IEEE (2009)
7. Elhamifar, E., Vidal, R.: Sparse subspace clustering: algorithm, theory, and applications. IEEE Trans. Pattern Anal. Mach. Intell. (TPAMI) **35**(11), 2765–2781 (2013)
8. Liu, G., Lin, Z., Yan, S., Sun, J., Yu, Y., Ma, Y.: Robust recovery of subspace structures by low-rank representation. IEEE Trans. Pattern Anal. Mach. Intell. (TPAMI) **35**(1), 171–184 (2013)
9. Favaro, P., Vidal, R., Ravichandran, A.: A closed form solution to robust subspace estimation and clustering. In: Computer Vision and Pattern Recognition (CVPR 2011), pp. 1801–1807. IEEE (2011)
10. Saha, B., et al. Sparse subspace clustering via group sparse coding. In: SDM 2013: Proceedings of the Thirteenth SIAM International Conference on Data Mining. Society for Industrial and Applied Mathematics (2013)
11. Tron, R., Vidal, R.: A benchmark for the comparison of 3-D motion segmentation algorithms. In: Proceedings of IEEE Conference on Computer Vision and Pattern Recognition (2007)
12. Lee, K.-C., Ho, J., Kriegman, D.: Acquiring linear subspaces for face recognition under variable lighting. IEEE Trans. Pattern Anal. Mach. Intell. **27**(5), 684–698 (2005)
13. Boyd, S., Parikh, N., Chu, E., Peleato, B., Eckstein, J.: Distributed optimization and statistical learning via the alternating direction method of multipliers. Found. Trends Mach. Learn. **3**(1), 1–122 (2010)

Precise Image Matching:
A Similarity Measure Approach

Dan Yu, Zhipeng Ye, Wei Zhao, and Xianglong Tang[(✉)]

Pattern Recognition and Intelligent System Research Center,
School of Computer Science and Technology,
Harbin Institute of Technology, Harbin, China
tangxl@hit.edu.cn

Abstract. An algorithm that utilizes the similarity comparison is proposed to get more proper match result, which is easy to implement. SIFT depends on principal direction which will lead to low precision rate when the direction is incorrectly computed. In this paper, similarities are tested by cosine theorem of matched points in some area to find stable matches and exclude mismatches (push) at first. Part of correct matches in excluded points are revived (pull) through stable matches, which are located in cluster sets centered by stable matched points, thus shrink search field and boosting the algorithm. Sum of Square Distance (SSD) measurement function is tested and chosen as similarity function to accomplish the reviving step. Experimental results show that the proposed method exhibits improved performance compared with SIFT and other methods.

Keywords: Image processing · Image matching · Signal processing · Similarity measurement · Push-and-Pull SIFT

1 Introduction

Many machine learning and pattern recognition applications are related with image matching. Given two images, the purpose of match is to find the correspondence between them. Stable and high quality matching points are the bases for 3D reconstruction, shape retrieval, object recognition and so on. During the past score years, image matching problem has been investigated by many researches [1–4]. SIFT [5] has not only good scale and brightness invariance but also a certain robustness to affine distortion. Comparative study has been carried out for the main local descriptors [6]. The results show that SIFT descriptor has the best performance for object recognition. However, the quality of matching points generated by SIFT is often not high enough. Much research has been carried out to improve the precision of matching. These studies focus on combining all kinds of features [7, 8] or methods [9, 10] to improve matching quality. The limitation of these methods is trying to eliminate incorrect matches. However, there exists some correct match which is also eliminated. It leads to the final result that some correct features are missing in the final result.

In this paper, we utilize similarity measurement methods to improve matching quality. In the excluding (push mismatches out) step, cosine theorem is utilized to measure similarity of areas to each matched points detected by SIFT. Then, a reviving

© Springer International Publishing Switzerland 2015
X. He et al. (Eds.): IScIDE 2015, Part I, LNCS 9242, pp. 137–144, 2015.
DOI: 10.1007/978-3-319-23989-7_15

(pull correct matches back) step is proposed by testing similarity of areas between unmatched points, trying to revive the potential correct matches to improve precision. Experimental results on popular datasets demonstrate that, compared with SIFT and other methods, the proposed method could obtain much precise result for matching. The rest of paper is organized as follows: Sect. 2 describes each step of the proposed method in detail. In Sect. 3, we first decide the optimal parameter for our method, then compare with other SIFT-based state-of-art methods. We conclude our method in Sect. 4.

2   Proposed Method

The method proposed here is a two-step algorithm. Firstly, Cosine Theorem is utilized to clean set of matching points which are computed by SIFT, to make sure there are only correctly matched points, i.e. 'push' the error matches out of the matching set. In this step, some correct matches may also be excluded. To solve this problem, the second step uses area similarity measurement method to revive these points, i.e. 'pull' the correct matches back into the matching set. At last, the output of the two steps is accumulated as the final result. The procedure is given by Fig. 1.

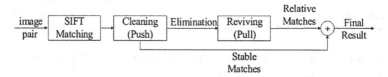

Fig. 1. Flow of the proposed algorithm

2.1   Push-to-Clean

Similarity-based methods have proven effective in many computer vision tasks [11–13]. A plain way to measure similarity is to match their features, and two images should be deemed similar if features in one image have matching features in the other. Cosine-theorem-based methods are popular in measuring text similarity. Here it is extended into image similarity matching. After SIFT is executed, matching set M and point set P are generated. M is the relationship index of each matched point pair and P is the set of matched points. Firstly, a histogram is composed to get each graylevel's occurring frequency. We segment grayscale value $0 \sim 255$ into n parts, the i-th part will count the number of pixels, whose value falls in $[i*(\lceil \frac{255}{n} \rceil + 1),(i+1)*\lceil \frac{255}{n} \rceil + i]$.

Then, the two areas to be matched could be expressed as $\mathbf{A} = (hisA_1, hisA_2, \ldots, hisA_n)^T$ and $\mathbf{B} = (hisB_1, hisB_2, \ldots, hisB_n)^T$, where $hisX_i$ is the number of aforementioned corresponding pixels in area X. Similarity value is given by (1):

$$\cos(\mathbf{A},\mathbf{B}) = \frac{\mathbf{A}^T\mathbf{B}}{\sqrt{(\mathbf{A}^T\mathbf{A})(\mathbf{B}^T\mathbf{B})}} \qquad (1)$$

The more the value is close to 1, the higher similarity the two areas have. The relationship of Match between two points detected by SIFT with low similarity will be released from M. Notice that the points still remains in P, waiting for reviving. Since there is possibility push-to-clean step may also release relationship of some correctly matched points, a pull-to-revive method is needed to recover them, which is discussed in next section.

2.2 Pull-to-Revive

After stable matching point set M_{stable}, named as seed set in this paper, is extracted, the Pull-to-Revive step uses it to recover some missed correct matches in first step. Firstly, we use seed points in M_{stable} as center and classify all of the unmatched points in point set P into cluster sets by Euclidean distance. Notice that the center is no need to be updated during the process since we need matching relationship of real existing points to be cluster center for further usage. As a result, we'll get the same number of cluster sets as the number of pairs in M_{stable}. For an unmatched point ^oP_i in the original image of cluster $C_{original}$ centered by $^oP_i^c$, to look for its match in target image, every point in corresponding cluster set C_{target} centered by $^tP_i^c$ will be checked. A pair of points $^oP_i^c$ and $^tP_i^c$ is written as $<^oP_i^c, {}^oP_i^c>$, which is the i-th pair of matching points in M_{stable}. If a point ^tP_i satisfies the similarity constrains, i.e. similarity of two neighborhood areas in which ^oP_i and ^tP_i located are checked by similarity measurement function and they are close enough, $<^oP_i,{}^tP_i>$ will be add back to M_{stable} and the match is revived successfully. Similarity measurement function (in this case the SSD) is given as following

$$SSD = \frac{1}{N}\sum_{i=1}^{N}|A(i) - B'(i)|^2, \forall i \in A \cap B', \qquad (2)$$

where N is the number of pixels in selected area. A and B are two areas to be compared. The smaller SSD value is, the more similar two areas are. Points in C_{target} and $C_{original}$ will be wiped out if there are no proper matches found between them, as illustrated in Fig. 2. Looking for matched points in corresponding cluster set will lower the possibility of mismatch again and reduce the complexity of computation to boost the reviving step at the same time. As a result of reviving, $p\%$, for example, $p = 30$, unmatched points in original image satisfy similarity constrains most will be revived. Here we get only one match pre pair of points according to the evaluation method to be described in Sect. 2.3, that correct-match is a match where the two key points correspond to the same physical location, while two real existing points in an image often stands for different physical location. Section 4 is the conclusion. The result of the algorithm is shown in Fig. 2.

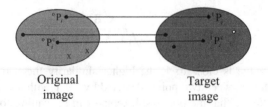

Original   Target
image   image

Fig. 2. Locating potential matching points. Solid dots are successfully revived matches. Hollow dot and X marks in original image stands for the points without proper matches.

2.3   Algorithm

A summarization of the proposed algorithm is given below. There are several ways to evaluate matching features. All of them were operated under datasets which ground-truth is known. In this paper, Ke's classical method [14] is used to evaluate every method. We could see the effectiveness of each step. The cleaning step pushes mismatches out, however, at the cost of losing correct matches. After the reviving step, many correct matches are recovered so that the precision is improved. The complexity of our algorithm is approximately $O(n^3)$.

Algorithm 1. Push-and-Pull Matching

1. Detect matching feature set P and their matching relationship M by SIFT.
2. For $i = 1, \cdots n$, match $P_i \in P$,

   If $!isCosSimilar(neibr(P^i_{original}), neibr(P^i_{target}))$ then Remove M_i from M

3. Let remaining M be a new set M_{stable}, marking corresponding points as the center of clustering. Cluster the points which matching relationship are removed in 2, forming two cluster sets: $C_{original}$ and C_{target}.

4. For $i = 1, \cdots, sizeof(M_{stable})$

   For $j = 1, \quad n$,

   For $k = 1, \cdots n$

   If $isSim(neibr(^oP^j_i), neibr(^tP^k_i))$ then

   add $< {}^oP^j_i, {}^tP^k_i >$ to M_{revive}

5. Sort M_{revive} by SSD value and put first 30% into M_{stable}.

6. Output M_{stable}

3   Experimental Results and Analysis

In this section, we present comparative results of the proposed method with other state-of-the-art methods in both public datasets (Middlebury [15], CMU-House) and our dataset, named as WALL-E. The first experimental evaluation has been carried out

Table 1. Comparison of similarity functions

	Middlebury	CMU-house	WALL-E
SAD	45.8 %	83.7 %	45.6 %
ZSAD	42.7 %	81.4 %	35.6 %
SSD	65.3 %	86.8 %	48.9 %
ZSSD	64 %	85.1 %	46.7 %
NCC	37.3 %	60.3 %	43.3 %
ZNCC	27.6 %	57.3 %	32.2 %

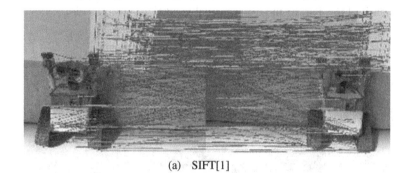

(a) SIFT[1]

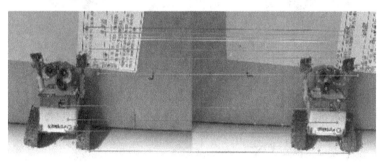

(b) Cleaning(push)

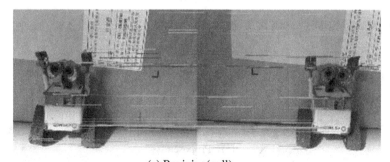

(c) Reviving(pull)

Fig. 3. Results of each procedure of proposed algorithm.

to identify the best similarity function in the second step of our method. Several mainstream methods are compared, including sum of average differences (SAD), zero mean sum of average differences (ZSAD), sum of squared differences (SSD), zero mean sum of squared differences (ZSSD), normalized cross-correlation (NCC) and zero mean normalized cross-correlation (ZNCC). Each strategy is tested as similarity function on all datasets and the best performance is chosen as the measurement function of our method.

In these experiments, $p\% = 30\%$. As shown in Table 1, SSD gets the highest precision rate and is chosen as similarity function of the proposed method. Local algorithms SAD/ZSAD and NCC/ZNCC give poor results for both rotation and translation situations. Their mismatches are mainly due to insufficient quantization in image intensity [16]. Part of the experimental results are given in Fig. 3.

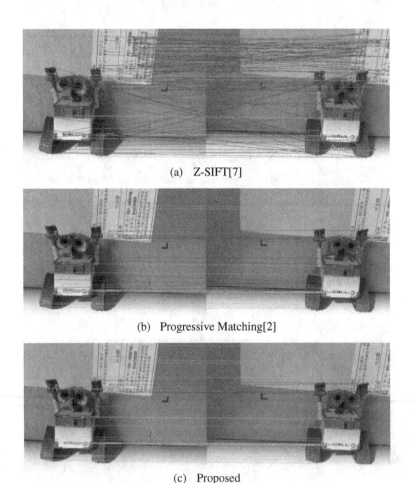

(a)   Z-SIFT[7]

(b)   Progressive Matching[2]

(c)   Proposed

Fig. 4. Experiments on images from WALL-E dataset. Correct and incorrect matches are denoted as cyan and red lines, respectively. These figures are best viewed in color.

To further demonstrate how well the algorithm works, we compared the results with other matching methods with various images and summarize the quantitative analysis in Table 2. Cho [2] is an improvement of method described in [10] utilizing area feature points structure matching. The number of candidate input matches was always fixed to the same for every method. Thus, the performance increase in this experiment is not due to the increase or decrease in the sizes of detected feature points, but to the push-to-clean step cleaning error matches and pull-to-revive step recovering missed correct matches of the proposed method. We can see from Fig. 4 and Table 2, the precision ratios of our method outperforms others on all the test cases.

Table 2. Comparison of matching accuracy on different methods

	Middlebury	CMU-house	WALL-E
SIFT [1]	56.4 %	73.9 %	21.4 %
Z-SIFT [3]	57.8 %	76.5 %	25 %
Progressing matching [11]	61.4 %	79.7 %	39.3 %
Proposed	65.3 %	86.8 %	48.2 %

As shown in Figs. 3 and 4, many of the matches generated by SIFT and ZSIFT are repeatedly and incorrect. Points under same depth are matched despite their relativity. For example, the hand and wheel of robot are in same depth and are matched incorrectly. Progressive Matching improves the matching quality, however, not working properly in low texture (the words on the box in white background) or viewpoint change (the left wheel, eyes and body of the robot) areas. Although the reviving method of the proposed method may introduce some outliers, it still perform better in aforementioned areas than other methods because the Pull-to-Revive is operating under reliable match offered by Push-to-Clean.

4 Conclusions

Based on SIFT, we introduced a novel matching algorithm which overcomes the limitations of conventional graph matching and achieves impressive performance in image matching experiments. Two similarity measurement methods are utilized: Cosine theorem and SSD, to clean (push) the original matching set and revive (pull) correct matches which is excluded at first. K-means clustering is used to diminish the scope of comparisons. The experimental results demonstrate the effectiveness in both visual and quantitative aspects. The proposed framework will contribute to a wide range of graph-based approaches and applications.

Acknowledgments. The research is supported by National Natural Science Foundation of China (61171184, 61201309).

References

1. Thirion, J.-P.: Image matching as a diffusion process: an analogy with Maxwell's demons. Med. Image Anal. **2**(3), 243–260 (1998)
2. Cho, M., Lee, K.M.: Progressive graph matching: making a move of graphs via probabilistic voting. In: 2012 IEEE Conference on Computer Vision and Pattern Recognition (CVPR), pp. 398–405. IEEE (2012)
3. Berg, A.C., Berg, T.L., Malik, J.: Shape matching and object recognition using low distortion correspondences. In: IEEE Computer Society Conference on Computer Vision and Pattern Recognition, 2005, CVPR 2005. IEEE (2005)
4. Hirschmuller, H., Scharstein, D.: Evaluation of cost functions for stereo matching. In: IEEE Conference on Computer Vision and Pattern Recognition, CVPR 2007. IEEE (2007)
5. Lowe, D.G.: Distinctive image features from scale-invariant keypoints. Int. J. Comput. Vis. **60**(2), 91–110 (2004)
6. Mikolajczyk, K., Schmid, C.: A performance evaluation of local descriptors. IEEE Trans. Pattern Anal. Mach. Intell. **27**(10), 1615–1630 (2005)
7. He, L., Wang, S., Pappas, T.N.: 3D surface registration using Z-SIFT. In: ICIP 2011, pp. 1985–1988 (2011)
8. Mazin, B., Delon, J., Gousseau, Y.: Combining color and geometry for local image matching. In: 21st International Conference on Pattern Recognition (ICPR), pp. 2667–2680. IEEE. (2012)
9. Ishii, J., et al.: Wide-baseline stereo matching using ASIFT and POC. In: 19th IEEE International Conference on Image Processing (ICIP), pp. 2977–2980. IEEE (2012)
10. Lee, J., Cho, M., Lee, K.M.: Hyper-graph matching via reweighted random walks. In: 2011 IEEE Conference on Computer Vision and Pattern Recognition (CVPR), pp. 1633–1640. IEEE (2011)
11. Collins, R.T., Beveridge, J.R.: Matching perspective views of coplanar structures using projective unwarping and similarity matching. In: 1993 IEEE Computer Society Conference on Computer Vision and Pattern Recognition, Proceedings CVPR 1993. IEEE (1993)
12. Adjeroh, D.A., Lee, M.-C., King, I.: A distance measure for video sequence similarity matching. In: International Workshop on Multi-Media Database Management Systems, Proceedings. IEEE (1998)
13. Lim, J.-H., et al.: Learning similarity matching in multimedia content-based retrieval. IEEE Trans. Knowl. Data Eng. **13**(5), 846–850 (2001)
14. Ke, Y., Sukthankar, R.: PCA-SIFT: a more distinctive representation for local image descriptors. In: Proceedings of the 2004 IEEE Computer Society Conference on Computer Vision and Pattern Recognition, CVPR 2004, pp. 506–513. IEEE (2004)
15. Scharstein, D., Szeliski, R.: High-accuracy stereo depth maps using structured light. In: 2003 IEEE Computer Society Conference on Computer Vision and Pattern Recognition, 2003, Proceedings. IEEE (2003)
16. Lin, H.-Y., Chou, X.-H.: Stereo matching on low intensity quantization images. In: 2012 21st International Conference on Pattern Recognition (ICPR), pp. 2618–2621. IEEE (2012)
17. Middlebury. http://vision.middlebury.edu/stereo
18. CMU-House. http://vasc.ri.cmu.edu/idb/html/motion/house/

Assigning PLS Based Descriptors by SVM in Action Recognition

Jiayu Sheng[1,2], Biyun Sheng[1,2(✉)], Wankou Yang[1,2,3], and Changyin Sun[1,2]

[1] School of Automation, Southeast University, Nanjing 210096, China
hisby@126.com
[2] Key Laboratory of Measurement and Control of Complex Systems of Engineering, Ministry of Education, Southeast University, Nanjing 210096, China
[3] Key Laboratory of Child Development and Learning Science of Ministry of Education, Southeast University, Nanjing 210096, China

Abstract. In this paper, we propose assigning PLS based descriptors by SVM to obtain the representations of human action videos. First, in addition to the spatially gradient orientation, we add spatio-temporal gradient statistic to generate the extended Histogram of Oriented Gradient (HOG). Second, different from requently-used cuboid descriptors in which Principal Component Analysis (PCA) is applied for dimension reduction, the proposed features utilize the Partial Least Squares (PLS) method for better performance. Then, we apply a multi-class SVM for assignment instead of assigning descriptors to the nearest (Euclidean distance) visual word in traditional Bag of Visual Words (BOVW) framework. Finally, the K-nearest neighbor algorithm is used to classify the histogram of visual words. The experimental results on the facial expression dataset and KTH human activity dataset validate the effectiveness of our proposed method.

1 Introduction

Recently, there is a growing trend for researchers to work on action recognition because of its applications in navigation, surveillance and video indexing [1–4]. The task is challenging because there exists variations in motion performance, recording settings and inter-personal differences [3]. For demanding recognition systems in unconstrained scenarios, research has been transferred from recognizing simple human actions under controlled conditions to more complex activities and events "in the wild" [4].

Among all the action recognition methods, spatio-temporal interest points (STIPs) methods are most popular for the reason that the yielded histogram of visual words is robust to noise, occlusion and geometric variation, without requiring reliable tracks on a particular subject [1,2,11,12]. In general, points with non-constant motion correspond to accelerating local image structures that may correspond to accelerating objects in the world, so that those points can be expected to contain information about the forces acting in the physical environment and changing its structure [1]. Recent work has shown promising results

© Springer International Publishing Switzerland 2015
X. He et al. (Eds.): IScIDE 2015, Part I, LNCS 9242, pp. 145–153, 2015.
DOI: 10.1007/978-3-319-23989-7_16

using local space-time features together with bag of visual words (BOVW) models [11], in which a video clip is represented by the histogram of its local features followed by interest point's extraction.

In order to characterize action videos effectively, researchers have been concentrated on discriminative representations from low-level to high-level. Based on high level concepts detection [5] or low-level building blocks [6–9], high-level approaches rely on expensive computation cost despite its superiority on recognition rate. Besides, high-level representations are sensitive to local geometric disturbances such as occlusion and are consequently less scalable [5,18]. Low-level approaches are widely applied for its advantages of easy implement, sparse and efficient property, and robustness to geometric disturbances such as occlusion and viewpoint changes. However, the features such as cuboid descriptors only describe the spatially gradient orientation while its complementary information of space to time is ignored. Besides, the dimension reduction way PCA in flatted vectors of cuboid descriptors doesn't take the correlation between the dependent variables and the independent ones into consideration. Furthermore, the hard-assignment method in traditional BOVW leads to major errors. Thus, we modify the stages mentioned above and aim at proposing a method with superiorities of easy implementation in low-level approaches and high recognition accuracy in high-level ones.

In this paper, we follow the framework of STIPs and BOVW, and obtain more discriminative video representations with the method of assigning PLS_based extended HOG descriptors by a multi-class SVM. Specifically, for each voxel in extracted cuboids, the gradient orientations of horizontal and vertical axis to the time axis are simultaneously aggregated besides that of vertical to horizontal axis in original HOG. As the operations made in [8], the local descriptor is the concentration of extended HOG in each block of the divided cuboid. With action types as labels, a supervised dimension reduction PLS is utilized instead of PCA. Finally we classify the descriptors of a video to the visual words by a multi-class SVM. The flowchart of our algorithm is shown in Fig. 1.

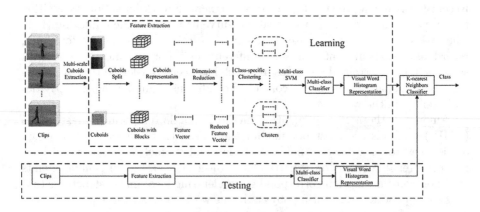

Fig. 1. Flowchart of our algorithm.

2   Proposed Method

As the classical machine learning procedure, our algorithm is designed in two stages. In the learning stage, after representing all the extracted cuboids by PLS_based extended HOG, we first execute class-specific clustering by the K-means algorithm to build a vocabulary. Given N action classes and K centers in each class, the final number of words in vocabulary is $N \times K$. With the proposed features and corresponding action labels as training samples, a multi-class SVM is learned to classify the cuboids into different clusters. The clips are represented by calculating the number of descriptors in each visual word. In the testing stage, the testing clip is also represented by the visual word histogram based on the learned multi-class SVM. Finally K-nearest neighbor classifier is used to classify the clip to a specific class.

2.1   Interest Points Detector

We use the interest points detector in [12]. The response function has the form

$$R = (I * g * h_{ev})^2 + (I * g * h_{od})^2 \tag{1}$$

where I is the video sequence, $g(x, y; \sigma)$ is the 2D Gaussian smoothing kernel, applied only along the spatial dimensions, and h_{ev} and h_{od} are a quadrature pair of 1D Gabor filters used temporally. They are defined as

$$\begin{cases} h_{ev}(t; \tau, w) = -\cos(2\pi t w)e^{-t^2/\tau^2} \\ h_{od}(t; \tau, w) = -\sin(2\pi t w)e^{-t^2/\tau^2} \end{cases} \tag{2}$$

Interest points are local maximum points of R after non-maximum suppression. In order to avoid scale selection and detect more effective interest points, here we use different combinations of spatial scales and temporal scales $(\sigma_i, \tau_i), i = 1, 2, \cdots, n$.

2.2   Extended HOG of Extracted Cuboids

In spatial pyramid matching, the image is subdivided at different levels of resolution and the features falling in each spatial bin are calculated for each level of resolution and channel [13,14]. Finally, the spatial histogram is weighted to construct the final features so that those unmatched at low levels of resolution can be matched at high levels of resolution. Inspired by this technique and the spatio-temporal method used in [8], we divide the extracted cuboids into different combinations of blocks, features of which are then concatenated to get the final vectors of cuboids.

The extracted cuboids with the same size of (L_X, L_Y, L_T) is averagely split into $n_{xi} \times n_{yi} \times n_{ti}$ blocks. As shown in Fig. 2, the block has some overlaps. The lengths of each block in each dimension are

$$\begin{cases} l_{BX} = ceil(\dfrac{L_X}{n_{xi}}) \\[2mm] l_{BY} = ceil(\dfrac{L_Y}{n_{yi}}) \\[2mm] l_{BT} = ceil(\dfrac{L_T}{n_{ti}}) \end{cases} \tag{3}$$

where $ceil(\cdot)$ rounds the element toward positive infinity. The initial point of each block is P_X, P_Y, P_T, which can be computed as follows:

$$\begin{cases} P_X = 1 + (i-1)\dfrac{L_X - l_{BX}}{n_{xi}}, i = 1, 2, \cdots, n_{xi} \\[2mm] P_Y = 1 + (i-1)\dfrac{L_Y - l_{BY}}{n_{yi}}, i = 1, 2, \cdots, n_{yi} \\[2mm] P_T = 1 + (i-1)\dfrac{L_T - l_{BT}}{n_{ti}}, i = 1, 2, \cdots, n_{ti} \end{cases} \tag{4}$$

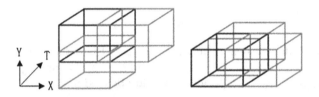

Fig. 2. Blocks splitting with overlap.

In original HOG, the gradient orientation of G_Y relative to G_X is used with temporal information in 3D space ignored. Here, the combinations of channels in G_Y/G_T, G_X/G_T are computed at the same time besides the gradient orientation G_Y/G_X. All the gradient orientations are equally projected into n_{bin} orientations, which leads to a $3 \times n_{bin}$-dimensional vector of a block and a $3 \times n_{bin} \times n_{xi} \times n_{yi} \times n_{ti}$-dimensional feature vector of the extracted cuboid.

2.3   Dimension Reduction with PLS

Principal Component Analysis (PCA) is usually applied to reduce the dimension of descriptors, which aims at maximizing the covariance of dependent variables (namely extend HOG descriptors). Different from PCA, Partial Least Squares (PLS) is modeling relations between sets of observed variables by means of latent variables [15–17]. The basic idea of PLS is to construct new predictor variables, latent variables, as linear combinations of the original variables summarized in descriptor variables and class labels. PLS decomposes the descriptors matrix X and the class label vector y into

$$\begin{cases} X = TP^T + E \\ y = Uq^T + f \end{cases} \tag{5}$$

where T and U are matrices containing extracted latent vectors, the matrix P and the vector q represent the loadings, and the matrix E and the vector f are the residuals. The PLS method, using the nonlinear iterative partial least squares (NIPALS) algorithm, constructs a set of weight vectors (or projection vectors) $W = \{w_1, w_2, \ldots, w_p\}$ such that

$$[cov(t_i, u_i)]^2 = \max_{|w_i|=1} [cov(Xw_i, y)]^2 \tag{6}$$

where t_i is the ith column of matrix T, u_i is the ith column of matrix U and $cov(t_i, u_i)$ is the sample covariance between latent vectors t_i and u_i. The dimensionality reduction is performed by projecting the features vector onto the weight vectors $W = \{w_1, w_2, \ldots, w_p\}$, obtaining the latent vector $z_i \in \mathbb{R}^{1 \times p}$ as the spatio-temporal features.

2.4 Feature Assignment by Multi-class SVM

In the standard Bag of Visual Words (BOVW) framework, spatio-temporal features are assigned to the nearest visual word of the dictionary. Simply comparing the Euclidean distance for feature assignment may lead to major errors. In this paper, each visual word of a class-specific dictionary is associated with a label and spatio-temporal features are classified to the corresponding dictionary element by a multi-class SVM [19]. The procedure of feature assignment is shown in Fig. 3.

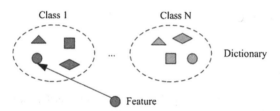

Fig. 3. A toy example of feature assignment. The symbols in each class are the class-specific dictionary elements and the feature is assigned to the visual word with the same label.

The video clips are represented by the histogram of visual word. And the K-nearest neighbors classifier is utilized on the visual word histogram to classify.

3   Experiments

We evaluate our algorithm on two benchmark datasets: facial expression dataset [12] and KTH human action dataset [10].

3.1   Facial Expression Dataset

The facial expression dataset involves 2 individuals, each expressing 6 different emotions under 2 lighting setups. The expressions are anger, disgust, fear, joy, sadness and surprise. Certain expressions are quite distinct, such as sadness and joy while others are fairly similar, such as fear and surprise. Under each lighting setup, each individual was asked to repeat each of the 6 expressions 8 times.

In this experiment the block splitting combinations (n_{xi}, n_{yi}, n_{ti}) are: [3,3,2], [4,4,4] and [5,5,6]. The scale combinations of (σ_i, τ_i) for detecting interest points are: [4,3], [4,4], [8,3] and [8,4], and the dictionary size of each class is set to 50. The recognition rate is 98.5 % while the standard BOVW model can only reach around 72 % [12]. The confusion matrix is shown in Fig. 3.

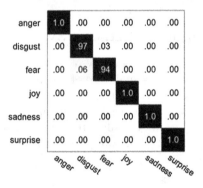

Fig. 4. Confusion matrix of facial expression dataset.

3.2   KTH Dataset

The KTH dataset contains six types of human actions (walking, jogging, running, boxing, hand waving and hand clapping) performed several times by 25 subjects in four different scenarios: outdoors s1, outdoors with scale variation s2, outdoors with different clothes s3 and indoors s4 as illustrated below. Currently the database contains 2391 sequences, all of which were taken over homogeneous backgrounds with a static camera with 25 fps frame rate.

In this experiment the scale combinations of (σ_i, τ_i) and block splitting combinations are set the same as the facial expression dataset. The overall dictionary size is set to 100 × 6, and the recognition rate reaches 93.16 %. The confusion matrix is shown in Fig. 4 and the comparison of our algorithm with previous work is shown in Table 1. Although the accuracy isn't comparable to all of the listed methods, our method is only constructed on a basic low-level approach with less computation complexity and time consumption.

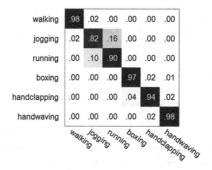

Fig. 5. Confusion matrix of KTH human action dataset.

Table 1. Comparison with previous work on the KTH dataset

Approach	Accuracy(%)
Laptev et al. [8]	91.8
Bregonzio et al. [20]	93.17
Niebles et al. [21]	91.3
Li et al. [22]	93.60
Ours	**93.16**

4 Conclusion

Along the basic BOVW framework, the proposed method improves the discriminative power of the final representations. In order to capture more spatiotemporal information in 3D space, we extend original HOG by aggregating the gradient orientations of space axis to time axis additionally. Then PLS is applied to reduce the dimension of features, which takes the class labels of features into consideration. Instead of comparing the Euclidean distance of features to clusters, a multi-class SVM is used to assign the descriptors. The experimental results on facial expression dataset and KTH human activity dataset show the effectiveness of the proposed method (Fig. 5).

Acknowledgments. The work was supported in part by National Natural Science Foundation of China under Grant No. 61305058, No. 61375001, Natural Science Foundation of Jiangsu Province of China under Grant No. BK20130471 and No. BK20140638, China Postdoctoral Science Foundation under grant No.2013M540404, Jiangsu Planned Projects for Postdoctoral Research Funds under grant No.1401037B, open fund of Key Laboratory of Measurement and Control of Complex Systems of Engineering, Ministry of Education under Grant No.MCCSE2013B01, the Open Project Program of Key Laboratory of Child Development and Learning Science of Ministry of Education, Southeast University (No. CDLS-2014-04), and A Project Funded by the Priority Academic Program Development of Jiangsu Higher Education Institutions (PAPD), and the Fundamental Research Funds for the Central Universities.

References

1. Laptev, I.: On space-time interest points. Int. J. Comput. Vis. **64**(2), 107–123 (2005)
2. Wang, H., Yuan, C., Luo, G., Weiming, H., Sun, C.: Action recognition using linear dynamic systems. Pattern Recogn. **46**(6), 1710–1718 (2013)
3. Poppe, R.: A survey on vision-based human action recognition. Image Vis. Comput. **28**(6), 976–990 (2010)
4. Liu, J., Luo, J., Shah, M.: Recognizing realistic actions from videos "in the wild". In: Proceedings of the IEEE Conference on Computer Vision and Pattern Recognition, pp. 1996–2003 (2009)
5. Sadanand, S., Corso, J.J.: Action bank: a high-level representation of activity in video. In: Proceedings of the IEEE Conference on Computer Vision and Pattern Recognition, pp. 1234–1241 (2012)
6. Tamrakar, A., Ali, S., Yu, Q., Liu, J., Javed, O., Divakaran, A., Cheng, H., Sawhney, H.: Evaluation of low-level features and their combinations for complex event detection in open source videos. In: Proceedings of the IEEE Conference on Computer Vision and Pattern Recognition, pp. 3681–3688 (2012)
7. Klaser, A., Marszalek, M., Schmid, C.: A spatio-temporal descriptor based on 3D-gradients. In: Proceedings of the British Machine Vision Conference, pp. 995–1004 (2008)
8. Laptev, I., Marszalek, M., Schmid, C., Rozenfeld, B.: Learning realistic human actions from movies. In: Proceedings of the IEEE Conference on Computer Vision and Pattern Recognition, pp. 1–8 (2008)
9. Wang, H., Ulla, M.M., Klaser, A., Laptev, I., Schmid, C.: Evaluation of local spatio-temporal features for action recognition. In: Proceedings of the British Machine Vision Conference **124**(11), pp. 1–124 (2009)
10. Schuldt, C., Laptev, I., Caputo, B.: Recognizing human actions: a local SVM approach. In: Proceedings of the International Conference on Pattern Recognition, pp. 32–36 (2004)
11. Wang, H., Yuan, C., Weiming, H., Sun, C.: Supervised class-specific dictionary learning for sparse modeling in action recognition. Pattern Recog. **45**(11), 3902–3911 (2012)
12. Dollar, P., Rabaud, V., Cottrell, G., Belongie, S.: Behavior recognition via sparse spatiotemporal features. In: IEEE International Workshop on Visual Surveillance and Performance valuation of Tracking and Surveillance, pp. 65–72 (2005)
13. Lazebnik, S., Schmid, C., Ponce, J.: Beyond bags of features: spatial pyramid matching for recognizing natural scene categories. In: Proceedings of the IEEE Conference on Computer Vision and Pattern Recognition, pp. 2169–2178 (2006)
14. Yang, J., Yu, K., Gong, Y., Huang, T.: Linear spatial pyramid matching using sparse coding for image classification. In: Proceedings of the IEEE Conference on Computer Vision and Pattern Recognition, pp. 794–1801 (2009)
15. Schwartz, W.R., Davis, L.S.: Learning discriminative appearance-based models using partial least squares. In: 2009 XXII Brazilian Symposium on Computer Graphics and Image Processing, pp. 322–329 (2009)

16. Schwartz, W.R., Kembhavi, A., Harwood, D., Davis, L.S.: Human detection using partial least squares analysis. In: IEEE 12th International Conference on Computer vision, pp. 24–31 (2009)

17. Hu, Y.-G., Ren, C.-X., Yao, Y.-F., Li, W.-Y., Feng-Wang, : Face recognition using nonlinear partial least squares in reproducing kernel hilbert space. In: Liu, C.-L., Zhang, C., Wang, L. (eds.) CCPR 2012. CCIS, vol. 321, pp. 316–323. Springer, Heidelberg (2012)

18. Everts, I., van Gemert, J.C., Gevers, T.: Evaluation of color STIPs for human action recognition. In: Proceedings of the IEEE Conference on Computer Vision and Pattern Recognition, pp. 2850–2857 (2013)

19. Crammer, K., Singer, Y.: On the algorithmic implementation of multi-class SVMs. J. Mach. Learn. Res. $2(2)$, 265–292 (2001)

20. Bregonzio, M., Gong, S., Xiang, T.: Recognising action as clouds of space-time interest points. In: Proceedings of the IEEE Conference on Computer Vision and Pattern Recognition, pp. 1948–1955 (2009)

21. Niebles, J.C., Chen, C.-W., Fei-Fei, L.: Modeling temporal structure of decomposable motion segments for activity classification. In: Daniilidis, K., Maragos, P., Paragios, N. (eds.) ECCV 2010, Part II. LNCS, vol. 6312, pp. 392–405. Springer, Heidelberg (2010)

22. Li, B., Ayazoglu, M., Mao, T., Camps, O., Sznaier, M.: Activity recognition using dynamic subspace angles. In: Proceedings of the IEEE Conference on Computer Vision and Pattern Recognition, pp. 3193–3200 (2011)

Coupled Dictionary Learning with Common Label Alignment for Cross-Modal Retrieval

Xu Tang, Yanhua Yang, Cheng Deng$^{(\boxtimes)}$, and Xinbo Gao

School of Electronic Engineering, Xidian University, Xi'an 710071, China
tangxu@stu.xidian.edu.cn, {yangyanhua.xd,chdeng.xd}@gmail.com,
xbgao@mail.xidian.edu.cn

Abstract. Cross-modal retrieval has been an active research topic in recent years. However, most existing methods ignored discovering the common semantic relationship among different modalities so as to seriously reduce the retrieval accuracy. To cope with this problem, we propose a novel cross-modal retrieval method based on coupled dictionary learning with common label alignment. Concretely, our method first conducts coupled dictionary learning on the data from different modalities separately and then projects them into a common space, where the correlation between these modalities is encouraged by using common label alignment. Experimental results on two public datasets demonstrate that our method outperforms several state-of-the-art methods.

Keywords: Cross-modal retrieval · Common space · Coupled dictionary learning · Label alignment

1 Introduction

Cross-modal retrieval has attracted extensively attention due to the explosive growth of multimedia data, such as image, video and text. Given a text query, the task of cross-modal retrieval is to search the most relevant images from an image dataset, and vice versa. Different from the unimodal case, the data from different modalities reside in different feature space, and exist a natural semantic gap between them. Hence, the challenging problem of cross-modal retrieval is that how to sufficiently represent the heterogeneous data and effectively narrow down the semantic gap.

Recently, several cross-modal retrieval approaches have been proposed from different perspectives. As a classic technique, Canonical Correlation Analysis (CCA) [4] is widely used in cross-modal retrieval task, which aims to obtain the maximally correlated subspace between two different modalities. Rasiwasia *et al.* [8] proposed to map the text and image from their original spaces to a CCA space. The work of [10] used a bilinear model (BLM) to learn a common space for cross-modal retrieval. Sharma *et al.* [9] and Chen *et al.* [2] utilized Partial Least Squares (PLS) for cross-modal retrieval by linearly mapping data in different modalities into the common space. Besides CCA, BLM and PLS, Sharma *et al.* [10] proposed Generalised Multiview Anlysis to extend Linear Discriminant Analysis

© Springer International Publishing Switzerland 2015
X. He et al. (Eds.): IScIDE 2015, Part I, LNCS 9242, pp. 154–162, 2015.
DOI: 10.1007/978-3-319-23989-7_17

(LDA) [1] and Marginal Fisher Analysis (MFA), named Generalized Multiview LDA (GMLDA) and Generalized Multiview MFA (GMMFA) for cross-modal retrieval. It is worth noting that, the foregoing methods always assume that both modalities can be projected into a common feature space where the maximized correlation is explored to narrow down the semantic gap.

On the other hand, to sufficiently represent the heterogeneous data resided in different feature spaces, dictionary learning has absorbed ever-increasing attention in recent years. Dictionary learning considers that a data sample can be reconstructed with a linear combination of a few atoms in a learned dictionary. Due to the intrinsic power of representing the heterogeneous features by generating different dictionaries for multi-modal data, dictionary learning technique has been adopted to deal with cross-style problems such as photo-sketch synthesis [5] and super-resolution [12]. Huang et al. [5] proposed a coupled dictionary learning (CDL) for cross-domain image recognition, where the different sparse coefficients are mapped into a common feature space. Recently Wu et al. [13] proposed coupled dictionary learning with group structure information for multi-modal retrieval (SLiM^2). Nevertheless, these methods also have very strict assumptions, which make them unsuitable for real multi-modal data and incapable to capture the inherent semantic relationship.

To cope with the aforementioned problems, this paper presents a novel cross-modal retrieval framework based on coupled dictionary learning with common label alignment. Specifically, we first separately obtain sparse coefficients to represent the heterogeneous features from different modalities by imposing dictionary learning into our model. Then, the data samples from different modalities are projected into a common space where the inherent relation between modalities can be well discovered. Moreover, label information is leveraged to align the cross-modal data sample pairs in the common space so as to encourage the inherent correlation across the different modalities. Our model minimizes representation error and label space projection error simultaneously. To solve the framework efficiently, an iterative optimization strategy is exploited. Experimental results on two public datasets show that the proposed model outperforms several state-of-the-art methods.

2 The Proposed Model

In this section, we begin with a brief introduction of dictionary learning. Then we formulate the objective function and its corresponding optimization algorithm.

2.1 Dictionary Learning

Dictionary learning assumes that a data sample could be reconstructed with a linear combination of a few atoms in a dictionary. Let $\mathbf{X} = [\mathbf{x}_1, \mathbf{x}_2, \ldots, \mathbf{x}_N] \in \mathbb{R}^{p \times N}$ be the data matrix, where N is the number of data and p is the feature dimension of the data. The sparse dictionary can be learned by minimizing the following formula:

$$\min_{\mathbf{D},\mathbf{A}} \|\mathbf{X} - \mathbf{DA}\|_F^2 + \lambda \|\mathbf{A}\|_1$$

$$s.t. \quad \|\mathbf{d}_i\| \leq 1, \quad \forall i, \tag{1}$$

where $\mathbf{D}$ is the learned dictionary and $\mathbf{d}_i$ is one of the dictionary atoms of $\mathbf{D}$. $\mathbf{A}$ is the sparse coefficient of $\mathbf{X}$, $\|\cdot\|_1$ denotes l_1-norm to enforce the sparsity, and the parameter λ controls the sparsity. The formulation minimizes the reconstruction error of the given set of data $w.r.t.$ a sparsity constraint.

2.2 Problem Formulation

Assume that there are M-modal datasets $\{\mathbf{X}_1, \mathbf{X}_2, \ldots, \mathbf{X}_M\}$. Suppose that $\mathbf{X}_m = [\mathbf{x}_1^m, \mathbf{x}_2^m, \ldots, \mathbf{x}_N^m] \in \mathbb{R}^{p_m \times N}$ from the m-th modality, where N denotes the number of data samples in $\mathbf{X}_m$, and p_m denotes the dimensionality of the m-th modality. Each sample pair $\{\mathbf{x}_i^1, \ldots, \mathbf{x}_i^M\}$ exclusively belongs to one of C classes and describes the relevant underlying content.

Let $\mathbf{D}_m \in \mathbb{R}^{p_m \times K}$ be the dictionary from $\mathbf{X}_m$ and let $\mathbf{A}_m \in \mathbb{R}^{K \times N}$ be its corresponding sparse coefficients, where K is the size of the dictionary. We also suppose that a common label matrix $\mathbf{Q} = [\mathbf{q}_1, \mathbf{q}_2, \ldots, \mathbf{q}_N] \in \{0,1\}^{C \times N}$, where $\mathbf{q}_i$ is the label vector of the i-th sample pair. For the i-th sample pair, if it belongs to the j-th class, $q_{ij} = 1$, otherwise $q_{ij} = 0$. As one sample only belongs to one class, $\sum_{j=1}^{C} q_{ij} = 1$ for the i-th sample pair.

Our model aims to separately obtain the sparse coefficients from different modalities by dictionary learning, and project them to a common label space where the data share the similar semantic concepts by label alignment. Therefore, the object function is formulated as a minimization problem as follows:

$$\min_{\mathbf{D}_m, \mathbf{A}_m, \mathbf{W}_m} \sum_{m=1}^{M} \|\mathbf{X}_m - \mathbf{D}_m \mathbf{A}_m\|_F^2 + \sum_{m=1}^{M} \|\mathbf{Q} - \mathbf{W}_m \mathbf{A}_m\|_F^2$$

$$+ \sum_{m=1}^{M} \lambda \|\mathbf{A}_m\|_1 + \sum_{m=1}^{M} \gamma \|\mathbf{W}_m\|_F^2 \tag{2}$$

where the first term minimizes the reconstruction error, and the second term aligns the common labels of the relevant data sample by minimizing the projection error. The regularization parameters λ and γ can balance the weight of the two terms in the object function. In Eq. (2), $\mathbf{W}_m \in \mathbb{R}^{C \times K}$ is the projection matrix, by which $\mathbf{A}_m$ is projected into the common label space. The term $\|\mathbf{W}_m\|_F^2$ can avoid over-fitting. In our model, different modalities data can be represented and correlated simultaneously.

2.3 Optimization

Since the objective function is not jointly convex $w.r.t.$ $\mathbf{D}_m$, $\mathbf{A}_m$ and $\mathbf{W}_m$, it is difficult to optimize them jointly. Therefore we use a iterative algorithm to update each variable when fixing the other two for each modal, respectively.

Algorithm 1. Iterative Algorithm for Our Proposed Model

Input: Data matrices $\mathbf{X}_m$ from the m-th modality; the label matrix $\mathbf{Q}$.

1: Initialize the dictionary $\mathbf{D}_m$ and the sparse coefficients $\mathbf{A}_m$ for m-th modality and initialize the projection matrix $\mathbf{W}_m$ by Eq. (7).

2: **while** not converge **do**

3: Update $\mathbf{A}_m$ of the m-th modality by Eq. (4) with $\mathbf{D}_m$ and $\mathbf{W}_m$ obtained from the previous iteration.

4: Update $\mathbf{D}_m$ of the m-th modality by Eq. (5) with $\mathbf{A}_m$ and $\mathbf{W}_m$ fixed.

5: Update $\mathbf{W}_m$ of the m-th modality by Eq. (7) with $\mathbf{D}_m$ and $\mathbf{A}_m$ fixed.

6: **end while**

Output: Dictionaries $\mathbf{D}_m$, and project matrices $\mathbf{W}_m$ of data from the m-th modality.

With the initialization of dictionaries $\mathbf{D}_m$ and projection matrices $\mathbf{W}_m$, we can update sparse coefficients $\mathbf{A}_m$ by considering $\mathbf{D}_m$ and $\mathbf{W}_m$ as constants, which can be formulated as follows:

$$\min_{\mathbf{A}_m} \|\mathbf{X}_m - \mathbf{D}_m\mathbf{A}_m\|_F^2 + \|\mathbf{Q} - \mathbf{W}_m\mathbf{A}_m\|_F^2 + \lambda\|\mathbf{A}_m\|_1 \tag{3}$$

To conduct the optimization above, it can be rewritten as the following problem:

$$\min_{\mathbf{A}_m} \left\| \begin{bmatrix} \mathbf{X}_m \\ \mathbf{Q} \end{bmatrix} - \begin{bmatrix} \mathbf{D}_m \\ \mathbf{W}_m \end{bmatrix} \mathbf{A}_m \right\|_F^2 + \lambda\|\mathbf{A}_m\|_1 \tag{4}$$

which is a l_1-norm lasso problem and can be solved by SPAMS Toolbox [7].

Then with $\mathbf{A}_m$, $\mathbf{W}_m$ fixed, we can update $\mathbf{D}_m$ as follow:

$$\min_{\mathbf{D}_m} \|\mathbf{X}_m - \mathbf{D}_m\mathbf{A}_m\|_F^2 \quad \text{s.t.} \quad \|\mathbf{d}_{m,i}\| \leq 1, \forall i \tag{5}$$

which is a quadratically constrained quadratic program (QCQP) problem $w.r.t.$ $\mathbf{D}_m$ and the solutions can be obtained by Lagrange dual techniques [6].

When $\mathbf{A}_m$ and $\mathbf{D}_m$ are fixed, we can calculate projection matrices $\mathbf{W}_m$ as follows:

$$\min_{\mathbf{W}_m} \|\mathbf{Q} - \mathbf{W}_m\mathbf{A}_m\|_F^2 + \gamma\|\mathbf{W}_m\|_F^2 \tag{6}$$

Finally, we update the projection matrices $\mathbf{U}_m$. When $\mathbf{D}_m$ and $\mathbf{A}_m$ are fixed, Eq. (2) is a ridge regression problem and the analytical solutions can be obtained as below:

$$\mathbf{W}_m = \mathbf{Q}\mathbf{A}_m^T(\mathbf{A}_m\mathbf{A}_m^T + \gamma\mathbf{I})^{-1} \tag{7}$$

The algorithm procedures are summarized in Algorithm 1.

2.4 Application to Cross-Modal Retrieval

After the optimization algorithm is completed, $\mathbf{D}_m$, $\mathbf{A}_m$ and $\mathbf{U}_m$ can be derived and we can use them to conduct cross-modal retrieval application.

Given a query q_m from m-th modality, the sparse coefficients a_m can be obtained by sparse coding as follows:

$$\min_{a_m} \|q_m - \mathbf{D}_i a_m\|_F^2 + \lambda \|a_m\|_1 \tag{8}$$

Then the sparse coefficients will be projected into the common label space by the projection matrix $\mathbf{W}_m$:

$$p_m = \mathbf{W}_m a_m \tag{9}$$

where p_m is the representation of q_m in the common label space. By using the projection matrices we can projected all data from different modalities into the common label space, in which the relevance of data from different modalities can be measured easily and the nearest neighbors of the query are returned as the retrieval results.

3   Experiments

In this section, we will evaluate the performance of the proposed cross-modal retrieval. We first elaborate the experiment setting and the evaluation metrics adopted in this paper, and then compare our approach with several state-of-the-art methods.

3.1   Experiment Setting

We evaluate the proposed method on two public image-text datasets, $i.e.$, Wiki Text-Image dataset [8] and the NUS-WIDE dataset [3]. And our experiments are conducted for two retrieval tasks: (1) image query in text database, (2) text query in image database.

The Wiki dataset consists of 2173/693 (training/test) image-text pairs which are generated from Wikipedia. Each pair is labeled by only one of 10 semantic classes. SIFT descriptors of images are extracted and quantized into Bag-of-Visual-Words (BoVW) by K-means clustering. For texts, we represent them with Bag-of-Words (BoW) by counting the word frequency. Considering the effect of feature dimensions, we finally obtain two datasets: one with 500 dimensions BoVW and 1,000 dimensions BoW, and the other with 1,000 dimensions BoVW and 5,000 dimensions BoW.

The NUS-WIDE dataset contains 269,648 images from Flickr, with a total number of 5,018 unique tags. It contains 81 labels and some image-tags pairs have two or more labels. In our experiments, image-tags pairs with only one label are selected. We use 500 dimensions BoVW based on SIFT as image features and 1000 dimensions tags as text feature. Finally 63,641 pairs are obtained as dataset. And we randomly take 3 % each class for training set and 1 % each class from the testing set.

In the paper, mean Average Precision (mAP) is utilized to evaluate the performance and the cosine distance is adopt to measure the similarity. The Average Precision (AP) is defined as $AP = \frac{1}{T} \sum_{r=1}^{N} P(r)\delta(r)$, where T is the number of

Table 1. Comparison of mAP performance of different methods on the Wiki dataset with 500-D BoVW and 1000-D BoW, for image query text task, text query image task and the average of two tasks. The best results are marked by bold font.

Methods	mAP		
	Image query	Text query	Average
CCA	0.1785	0.1784	0.1785
PLS	0.2850	0.3419	0.3135
BLM	0.2575	0.2806	0.2691
GMLDA	0.2770	0.3411	0.3091
GMMFA	0.2722	0.3241	0.2982
SLiM^2	0.2242	0.2334	0.2288
Proposed	**0.3094**	**0.3762**	**0.3428**

Table 2. Comparison of mAP performance of different methods on the Wiki dataset with 1000-D BoVW and 5000-D BoW, for image query text task, text query image task and the average of two tasks. The best results are marked by bold font.

Methods	mAP		
	Image query	Text query	Average
CCA	0.2454	0.2405	0.2430
PLS	0.2892	0.3258	0.3075
BLM	0.2623	0.3247	0.2935
GMLDA	0.2456	0.2259	0.2358
GMMFA	0.2020	0.2449	0.2235
SLiM^2	0.2647	0.2852	0.2750
Proposed	**0.3410**	**0.4253**	**0.3832**

retrieved data belonging to the same class of the query. $P(r)$ denotes the precision of the top r retrieved data, and $\delta(r) = 1$ if the rth retrieved data has the same label as the query and $\delta(r) = 0$ otherwise. In the experiments, we set $N = 50$. The average AP values over all the queries in the query set is obtained as mAP. In our experiments, both the parameters λ and γ are set to 0.1, and dictionary size $K = 200$ empirically.

3.2 Performance Comparisons

We evaluate the performance of our proposed method, and compare it with several related methods: CCA [4], PLS [9], BLM [11], GMMFA [10], GMLDA [10], and SLiM^2 [13] over Wiki Text-Image dataset and NUS-WIDE dataset.

Tables 1 and 2 show the mAP scores of different methods on the Wiki dataset. As shown in Tables 1 and 2, our method outperforms other approaches in both

Table 3. Comparison of mAP performance of different methods on the NUS-WIDE dataset with 500-D BoVW and 1000-D BoW for image query text task, text query image task and the average of two tasks. The best results are marked by bold font.

Methods	mAP		
	Image query	Text query	Average
CCA	0.3329	0.3393	0.3361
PLS	**0.4297**	0.4363	0.4330
BLM	0.4264	0.4249	0.4257
GMLDA	0.3775	0.4138	0.3957
GMMFA	0.3795	0.4067	0.3931
SLiM^2	0.3652	0.3500	0.3576
Proposed	0.4217	**0.4903**	**0.4560**

two retrieval tasks on Wiki dataset. Moreover, we obtain the highest average mAP scores of 0.3428 and 0.3832 on two dataset, respectively. The reason is that our method not only effectively represent heterogenous features by sparse coefficients and also sufficiently explore the common label information.

The comparison of mAP results on NUS-WIDE dataset are shown in Table 3. As we have seen, our method achieves mAP score of 0.4903 on image-query-text task, which is much better than other methods. Although the PLS method outperforms slightly on text-query-image task, our method obtains the best mAP score on average.

Figure 1 illustrates a visual case of cross-modal retrieval by our method. The top row shows the result of text-query-images task and the bottom row is the image-query-texts task. It is worth noting that our method can retrieve the semantically related results.

Like most seabirds, the majority of procellariids breed once a year. There are exceptions; many individuals of the larger species, such as the White-headed Petrel, will skip a breeding season after successfully fledging a chick......

text query   Top images retrieved

image query   The corresponding images to the top texts

Fig. 1. Top: An example of an text query and the top four images retrieved by our method. Bottom: An example of an image query and the corresponding images of the top four texts retrieved by our method.

4   Conclusions

In this paper, we propose a cross-modal retrieval method based on coupled dictionary learning with common label alignment. The main contributions of our method include: (1) we exploit coupled dictionary learning to project the data from different modalities into a common space where the inherent relation between modalities can be well discovered; (2) we utilize the shared label information in the common space to further encourage the correlation across the different modalities. Experiment results on two public datasets confirm that our method outperforms several state-of-the-art approaches.

Acknowledgments. This work is supported by the National High Technology Research and Development Program of China (2013AA01A602), the Program for New Century Excellent Talents in University (NCET-12-0917), the Fundamental Research Funds for the Central Universities (No. K5051302019), the Key Science and Technology Program of Shaanxi Province, China (2014K05-16).

References

1. Blei, D.M., Ng, A.Y., Jordan, M.I.: Latent dirichlet allocation. J. Mach. Learn. Res. **3**, 993–1022 (2003)
2. Chen, Y., Wang, L., Wang, W., Zhang, Z.: Continuum regression for cross-modal multimedia retrieval. In: Proceedings of IEEE Conference on Image Processing, pp. 1949–1952 (2012)
3. Chua, T.S., Tang, J., Hong, R., Li, H., Luo, Z., Zheng, Y.: Nus-wide: a real-world web image database from national university of singapore. In: Proceedings of the ACM International Conference on Image and Video Retrieval, pp. 48:1–48:9 (2009)
4. Hardoon, D., Szedmak, S., Shawe-Taylor, J.: Canonical correlation analysis: an overview with application to learning methods. Neural Comput. **16**(12), 2639–2664 (2004)
5. Huang, D.A., Wang, Y.C.F.: Coupled dictionary and feature space learning with applications to cross-domain image synthesis and recognition. In: Proceedings of IEEE International Conference on Computer Vision, pp. 2496–2503 (2013)
6. Lee, H., Battle, A., Raina, R., Ng, A.Y.: Efficient sparse coding algorithms. In: Advances in Neural Information Processing Systems, pp. 801–808 (2006)
7. Mairal, J., Bach, F., Ponce, J., Sapiro, G.: Online dictionary learning for sparse coding. In: Proceedings of the 26th International Conference on Machine Learning, pp. 689–696 (2009)
8. Rasiwasia, N., Pereira, J.C., Coviello, E., Doyle, G., Lanckriet, G.R., Levy, R., Vasconcelos, N.: A new approach to cross-modal multimedia retrieval. In: Proceedings of the ACM International Conference on Multimedia, pp. 251–260 (2010)
9. Sharma, A., Jacobs, D.W.: Bypassing synthesis: PLS for face recognition with pose, low-resolution and sketch. In: Proceedings of IEEE Conference on Computer Vision and Pattern Recognition, pp. 593–600 (2011)
10. Sharma, A., Kumar, A., Daume, H., Jacobs, D.W.: Generalized multiview analysis: a discriminative latent space. In: IEEE Conference on Computer Vision and Pattern Recognition, pp. 2160–2167 (2012)

11. Tenenbaum, J.B., Freeman, W.T.: Separating style and content with bilinear models. Neural Comput. **12**(6), 1247–1283 (2000)
12. Wang, S., Zhang, L., Liang, Y., Pan, Q.: Semi-coupled dictionary learning with applications to image super-resolution and photo-sketch synthesis. In: Proceedings of IEEE Conference on Computer Vision and Pattern Recognition, pp. 2216–2223 (2012)
13. Zhuang, Y., Wang, Y.F., Wu, F., Zhang, Y., Lu, W.: Supervised coupled dictionary learning with group structures for multi-modal retrieval. In: Proceedings of the Twenty-seventh AAAI Conference on Artificial Intelligence, pp. 1070–1076 (2013)

Combining Active Learning and Semi-Supervised Learning Based on Extreme Learning Machine for Multi-class Image Classification

Jinhua Liu[1,2], Hualong Yu[1,2,3], Wankou Yang[1,2,4],
and Changyin Sun[1,2(✉)]

[1] School of Automation, Southeast University, Nanjing 210096, China
cysun@seu.edu.cn
[2] Key Lab of Measurement and Control of Complex Systems of Engineering,
Ministry of Education, Southeast University, Nanjing 210096, China
[3] School of Computer Science and Engineering, Jiangsu University of Science
and Technology, Zhenjiang 212003, Jiangsu, China
[4] Key Laboratory of Child Development and Learning Science of Ministry
of Education, Southeast University, Nanjing 210096, China

Abstract. An accurate image classification system often requires many labeled training instances to train the classification models, which is expensive and time-consuming. Therefore, machine learning technologies which could utilize unlabeled instances to promote classification accuracy attract more attentions in the image classification field. Active learning and semi-supervised learning could both automatically discovery the hidden useful information from unlabeled instances. In this article, we try to combine active learning and semi-supervised learning to improve the classification performance of multi-class images. Specifically, extreme learning machine (ELM) is adopted as baseline classifier to accelerate the learning procedure, and an uncertainty estimation strategy is used to evaluate the information of each unlabeled instance. The experimental results on five multi-class image data sets show that the proposed method outperforms both random sampling and active learning. Meanwhile, we found that contrast with support vector machine (SVM), ELM could save much training time without obvious loss of performance.

Keywords: Extreme learning machine · Multi-class image classification · Active learning · Semi-supervised learning · Uncertainty estimation

1 Introduction

In the past decade, with the development of multimedia technology and Internet, image classification problem has received considerable attention [1]. Most existing image classification methods, however, often require a mass of labeled instances to train an accurate classification model. Some researchers considered to adopt the machine learning technologies which could make use of the information of the unlabeled examples to rapidly promote the performance of the image classification systems [2–5].

© Springer International Publishing Switzerland 2015
X. He et al. (Eds.): IScIDE 2015, Part I, LNCS 9242, pp. 163–175, 2015.
DOI: 10.1007/978-3-319-23989-7_18

As we know, both active learning and semi-supervised learning can improve the equality of classification model by iteratively inserting unlabeled instances into the training set. Active learning extracts the most uncertain unlabeled instances on each round to submit to the human annotator to label, which guarantees the classification model could be improved to the maximum extent [6]. And semi-supervised learning often selects the most confident samples to label and then to change the classification model, thus it does not acquire the participation of human annotators, as well its performance promotion is relatively slow [7].

As for image classification, Tong and Chang [2] adopted support vector machine (SVM) to perform active learning, which extracts the closest images to the classification hyperplane as the uncertain examples. Gu et al. [3] proposed an active learning-based multi-class image classification method which combines both uncertainty metric and diversity metric to extract samples. Specifically, in their method, one-versus-one SVM is adopted to solve multi-class image classification, Gaussian kernel is used to estimate the diversity among the selected uncertain samples, and a dynamic programming method is used to determine the final instances that should be submitted to the human annotators for labeling. Li et al. [4] presented a semi-supervised image classification algorithm based on random feature subspace. The method maps the original feature vector into three different random feature subspaces and adopts Tri-training learning algorithm [8] to improve the classification performance. Chen et al. [5] integrated active learning and semi-supervised learning into a unified framework to classify multi-class images. In particular, Best-Versus-Second-Best (BVSB) strategy [9] is used to estimate the uncertainty of each unlabeled instance, and a constrained self-training approach is proposed to find a trade-off between confidence and information in semi-supervised learning. The experimental results indicated that Chen's method outperforms to merely using active learning algorithm with the same manually labeled instances. However, its drawback lies in the training procedure is excessively time-consuming as the use of SVM.

In this article, we try to adopt extreme learning machine (ELM) [10–12] to implement active learning and semi-supervised learning, and further combine them to improve the classification performance of multi-class images. First, we use a few labeled images to train an initial ELM classifier. Then the classifier is used to classify all unlabeled instances. Next, the actual outputs of each unlabeled sample are transferred to approximate posterior probabilities by a sigmoid function, and further BVSB strategy is adopted to estimate the uncertainty and/or confidence for each instance. At last, several most uncertain samples are labeled by human annotators and several confident examples are automatically labeled by the classifier to expend the training set and update the classifier. Specifically, considering the most confident samples often include little useful information to improve the classification model, we profit from the idea of Chen et al. [5] to propose a constrained sampling strategy which could satisfy correctness of labeling and information of sampling, simultaneously. The proposed method is compared with random sampling and active learning on three handwriting digit data sets, an object recognition data set and a scene categorization data set, showing more superior performance. The results indicate that semi-supervised learning could help accelerate the progress of active learning. Moreover, we found that in contrast with support vector machine (SVM) which is adopted in the method of Chen et al. [5], ELM could save much training time without obvious loss of performance.

The rest of this article is organized as follows. In Sect. 2, we describe the method used in this paper, including the basic ELM algorithm, the uncertainty estimation strategy based on ELM, active learning and semi-supervised learning algorithms taking advantage of the uncertainty estimation strategy, and the proposed combined algorithm. Section 3 first reports our experimental results on three handwritten character data sets, an object recognition data set and a scene categorization data set, then presents the application conditions of the learning algorithm. Finally, Sect. 4 concludes the article and indicates our future research.

2 Methods

2.1 Extreme Learning Machine

ELM is a new learning algorithm to train single-hidden layer feedforward network (SLFN) [10–12]. The basic structure of SLFN is described in Fig. 1. Unlike gradient decent-based back propagation (BP) method [13], ELM doesn't need to iteratively tune the parameters between the input layer and the hidden layer, but denotes their values randomly, then directly calculates the weights connecting the hidden layer and the output layer by lease-square method [10] or optimization method [12]. It has been found that ELM often provides better or at least comparable generalization ability at a much faster learning speed than those traditional classification approaches, including BP neural network (BPNN), support vector machine (SVM) and least-square support vector machine (LS-SVM) [10–12].

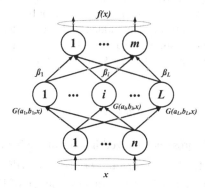

Fig. 1. The basic structure of one SLFN

Suppose there are N arbitrary distinct training instances $(x_i, t_i) \in R^n \times R^m$, where x_i is one $n \times 1$ input vector and t_i is one $m \times 1$ target vector. If an SLFN with L hidden nodes can approximate these N instances with zero error, it then implies that there exist β_i, a_i and b_i, such that:

$$f_L(x_j) = \sum_{i=1}^{L} \beta_i G(a_i, b_i, x_j) = t_j, \quad j = 1, \ldots, N \tag{1}$$

where a_i and b_i denote the weights and biases between the input layer and the hidden layer, β_i is the weight vector connecting the ith hidden node to the output nodes. Then Eq. (1) can be written compactly as:

$$H\beta = T \tag{2}$$

where

$$H(a_1, \ldots, a_L, b_1, \ldots, b_L, x_1, \ldots, x_N) = \begin{bmatrix} G(a_1, b_1, x_1) & \cdots & G(a_L, b_L, x_1) \\ \vdots & \cdots & \vdots \\ G(a_1, b_1, x_N) & \cdots & G(a_L, b_L, x_N) \end{bmatrix} \tag{3}$$

$$\beta = \begin{bmatrix} \beta_1^T \\ \vdots \\ \beta_L^T \end{bmatrix}_{L \times m} \text{ and } T = \begin{bmatrix} t_1^T \\ \vdots \\ t_N^T \end{bmatrix}_{N \times m} \tag{4}$$

Here, $G(a_i, b_i, x_j)$ denotes activation function which is used to calculate the output of the ith hidden node for the jth training instance. In ELM, many nonlinear activation functions can be used, like sigmoid (sig), sine (sin), hardlimit ($hardlim$) and radial basis functions (rbf) [12]. H is called hidden layer output matrix of the network, where its ith column denotes the ith hidden node's output vector with respect to inputs $x_1, x_2 \ldots x_N$ and its jth line represents the output vector of the hidden layer with respect to input x_j.

In SLFN, the number of hidden nodes, L, would always be less than the number of training samples, N, and, hence, the training error cannot be made exactly zero but can approach a nonzero training error ε. The hidden node parameters a_i and b_i need not be tuned during training and may simply be assigned with random values according to any continuous sampling distribution [10, 11]. Equation (2) then becomes a linear system and the output weights β are estimated as:

$$\hat{\beta} = H^\dagger T \tag{5}$$

where $H^\dagger$ is the Moore-Penrose generalized inverse of the hidden layer output matrix H. $H^\dagger = (H^T H)^{-1} H^T$ if $H^T H$ is nonsingular or $H^\dagger = H^T (H H^T)^{-1}$ if $H H^T$ is nonsingular. Here, $\hat{\beta}$ is the minimum-norm least squares solution of Eq. (2).

ELM could also be explained in the optimization view. As we know, the norm of output weights $\|\beta\|$ is closely related with the generalization ability of a neural network, thus ELM may try to minimize both $\|H\beta - T\|$ and $\|\beta\|$ simultaneously. Then the solution of Eq. (2) could be obtained by [12]:

$$\text{Minimize: } Lp_{ELM} = \frac{1}{2}||\beta||^2 + C\frac{1}{2}\sum_{i=1}^{N}\varepsilon_i^2 \tag{6}$$

$$\text{Subject to: } h(x_i)\beta = t_i - \varepsilon_i$$

where $\varepsilon_i = [\varepsilon_{i,1}, \ldots, \varepsilon_{i,m}]$ is the training error vector of the m output nodes corresponding to training instance x_i, C is the trade-off regularization parameter between the minimization of training errors and the maximization of the marginal distance. The solution of Eq. (6) could be obtained based on KKT theorem [14]. Given a new instance x, the output function of ELM is obtained by [12]:

$$f(x) = \begin{cases} h(x)H^T(\frac{I}{C} + HH^T)^{-1}T, & \text{when } N < L \\ h(x)(\frac{I}{C} + HH^T)^{-1}H^T T, & \text{when } N \geq L \end{cases} \tag{7}$$

where $f(x) = [f_1(x), \ldots, f_m(x)]$ is the output function vector. Then users may use the following equation to find out the predication label of x:

$$label(x) = \underset{i}{\arg\max} f_i(x), i \in [1, \ldots, m] \tag{8}$$

In this article, we use the optimization version of ELM to implement active learning.

2.2   Uncertainty Estimation Based on the Outputs of ELM

As we know, both active learning and semi-supervised learning need to estimate the uncertainty and/or confidence of each unlabeled instance. ELM, however, merely provides real-value outputs. Generally, the class corresponding to the maximum output is assigned as the class label for one unlabeled test instance. Here, we wish to make use of these real-value outputs to quantificationally estimate the uncertainty/certainty for each example in the unlabeled pool. Profiting from the idea of Platt [15] which transforms the real value outputs of SVM as approximately accurate posteriori probabilities by solving the following sigmoid model:

$$P(y = 1|f(x)) = \frac{1}{1 + \exp(Af(x) + B)} \tag{9}$$

where A and B are two unknown variances which need to be solved by using all training instances. Therefore, the transformation is an optimization procedure that is much time-consuming. Active learning and semi-supervised learning, however, does not need to accurately approximate the posteriori probabilities, but only needs to provide the sorting of all instances according to their uncertainties and/or confidences. Therefore, we adopt the basic sigmoid function to construct a mapping relationship between real outputs of ELM and posteriori probabilities, which is described as follows:

$$P(y = 1|f_i(x)) = \frac{1}{1 + \exp(-f_i(x))} \qquad (10)$$

where $f_i(x)$ denotes the real output of the ith output node for the sample x. In Eq. (10), with the increase of the output value, the posteriori probability would gradually approximate as 1, while when the output is a very small negative value, the posteriori probability would approximate as 0. The mapping relationship between actual output value of ELM and the approximated posteriori probability is described in Fig. 2.

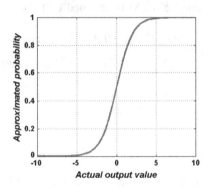

Fig. 2. The mapping relationship between actual output value and the approximated probability

It is difficult to prove that for binary classification problem, the sum of posteriori probabilities is exactly as 1 by using Eq. (10). Multiclass problem, however, could not guarantee the sum of posteriori probabilities is exactly as 1. In fact, it is often larger than 1. Therefore, we adopted a normalized method to transform them. The normalized posteriori probabilities are calculated using the following function:

$$p'(y = 1|f_i(x)) = \frac{p(y = 1|f_i(x))}{\sum_{j=1}^{m} p(y = 1|f_j(x))} \qquad (11)$$

where $p(y = 1|f_i(x))$ and $p'(y = 1|f_i(x))$ are original and normalized posteriori probabilities of the ith class. Then, ELM could be transformed as probabilistic classifier, as well we could use the probabilities to estimate the uncertainty and/or confidence for each unlabeled instance. BVSB, which was proposed by Joshi et al. [9], is adopted as the uncertainty estimation measure. BVSB calculates the difference between the largest posteriori probability and the second largest posteriori probability. It is obvious that the larger the difference is, the more confident the corresponding instance is, while the smaller the difference is, the more uncertain the corresponding instance is.

2.3   Active Learning and Semi-Supervised Learning Based on ELM

Next, we wish to take advantage of the results of uncertainty estimation to implement active learning and semi-supervised learning, simultaneously. As we know, active learning often extracts those most uncertain instances as they contain more useful information to improve the classification model. While semi-supervised learning usually automatically labels those most confident instances which correspond the ones having the least uncertainties. Actually, the semi-supervised learning method used in this article is called self-training method [7]. Compared with active learning, semi-supervised learning does not need the participation of human annotators, but its performance promotion is quite limited.

Considering the most confident unlabeled instances are often far from the classification hyperplane, thus they are usually useless to improve the quality of the classification model. Therefore, we present an improved sampling strategy for semi-supervised learning. First, the strategy samples a half of most confident instances in the unlabeled set. Then the extracted unlabeled instances are ranked in descending order according to the uncertainty measures. Next, we find the nearest neighbor of each extracted unlabeled instance in labeled training set. If one unlabeled instance has the same class label as its nearest neighbor, we believe that the instance is absolutely confident, otherwise, the instance would be deserted. At last, a batch of instances are extracted in order, as well are automatically labeled and feed into the training set. It is clear that the proposed strategy could find a trade-off between accuracy of labeling and usability of instances, consequently improving the effect of semi-supervised learning.

2.4 The Proposed Integrated Learning Algorithm for Multi-class Image Classification

The learning procedure is described as follows:

Input: Initial labeled training set L, unlabeled set U, the times of iteration T and the batch size N
Output: The final learned ELM classifier C
Procedure:
1. Train an initial ELM classifier C_0 using L;
2. for $i=1:T$
3. Adopt the classifier C_{i-1} to estimate the uncertainty of each instance in U (using Eq.(10) and Eq.(11));
4. Extract N most uncertain instances from U to submit to the human annotators to label;
5. Extract a half of most certain instances from U to rank them in descending order according to the uncertainty measures;
6. Find the nearest neighbor of each extracted unlabeled instance in labeled training set L, if they have the same class label, we reserve it, otherwise, we desert it;
7. Extract N instances in order and automatically label them by the classifier C_{i-1};
8. Move the extracted $2*N$ instances from U to L;
9. Use the extended L to train a new ELM classifier C_i;
10. end for
11. Output the final learned ELM classifier $C=C_T$.

A simple schematic diagram about the learning procedure is presented in Fig. 3.

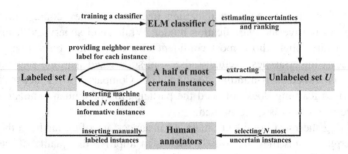

Fig. 3. The flow chart of the proposed learning algorithm

3 Results and Discussions

3.1 Results on Three Handwriting Digit Data Sets

First, we used three handwriting digit data sets acquired from Keel data repository [16]. For each data set, the instance is one image denoting one of the digits between 0 to 9. The detailed information about these three data sets are described in Table 1.

Table 1. The details of the three used handwriting digit data sets

Dataset	Data types	Attribute types	# Instances	# Attributes
Optidigits	Multivariate	Integer	5,620	64
Penbased	Multivariate	Integer	10,992	16
USPS	Multivariate	Real	11,000	256

For each data set, we randomly extracted a few instances as initial labeled training set, a mount of examples as initial unlabeled set, and the rest samples were used as testing set. We compared the proposed method (Active-Semi) with merely active learning (Active) and random sampling (Random) with batch-mode learning style, where the batch size was assigned as 5. For ELM, Radial Basis Kernel Function (RBF) was adopted to calculate the outputs of hidden nodes as it had revealed better performance than other kernel functions [12]. The other parameters, such as the number of hidden nodes L, the penalty factor C and the times of iteration in learning procedure could be observed in Table 2. To obtain the real results, each experiment was executed 50 times, then the average results were provided.

Table 2. Parameter settings, where #iteration denotes the times of iteration

Dataset	Labeled percentage	Unlabeled percentage	L	C	# iteration	Batch size
Optical	10 %	40 %	300	100,000	100	5
Penbased	5 %	40 %	300	100,000	100	5
USPS	10 %	40 %	300	100,000	150	5

The corresponding learning curves of the three learning algorithms on the three handwriting digit data sets are presented in Fig. 4, which indicates:

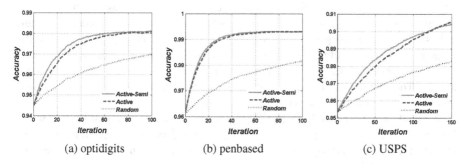

Fig. 4. Learning curves of the three learning algorithms on the three handwriting digit data sets, (a) optidigits, (b) penbased, (c) USPS

1. The proposed uncertainty estimation approach is effective as active learning could rapidly promote the classification performance compared with random sampling. That means the uncertainty measure is helpful to finding those informative instances which are generally close to the borderline.
2. In contrast with active learning, semi-supervised learning helps accelerating the learning procedure without more manual labels. Furthermore, we found that Active-Semi is superior to Active only during the initial learning phases. Since later in the learning procedure, more and more falsely labeled instances by semi-supervised learning would be inserted into the training set, causing accumulation of errors and degeneration of classification performance.

3.2   Results on the Object and Scene Recognition Data Sets

We have also conducted experiments on one object recognition data set (Caltech-101 data set) and one scene categorization data set (15 natural scene categories data set), respectively. We firstly take advantage of the pipeline of Fig. 5 to extract features of images [17].

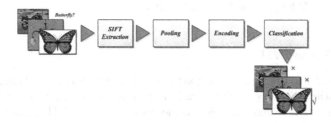

Fig. 5. The image classification pipeline using the typical components of a bag of visual words

We randomly extracted 10 classes (1761 images in total) in Caltech-101 data set and 15 classes (2850 images in total) in 15 natural scene categories data set. An improved dense-sift descriptor was applied to extract local image features, which is very similar with Lowe's original sift descriptor [18] but much faster. The encoding method divides local descriptor space into informative regions whose internal structure could be disregarded or parameterized linearly. These regions are also called visual words [17]. K-means clustering was used to construct visual vocabularies. K is assigned as 100 on the Caltech-101 data set and 150 on the 15 natural scene categories data set. Then the basic histogram encoding was used to make the local descriptors quantitative and constitute the baseline of each image.

We randomly extracted 35 % and 20 % original instances as initial labeled set on Caltech-101 data set and 15 natural scene categories data set, respectively. For both data sets, the initial percentage of unlabeled instances were assigned as 40 %. The other parameters stayed the same as the ones in Table 2. Then the average learning curves are presented in Fig. 6.

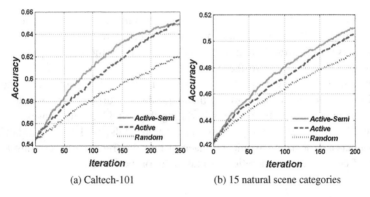

(a) Caltech-101   (b) 15 natural scene categories

Fig. 6. Learning curves of the three learning algorithms on Caltech-101 and 15 natural scene categories sets, (a) Caltech-101, (b) 15 natural scene categories

Figure 6 shows the same phenomenon as Fig. 4, i.e., our proposed Active-Semi algorithm has more rapid learning speed than active learning and random sampling, which verifies its effectiveness and feasibility again. In addition, we found the classification accuracies of various learning algorithms are relatively low as the use of a simple feature descriptor. We believe that if better feature descriptors were adopted, the performance could be promoted to a large extent.

3.3   ELM Versus SVM

As we know, SVM often produces good classification performance. Here, we compared ELM with SVM beyond the same learning frameworks [5] with examining two measures: running time and area below the learning curve (ALC) [19] which is used to detect both learning speed and classification performance. Obviously, the larger the

ALC is, the better the learning algorithm is. In particular, considering SVM could be only directly constructed on binary-class problem, one-versus-rest coding strategy of SVM was adopted to solve multi-class problem. Moreover, SVM also used RBF kernel function and the optimal combination of parameters was determined by grid search, where $C \in [2^{-8}, 2^{-7}, ..., 2^8]$, $\sigma \in [2^{-8}, 2^{-7}, ..., 2^8]$. Taking optidigits data set as an example, the comparable results are provided in Table 3.

Table 3. Comparison between ELM and SVM

Classifier	Active-semi		Active		Random	
	Time (s)	ALC	Time (s)	ALC	Time (s)	ALC
ELM	6.6651	0.9750	5.1309	0.9732	3.237	0.9621
SVM	363.5603	0.9847	283.1574	0.9840	178.6055	0.9753

Table 3 shows that SVM only performs a little better than ELM, but consumes about 50 times more running time for any one learning style. We believe that with the increase of training instances and the number of classes, the difference of running time between SVM and ELM would be further enlarged. The adoption of ELM provides one trade-off between classification performance and running time.

3.4 Application Conditions of the Proposed Learning Algorithm

We determined the influence of percentage of initial labeled instances to the learning performance. We still take the optidigits data set as an example. We respectively extracted 5 %, 10 %, 20 % and 30 % instances as initial labeled set, as well extracted 40 % instances as unlabeled set. The results are presented in Fig. 7.

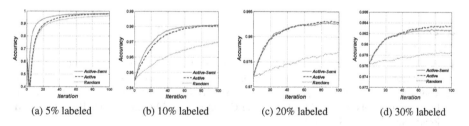

(a) 5% labeled   (b) 10% labeled   (c) 20% labeled   (d) 30% labeled

Fig. 7. Learning curves with different percentages of initial labeled instances, (a) 5 % labeled, (b) 10 % labeled, (c) 20 % labeled, (d) 30 % labeled

Figure 7 shows that when initial labeled instances are quite limited, the proposed learning algorithm is superior to others, while with the increase of initial labeled instances, the difference between Active-Semi and Active would be less and less, sometimes active learning even outperforms to Active-Semi learning. Therefore, the proposed learning algorithm is more appropriate to be applied in the scenario that only a few initial labeled samples could be available.

4 Conclusions

In this article, we proposed a combined active learning and semi-supervised learning algorithm based on ELM classifier for classifying multi-class images. Specifically, one uncertainty measure based on the actual outputs of ELM is presented to estimate the uncertainty and/or confidence of each unlabeled instance. Meanwhile, the uncertainty-based active learning algorithm and an improved self-training semi-supervised algorithm are integrated into an unified learning framework. Experimental results on five different multi-class image data sets indicated that semi-supervised learning could accelerate the progress of active learning without more manually labels. Furthermore, we found that in contrast with SVM, ELM could save the time of learning to a large extent without obvious loss of performance.

In future work, we wish to verify the effectiveness of the proposed method on more multi-class image data sets. Moreover, the possibility of combining multiple semi-supervised ELMs to implement the integration of query-based committee (QBC) active learning to classify multi-class images would be investigated, too.

Acknowledgments. The work was supported in part by National Natural Science Foundation of China under Grant No. 61305058, No. 61473086, No. 61375001, Natural Science Foundation of Jiangsu Province of China under Grant No. BK20130471 and No. BK20140638, China Postdoctoral Science Foundation under grant No. 2013M540404, Jiangsu Planned Projects for Postdoctoral Research Funds under grant No. 1401037B, open fund of Key Laboratory of Measurement and Control of Complex Systems of Engineering, Ministry of Education under Grant No. MCCSE2013B01, the Open Project Program of Key Laboratory of Child Development and Learning Science of Ministry of Education, Southeast University (No. CDLS-2014-04), and A Project Funded by the Priority Academic Program Development of Jiangsu Higher Education Institutions (PAPD), and the Fundamental Research Funds for the Central Universities.

References

1. Alajlan, N., Pasolli, E., Melgani, F., Franzoso, A.: Large-scale image classification using active learning. IEEE Geosci. Remote Sens. Lett. **11**, 259–263 (2014)
2. Tong, S., Chang, E.: Support vector machine active learning for image retrieval. In: Proceedings of the 9th ACM International Conference on Multimedia, pp. 107–118. ACM Press, New York, USA (2001)
3. Gu, Y.J., Jin, Z., Chiu, S.C.: Active learning combining uncertainty and diversity for multi-class image classification. IET Comput. Vis. (2014). doi:10.1049/iet-cvi.2014.0140
4. Li, L., Huaxiang, Z., Xiaojun, H., Feifei, S.: Semi-supervised image classification learning based on random feature subspace. In: Li, S., Liu, C., Wang, Y. (eds.) Pattern Recognition, pp. 237–242. Springer, Berlin, Heidelberg (2014)
5. Chen, R., Cao, Y.F., Sun, H.: Multi-class active learning and a semi-supervised learning for image classification (in Chinese). Acta Automatica Sinica **37**, 954–962 (2011)
6. Settles, B.: Active learning literature survey. Univ. Wis. Madison **52**, 55–66 (2010)
7. Chapelle, O., Scholkopf, B., Zien, A. (eds.): Semi-Supervised Learning. MIT Press, Cambridge (2006)

8. Zhou, Z.H., Li, M.: Tri-training: exploiting unlabeled data using three classifiers. IEEE Trans. Knowl. Data Eng. **17**, 1529–1541 (2005)

9. Joshi, A.J., Porikli, F., Papanikolopoulos, N.: Multi-class active learning for image classification. In: Proceedings of 2009 IEEE Computer Society Conference on Computer Vision and Pattern Recognition, pp. 2372–2379. IEEE press, Miami, USA (2009)

10. Huang, G.B., Zhu, Q.Y., Siew, C.K.: Extreme learning machine: theory and applications. Neurocomputing **70**, 489–501 (2006)

11. Huang, G.B., Wang, D.H., Lan, Y.: Extreme learning machine: a survey. Int. J. Mach. Learn. Cybernet. **2**, 107–122 (2011)

12. Huang, G.B., Zhou, H., Ding, X., Zhang, R.: Extreme learning machine for regression and multiclass classification. IEEE Trans. Syst. Man Cybern. B Cybern. **42**, 513–529 (2012)

13. Rumelhart, D.E., Hinton, G.E., Williams, R.J.: Learning representations by back-propagation errors. Nature **323**, 533–536 (1986)

14. Fletcher, R.: Practical Methods of Optimization, Constrained Optimization, vol. 2. Wiley, New York (1981)

15. Platt, J.C.: Probabilistic Outputs for Support Vector Machines and Comparisons to Regularized Likelihood Methods, Advances in Large-Margin Classifiers. MIT Press, Cambridge (2000)

16. Alcalá-Fdez, J., Fernandez, A., Luengo, J., Derrac, J., García, S., Sánchez, L., Herrera, F.: KEEL data-mining software tool: data set repository, integration of algorithms and experimental analysis framework. J. Multiple-Valued Logic Soft Comput. **17**, 255–287 (2011)

17. Chatfield, K., Lempitsky, V., Vedaldi, A., Zisserman, A.: The devil is in the details: an evaluation of recent feature encoding methods. In: British Machine Vision Conference (2011)

18. Lowe, D.G.: Object recognition from local scale-invariant features. In: Proceedings of the Seventh IEEE International Conference on Computer Vision, vol. 2, pp. 1150–1157. IEEE Press (1999)

19. Guyon, I., Gawley, G., Dror, G., Lemaire, V.: Results of active learning challenge. JMLR Workshop Conf. Proc. **16**, 19–45 (2011)

Multi-modal Retrieval via Deep Textual-Visual Correlation Learning

Jun Song, Yueyang Wang, Fei Wu[✉], Weiming Lu, Siliang Tang, and Yueting Zhuang

College of Computer Science, Zhejiang University,
NO. 38 Zheda Road, Hangzhou 310027, Zhejiang, China
{songjun54cm,yywang,luwm,siliang}@zju.edu.cn,
{wufei,yuzhuang}@cs.zju.edu.cn

Abstract. In this paper, we consider multi-modal retrieval from the perspective of deep textual-visual learning so as to preserve the correlations between multi-modal data. More specifically, We propose a general multi-modal retrieval algorithm to maximize the canonical correlations between multi-modal data *via* deep learning, which we call *Deep Textual-Visual correlation learning* (DTV). In DTV, given pairs of images and their describing documents, a convolutional neural network is implemented to learn the *visual* representation of images and a dependency-tree recursive neural network(DT-RNN) is conducted to learn compositional *textual* representations of documents respectively, then DTV projects the *visual-textual* representation into a common embedding space where each pair of multi-modal data is maximally correlated subject to being unrelated with other pairs by *matrix-vector* canonical correlation analysis (CCA). The experimental results indicate the effectiveness of our proposed DTV when applied to multi-modal retrieval.

Keywords: Multi-modal retrieval · Deep learning · CCA

1 Introduction

Nowadays, many real-world applications involve multi-modal data, where information inherently consists of data with different modalities, such as a web image with loosely related narrative text descriptions, and a historic news report with paired text and images. Therefore, multi-modal retrieval is imperative to many applications of practical interest, for examples, finding relevant textual documents of a tourist spot that best match a given image of the spot or finding a set of images that visually best illustrate a given text description.

However, The *heterogeneity-gap* between multi-modal data has been widely understood as a fundamental barrier to hinder multi-modal retrieval. To bridge this gap, one straightforward way is to map the multi-modal data into a common low-dimensional embedding space, with then the multi-modal retrieval can be conducted in the newly mapping space.

© Springer International Publishing Switzerland 2015
X. He et al. (Eds.): IScIDE 2015, Part I, LNCS 9242, pp. 176–185, 2015.
DOI: 10.1007/978-3-319-23989-7_19

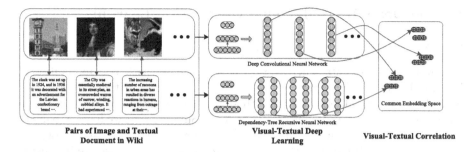

Fig. 1. The intuitive algorithmic flowchart of the proposed deep textual-visual correlation learning (DTV). Given pairs of images and their describing documents. Two neural networks(i.e. the convolutional neural network and the dependency-tree recursive neural network) are conducted to generate visual and textual representations. In the end, DTV projects them into a low-dimensional embedding space where the correlations between the paired multi-modal data are preserved.

To the best of our knowledge, there are generally two kinds of approaches to boost multi-modal retrieval: one is Canonical Correlation Analysis (CCA) [8] and its variants. CCA tends to map the multi-modal data into a common (or shared) space such that the distance between two similar objects is minimized, while the distance between two dissimilar objects is maximized. For example, after the maximally correlated subspaces of text and image features are obtained by CCA, a logistic regression is employed to multi-modal retrieval in [9]. Another kind of approach is Latent Dirichlet Allocation (LDA) [5] and its extensions. The LDA-based approaches attempts to model correlations among multi-modal data from a latent semantic (topic) level across different modalities, e.g. correspondence LDA [10].

Motivated by the recent advance that deep learning can identify and disentangle the underlying explanatory factors hidden in the observed milieu of low-level sensory data, this paper is interested in the discovery of multi-modal correlations for multi-modal retrieval *via* deep learning. We particularly consider multi-modal retrieval from the perspective of deep textual-visual learning so as to preserve the multi-modal correlations. We propose a general multi-modal retrieval algorithm to maximize the canonical correlations between multi-modal data via deep learning, which we call *Deep Textual-Visual correlation learning* (DTV). In DTV, given pairs of images and their describing documents, a convolutional neural network is implemented to learn the *vectorized visual* representation of images and a dependency-tree recursive neural network(DT-RNN) is conducted to learn compositional *matrix-based textual* representations of documents respectively, then DTV projects the *visual-textual* representation into a common embedding space where each pair of multi-modal data is maximally correlated subject to being unrelated with other pairs by *matrix-vector* canonical correlation analysis (CCA) (the intuitive flowchart of the proposed DTV is given in Fig. 1).

2   The Deep Textual-Visual Correlation Learning

2.1   Notations

To simplify our presentation, we use the special case with two modalities of data in this paper. However, our DTV has an inherent extension ability to more than two modalities. We name these two modalities $\mathcal{V}$ and $\mathcal{T}$ (e.g. visual images versus textual documents in this paper). Suppose that we have N pairs of visual images $\mathcal{V}$ and their describing textual documents $\mathcal{T}$,which are denoted as $\mathbf{V} = \{v_1, v_2, ..., v_N\}$ and $\mathbf{T} = \{t_1, t_2, ..., t_N\}$ respectively. Since we are interested with the representation of $\mathcal{V}$ and $\mathcal{T}$ *via* deep learning, we assume that the i-th image $v_i(1 \leq i \leq N)$ come from the original visual space (bag-of-visual-words (BoVW)used in this paper) and the i-th documents $t_i(1 \leq i \leq N)$ comes from the text space (bag-of-words (BoW)used in this paper).

2.2   Visual Feature Representation

Deep Convolutional Networks (DCN) have a long history in computer vision. More recently, DCN have achieved attractive competition-winning on large benchmark data sets such as ImageNet [18]. The deep convolutional networks consist of several convolutional filtering, local contrast normalization, and max-pooling layers, followed by several fully connected neural network layers trained using the dropout regularization technique. In this paper, the deep convolutional activation features (DeCAF) [3] are conducted to obtain the feature we used to represent each image. The idea of DeCAF is to transform learning from a well learned task to a generic task.

DeCAF is adopted from the deep convolutional neural network architecture proposed by [4]. Specifically, DeCAF contains 5 convolutional layers (with pooling and ReLU non-linearities) and 3 fully-connected layers. Different from [4], the images in DeCAF are warped into 256×256 resolution and the data augmentation trick of adding random multiples of the principle components of the RGB pixel values is removed.

In the end, the i-th image v_i is represented by a n_v-dimensional feature vector $V_i \in \mathbb{R}^{1 \times n_v}$, where n_v is the dimension of feature representation for each image via DCN (n_v equals to 4096 in this paper).

2.3   Textual Feature Representation

Given the i-th document t_i which is used to describe its corresponding image. We first select out the first 20 sentences from $t_i(t_i = \{t_1^i, \cdots, t_{20}^i\})$, where t_j^i ($1 \leq j \leq 20$) is the j-th sentence in t_i(For document which has less than 20 sentences, we first select all its sentences and then randomly select sentences from the document repeatedly, until 20 sentences are obtained). We assume t_j^i consists of m words, namely $t_j^i = (t_{j,1}^i, \cdots, t_{j,m}^i)$. For the k-th word $t_{j,k}^i$ ($1 \leq k \leq m$) in t_j^i, we represent $t_{j,k}^i$ in a vector by the word2vec model [1], which can generate a vector for each word keeping the semantic similarities between words. Moreover, in order to capture the *syntactic structure* in t^i_j, the dependency-tree recursive neural network(DT-RNN) [2] is conducted to transform each sentence t^i_j into a sentence-vector. In DT-RNN, each sentence t^i_j, after being parsed by using existing NLP methods [17] and a dependence tree of t^i_j, is thus denoted as an ordered list of (child,parent) indices. In the following, the corresponding vector representation of each word $t^i_{j,k}$ by [1] is denoted as $t^i_{j,k}$.

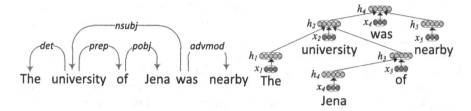

Fig. 2. Example of a dependency tree for the sentence "The university of Jena was nearby." On the left, we show the full dependency tree of the sentence. On the right, every word is firstly represented as a vector x_i. Then the parent vectors h_i are generated via a bottom up manner. The final representation is computed at the root node ('was')

Given the sentence t^i_j, using the word-vector $t^i_{j,k}$ and dependence list of t^i_j as input, the DT-RNN model will compute parent vectors at each node that include all the dependent (children) nodes in the dependence tree generated from t^i_j *via* a bottom up manner using the composition function g and a nonlinearity function $f = \tanh$ as follows:

$$h^i_{j,k} = f(g(t^i_{j,k})) \tag{1}$$

$$g(t^i_{j,k}) = \frac{1}{\ell(k)}(W_v t^i_{j,k} + \sum_{p \in C(k)} \ell(p) W_{pos(k,p)} h^i_{j,p}) \tag{2}$$

The hidden vector $h^i_{j,k}$ is the hidden layer representation of the word with index k in the j-th sentence of the i-th document for word vector $t^i_{j,k}$. The composition function g is a linear function parameterized with W_v and $W_{pos(k,p)}$. Matrix $W_{pos(k,p)}(p \in C(k))$ are used for composing with hidden child from the right and left sides: $W_{pos(k,p)} = W_{r_q}$, if p is the q-th child from right side of k; $W_{pos(k,p)} = W_{l_q}$, if p is the q-th child from left side of k. $C(k)$ is the set of children of node k, and $\ell(k)$ is the number of leaf nodes under node k. Here g is a linear function and its complexity depends on the number of children of the current node.

The calculation of hidden vector h_4 of an leaf node and h_2 of a inner node regarding to the example sentence in Fig. 2 are show below:

$$h_4 = f(g(x_4)) = f(W_v x_4) \tag{3}$$

$$h_2 = f(g(x_2)) = f(\frac{1}{\ell(2)}(W_v x_2 + \frac{1}{\ell(1)} W_{l_1} h_1 + \frac{1}{\ell(3)} W_{r_1} h_3)) \tag{4}$$

Finally, the n_t-dimensional hidden vector of the root node is used to represent the whole sentence t_j^i (n_t equals to 500 in this paper). Therefore, the hidden vector of the root node of the 20 sentences in t_i are stacked to generate the textual representation of the document, namely t_i is represented as $T_i \in \mathbb{R}^{20 \times n_t}$.

2.4   Deep Multi-Modal Mapping

Since each image is represented as one *vector* and each document is represented as one *matrix*, we conduct *matrix-vector* canonical correlation analysis, which can be regarded as a special case of 2D-CCA [6], to learn the maximum correlation between pairs of images and documents.

Given the pairs of document $t_i \in \mathbf{T}$ denoted as $X_i \in \mathbb{R}^{20 \times n_t}$ and image $v_i \in \mathbf{V}$ denoted as $Y_i \in \mathbb{R}^{1 \times n_v}$, the *matrix-vector*-CCA seeks left transforms l_t and l_v and right transforms r_t and r_v, which maximize the correlations between $l_t^T X r_t$ and $l_v^T Y r_v$ (here X and Y are the representation of all of pairs of documents and images respectively). The objective function to be maximized is given as follows:

$$\arg\max_{l_t, r_t, l_v, r_v} cov\left(l_t^T X r_t,\ l_v^T Y r_v\right)$$
$$s.t.\ var\left(l_t^T X r_t\right) = 1,\ var\left(l_v^T Y r_v\right) = 1. \tag{5}$$

Let $\tilde{X}_i$ and $\tilde{Y}_i$ equal to the following equations:

$$\tilde{X}_i = X_i - \frac{1}{N}\sum_{j=1}^N X_i$$
$$\tilde{Y}_i = Y_i - \frac{1}{N}\sum_{i=1}^N Y_i \tag{6}$$

Σ_{tv}^r and Σ_{tv}^l can be rewritten as the following equations, as well as Σ_{tt}^r, Σ_{vv}^r, Σ_{tt}^l and Σ_{vv}^l:

$$\Sigma_{tv}^r = \left\langle \tilde{X} r_t r_v^T \tilde{Y}^T \right\rangle = \frac{1}{N}\sum_{i=1}^N \tilde{X}_i r_t r_v^T \tilde{Y}_i^T$$
$$\Sigma_{tv}^l = \left\langle \tilde{X} l_t l_v^T \tilde{Y}^T \right\rangle = \frac{1}{N}\sum_{i=1}^N \tilde{X}_i l_t l_v^T \tilde{Y}_i^T \tag{7}$$

$$\Sigma_{tt}^r = \left\langle \tilde{X} r_t r_t^T \tilde{X}^T \right\rangle;\qquad \Sigma_{vv}^r = \left\langle \tilde{Y} r_v r_v^T \tilde{Y}^T \right\rangle;$$
$$\Sigma_{tt}^l = \left\langle \tilde{X}^T l_t l_t^T \tilde{X} \right\rangle;\qquad \Sigma_{vv}^l = \left\langle \tilde{Y}^T l_v l_v^T \tilde{Y} \right\rangle, \tag{8}$$

the optimization of Eq. 3 is equivalent to the following two eigenvalue problems:

$$\begin{bmatrix} 0 & \Sigma_{tv}^r \\ \Sigma_{vt}^r & 0 \end{bmatrix}\begin{bmatrix} l_t \\ l_v \end{bmatrix} = \lambda \begin{bmatrix} \Sigma_t^r t & 0 \\ 0 & \Sigma_{vv}^r \end{bmatrix}\begin{bmatrix} l_t \\ l_v \end{bmatrix}. \tag{9}$$

$$\begin{bmatrix} 0 & \Sigma_{tv}^l \\ \Sigma_{vt}^l & 0 \end{bmatrix}\begin{bmatrix} r_t \\ r_v \end{bmatrix} = \lambda \begin{bmatrix} \Sigma_t^l t & 0 \\ 0 & \Sigma_{vv}^l \end{bmatrix}\begin{bmatrix} r_t \\ r_v \end{bmatrix}. \tag{10}$$

After the convergence of Equation (9) and Equation (10) in a iterative manner, the left transforms(l_t and l_v) and the right transforms(r_t and r_v) can be determined (see [6]) for more detailed explanations). After optimizing the transforms above, we are able to project multi-modal data into a common embedding space. Algorithm 1 summarized our proposed DTV.

Algorithm 1. Algorithm 1 The multi-modal retrieval by DTV

Input: The two well trained neural networks, pairs of multi-modal data as training data and query data x_q.

1 Represent x_q as matrix T_x if x_q is a document query. Otherwise, represent x_q as image feature vector V_x.;

2 Optimize Equation 9 and Equation 10 by using the training data to get the left transforms and the right transforms.;

3 Project the multi-modal data into the common embedding space.;

4 Rank the neighbors of the query data in the common embedding space.;

Output: The retrieved similar data in response to x_q.

---

3 Experiments

We evaluate the performance of our proposed DTV on the Wiki data set [9], which is the largest available multi-modal dataset that are fully paired to the best of our knowledge. The Wiki data set [9] consists of 2866 images, each with a short paragraph describing the image. The images are labeled with exactly one of the 10 different semantic classes, such as art and geography. In the originally provided dataset, the text comes with a 10 dimensional feature vector representing the probabilistic proportions over the 10 topics. We extract bag-of-word feature vectors using the TF-IDF weighting scheme. For images, we first extract SIFT points from each image in the dataset. The randomly selected SIFT points are clustered by K-means to generate the visual dictionary. Then each image is quantized into histogram feature vector using the bag-of-visual words (BoVW) model.

The data set has been randomly reorganized to 2000/866 (training/testing) text-image pairs from different categories. In this paper, each document is originally denoted as 500-dimensional bag of textual words and each image is originally denoted as 1000-dimensional bag of visual words [14]. We use two different evaluation criteria as follows:

- Mean average precision (MAP), which is based on the retrieved ranking list of queries, is conducted as the evaluation criterion for our experiments. MAP is defined here to measure whether the retrieved data belongs to the same class as the query (*relevant*) or does not belong to the same class (*irrelevant*). Given a query (one image or one text) and a set of its corresponding R retrieved data, the Average Precision is defined as

$$AP = \frac{1}{L} \sum_{r=1}^{R} prec(r)\delta(r), \tag{11}$$

where L is the number of relevant data in the retrieved set, $prec(r)$ represents the precision of the r retrieved data. $\delta(r) = 1$ if the r-th retrieved datum is relevant to the query and $\delta(r) = 0$ otherwise. MAP is defined as the average AP of all the queries. In this paper we set R = 866, same as the number of test data.

- Since there is only one ground-truth match for each image/text, to evaluate the multi-modal performance we can resort to the position of the ground-truth text/image in the ranked list obtained. In general, one image/text is considered correctly retrieved if it appears in the first t percent of the ranked list of its corresponding retrieved texts/images [19]. We set $t = 0.2$ in our experiments.

The following state-of-art algorithms are chosen for comparisons:

- **CCA**: CCA maps pairs of documents(denoted as bag of textual words) and images (denoted as bag of visual words) into a latent space and therefore the latent representation of images and texts is obtained. After that, CCA performs multi-modal retrieval. CCA is compared here to validate whether the deep learning is effective.
- DTV^1: In DTV^1, each image is denoted as bag of visual words, same as CCA. However, each document is denoted as the feature representation by DT-RNN, same as DTV.
- DTV^2: In DTV^2, each image is denoted as the feature representation by DeCAF, same as DTV. However, each document is denoted as bag of textual words, same as CCA.

In our experiments, we can submit one image to retrieve documents (*image-query-texts*), or submit one document to retrieve images (*text-query-images*). We report the performance results on both directions of retrieving images from text queries (text-query-image) and retrieving text documents from image queries (image-query-text).

Table 1. The performance comparison in terms of MAP. The results shown in boldface are the best results.

Methods	*text-query-image*	*image-query-text*
CCA	0.125	0.118
DTV^1	0.128	0.124
DTV^2	0.125	0.118
DTV	**0.145**	**0.129**

Table 2. The performance comparison in terms of Percentage while $t = 0.2$. The results shown in boldface are the best results.

Methods	*text-query-image*	*image-query-text*
CCA	0.201	0.214
DTV^1	0.239	0.219
DTV^2	0.222	0.214
DTV	**0.255**	**0.226**

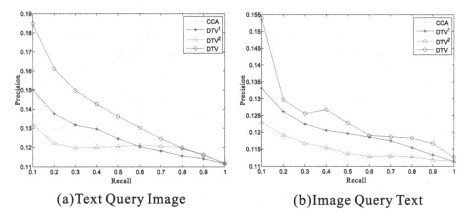

(a)Text Query Image   (b)Image Query Text

Fig. 3. Precision-Recall curves on different algorithms.

Fig. 4. Examples of *text-query-image* and *image-query-text* by CCA (top row) and DTV (bottom row). For the *text-query-image*, the query text comes from "Warfare" category and three of the top four retrieved images by DTV are from "warfare", much better than by CCA. For the example of *image-query-text*, the query image comes from the "Biology" category and the images retrieved by DTV preserve stronger multi-modal correlation, compared with the images retrieved by CCA.

Besides, to give an pictorial demonstration of each algorithms performance, the Precision-Recall curves are also reported on all the approaches. The performance in terms of MAP and Percentage by each algorithm are given in Table 1 and Table 2, and the Precision-Recall curves are reported in Fig. 3.

For both text-query-image retrieval and image-query-text retrieval, DTV outperforms other comparative algorithms. The reason is probably that DTV can capture the semantics of text queries (even long text queries) and boost the performance of semantic representation due to the consideration of syntactic structure in sentences.

Figure 4 illustrates one example of *image-query-text* and one example of *text-query-image* by CCA and our proposed DTV. From the experiments, we can observe that our proposed DTV can achieve the best performance due to its aptitude to preserve the multi-modal correlations *via* deep learning.

4   Conclusion

DTV is proposed in this paper for multi-modal retrieval. DTV can maximize the canonical correlations between multi-modal data *via* deep learning. In DTV multi-modal data are projected into a common low-dimensional embedding space, where multi-modal retrieval can easily be achieved. We have demonstrated the superior performance of DTV in terms of MAP and Percentage.

Acknowledgments. This work is supported in part by 973 Program (2012CB316400), NSFC (61402401), Zhejiang Provincial Natural Science Foundation of China (LQ14F01 0004), Chinese Knowledge Center of Engineering Science and Technology (CKCEST).

References

1. Mikolov, T., et al. Efficient estimation of word representations in vector space (2013). arXiv preprint arXiv:1301.3781
2. Socher, R., et al. Grounded Compositional Semantics for Finding and Describing Images with Sentences. NIPS Deep Learning Workshop (2013)
3. Donahue, J., et al. Decaf: A deep convolutional activation feature for generic visual recognition (2013). arXiv preprint arXiv:1310.1531
4. Krizhevsky, A., Ilya, S., Geoffrey, E.H.: Imagenet classification with deep convolutional neural networks. In: Advances in Neural Information Processing Systems (2012)
5. Blei, D.M., Andrew, Y.N., Michael, I.J.: Latent dirichlet allocation. J. mach. Learn. Res. **3**, 993–1022 (2003)
6. Lee, S.H., Seungjin, C.: Two-dimensional canonical correlation analysis. IEEE Signal Process. Lett. **14**(10), 735–738 (2007)
7. Siagian, C., Itti, L.: Rapid biologically-inspired scene classification using features shared with visual attention. IEEE Trans. Pattern Anal. Mach. Intell. **29**(2), 300–312 (2007)
8. Hotelling, H.: Relations between two sets of variates. Biometrika **28**, 321–377 (1936)

9. Rasiwasia, N., et al. A new approach to cross-modal multimedia retrieval. In: Proceedings of the International Conference on Multimedia. ACM (2010)
10. Blei, D.M., Jordan, M.I.: Modeling annotated data. In: Proceedings of the 26th Annual International ACM SIGIR Conference on Research and Development in Informaion Retrieval, pp. 127–134. ACM (2003)
11. Salomatin, K., Yiming, Y., Abhimanyu, L.: Multi-field Correlated Topic Modeling. SDM (2009)
12. Putthividhy, D., Hagai, T.A., Srikantan, S.N.: Topic regression multi-modal latent dirichlet allocation for image annotation. In: 2010 IEEE Conference on Computer Vision and Pattern Recognition (CVPR). IEEE (2010)
13. Zhuang, Y., et al. Supervised Coupled Dictionary Learning with Group Structures for Multi-modal Retrieval. AAAI (2013)
14. Fei-Fei, L., Fergus, R., Perona, P.: Learning generative visual models from few training examples: an incremental bayesian approach tested on 101 object categories. Comput. Vis. Image Underst. **106**(1), 59–70 (2007)
15. Zhen, Y., Dit-Yan, Y.: A probabilistic model for multimodal hash function learning. In: Proceedings of the 18th ACM SIGKDD International Conference on Knowledge Discovery and Data Mining. ACM (2012)
16. Lowe, D.G.: Distinctive image features from scale-invariant keypoints. Int. J. Comput. Vis. **60**(2), 91–110 (2004)
17. De Marneffe, M.-C., MacCartney, B., Manning, C.D.: Generating typed dependency parses from phrase structure parses. In: Proceedings of LREC, vol. 6 (2006)
18. Deng, Jia., et al. Imagenet: A large-scale hierarchical image database. In: IEEE Conference on Computer Vision and Pattern Recognition, CVPR 2009. IEEE (2009)
19. Jia, Y., Mathieu, S., Trevor, D.: Learning cross-modality similarity for multinomial data. In: 2011 IEEE International Conference on Computer Vision (ICCV). IEEE (2011)

Bidirectional Covariance Matrices: A Compact and Efficient Data Descriptor for Image Set Classification

Jieyi Ren[1] and Xiaojun Wu[2(✉)]

[1] School of Digital Media, Jiangnan University, Wuxi 214122, China
alvisland@gmail.com
[2] School of IOT Engineering, Jiangnan University, Wuxi 214122, China
xiaojun_wu_jnu@163.com

Abstract. Symmetric Positive Definite (SPD) matrices have been widely used in many computer vision tasks. Recently, there are growing interests in applying covariance matrices to image set classification due to their benefit of encoding image features as a data descriptor. Since SPD matrices follow a non-linear Riemannian geometry, exploiting an appropriate Riemannian metric is the key to successful classification. Adopting Riemannian metrics to classify covariance matrices of image sets is nontrivial, since such matrices are usually singular matrices. Besides, the computational complexity is intolerable while dealing with high dimensional covariance matrices. This paper proposes to use bidirectional covariance matrices instead of covariance matrices as a data descriptor. We model image sets both from the row and column directions of images and these bidirectional covariance matrices are proved to be compact and efficient. Improved accuracy and efficiency are obtained through experiments on standard datasets for comparing bidirectional covariance matrices with covariance matrices.

Keywords: Riemannian manifold · Covariance matrices · Image set classification

1 Introduction

Besides being described in terms of feature vectors, images can be represented as many mathematical entities which do not form vector spaces, but reside on non-linear manifolds in the field of computer vision. For instance, Symmetric Positive Definite (SPD) matrices have recently obtained significant results for several computer vision tasks such as object recognition [1], face recognition [2], texture classification [3], visual surveillance [4], human tracking [5], etc.

As a data descriptor, covariance matrices model the second-order statistics of image features and result in SPD matrices, which can be applied to various applications. Compared with other vectorial data descriptors such as Bag-of-Words (BoW) and Fisher Vectors, covariance matrices are superior in many aspects. For example, different features can be fused into a covariance matrix compactly, ignoring the huge

X. He et al. (Eds.): IScIDE 2015, Part I, LNCS 9242, pp. 186–195, 2015.
DOI: 10.1007/978-3-319-23989-7_20

number of data samples. This kind of data fusion is robust to noise, illumination changes and rigid or non-rigid deformations of images [6].

Learning methods based on SPD matrices usually suffer from the problem of how to model and compute the SPD matrices efficiently. It is well known that the space of a $d \times d$ SPD matrix (denoted by Sym_d^+) is not a linear space but a non-linear Riemannian manifold [7] so that conventional mathematical modeling methods in the Euclidean space are not suitable [8]. In image set classification, previous works mainly model image sets in several prevalent ways: Gaussian [9] or Gaussian mixture models (GMM) [10], subspace learning methods [11] and nearest pair of points matching methods [12]. More recently, some researches have shown encouraging results by modelling image sets with their global covariance matrices [13–15], thus the classification problem can be formulated as classifying points on a Riemannian manifold. These points are not connected with straight lines, but geodesics along the curvature of the manifold. Therefore, appropriate Riemannian metrics are needed for calculating distance or similarity between different image sets. In most of the cases, covariance matrices of image sets are singular matrices, which means it is unable to apply Riemannian metrics directly. Moreover, the computational complexity of high dimensional covariance matrices is expensive. By simultaneously considering the row and column directions of images, we propose to replace covariance matrices with bidirectional covariance matrices for overcoming their drawbacks. The benefits of bidirectional covariance matrices as a data descriptor are demonstrated through experiments in image set classification. Our experimental results show that the bidirectional covariance matrices have a clear superiority over the covariance matrices.

2   Background

This section provides an overview on Riemannian geometry of the SPD matrices manifold and some popular Riemannian metrics.

2.1   Riemannian Manifold of SPD Matrices

In mathematics, a manifold is a topological space that resembles Euclidean space near each point. More precisely, each point of the manifold has a tangent space, which is a vector space that consists tangent vectors of all possible curves passing through the corresponding point. A Riemannian manifold is a smooth manifold equipped with an inner product on each tangent space. The family of these inner products is called as Riemannian metric. A Riemannian metric makes it possible to define various geometric notions on a Riemannian manifold, such as angles, lengths of curves, etc.

A SPD matrix is a symmetric matrix with the property that all its eigenvalues are positive. As studied in [1], Sym_d^+ is a specific Riemannian manifold. The geodesics between points on the Riemannian manifold specify the length of the shortest curves that connect these points, which are similar to straight lines in $\mathbb{R}^n$. Obviously, the geodesic induced by Riemannian metric is a more proper measurement for comparing similarity between two SPD matrices than the Euclidean distance.

2.2   Riemannian Metric

Although a number of metrics on Sym_d^+ have been proposed, not all of them define a true geodesic on the manifold [16]. There are two most popular ones: Affine-Invariant Distance (AID) [17] and Log-Euclidean Distance (LED) [7].

The first metric AID is defined with two SPD matrices X_1 and X_2 as:

$$d_{AID}(X_1, X_2) = \sqrt{\sum_{i=1}^{d} \ln^2 \lambda_i(X_1, X_2)}, \tag{1}$$

where $\lambda_i(X_1, X_2)(i = 1, \ldots, d)$ are the eigenvalues computed from $|\lambda X_1 - X_2| = 0$. This metric is invariant to affine transformations and inversions [1] and it defines a true geodesic on Sym_d^+, but with high computational costs in practical applications.

Another metric LED for Sym_d^+ is expressed by classical Euclidean computations as:

$$d_{LED}(X_1, X_2) = \|\log(X_1) - \log(X_2)\|_F, \tag{2}$$

where $X_1, X_2 \in Sym_d^+$ and F denotes the matrix Frobenius form, log is the common matrix logarithm operator. For a SPD matrix X, its eigen-decomposition is given by $X = U \sum U^T$ and the logarithm is defined as:

$$\log(X) = U \log\left(\sum\right) U^T. \tag{3}$$

The LED metric also defines a true geodesic on Sym_d^+, and it is simple to use compared with AID metric. Nevertheless, both AID and LED are computationally expensive when the dimensionalities of SPD matrices are high. More detail information about the two metrics can be found in [7].

3   Proposed Bidirectional Descriptor

Covariance matrices have shown notable performance improvement over other modelling methods in image set classification [13, 14]. In this section, we propose bidirectional covariance matrices based learning strategy, which address the shortcomings of the covariance matrices.

3.1   Bidirectional Covariance Matrix

As mentioned in Sect. 1, an image set can be modelled with its global covariance matrix. Given an image set S with n samples, $S = [s_1, s_2, \ldots, s_n]$. S_i denotes the i-th samples with a d-dimensional feature vector obtained by lexicographic ordering of the pixel elements of the i-th sample as a $w \times h$ matrix, $d = w \times h$. Instead of pixel values, the vector may also contain feature values such as LBP or Gabor features. Thus, the image set S can be represented with its $d \times d$ covariance matrix:

$$C = \frac{1}{n-1} \sum_{i=1}^{n} (s_i - \bar{s})(s_i - \bar{s})^T, \tag{4}$$

and $\bar{s}$ is the mean of image samples. If the number of images n in image set S is less than the dimensionality d of the feature vector, C will be a semi positive definite matrix rather than a SPD matrix. This situation is commonly encountered in image set classification, so that both the two metrics introduced in Sect. 2.2 cannot be applied directly to calculate the distance between different image sets through Eqs. (1)–(3). To tackle this singularity problem, regularization can be applied to the original covariance matrix C as $C' = C + \lambda I$. I is an identity matrix of the same size as C and λ is set to $0.001 * trace(C)$. It is important to know that this additional perturbation could lead to deviations in the classification phase, especially when the size of C is large.

Motived by [18], we model an image set with its samples in the form of matrices rather than vectors. Consider an image set S with n samples, $S = [x_1, x_2, \ldots, x_n]$. x_i denotes the i-th samples with a $w \times h$ matrix. Without the matrix-to-vector conversion, the covariance matrix is defined as:

$$C_r = \frac{1}{n-1} \sum_{i=1}^{n} (x_i - \bar{x})^T (x_i - \bar{x}), \tag{5}$$

and $\bar{s}$ is the mean of image samples as a $w \times h$ matrix. From [19] we know that C_r is working in the row direction of images, and an alternative covariance matrix working in the column direction of image is defined as:

$$C_c = \frac{1}{n-1} \sum_{i=1}^{n} (x_i - \bar{x})(x_i - \bar{x})^T. \tag{6}$$

In contrast to the covariance matrix C in Eq. (4), the sizes of the covariance matrices C_r and C_c are quite smaller. C_r is a $h \times h$ matrix and C_c is a $w \times w$ matrix. The singularity problem here could be negligible, since $n \gg w$ and $n \gg h$ in most cases. As a result, C_r and C_c model the image set more accurately than C, with less time required to calculate the distance.

3.2 Learning with Proposed Descriptor

The fundamental thought of our learning strategy based on bidirectional covariance matrices could be summarized into two major points: (1) Represent image sets with bidirectional covariance matrices in the modelling phase; (2) Fuse bidirectional covariance matrices together in the classification phase. It is easy to extend existing image set classification methods with our bidirectional covariance matrices and just few changes are needed. Here we take [13] for example and introduce the main steps briefly.

Suppose there are m gallery image sets for training and l probe image sets for testing, which are all from c classes. First we compute the corresponding bidirectional covariance matrices through Eqs. (5) and (6). Two kernel matrices K_r and K_c are calculated with training image sets respectively by a Riemannian kernel based on LED metric, then fed into Kernel Discriminant Analysis (KDA) [20] to solve the optimization problem. In the testing phase, both training and testing image sets are projected to the $c - 1$ dimensional discriminant subspace. Nearest-Neighbor (NN) classification is then conducted based on Euclidean distance. Specially, the Euclidean distance here is the sum of distance both from the row and column directions, which is a fused bidirectional distance actually.

4   Experiments

To evaluate the proposed bidirectional data descriptor, extensive experiments are performed on four standard datasets for three computer vision tasks: face recognition, object categorization and scene classification. Face recognition experiments are performed on Honda/UCSD [21] and YouTube Celebrities [22] datasets. Experiments for object categorization are performed on ETH-80 [23] dataset. Fifteen Scene Categories [24] dataset is used for scene classification experiments. Literature [13] is chose as a benchmark method to compare both accuracy and execution time complexity of bidirectional covariance matrices and covariance matrices. For convenience, we denote the original method with covariance matrices by CDL and the improved method with bidirectional covariance matrices by BDCDL in the following sections.

4.1   Datasets and Presets

The Honda/UCSD dataset contains 59 video sequences of 20 different persons. Each video contains approximately 300–500 frames covering large variations in head pose and facial expression. The YouTube Celebrities has 1,910 video clips of 47 subjects collected from YouTube. Each clip contains hundreds of frames, which are mostly low resolution and highly compressed. For both datasets, a cascaded human face detector [25] was used to detect face in every frame of the videos automatically, and then each face images were cropped and converted to gray scale. Histogram equalization was the only pre-processing used to eliminate lighting effects. Each video generated an image set of faces. Sample face images from both datasets are shown in Fig. 1.

Fig. 1. Samples in Honda and YouTube datasets

ETH-80 dataset contains images of 8 object categories and each category has 10 objects. Each object has 41 images of different views (Fig. 2).

Fig. 2. 8 object categories in the ETH-80 dataset

The Fifteen Scene Categories dataset is a dataset of fifteen natural scene categories and each category has about 200–400 images. The major sources of the images in the dataset include the COREL collection, personal photographs, and Google image search. It should be noted that the results on this dataset are used to show the advance of our proposed descriptor, instead of compare with other state-of-the-art nature scene classification methods (Fig. 3).

Fig. 3. Scene images in the fifteen scene categories dataset

To allow fair comparison, ten-fold cross validation experiments (randomly selected gallery/probe image sets combinations) were conducted on all four datasets to obtain average results, and image resolutions from 16 × 16 to 24 × 24 (increase 2 pixels per time) were adopted to estimate performance under different image intensity.

For Honda dataset, each person had one image set as the gallery and the rest for probes. The single gallery set was randomly divided into two non-overlapping subsets and each subset was used as a full image set. For YouTube dataset, one person had 3 randomly chosen image sets for gallery and 6 for probes. For ETH-80 dataset, each category had 5 objects for gallery and other 5 for probes. For Fifteen Scene Categories dataset, each category was divided into five subsets and 3 for gallery and other 2 for probes. For covariance matrices, regularization was applied to avoid singularity, while there was no need to do that for bidirectional covariance matrices, as stated in Sect. 3.1.

4.2   Results and Analysis

The results of both two methods on all datasets are summarized in Table 1. In the face recognition experiments, BDCDL is better than CDL with all image resolutions. Furthermore, the accuracy of CDL keeps descending when the image resolution increases, while the accuracy of BDCDL remains stable. This result confirms our conjecture about the deviation in Sect. 3.1. CDL outperforms BDCDL on YouTube dataset, but the disparity is tiny. Since YouTube dataset contains a huge number of image sets, BDCDL might achieve the same or even better result than CDL when the image resolution is large enough. In the object categorization experiments on ETH-80 dataset, BDCDL reports higher accuracy than CDL too. CDL only obtains better result than BDCDL when the image resolution is 16×16. In the scene classification experiments on Fifteen Scene Categories dataset, BDCDL outperforms CDL again, which is similar to the situation in the face recognition experiments.

Table 1. Average recognition rates of both methods

Method	Honda	YouTube	ETH-80	Fifteen Scene
CDL (16px)	**0.987**	**0.642**	**0.921**	**0.692**
CDL (18px)	0.982	0.629	0.903	0.661
CDL (20px)	0.971	0.618	0.903	0.652
CDL (22px)	0.967	0.636	0.909	0.672
CDL (24px)	0.964	0.617	0.911	0.653
BDCDL (16px)	0.996	0.547	0.901	0.739
BDCDL (18px)	0.999	0.564	0.918	0.741
BDCDL (20px)	1	0.588	0.924	0.743
BDCDL (22px)	1	0.597	0.932	0.764
BDCDL (24px)	1	**0.602**	**0.942**	**0.794**

The computational complexities of both two methods were compared with You-Tube dataset. The time cost for each method is showed in Table 2. The superiority of BDCDL can be clearly observed, especially for high image resolution. The time cost by CDL is over 10 times slower than that in BDCDL. As stated in Sect. 3.1, smaller sizes of bidirectional covariance matrices than that of covariance matrices lead to this result.

Table 2. Execution times on YouTube dataset (one validation)

Method	Training (s)	Testing (s)
CDL	2.7	8.1
BDCDL	**0.19**	**0.49**

After comparing all results above, there are two following observations: (1) Among all experiments, bidirectional covariance matrices are generally superior to covariance matrices as a data descriptor both in classification accuracy and execution time complexity; (2) A proper data intensity is important for using bidirectional covariance matrices, extremely low image resolution may leads to an unsatisfied result.

5 Conclusion

In this paper, a compact and efficient data descriptor for image set classification has been proposed. Bidirectional covariance matrices consider both the row and column directions of images and model the image set more accurately than covariance matrices, with less time required for classification. We made extensive experiments on standard datasets and the results demonstrated the superiority of bidirectional covariance matrices over covariance matrices in terms of accuracy and efficiency, as well as the robustness to the varying image resolution.

Acknowledgments. This work was supported in part by the project of NSFC (No. 61373055) and the Research Project on Surveying and Mapping of Jiangsu Province (No. JSCHKY201109).

References

1. Tuzel, O., Porikli, F., Meer, P.: Region covariance: a fast descriptor for detection and classification. In: Leonardis, A., Bischof, H., Pinz, A. (eds.) ECCV 2006. LNCS, vol. 3952, pp. 589–600. Springer, Heidelberg (2006)
2. Pang, Y., Yuan, Y., Li, X.: Gabor-based region covariance matrices for face recognition. IEEE Trans. Circ. Syst. Video Technol. **18**, 989–993 (2008)
3. Sivalingam, R., Boley, D., Morellas, V., Papanikolopoulos, N.: Tensor sparse coding for region covariances. In: Daniilidis, K., Maragos, P., Paragios, N. (eds.) ECCV 2010, Part IV. LNCS, vol. 6314, pp. 589–600. Springer, Heidelberg (2010)
4. Cherian, A., Morellas, V., Papanikolopoulos, N., Bedros, S.J.: Dirichlet process mixture models on symmetric positive definite matrices for appearance clustering in video surveillance applications. In: IEEE Conference on Computer Vision and Pattern Recognition, pp. 3417–3424. IEEE (2011)
5. Porikli, F., Tuzel, O., Meer, P.: Covariance tracking using model update based on lie algebra. In: IEEE Conference on Computer Vision and Pattern Recognition, pp. 728–735. IEEE (2006)
6. Ma, B., Wu, Y., Sun, F.: Affine object tracking using kernel-based region covariance descriptors. In: Wang, Y., Li, T. (eds.) Foundations of Intelligent Systems, pp. 613–623. Springer, Heidelberg (2012)

7. Arsigny, V., Fillard, P., Pennec, X., Ayache, N.: Geometric means in a novel vector space structure on symmetric positive-definite matrices. SIAM J. Matrix Anal. Appl. **29**, 328–347 (2007)
8. Xie, Y., Vemuri, B.C., Ho, J.: Statistical analysis of tensor fields. In: Jiang, T., Navab, N., Pluim, J.P.W., Viergever, M.A. (eds.) MICCAI 2010, Part I. LNCS, vol. 6361, pp. 682–689. Springer, Heidelberg (2010)
9. Shakhnarovich, G., Fisher III, J.W., Darrell, T.: Face recognition from long-term observations. In: Heyden, A., Sparr, G., Nielsen, M., Johansen, P. (eds.) ECCV 2002, Part III. LNCS, vol. 2352, pp. 851–865. Springer, Heidelberg (2002)
10. Arandjelovic, O., Shakhnarovich, G., Fisher, J., Cipolla, R., Darrell, T.: Face recognition with image sets using manifold density divergence. In: IEEE Conference on Computer Vision and Pattern Recognition, pp. 581–588. IEEE (2005)
11. Hamm, J., Lee, D.D.: Grassmann discriminant analysis: a unifying view on subspace-based learning. In: 25th International Conference on Machine Learning, pp. 376–383. ACM (2008)
12. Hu, Y., Mian, A.S., Owens, R.: Sparse approximated nearest points for image set classification. In: IEEE Conference on Computer Vision and Pattern Recognition, pp. 121–128. IEEE (2011)
13. Wang, R., Guo, H., Davis, L.S., Dai, Q.: Covariance discriminative learning: a natural and efficient approach to image set classification. In: IEEE Conference on Computer Vision and Pattern Recognition, pp. 2496–2503. IEEE (2012)
14. Uzair, M., Mahmood, A., Mian, A., McDonald, C.: A compact discriminative representation for efficient image-set classification with application to biometric recognition. In: International Conference on Biometrics, pp. 1–8. IEEE (2013)
15. Huang, Z., Wang, R., Shan, S., Chen, X.: Hybrid Euclidean-and-Riemannian metric learning for image set classification. In: Cremers, D., Reid, I., Saito, H., Yang, M.-H. (eds.) ACCV 2014. LNCS, vol. 9005, pp. 562–577. Springer, Heidelberg (2015)
16. Jayasumana, S., Hartley, R., Salzmann, M., Li, H., Harandi, M.: Kernel methods on the Riemannian manifold of symmetric positive definite matrices. In: IEEE Conference on Computer Vision and Pattern Recognition, pp. 73–80. IEEE (2013)
17. Förstner, W., Moonen, B.: A metric for covariance matrices. In: Grafarend, E.W., Krumm, F.W., Schwarze, V.S. (eds.) Geodesy-The Challenge of the 3rd Millennium, pp. 299–309. Springer, Heidelberg (2003)
18. Yang, J., Zhang, D., Frangi, A.F., Yang, J.Y.: Two-dimensional PCA: a new approach to appearance-based face representation and recognition. IEEE Trans. Pattern Anal. Mach. Intell. **26**, 131–137 (2004)
19. Zhang, D., Zhou, Z.H.: (2D) 2PCA: two-directional two-dimensional PCA for efficient face representation and recognition. Neurocomputing **69**, 224–231 (2005)
20. Baudat, G., Anouar, F.: Generalized discriminant analysis using a kernel approach. Neural Comput. **12**, 2385–2404 (2000)
21. Lee, K.C., Ho, J., Yang, M.H., Kriegman, D.: Video-based face recognition using probabilistic appearance manifolds. In: IEEE Conference on Computer Vision and Pattern Recognition, pp. 313–320. IEEE (2003)
22. Kim, M., Kumar, S., Pavlovic, V., Rowley, H.: Face tracking and recognition with visual constraints in real-world videos. In: IEEE Conference on Computer Vision and Pattern Recognition, pp. 1–8. IEEE (2008)
23. Leibe, B., Schiele, B.: Analyzing appearance and contour based methods for object categorization. In: IEEE Conference on Computer Vision and Pattern Recognition, pp. II-409-15. IEEE (2003)

24. Lazebnik, S., Schmid, C., Ponce, J.: Beyond bags of features: spatial pyramid matching for recognizing natural scene categories. In: IEEE Conference on Computer Vision and Pattern Recognition, pp. 2169–2178. IEEE (2006)
25. Viola, P., Jones, M.J.: Robust real-time face detection. Int. J. Comput. Vision **57**, 137–154 (2004)

Image Denoising Using Modified Nonsubsampled Contourlet Transform Combined with Gaussian Scale Mixtures Model

Chunman Yan[1][✉], Kaibing Zhang[2], and Yunping Qi[1]

[1] College of Physics and Electronic Engineering, Northwest Normal University,
Lanzhou 730070, Gansu, China
yancha02@163.com, ypqi@nwnu.edu.cn
[2] School of Computer and Information Science, Hubei Engineering University,
Xiaogan 432000, China
kbzhang@hbeu.edu.cn

Abstract. Nonsubsampled contourlet transform (NSCT) combined with Gaussian scale mixtures model (GSM) has been recognized as an excellent method for image denoising. However, the processing performance of this method is highly relied on the performance of nonsubsampled directional filter bank (NSDFB) applied in NSCT. In this paper, we employ a lifting scheme to develop a new directional filter bank (DFB). The new DFB is adopted to improve the original NSDFB for a highly efficient NSCT. By combining with the GSM, the improved NSCT is particularly propitious for image denoising. The experimental results show that the modified NSCT significantly outperforms the traditional NSCT in processing performance while preserving good visual quality of denoised images.

Keywords: Image denoising · Nonsubsampled contourlet transform · Gaussian scale mixtures model

1 Introduction

During the process of capturing, processing, and storing an image, it is inevitable to introduce noise into the image. A noised image may affect the performance of the subsequent processing tasks such as image compression and feature extraction. Therefore, image denoising remains an important problem in image understanding and computer vision and has been becoming one of the most active topics on image processing.

Generally, conventional image denoising methods can be divided into space filtering and frequency filtering methods. The space filtering methods directly

C. Yan—This work is supported by the National Natural Science Foundation of China under Grant Nos. 61471161, 41461078, and 61367005, and by the Fundamental Research Funds for the Universities of Gansu Province.

© Springer International Publishing Switzerland 2015
X. He et al. (Eds.): IScIDE 2015, Part I, LNCS 9242, pp. 196–207, 2015.
DOI: 10.1007/978-3-319-23989-7_21

performis denoising on the neighborhood pixels of the target pixel in the original image; while the frequency filtering first transform the original image into the frequency representation and then carry out the filtering process on the high subbands based upon the principle that the noise coefficients are found mainly escaped into these subbands. Frequency filtering also belongs to the transform domain technique for image denoising. With the development of wavelet transform, a wide variety of denoising methods based on signal sparse representation on wavelet domain have been proposed (e.g. threshold shrinking algorithms [1–3]). Derived from wavelet transform, a new sparse representation theory called multiscale geometric analysis (MGA) is popular [4]. Due to the multidirection and the anisotropy of their basis function, MGA can provide more sparse representation for 2-D image and can achieve better denoising performance than wavelets. Besides the MGA using sparse representation over off-the-shelf dictionary, signal sparse representation based on online dictionary learning methods has attracted extensive attention [5,6]. The representative work is the KSVD algorithm for image denoising [7]. In addition, many image denoising algorithms that employ the statistical learning and spatial correlation have been presented and shown their excellent denoising performance, including the Gaussian scale mixtures model (GSM) in the Wavelet Domain [8,9], the non local mean filtering [10], and so on. Recently, combing sparse representation and image statistical model for image denoising has attracted great attention. In particular, Zhou et al. [11] presented an image denoising method by combining the nonsubsampled contourlet transform (NSCT) and the GSM, leading to the state-of-art results.

As a key preprocessing step, image denoising approaches should take the following factors into account: (1) denoised image quality; (2) processing efficiency; (3) compatibility of the algorithm. The denoised image quality evaluation is a systemic problem in that a credible evaluation system should be a comprehensive one which depends on both subjective results and the objective criteria. It is widely acknowledged that only one objective index (e.g. PSNR) is insufficient to evaluate image quality. Secondly, the processing efficiency, namely the speed of denoising algorithm should be considered while pursuing high denoised effect. Additionally, as a key preprocessing step of image processing, the denoising algorithm should fully consider the compatibility with the following procedures (e.g. compressing, feature extraction, information fusion).

Taking the above factors into consideration, this paper chooses the algorithm proposed in [11,12] for further study. As an excellent decomposition method of MGA for image sparse representation, the NSCT is shift-invariant. Moreover, the denoising algorithms based on NSCT have good compatibility. Because of the sparsity and the redundancy of the NSCT coefficients, the denoised coefficients can be directly used for image compressing, feature extraction, information fusion and image inpainting, etc. However, because of the high redundancy of NSCT coefficients are computationally intensive and therefore are not beneficial to its practical applications. To reduce the computational cost, this paper analyzes the structure of NSCT and adopts a lifting scheme for directional filter bank (DFB) to improve the nonsubsampled directional filter bank (NSDFB) of the NSCT. We combine the improved NSCT with the GSM for image denoising.

The experimental results show that our algorithm outperforms other algorithms in most cases, and maintains comparable quality to the original method while keeping the efficiency of about 11 times.

The remainder of the paper is organized as follows. Section 2 briefly reviews the NSCT and analyzes the factors that affect the performance of it, and then presents an optimized DFB to improve the original NSCT. Section 3 introduces the GSM and its implementation for image processing. Section 4 develops the application of the modified NSCT that combines GSM for image denoising. The experimental results and analysis are presented in Sect. 5, and Sect. 6 concludes the paper.

2   Analysis and Improvement of Nonsubsampled Contourlet Transform

2.1   Nonsubsampled Contourlet Transform (NSCT)

The contourlet transform associated with multiscale, multidirection, and anisotropy uses contour segment like function, to approximate a natural image at multiple scales and multiple directions [13]. There are two filter banks used for contourlet transform: the Laplacian pyramid (LP) transform for multiscale decomposition and the directional filter bank (DFB) for directional subbands at high scale. Therefore, the contourlet expansion is also called as pyramidal directional filter bank (PDFB).

The contourlet transform is constructed by cascade LP and DFB. In the process of decomposition and reconstruction, downsampling and upsampling may cause spectral aliasing and thus it is not shift-invariant. By contrast, the NSCT uses the nonsubsampled laplacian pyramid filter bank (NSPFB) and the nonsubsampled directional filter bank (NSDFB) to replace the old LP and DFB [14,15]. The NSCT is shift-invariant because of the nonsubsampling process. Accordingly, the NSCT employs a double filter bank structure, i.e., the NSPFB is used for multiscale decomposition, followed by NSDFB for directional subbands at high scale, when this process iterates at the coarse scale, then the multiscale and multidirection decomposition for an image can be obtained. Although the coefficients of the NSCT are redundant, they are very useful for many image processing tasks (e.g. image denoising).

2.2   Improvement for NSDFB of NSCT

The filter design of the NSCT comprises two filter banks: NSPFB and NSDFB. Because the NSDFB is the major factor determining the performance of the NSCT, this section will deal with the design and optimization of the NSDFB for an computationally efficient NSCT.

The NSDFB is efficiently implemented via an l level binary tree decomposition, which leads to 2^l subbands with wedge-shaped frequency partitioning. The seminal work of the DFB which uses quincunx filter banks (QFB) with diamond-shaped filters was proposed by Bamberger *et al.* [15]. For simplifying and improving the performance of frequency partitioning, Do *et al.* [13,16] proposed the improved DFB which uses the QFB and obtains an ideal frequency partition while not modulating the input image. In this paper, we adopt an optimized QFB [17] to improve the NSDFB for a fast NSCT. For a QFB, diamond filter pair and fan filter pair are used for signal frequency partition, where the first one splits the frequency into low and high subbands and the second one obtains vertical and horizontal subbands. We can obtain one filter pair from the other by simply modulating. Thus there is only one problem for the filter design.

For a high performance QFB, it should be perfect reconstruction (PR), high coding gain, good frequency selectivity, certain prescribed vanishing moments properties, and linear-phase. The PR property ensures high PSNR of reconstructed image or lossless compression. The linear-phase property helps to avoid phase distortion. The presence of vanishing moments can reduce the number of nonzero coefficients in the highpass subbands and get smoother synthesis basis functions. Good frequency selectivity serves to minimize aliasing in the subband signals. Compared with other properties, designing nonseparable two-dimensional filter banks is an extremely challenging task [17].

Lifting scheme is valuable for filter design [18–20]. The PR condition can be naturally satisfied by its parameterization. The linear phase condition can easily be imposed, and this reversible integer-to-integer transform is useful for image processing. Thus, unlike the traditional structure, this paper adopts the lifting scheme proposed in [17] .

A QFB consists of the lowpass and highpass analysis filters $\{H_0(z), H_1(z)\}$ and lowpass and highpass synthesis filters $\{G_0(z), G_1(z)\}$, respectively. Given the coefficients of lifting filter $\{A_k(z)\}$, the corresponding transfer function of the analysis filter can be calculated, and the transfer function of the synthesis filters can obtained by the PR condition, calculating from the function $G_k(z) = (-1)^{1-k} z_0^{-1} H_{1-k}(-z)$. Thus, the task of filter bank design is only to construct $\{H_0(z), H_1(z)\}$, and calculate the coefficients of lifting filter $\{A_k(z)\}$. We need to parameterize the quincunx filter bank and the analysis side of the filter bank is only considered as follows.

In order to parameterize the QFB, two important constraint conditions should be considered: one is frequency response and the other is vanishing moments. According to the lifting property, the needs for additional constraints during optimization for PR and linear phase can be eliminated.

(1) Frequency Response. Let x be a vector consisting of all the independent coefficients of lifting filter $\{A_k\}$, $x = \begin{bmatrix} a_1^T & a_2^T & ... & a_{2\lambda}^T \end{bmatrix}^T$, and the number of independent coefficients in $\{A_k\}$ is $2l_{k,0}l_{k,1}$. The frequency responses of the analysis filters is defined as:

$$\begin{bmatrix} \hat{h}_0(\omega) \\ \hat{h}_1(\omega) \end{bmatrix} = \left(\sum_{k=1}^{\lambda} \left(\begin{bmatrix} 1 & e^{-j\omega_0} a_{2k}^T V_{2k} \\ 0 & 1 \end{bmatrix} \right. \right. \\ \left. \left. \times \begin{bmatrix} 1 & 0 \\ e^{j\omega_0} a_{2k-1}^T V_{2k-1} & 1 \end{bmatrix} \right) \right) \begin{bmatrix} 1 \\ e^{j\omega_0} \end{bmatrix} \quad (1)$$

where V_{2k} is a vector with $2l_{2k,0}l_{2k,1}$ elements, V_{2k-1} is a vector with $2l_{2k-1,0}$ $l_{2k-1,1}$ elements, and the n-th element of them is given by:

$$V_{2k}[n] = 2\cos\left[\omega_0(n_0 + n_1 - 1) + \omega_1(n_0 - n_1)\right] \tag{2}$$

$$V_{2k-1}[n] = 2\cos\left[\omega_0(n_0 + n_1 + 1) + \omega_1(n_0 - n_1)\right] \tag{3}$$

with $n = \begin{bmatrix} n_0 & n_1 \end{bmatrix}^T$, and $n_0, n_1 \in Z$.

By expanding the above equation, each of the analysis filter frequency responses can be viewed as a polynomial of x, the order of which depends on the number of lifting steps.

(2) Vanishing Moments. For a QFB, the number of vanishing moments is equivalent to the order of zero at $\begin{bmatrix} 0 & 0 \end{bmatrix}^T$ or $\begin{bmatrix} \pi & \pi \end{bmatrix}^T$ in the frequency response respectively. Given a linear-phase filter H with group delay $d \in Z^2$, its frequency response is defined as $\hat{h}(\omega)$ and its signed amplitude response as $\hat{h}_a(\omega)$, and then the m-th order partial derivative of $\hat{h}_a(\omega)$ in the Fourier domain is given by

$$\frac{\partial^{m_0+m_1}\hat{h}_a(\omega)}{\partial\omega_0^{m_0}\partial\omega_1^{m_1}} = \begin{cases} \displaystyle\sum_{n\in Z^2} h[n](n-d)^m \cos\left[\omega^T(n-d)\right]; & \text{for } |m| \in Z_{even} \\ -\displaystyle\sum_{n\in Z^2} h[n](n-d)^m \cos\left[\omega^T(n-d)\right]; & \text{otherwise} \end{cases} \tag{4}$$

where $m = \begin{bmatrix} m_0 & m_1 \end{bmatrix}^T$. From (4), when $|m| \in Z_{even}$, the mth-order partial derivative is zero at $\begin{bmatrix} 0 & 0 \end{bmatrix}^T$ and $\begin{bmatrix} \pi & \pi \end{bmatrix}^T$. Therefore, in order to obtain an Nth-order zero at $\begin{bmatrix} 0 & 0 \end{bmatrix}^T$, the coefficients of the filter only should be satisfied with the following condition:

$$\sum_{n\in Z^2} h[n](n-d)^m = 0 \text{ for } |m| \in Z_{even}, |m| < N \tag{5}$$

Similarly, in order to obtain an Nth-order zero at $\begin{bmatrix} \pi & \pi \end{bmatrix}^T$, the filter coefficients need satisfying with (6)

$$\sum_{n\in Z^2} (-1)^{|n-d|} h[n](n-d)^m = 0 \text{ for } |m| \in Z_{even}, |m| < N. \tag{6}$$

Since we use a lifting-based parameterization, the relationships need to be represented in terms of the lifting filter coefficients. For a filter bank to $N, \tilde{N}$ primal and dual vanishing moments, the analysis filter coefficients are required to satisfy (7):

$$Ax = b. \tag{7}$$

Here, we briefly give an example of a filter with two lifting steps. For (7), $A = \begin{bmatrix} A_1 & 0 \\ 0 & A_2 \end{bmatrix}$, $x = \begin{bmatrix} a_1 \\ a_2 \end{bmatrix}$, $b = \begin{bmatrix} b_1 \\ b_2 \end{bmatrix}$. The number of equations is $\left[\tilde{N}/2\right]^2 +$ $[N/2]^2$. b_1 is a vector with $\left[\tilde{N}/2\right]^2$ elements, and each element of it takes the form of $-2^{-|m|}$. b_2 is a vector with $[N/2]^2$ elements, each element of it takes the form of $-(-2)^{-|m|-1}$. Let the singular value decomposition (SVD) of A be $A = USV^T$, then all of the solutions to (7) can be parameterized as (8):

$$x = x_s + V_r\varphi \tag{8}$$

where V_r is a matrix composed of the last $(n - r)$ columns of V, and φ is an arbitrary $(n - r)$-dimensional vector. Henceforth, φ can be used as the design vector instead of x.

From the above discussion, the vanishing moment condition is the solution to a system of $[N/2]^2 + [N/2]^2$ linear equations, and the frequency condition is an optimizing problem for the coefficients of the lifting filter, which makes it as close as possible to ideal frequency response. Then the design objective is to maximize the coding gain G_c subject to a set constrains, which are chosen to ensure that the desired vanishing moment and frequency selectivity condition are met. Since the coding gain G_c can be expressed as a nonlinear function of vector φ, let $G = -10\log_{10} G_c$, for a given parameter vector φ, we should seek a small perturbation δ_φ such that $G(\varphi + \delta_\varphi)$ is reduced relative to $G(\varphi)$. Because $\|\delta_\varphi\|$ is small, the linear approximations of $G(\varphi + \delta_\varphi)$ can be written as (9):

$$G(\varphi + \delta_\varphi) = G(\varphi) + g^T\delta_\varphi \tag{9}$$

where g is the derivative of G at φ point. Then the design task is to search δ_φ, until $|G(\varphi + \delta_\varphi) - G(\varphi)|$ becomes less than a prescribed tolerance ε. And x^* is the desired output which contains the optimized coefficients of lifting filter. Due to space limitations, details about the optimization process can be referred to [17].

3   Gaussian Scale Mixtures Model

The Gaussian scale mixtures model (GSM) can characterize both the marginal and joint distributions of wavelet coefficients of natural image. This model is promising for applications in image processing [8,9].

For the coefficients of natural image in transform domain, the observed sample in each neighborhood can be represented by $y = x + w = \sqrt{z}u + w$, where x stands for the coefficients to be estimated, and w represents the noise coefficients. w, x, and y are random vectors. z is a positive scalar random variable, w and u are zero-mean Gaussian vectors, C_w and C_u are the corresponding covariance, respectively. The density of x is Gaussian conditioned on z and y is an infinite mixture of Gaussian vector. The density of the observed neighborhood vector conditioned on z is zero-mean Gaussian with covariance $C_{y|z} = zC_u + C_w$ as below:

$$p(y|z) = \frac{\exp(-y^T(zC_u + C_w)^{-1}y/2)}{\sqrt{(2\pi)^N |zC_u + C_w|}} \tag{10}$$

where N is the number of observed coefficients namely the size of neighborhood window. The noise PSD is assumed known, then C_w can be easily estimated.

x_c is estimated from the neighbors of y. The estimate of the Bayesian least squares (BLS) is given by

$$E\{x_c | y\} = \int x_c p(x_c | y) dx_c = \int_0^\infty p(z|y) E\{x_c | y, z\} dz \tag{11}$$

Therefore, the solution is an average of the least squares estimate of x_c when conditioned on z, weighted by the posterior density of the multiplier $p(z|y)$. As z is zero-mean Gaussian, (11) is a local Wiener solution as

$$E\{x|y, z\} = z C_u (z C_u + C_w)^{-1} y \tag{12}$$

We can simplify the dependence of this expression on z by the diagonalization of the matrix $z C_u + C_w$. Let S be the square root of the positive definite matrix C_w, $C_w = SS^T$, Q and Λ are the eigenvector and eigenvalue of matrix $S^{-1} C_u S^{-T}$, respectively, then

$$z C_u + C_w = SQ(z\Lambda + I) Q^T S^T \tag{13}$$

Let $M = C_u S^{-T} Q v = Q^T S^{-1} y$. From (12), we can get the reference x_c:

$$E\{x_c | y, z\} = \sum_{n=1}^N \frac{z m_{cn} v_n}{z\lambda_n + 1} \tag{14}$$

where m_{cn} is an element of matrix M, λ_n is element of diagonal matrix Λ, v_n is element of matrix v, and c is the order of reference coefficient x. The posterior distribution of the multiplier can be computed by Bayesian rule:

$$p(z|y) = \frac{p(y|z) p_z(z)}{\int_0^\infty p(y|\alpha) p_z(\alpha) d\alpha} \tag{15}$$

where $p_z(z)$ is the density of z, $p_z(z) \propto \frac{1}{z}$. Using the relationship in (13) and the definition of v, the condition density $p(y|z)$ in (10) can be simplified as

$$p(y|z) = \frac{\exp(-\frac{1}{2} \sum_{n=1}^N \frac{v_n^2}{z\lambda_n+1})}{\sqrt{(2\pi)^N |C_w| \prod_{n=1}^N (z\lambda_n + 1)}} \tag{16}$$

4   Image Denoising Method

In order to validate the effectiveness of our improved NSCT, based on [11], we combined it with GSM to perform image denoising. Our algorithm is summarized as follows:

Step 1: Image decomposition. Use our modified NSCT to decompose the noisy image to coarse subband and the directional subbands of finer scale.

Step 2: For each subband (except the coarser)

(1) Calculate the covariance C_w of noise; calculate the covariance C_y of coefficients of neighbourhood; calculate the covariance C_u from C_w and C_y, suppose $E\{z\} = 1$, then $C_u = C_y - C_w$.

(2) Because C_w has diagonalized, then $C_w = SS^T$, and S is a positive definite matrix, calculate the eigenvalue expansion Λ and the eigenvector expansion Q of diagonal matrix $S^{-1}C_uS^{-T}$; calculate v and M, $v = Q^TS^{-1}y$, $M = C_uS^{-T}Q$.

(3) For each neighborhood
   i) According (14), calculate $E\{x_c|y, z\}$.
   ii) According (15), (16) calculate the posterior distribution of the multiplier $p(z|y)$, and the condition density $p(y|z)$.
   iii) According Eq. 10, estimate x_c in each neighborhood$y$, calculate $E\{x_c|y\}$.

Step 3: Gather the original coarse subband and the filtered directional subbands, use the improved inverse NSCT to reconstruct the denoised image.

5   Experimental Results and Analysis

In this section, we carry out comparison with the original NSCT combined with GSM [11], the NLM [10], the BLS-GSM [8], and the KSVD [7] algorithm to show the effectiveness of proposed method. In our experiments, we showed the denoised image for visual analysis and subjective assessment. We also compare the PSNR of denoised image for objective evaluation, and the statistic of the elapsed times of each algorithm is used for evaluating the speed of them. For convenient comparison, we denote the method that combines the original NSCT and the GSM as NSCT-GSM.

To test the denoising effectiveness, five representative images shown in Fig. 1 are chosen as benchmark images. The test environment is matlab2009, and the computer is with 2.33 GHz CPU, 4 Cores, and 4 GB DDR2 RAM.

We mimic noised images by adding the Gaussian white noise with different noise variance. The compared results of PSNR and elapsed times are reported in Table 1.

From Table 1, we can find that the PSNR values of NLM are the lowest. In the case of lower noise variance (e.g., $\sigma = 5$), the KSVD produces the highest

(a)   (b)   (c)   (d)   (e)

Fig. 1. The Original Image. (a) Lena,(b) House, (c) Barbara, (d) Mandrill, (e) Fingerprint.

denoising performance. However, with the increasing of the noise variance, the PSNR value begins to decrease, especially when $\sigma > 20$, the KSVD cannot function well. When $\sigma < 30$, the PSNR values of NSCT-GSM are constantly higher than those of the BLS-GSM. However, when a higher noise variance is added, the BLS-GSM outperforms the NSCT-GSM. For image Lena and image House, whose block smooth region is relative major, the results of the proposed method are comparable to those of the NSCT-GSM. For the images Barbara,

Table 1. The denoised results comparison of different methods

σ	5	10	15	20	25	30	50	70
Input PSNR	34.15	28.13	24.61	22.11	20.17	18.59	14.15	11.23
Method	PSNR/T (dB/Second) Lena 512×512							
NLM	37.16/293.76	33.80/249.09	31.49/294.54	29.89/293.42	28.79/293.28	27.95/293.53	25.55/294.65	23.97/294.70
KSVD	38.62/900.85	35.47/469.56	33.67/312.53	32.36/220.59	31.33/173.46	30.46/137.93	27.80/88.01	26.13/68.65
BLS-GSM	38.19/18.3	35.23/18.37	33.50/18.39	32.25/18.42	31.26/18.39	30.46/18.34	28.21/18.90	26.76/19.42
NSCT-GSM	38.53/930.57	35.67/935.98	33.91/927.71	32.59/930.06	31.55/928.70	30.67/927.89	28.17/926.57	26.55/922.71
Proposed	38.41/73.60	35.78/73.32	33.54/73.10	32.18/73.23	31.09/72.95	30.19/73.59	27.71/75.01	26.19/75.39
Method	PSNR/T (dB/Second) House 256×256							
NLM	37.35/60.20	34.61/61.34	32.63/60.48	30.74/57.5	29.23/58.64	27.95/57.8	24.67/58.48	22.94/58.79
KSVD	39.42/566.01	35.95/392.15	34.3/236.87	33.16/174.81	32.03/142.21	31.33/116.39	27.86/75.12	25.64/61.34
BLS-GSM	38.23/4.34	35.32/4.31	33.73/4.35	32.54/4.32	31.59/4.28	30.78/4.26	28.36/4.40	26.74/4.60
NSCT-GSM	38.81/230.26	35.46/230.95	33.68/230.04	32.37/230.57	31.33/230.30	30.45/230.56	27.84/230.73	26.10/231.01
Proposed	38.49/17.96	35.04/18.06	33.16/17.93	31.79/18.50	30.70/18.01	29.80/18.10	27.23/18.59	25.60/18.54
Method	PSNR/T (dB/Second) Barbara 512×512							
NLM	36.68/304.37	32.93/304.46	30.31/305.35	28.32/305.96	26.81/305.5	25.69/304.40	23.01/305.37	21.59/290.35
KSVD	38.08/1701.7	34.39/1056.7	32.38/585.62	30.89/391.71	29.61/289.43	28.57/225.09	25.45/126.03	23.32/87.92
BLS-GSM	37.19/18.82	33.13/18.70	30.76/18.60	29.08/18.57	27.81/18.59	26.79/18.64	24.33/18.50	23.17/18.53
NSCT-GSM	37.86/921.43	34.17/939.81	31.99/943.03	30.42/944.15	29.20/934.51	28.21/939.26	25.45/949.90	23.76/936.10
Proposed	37.59/73.95	33.64/73.95	31.32/74.18	29.66/73.93	28.40/74.87	27.38/74.93	24.69/74.34	23.18/75.26
Method	PSNR/T (dB/Second) Mandrill 512×512							
NLM	34.41/315.51	29.25/316.21	26.04/315.62	23.52/313.70	22.32/313.26	21.65/313.71	22.44/313.97	19.91/314.43
KSVD	35.27/4811.4	30.59/2818.1	28.11/1487.1	26.52/924.35	25.33/610.23	24.46/421.20	22.13/159.96	21.01/98.21
BLS-GSM	35.34/7.37	30.70/7.71	28.31/8.12	26.75/8.60	25.62/8.79	24.75/9.01	22.42/13.12	21.34/12.59
NSCT-GSM	35.28/937.95	30.66/931.43	28.25/955.53	26.67/950.51	25.52/948.21	24.63/946.35	22.42/944.54	21.28/929.60
Proposed	35.22/75.64	30.53/73.29	28.04/72.81	26.39/72.62	25.19/72.75	24.28/72.68	22.12/73.04	21.06/73.45
Method	PSNR/T (dB/Second) Fingerprint 512×512							
NLM	34.43/321.34	30.54/322.93	27.65/328.92	25.64/315.75	24.29/312.48	23.21/313.46	19.96/313.29	18.23/314.15
KSVD	36.63/2532.5	32.41/1682	30.09/1044.0	28.47/726.43	27.27/522.26	26.29/420.82	23.28/238.37	20.53/158.75
BLS-GSM	36.36/19.17	32.19/18.98	29.92/18.82	28.34/18.65	27.09/18.65	26.07/18.70	23.23/19.07	21.48/18.96
NSCT-GSM	36.56/917.73	32.33/918.43	30.06/926.09	28.52/918.43	27.35/919.21	26.42/919.07	23.85/918.14	22.14/912.46
Proposed	36.63/74.64	32.33/75.67	29.99/74.10	28.38/73.90	27.16/73.18	26.19/73.73	23.53/74.67	21.72/75.68

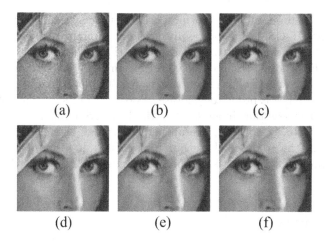

Fig. 2. The denoised detail comparison of image Lena ($\sigma = 10$). (a) Noised, (b) NLM, (c) KSVD, (d) BLS-GSM, (e) NSCT-GSM, (f) Proposed.

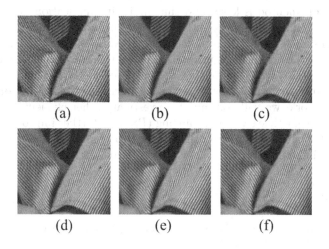

Fig. 3. The denoised detail comparison of image Barbara ($\sigma = 10$). (a) Noised, (b) NLM, (c) KSVD, (d) BLS-GSM, (e) NSCT-GSM, (f) Proposed.

Mandrill, and Fingerprint, which have plenty of textures, and the detail keeping ability of denoising algorithm is hoped, although the result of proposed method is slightly low than the NSCT-GSM's, the proposed method still outperforms BLS-GSM in most cases. When comparing the elapsed times as presented in Table 1, we can claim that: (1) the elapsed times of BLS-GSM is the least; (2) although the PSNR of KSVD is the highest in the case of low noise variance, the KSVD is the most time-consuming; (3) the PSNR values of NSCT-GSM are higher than those of the BLS-GSM in most cases, but it is highly time-consuming; (4) although the PSNR results of our method is slightly lower than the original

NSCT-GSM, the processing time of proposed method is significantly low than the original NSCT-GSM. Therefore, in terms of comprehensive evaluation, the proposed method shows a noticeable advantage than others.

Visually, the denoised results of local regions in Lena and Barbara are shown in Figs. 2 and 3, respectively. As contrastive results shown in Figs. 2 and 3, we can see that the NLM is prone to generate smoother results and cannot preserve fine details in the resultant images, while the BLS-GSM and KSVD somewhat outperform the NLM. However, overall performance of them is inferior to the NSCT-GSM and the proposed algorithm. In particular, in terms of the subjective quality, the proposed method is comparable to the NSCT-GSM while keeping high efficiency of processing.

6 Conclusions

The NSCT is one of the most excellent tools of the multiscale geometric analysis (MGA). The NSCT combined with GSM can achieve appealing denoising performance with good visual quality and high PSNR. However, the computational efficiency of this mehtod is lower and thus limits its application in practice. To reduce this difficulty, this paper introduces an optimized quincunx filter bank (QFB) to construct the NSDFB in the original NSCT to improve the runtime. Combined with the GSM, the improved NSCT is used for image denoising. The experimental results demonstrate that the proposed method can achieve noticeable performance in terms of both denoising results and spending of runtime. In the future work, we will continue to investigate other efficient optimization methods such as particle swarm optimizing (PSO) to further improve the performance of NSCT and extend it to other image processing tasks, including image fusion, image inpainting, and image superresolution, etc.

References

1. Donoho, D.L., Johnstone, I.M.: Ideal spatial adaptation by wavelet shrinkage. Biometrika **81**(3), 425–455 (1994)
2. Donoho, D.L.: De-noising by soft-threshoding. Ann. Stat. **41**(3), 613–627 (1995)
3. Blu, T.: The SURE-LET approach to image denoising. IEEE Trans. Image Process. **16**(11), 2778–2786 (2007)
4. Jiao, L.C., Tan, S.: Development and prospect of image multiscale geometric analysis. Acta Electronic Sinca **31**(12), 1975–1981 (2003)
5. LeCun, Ponce, J.: Learning midlevel features for recognition. In: IEEE International Conference on Computer Vision and Pattern Recognition, pp. 2559–2566 (2010)
6. Mairal, J., Bach, F., Ponce, J., Sapiro, G.: Online learning for matrix factorization and sparse coding. J. Mach. Learn. Res. **11**, 19–60 (2006)
7. Aharon, M., Elad, M., Bruckstein, A.M.: The K-SVD: an algorithm for designing of overcomplete dictionaries for sparse representation. IEEE Trans. Signal Process. **54**(11), 4311–4322 (2006)

8. Portilla, J., Strela, V., Wainwright, M., Simoncelli, E.P.: Image denoising using scale mixtures of gaussians in the wavelet domain. IEEE Trans. Image Process. **12**(11), 1338–1351 (2003)
9. Wainwright, M.J., Simoncelli, E.P.: Scale mixtures of Gaussians and the statistics of natural images. In: Advance in Neural Information Processing Systems, vol. 12, pp. 855–861 (2000)
10. Tasdizen, T.: Principal neighborhood dictionaries for nonlocal means image denoising. IEEE Trans. Image Process **18**(12), 2649–2660 (2009)
11. Zhou, H.F., Wang, X.T., Xu, X.G.: Image denoising using gaussian scale mixture model in the nonsubsampled contourlet domain. J. Electron. Inf. Technol. **31**(8), 1796–1800 (2009)
12. Yan, C.M., Guo, B.L., Yi, M.: Fast algorithm for nonsubsampled contourlet transform. Acta Automatica Sinica **44**(4), 757–762 (2014)
13. Do, M.N., Vetterli, M.: The contourlet transform: an efficient directional multiresolution image representation. IEEE Trans. Image Process. **14**(2), 2091–2106 (2005)
14. Cunha, A.L., Zhou, J.P., Do, M.N.: The nonsubsampled Contourlet transform: theory, design and application. IEEE Trans. Image Process. **15**(10), 3089–3101 (2006)
15. Bamberger, R.H., Smith, M.J.T.: A filter bank for the directional decomposition of images: Theory and design. IEEE Trans. Signal Process. **40**(4), 882–893 (1992)
16. Po, D.D.-Y., Do, M.N.: Directional multiscale modeling of images using the contourlet transform. IEEE Trans. Signal Process. **15**(6), 1610–1620 (2006)
17. Yi, C., Michael D., Adams, M. D., Lu, W.S.: Design of optimal quincunx filter banks for image coding. Journal on Advances in Signal Processing, Article ID 83858 (2007)
18. Sweldens, W.: The lifting scheme: a custom-design construction of biorthogonal. Appl. Comput. Harmon. Anal. **3**(2), 186–200 (1996)
19. Tran, T.D., de Queiroz, R.L., Nguyen, T.Q.: Linear-phase perfect reconstruction filter bank: Lattice structure, design, and application in image coding 48(1), 133–147 (2000)
20. Gouze, A., Antonini, M., Barlaud, M.: Quincunx lifting scheme for lossy image coding. In: IEEE International Conference on Image Processing, pp. 665–668 (2000)

Fast Correction Visual Tracking via Feedback Mechanism

Tianyang Xu and Xiaojun Wu$^{(\boxtimes)}$

School of IoT Engineering, Jiangnan University, Wuxi 214122, China
{tianyang_xu,xiaojun_wu_jnu}@163.com
http://www.springer.com/lncs

Abstract. Visual tracking is a fundamental problem in computer vision field. Most online visual trackers focus on the appearance information and inference theory to realize tracking frame by frame. However, enough attention has not been paid to the correction ability of a tracking system, which leads to drift problems or tracking failures in previous works. This paper investigates the contribution of feedback mechanism in a tracking-by-detection framework. Results indicate that the changing values of the target state's posterior distributions provide superior information to the connection between tracking result and the ground truth. We further analyse the spatial appearance information and propose an adaptive feedback tracking method using Discrete-Quaternion-Fourier-Transform (DQFT). Taking advantages of the stability of closed-loop control and the efficiency of DQFT, the proposed tracker can make a distinction between the easy-tracking frames and the hard-tracking frames, and then re-track hard-tracking frames using further temporal information to realize the correction ability. Experiments over 50 challenging videos demonstrate the effectiveness and robustness of the tracker, and the resulting tracker outperforms the existing state-of-the-art methods.

Keywords: Visual tracking · Feedback mechanism · Discrete quaternion fourier transform

1 Introduction

Visual tracking is a significant research task in computer vision, with practical applications in video surveillance, robot perception and human-computer interaction. Although a large variety of trackers have been proposed in the literature with a certain sense of success, it still suffers from several appearance challenges such as illumination variation, partial occlusion, shape deformation, and camera motion. In this paper we investigate to what extent the usage of feedback mechanism can alleviate some of these issues.Most state-of-the-art tracking methods rely on the appearance information and inference theory to realize tracking. Almost all of the previous methods have one thing in common: a fixed tracking strategy. In each frame, the same pattern — an updating appearance model and a corresponding searching strategy — is applied to achieve the tracking objective

© Springer International Publishing Switzerland 2015
X. He et al. (Eds.): IScIDE 2015, Part I, LNCS 9242, pp. 208–219, 2015.
DOI: 10.1007/978-3-319-23989-7_22

without considering the diversity in different frames. Due to the sensitive and unsupervised nature of online tracking, results directly depend on the quality of the pattern. On the other hand, the biological vision system has the ability to adaptively adjust different circumstances. For example, it is easy to track a running horse in a meadow while more attention should be paid to track a certain person in a crowded street. Inspired by the facts mentioned above, we propose a novel tracking method using feedback mechanism to make a distinction between the easy-tracking frames and the hard-tracking frames.

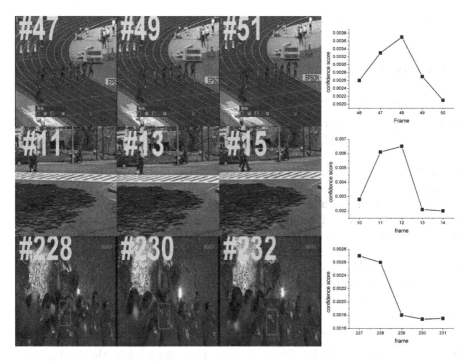

Fig. 1. The connection between a tracking failure and its corresponding changing values of the maximum a posteriori of the target state. From experimental results we find that all the tracking failures are accompanyed with a dramatic decline with respect to the confidence scores, and we consider the changing confidence scores as a necessary condition for a potential failure to realize feedback mechanism.

In order to realize feedback in an online tracking proceeding, we analyse the tracking results of different popular methods mentioned in [1,2]. Interestingly, we find that all the tracking failures are accompanyed with a dramatic decline in the values of the maximal posterior (we denote it as "confidence score", shown in Fig. 1). And according to this connection we design a feedback strategy to correct tracking results. The proposed method focuses on improving the processing capacity to build a robust tracker, and a flexible tracking strategy is applied to handle different challenges of the targets and scenarios. It achieves preliminary tracking via a normalized cross correlation (NCC) based spatial appearance model, which is calculated using Discrete-Quaternion-Fourier-Transform (DQFT). And then it exploits the inherent connection between tracking result and the ground truth to generate a feedback mechanism, which realizes a distinctive estimation between easy-tracking frames and hard-tracking frames. After that, a more effective temporal appearance model is established to re-track the hard-tracking-frames.

2   Related Works

Numerous tracking methods have been proposed in recent decades, a comprehensive introduction of different tracking methods is presented in recent surveys [2]. Among the various tracking algorithms, two main strategies– generative methods and discriminative methods– have been widely researched.

Generative methods. Generative methods aim to establish a model to describe the target without considering surrounding information. Since the Lucas-Kanade algorithm [3,4] was proposed in 1981, generative methods have become one of the most widely used techniques in computer vision, and its application in visual tracking is one of the pioneering work in this field. Subsequently, the famous mean-shift tracking method [6] was proposed to realize tracking by histograms matching, the spatial information is integrated by statistics theory using histograms and then the tracker maximizes the Bhattacharyya distance iteratively to achieve final results. Furthermore, Adam proposed fragments-based robust tracking method [5] based on matching the ensemble of patches to improve the description ability of histograms, more spatial information is exploited by these fragments. Then Subspace-based tracking approaches such as [7] make a better explanation for the appearance model. More recently, Trackers based on sparse theory, such as [8], also achieve robust results.

Discriminative methods. Discriminative methods exploit the background information to enhance the description ability of the target model. On-line boosting tracker [13] was proposed by Grabner to utilize the classification capacity of SVM. Then, Multiple instance learning tracker [9,10] was proposed to learn a discriminative classifier from positive and negative samples. In addition, Tracking-learning-detection method [11] was proposed to handle short-term occlusion using a double-tracking strategy. In order to exploit the labeled data more effectively, Struck [12] was proposed to train a structured supervised classifier to realise tracking.

In this paper, we extend the normalized cross correlation method [21] to realize fast preliminary tracking in all frames. And then a feedback mechanism is applied to distinguish potential failed frames. In the re-tracking stage, the temporal discriminative appearance information is integrated to correct the tracking results. The main contributions of the work are summarized as follows:

(1) A novel feedback mechanism model based tracker is proposed, which utilizing the confidence scores to make a judgement and re-track potential failed frames.
(2) A temporal-spacial appearance information based model is proposed to re-track the hard-tracking frames, which integrate the spacial information and the temporal information together to realize fine-grained robust tracking.
(3) Experimental results over 50 challenging videos indicate that the proposed tracking method outperforms previous state-of-the-art trackers.

3   Problem Formulation

Visual tracking problem is formulated by a Markov model with an inference task to solve the maximum a posteriori (MAP) problem:

$$p\left(X_t|Z_{1:t}\right) \propto p\left(Z_t|X_t\right) \times \int p\left(X_t|X_{t-1}\right) p\left(X_{t-1}|Z_{1:t-1}\right) dX_{t-1}. \tag{1}$$

where X_t is the state variable describing the target's location and size in the corresponding frame t, $p\left(Z_t|X_t\right)$ is the observation model. In order to estimate the likelihood of the target model and a candidate, an accurate and efficient appearance model should be established. Here we apply and extend the normalized cross correlation method to achieve this objective.

3.1   Normalized Cross Correlation Filter

Normalized cross correlation filter [21] is one of the most efficient methods in image matching. It provides a fast measure of the similarity between two images, and when it applied to visual tracking, we can calculate the similarity between the target template model T and a candidate I as follows (T and I have the same size, M × N, with corresponding mean value $\bar{T}$ and $\bar{I}$):

$$s = \frac{\sum\limits_{x=0,y=0}^{M-1,N-1} \left(I\left(x,y\right) - \bar{I}\right)\left(T\left(x,y\right) - \bar{T}\right)}{\sqrt{\sum\limits_{x=0,y=0}^{M-1,N-1} \left(I\left(x,y\right) - \bar{I}\right)^2 \sum\limits_{x=0,y=0}^{M-1,N-1} \left(T\left(x,y\right) - \bar{T}\right)^2}}. \tag{2}$$

3.2   The Definition of Discrete-Quaternion-Fourier-Transform

Quaternion multiplication [14] is non-commutative, which leads to different results by the left and the right multiplication of the exponential to the quaternion expression. In this paper, we use the left multiplication defined as follows:

$$F_{\mathrm{L}}\left(u,v\right) = \sum_{m=0}^{M-1} \sum_{n=0}^{N-1} e^{-\mu 2\pi\left(\frac{mu}{M} + \frac{nv}{N}\right)} f\left(m,n\right). \tag{3}$$

$$f(m, n) = \frac{1}{MN} \sum_{u=0}^{M-1} \sum_{v=0}^{N-1} e^{\mu 2\pi \left(\frac{mu}{M} + \frac{nv}{N}\right)} F_L(u, v). \tag{4}$$

where the subscript L stands for the left multiplication form of the DQFT, μ denotes the direction of a unit pure quaternion which represents one component in RGB color space here.

An RGB color image I can be represented in quaternion form as follows:

$$f(m, n) = r(m, n) \cdot i + g(m, n) \cdot j + b(m, n) \cdot k. \tag{5}$$

where $r(m, n)$, $g(m, n)$ and $b(m, n)$ are the red, green and blue components of pixel $I(m, n)$. This representation offers a holistical mode to handel different color components. Given an image I which is represented via quaternion form $f(m, n)$ and a translation shift vector (m_0, n_0), we can denote the new image I' as $f'(m, n) = f(m - m_0, n - n_0)$. From the translation property of the DQFT, we can derive that:

$$F_L'(u, v) = e^{-\mu 2\pi \left(\frac{m_0 u}{M} + \frac{n_0 v}{N}\right)} F_L. \tag{6}$$

Taking the modulus of both sides we can get that $|F_L'(u, v)| = |F_L(u, v)|$, and we can draw the conclusion that DQFT is shift invariant.

3.3 Normalized Cross Correlation Based Spatial Appearance Model Using DQFT

Bolme extended the normalized cross correlation filter to a MOSSE (Minimum Output Sum of Squared Error) form [17]. Henriques exploited the circulant structure of a tracking target and realized a fast tracking method with kernels [16,18]. Danelljan extended the learning scheme for the normalized cross correlation tracker to multichannel color features and propose a low dimensional adaptive extension of color attributes [19]. Zhang proposed a dense spatio-temporal context tracking scheme using the normalized cross correlation filter [20]. All these extensions mentioned above take advantage of the efficiency of NCC filter. Meanwhile, The latest survey [2] analyses the current state-of-the-art trackers and draws the conclusion that the NCC based tracker is one of the top 6 trackers in respect of the overall performance. And it is remarkable that NCC based trackers use only a plain matching without any complicated control strategy.

Due to the fact that DQFT is shift invariant (Sect. 3.2), we can exploit the convenience of the circulant structure proposed by Henriques and establish a robust appearance model with the highest speed. Back to visual tracking, we select the best candidate as tracking result in current frame. In order to evaluate a candidate, we train a classifier f (x) that minimizes the squared error bellow:

$$\min_{w} \sum_{i} (f(x_i) - y_i) + \lambda \langle w, w \rangle. \tag{7}$$

Here, λ is a regularization parameter. and the sample set $\{x_i\}, i = (m, n) \in \{(0, 0), \ldots, (M - 1, N - 1)\}$ represents all shift transformations of the original sample x. Inspired by tracking-by-detection methods, it is reasonable to assume that these transformations with less shift should get higher classifier scores and vice versa. As a result, the regression value $\{y_i\}$ is defined by a normalized Gaussian distribution with the center located at $x_{0,0}$.

We introduce the kernel method to map x to a Hilbert space by κ [22], and the classifier can be denoted as $f(x) = \langle \phi(x), w \rangle$. The kernel κ defines the inner product as $\langle \phi(\xi_1), \phi(\xi_2) \rangle = \kappa(\xi_1, \xi_2)$, and the optimization problem can be minimized by $w = \sum_i \alpha_i \phi(x_i) = \sum_{m,n} \alpha(m,n) \phi(x_{m,n})$ where the coefficients $\{\alpha\}$ can be directly expressed in matrix form as:

$$A = \mathscr{F}\{\alpha\} = \frac{Y}{K_{xx} + \lambda}. \tag{8}$$

Here the capital letters represent the DQFT of their corresponding lower-case letters, such that $Y = \mathscr{F}\{y\}$ and $K_{xx} = \mathscr{F}\{k_{xx}\}$, and $\mathscr{F}$ is the DQFT operator. The kernel matrix $\{k_{xx}\}$ is defined by the output of κ, $k_{xx}(m,n) = \kappa(x_{m,n}, x_{0,0})$. Thus, the classifier f can be determined by the valid example x.

After the training step, we can evaluate the candidates by f. We consider all the shift transformations of the new region-of-interest as our candidates, which can be denoted as $\{x_{m,n}^{\text{new}}\}$. For the same reason, we can evaluate $\{x_{m,n}^{\text{new}}\}$ expediently:

$$\{f(x_{m,n}^{\text{new}})\} = \mathscr{F}^{-1}\{A K_{xx^{\text{new}}}\}. \tag{9}$$

where $\{f(x_{m,n}^{\text{new}})\}$ is the results of candidates, and it is obvious that we can evaluate all candidates with only one operation. $K_{xx^{\text{new}}}$ represents the kernel output between $\{x\}$ and $\{x^{\text{new}}\}$: $k_{xx^{\text{new}}} = \kappa(x, x_{m,n}^{\text{new}})$.

3.4   Fast Calculation of DQFT Using FFT

We exploit the so-called reduced biquaternions [15] to achieve fast calculation of DQFT. Reduced biquaternions is based on double-complex algebra and it defines imaginary units as: $ij = ji = k, jk = kj = i, ik = ki = -j$. Given an RGB color image I and its reduced biquaternions form: $f = r \cdot i + g \cdot j + b \cdot k$, we can further represent it as:

$$\begin{cases} f = f_1 e_1 + f_2 e_2 \\ f_1 = g + (r+b) \cdot i \\ f_2 = -g + (r-b) \cdot i. \\ e_1 = \frac{1+j}{2} \\ e_2 = \frac{1-j}{2} \end{cases} \tag{10}$$

Based on the e_1-e_2 form of f, we can calculate the Fourier Transform as follows:

$$\begin{aligned} \mathscr{F}\{f\} &= \mathscr{F}\{f_1\}e_1 + \mathscr{F}\{f_2\}e_2 \\ &= \text{FFT}\{f_1\}e_1 + \text{FFT}\{f_2\}e_2 \end{aligned} \tag{11}$$

Finally we can realize fast calculation of DQFT using two 2-D FFT and guarantee the high speed of our appearance model, which is important to the normalized cross correlation based trackers.

4   Proposed Tracking Algorithm with Feedback Mechanism

According to confidence scores, we exploit feedback mechanism to reduce tracking failures in recent frames. The proposed tracking algorithm can be described as follows (Table 1):

Table 1. Proposed tracking algorithm

Feedback Mechanism based Tracking Algorithm:

Input: coefficient matrix A, new frame I_k, target position p_{k-1} in the last frame, confidence scores set S_{k-1}

Output: current coefficient matrix A, target position p_k (tracking result) in current frame, current confidence scores set S_k

---

Preliminary tracking:

1. Extract the search region R_k in current frame I_k, which is a padding of the target size at position p_{k-1}
2. Evaluate R_k using A and calculate the results of the classifier by equation (9)
3. Determine the maximal value s_k in step.2 as the final confidence score in current frame, and then the corresponding position p_k is identified as the tracking result
4. Update $S_k = S_{k-1} \bigcup s_k$

Feedback correction tracking:

5. If S_k satisfies feedback conditions (Sect. 4.1), we define the corresponding frames as hard-tracking frames and re-track them using Re-tracking Algorithm (Sect. 4.2)
6. Else, update A

4.1   Feedback Conditions

Based on earlier experimental results, we can obtain the inherent relationship between a tracking failure and the values of recent confidence scores. As shown in Fig. 2, a failure is always accompanyed with a dramatic decline with respect to the confidence scores and it is reasonable to exploit this necessary condition to correct tracking results. We define feedback conditions as:

1. s_{k-N} is the nearest peak value away from the current frame k.
2. s_{k-1} is a valley value.
3. The number of the declined frames: $N \geq 5$.

If the three above conditions are satisfied, then we denote frame $k-N$, $k-1$ and k as starting point, end point and trigger point respectively. Meanwhile, frames between starting point and end point are defined as hard-tracking frames which should be re-tracked via further processing.

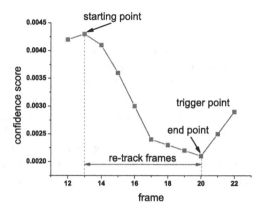

Fig. 2. An example of feedback conditions.

4.2   Re-Track Hard-Tracking Frames via Exploiting Temporal Information

After the hard-tracking frames are defined, we re-track them using a 3-step backward search strategy.

Step 1: Based on the preliminary tracking result at the end point, we calculate an auxiliary coefficient matrix A' by Eq. 8 and track the hard-tracking frames backward (from frame $k-1$ to frame $k-N$) using the same preliminary tracking method.

Step 2: Denote the tracking results in Step.1 as $\{p'_i\}$, $i = 1, \ldots, N$ corresponding to the frames between frame $k-1$ and frame $k-N$, we extract the image region R'_{k-N} which is positioned by $p'(N)$ in frame $k-N$. Then we calculate the overlap score between R'_{k-N} and R_{k-N}: $OS = \frac{|R'_{k-N} \cap R_{k-N}|}{|R'_{k-N} \cup R_{k-N}|}$. If $OS \geq 0.5$, we can conclude that the preliminary tracking is successful and the dramatic decline of confidence scores is caused by the target's internal transformation. On the contrary, if $OS < 0.5$, the preliminary tracking is failed and we define R'_{k-N} as a fake target.

Step 3: Exploiting the temporal discriminative information between the fake target R'_{k-N} and the real target R_{k-N}, we re-track hard-tracking frames by adding a penalty factor in Eq. 9:

$$\{f(x_{m,n}^{new})\} = \mathscr{F}^{-1}\{AK_{xx^{new}}\} - \beta\mathscr{F}^{-1}\{A'K_{x'x^{new}}\} . \tag{12}$$

where A' is the coefficient matrix trained by the fake target, $K_{x'x^{new}}$ is the kernel matrix calculated between the fake target region and the search region. It is obvious that this feedback method take advantages of the discriminative information to correct tracking results when a fake target is detected.

5   Experiments

We evaluate the proposed feedback correction tracker (FCT) using 50 publicly available videos which consist of the CVPR 2013 benchmark dataset [1]. We compare the proposed FCT method with 8 state-of-the-art methods, and the parameters of the proposed algorithm are fixed for all the experiments. The 8 trackers we compare with are: adaptive structural local sparse appearance model tracker(ASLA)[23], circulant structure kernels tracker (CSK)[16], Compressive Tracking (CT)[24], distribution fields tracking (DFT)[25], incremental visual tracking (IVT)[7], L1 tracker using accelerated proximal gradient (L1APG)[26], multi-task sparse learning tracker (MTT)[28], and online robust image alignment tracker(ORIA)[27].

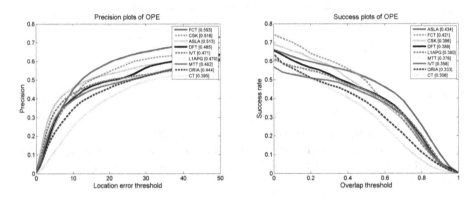

Fig. 3. Precision plot and success plot of overall performance

5.1   Overall Performance

We evaluate tracking performance using two criteria — precision plot and success plot. Precision plot is a widely used evaluation metric described the center location error, which is defined as the average distance between the center location of the tracking results and the ground truths. While success plot is also an important evaluating indicator, it counts the number of successful frames, a tracking result is successful only when the overlap score (defined as $OS = \frac{|r_t \bigcap r_g|}{|r_t \bigcup r_g|}$) is greater than a threshold (from 0 to 1 in our experiments). As shown in Fig. 3, our proposed FCT method achieves the best and the second best performance in terms of the center location error (precision) and success rate respectively. In addition, the FCT method is the second efficient algorithm (average 250 FPS) among all the evaluated methods, just behind the highest speed algorithm CSK (average 320 FPS). Furthermore, When location error threshold becomes larger (in precision plot) or overlap threshold becomes smaller (in success plot), the performance of FCT is getting more prominent.

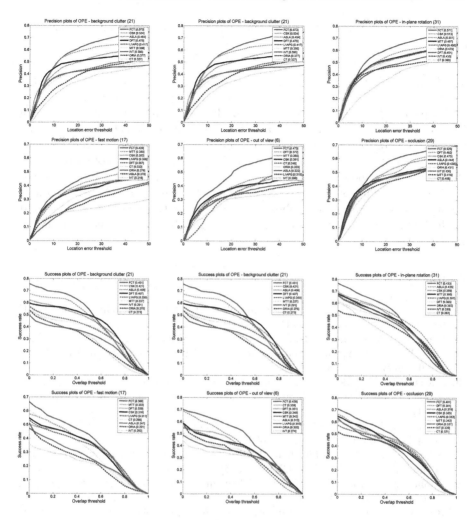

Fig. 4. Precision plot and Success plot on different attributes

5.2 Performance on Different Attributes

We divide video attributes into 6 categories: background clutter, motion blur, in-plane rotation, fast motion, out-of-view and occlusion. Then we evaluate and compare our FCT method on different attributes. As shown in Fig. 4, it is obvious that our feedback tracking mechanism is effective to correct the potential errors, it outperforms state-of-the-art trackers in many attributes, especially in motion blur, out-of-view and fast motion.**Motion blur, out-of-view and fast motion**: these three attributes are caused by external appearance changing, which can be better solved by FCT method. **Background clutter, in-plane rotation and occlusion**: these three attributes directly influence the region-of-interest, which reduces the description ability of our temporal discriminative model.

6   Conclusion

In this paper, we present an efficient feedback mechanism to correct our tracking results. A two-stage FCT tracking algorithm is proposed to achieve a coarse-to-fine tracking. In addition, the FCT method is robust to appearance variations introduced by occlusion, illumination changes, pose variations, motion blur, out-of-view and fast motion. Experiments with state-of-the-art methods on challenging videos demonstrate that the proposed FCT method achieves favorable results in terms of accuracy, robustness, and speed. Our research is supported by Research Project on Surveying and Mapping of Jiangsu Province (Grant No. JSCHKY201109).

References

1. Wu, Y., Lim, J., Yang, M.-H.: Online object tracking: A benchmark. In: IEEE CVPR, pp. 2411–2418. IEEE Press, Portland (2013)
2. Smeulders, A.W.M., Chu, D.M., Cucchiara, R., Calderara, S., Dehghan, A., Shah, M.: Visual tracking: an experimental survey. IEEE Trans. Pattern Anal. Mach. Intell. **36**(7), 1442–1468 (2014)
3. Lucas, B.D., Kanade, T.: An iterative image registration technique with an application to stereo vision. In: International Joint Conference on Artificial Intelligence (1981)
4. Baker, S., Matthews, I.: Lucas-Kanade 20 Years on: a unifying framework. Int. J. Comput. Vision **56**(3), 221–255 (2004)
5. Adam, A., Rivlin, E., Shimshoni, I.: Robust fragments-based tracking using the integral histogram. In: IEEE CVPR (2006)
6. Comaniciu, D., Ramesh, V., Meer, P.: Real-time tracking of non-rigid objects using mean shift. In: IEEE CVPR (2000)
7. Ross, D.A., Lim, J., Lin, R.S.: Incremental learning for robust visual tracking. Int. J. Comput. Vision **77**, 125–141 (2008)
8. Mei, X., Ling, H.: Robust visual tracking using L1 minimization. In: IEEE ICCV (2009)
9. Babenko, B., Yang, M.-H., Belongie, S.: Visual tracking with online multiple instance learning. In: IEEE CVPR (2009)
10. Dietterich, T.G., Lathropand, R.H., Lozano-Pez, T.: Solving the multiple instance problem with axis-parallel rectangles. Artif. Intell. **89**, 31–71 (1997)
11. Kalal, Z., Matas, J., Mikolajczyk, K.: P-N learning: Bootstrapping binary classifiers by structural constraints. In: IEEE CVPR (2010)
12. Hare, S., Saffari, A., Torr, P.H.S.: Struck: Structured output tracking with kernels. In: IEEE ICCV (2011)
13. Grabner, H., Grabner, M., Bischof, H.: Real-time tracking via on-line boosting. In: BMVC (2006)
14. Ell, T.A.: Quaternion Fourier transforms for analysis of 2-dimensional linear-time-invariant partial-diffenrential systems. In: IEEE Conference on Decision and Control, pp. 1830–1841 (1993)

15. Dimitrov, V.S., Cooklev, T.V., Donevsky, B.D.: On the multiplication of reduced biquaternions and applications. Inf. Process. Letters **43**, 161–164 (1992)
16. Henriques, J.F., Caseiro, R., Martins, P., Batista, J.: Exploiting the circulant structure of tracking-by-detection with kernels. In: Fitzgibbon, A., Lazebnik, S., Perona, P., Sato, Y., Schmid, C. (eds.) ECCV 2012, Part IV. LNCS, vol. 7575, pp. 702–715. Springer, Heidelberg (2012)
17. Bolme, D.S., Beveridge, J.R., Draper, B.A., Lui, Y.M.: Visual object tracking using adaptive correlation filters. In: IEEE CVPR (2010)
18. Henriques, J.F., Caseiro, R., Martins, P.: High-speed tracking with kernelized correlation filters. IEEE Trans. Pattern Anal. Mach. Intell. **37**(3), 583–596 (2014)
19. Danelljan, M., Khan, F.S., Felsberg, M., Weijer, J.: Adaptive color attributes for real-time visual tracking. In: IEEE CVPR, pp. 1090–1097 (2014)
20. Zhang, K., Zhang, L., Liu, Q., Zhang, D., Yang, M.-H.: Fast visual tracking via dense spatio-temporal context learning. In: Fleet, D., Pajdla, T., Schiele, B., Tuytelaars, T. (eds.) ECCV 2014, Part V. LNCS, vol. 8693, pp. 127–141. Springer, Heidelberg (2014)
21. Briechle, K., Hanebeck, U.D.: Template matching using fast normalized cross correlation. In: SPIE, vol. 4387, pp. 95–102 (2001)
22. Scholkopf, B., Smola, A.J.: Learning with kernels: Support vector machines, regularization, optimization, and beyond, pp. 125–148. MIT Press (2002)
23. Jia, X., Lu, H., Yang, M.-H.: Visual tracking via adaptive structural local sparse appearance model. In: IEEE CVPR (2012)
24. Zhang, K., Zhang, L., Yang, M.-H.: Real-time compressive tracking. In: Fitzgibbon, A., Lazebnik, S., Perona, P., Sato, Y., Schmid, C. (eds.) ECCV 2012, Part III. LNCS, vol. 7574, pp. 864–877. Springer, Heidelberg (2012)
25. Sevilla-Lara, L., Learned-Miller, E.: Distribution fields for tracking. In: IEEE CVPR (2012)
26. Bao, C., Wu, Y., Ling, H., Ji, H.: Real Time robust L1 tracker using accelerated proximal gradient approach. In: IEEE CVPR (2012)
27. Wu, Y., Shen, B., Ling, H.: Online robust image alignment via iterative convex optimization. In: IEEE CVPR (2012)
28. Zhang, T., Ghanem, B., Liu, S., Ahuja, N.: Robust visual tracking via multi-task sparse learning. In: IEEE CVPR (2012)

High Dynamic Range Image Rendering with a Luminance-Chromaticity Independent Model

Shaobing Gao, Wangwang Han, Yanze Ren, and Yongjie Li[✉]

University of Electronic Science and Technology of China, Chengdu 610054, China
gao_shaobing@163.com, liyj@uestc.edu.cn

Abstract. High dynamic range image (HDR) is widely used since it is capable of capturing more fine information. However, problems remain in its display. A good rendering of HDR color images requires careful treatment of both the brightness and chromaticity information. In this work, we first prove that the global logarithmic mapping of the R, G, B channels may result in desaturation. We then propose an improved way for HDR image rendering. Specifically, by keeping the chromaticity fixated, we use a global transformation and the Retinex-based adaptive filter only in the brightness channel. We finally transfer them back to the RGB space after combining the new brightness and the original chromaticity together. Our model works well in keeping the chromaticity information. Global mapping only in the brightness channel is a good way to avoid desaturation. In addition, our model ensures a good independence between brightness and chromaticity. By applying our method on HDR images, the details in both dark areas and bright areas can be well displayed with better appearance in hue and saturation.

Keywords: HDR · Tone mapping · Retinex · Color constancy

1 Introduction

The dynamic range of an image indicates its luminance range. High dynamic range image (HDR) indicates that the dynamic range of an image exceeds the range of reproducing ability of a display device. The human visual system (HVS) has nice adaptation to HDR scenes. In contrast, the traditional display devices have the limited performance on adapting to HDR scenes. When rendering a HDR image directly on a relative low dynamic range (LDR) device, details of the scene may suffer the loss. Thus, how to render an image on traditional LDR display devices is a nontrivial task [11]. To solve this problem, many HDR rendering algorithms with various motivations have been proposed [2].

Simple rendering algorithms globally compress the dynamic range utilizing a logarithmic function, gamma function or piecewise linear function [16]. These

---

S. Gao and W. Han—Contribute equally to this work.

© Springer International Publishing Switzerland 2015
X. He et al. (Eds.): IScIDE 2015, Part I, LNCS 9242, pp. 220–230, 2015.
DOI: 10.1007/978-3-319-23989-7_23

global mapping functions are mathematically simple and computationally fast. However, globall mapping may cause a serious loss of contrast, specifically the loss of detail visibility on both bright and dark areas of images. In order to overcome the insufficiency of direct global rendering function, a local processing is necessary for compensating the loss of contrast and thus conserving the more visually appealing details of scenes [2,3,17,20]. For example, Reinhard et al. [19] developed a local method based on the photographic dodging and burning technique. Fattal et al. compacted the dynamic range of HDR image using a gradient attenuation function defined by a multiresolution edge detection scheme [4]. Both of these two methods provide an efficient way of compressing the dynamic range while preserving the local contrast information of scenes. However, in order to obtain good performance, these approaches explicitly require careful parameter tuning. In contrast, HVS has the ability of adaptively dealing with the HDR scenes, which has motivated many algorithms to functionally imitate the mechanisms of HVS to reproduce the HDR images [1,5,6,12,14].

Inspired by the color perception of HVS, the Retinex theory proposed by Land [15] has been considered as the first attempt at designing computational model for lightness image recovering. Various versions of Retinex have been implemented for different applications. The simplest Retinex algorithm is max-RGB [7,8,15], which aims at recovering the true color of scenes [9,10,21]. Horn reformulated the Retinex as a two-dimensional Laplacien operator in logarithmic space [12], which can effectively remove the local uniform color cast of illuminant of scenes and enhance the contrast of images. However, directly applying the single-scale Retinex operator on images may introduce the problem of halo artifacts. In order to reduce this problem, Rahman et al. [13,18] suggested to use multiscale Retinex for color restoration (MSRCR) by averaging three single-scale Retinex. This way partially suppresses the halo artifacts but does not ensure a correct rendering of colors [1]. Although some methods [1,5,14] have been proposed to enhance color image by specifically applying the Retinex filter on luminance component of an image that is explicitly expressed in perceptually uniform color space (e.g., HSV/HSL), these color space are not truly luminance-chromaticity independent. Thus, such strategy may also cause color artifacts that magnify color shifts, or lead to color desaturation when processing the luminance.

In 2006, Meylan and Sabine Ssstrunk [16] proposed another Retinex-based algorithm for rendering HDR images. This method is characterized by using principle component analysis (PCA) as a color model and adopting the Retinex-based adaptive filter for local processing. This operator allows better compression in high-contrast areas while increasing the visibility in low contrast areas and can effectively avoid the halo artifacts in dark areas. However, its drawback is the introducing of quite obvious color distortion due to the global compressing of the HDR images.

In this work, we start by probing into the Meylan's model by first proving that the global logarithmic mapping on all of the R, G, B channels in Meylan's method results in desaturation. Then we prove that there are serious problems

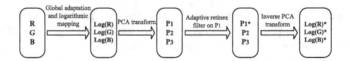

Fig. 1. The flowchart of RBAF algorithm. Adapted from [16].

when using PCA as a model to encode the luminance and chromaticity of image. According to these analyses, we then propose improved models for brightness and chromaticity information. Specifically, by keeping the chromaticity fixated, we use a global transformation and the Retinex-based adaptive filter only in the brightness channel. Since our new pipeline is truly luminance-chromaticity independent, the details in both dark and bright areas of HDR images can be well rendered with better appearance of both hue and saturation information.

The rest of this paper is organized as follows. In Sect. 2, the processing pipeline of Meylan's method [16] is briefly introduced, then we demonstrate that the global logarithmic transformation and PCA-based color model in Meylan's method are the main reasons for producing serious color distortion. Our improved model is introduced and comprehensively evaluated in Sect. 3. Finally, some discussion and concluding remarks are given in Sect. 4.

2   RBAF and It's Color Distortion Problem in Rendering HDR Image

For simplicity, we call Meylan's rendering method [16] as Retinex-Based Adaptive Filter (RBAF). Figure 1 shows the framework of RBAF, which could be primarily described as four steps. There is a power function based global adaptation on the input HDR images before a global logarithmic transformation. However, we have experimentally found that the power function based global adaptation has little effect on the final reproducing results of RBAF. Thus, we ignore such step in RBAF. The input HDR image thus is first undergone a global logarithmic transformation. Then PCA is used to decorrelate the logarithmic RGB representation of the input image into three principal components. Then, RBAF takes the first principal component (P1) as the luminance, and a Retinex-based adaptive filter is used to enhance the visibility of luminance while keeping the chromatic components (P2 and P3) unchanged. Finally, the original luminance is replaced by the filtered luminance (P1*) and the obtained image is transferred back to logarithmic RGB through inverse PCA to obtain the final output.

The key idea of RBAF is to process only the luminance component of images while keeping the chromatic component unchanged. However, from the experimental analysis we found that RBAF has poor ability of keeping the chromatic component unchanged, thus the final output of RBAF introduces serious color distortion. We found two steps in the pipeline of RBAF that lead to this problem. One is the initial logarithmic transformation imposed on HDR images, which results in desaturation. Another one is that the PCA is not a good chromatic model for separating the luminance and chromaticity components of images. The qualitative analysis and mathematical demonstration are as follows.

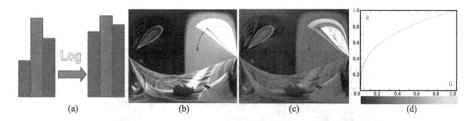

Fig. 2. (a) The histogram of an image before and after log transformation. (b) A HDR scene captured with a normal (non-HDR) camera, which is roughly regarded as the ground truth of a rendered HDR image. (c) The result of a HDR image after log transformation. (d) The relationship between R and G (Color figure online).

Figure 2(a) shows that directly using the logarithmic mapping on R, G, and B channels will reduce the contrast among color channels, thus resulting in desaturation. Figure 2(b) and (c) further indicate that although the global logarithmic mapping on the original HDR images (not shown here) can compress the high intensity areas in images while enhancing the visibility of dark areas, the color of HDR images after logarithmic mapping is seriously desaturated (Fig. 2(c)) in comparison to the color of the ground truth (Fig. 2(b)).

Figure 3 shows two examples of images, which show the color distortion introduced by PCA transformation in RBAF. Although one of the novelties of RBAF is to only process the luminance component of images while keeping the chromatic components unchanged, the experimental results show that the RBAF results in color distortion while achieving good contrast enhancement on luminance. For example, in comparison to the color of ground truth, the color of objects labeled with an arrow on the bottom row of Fig. 3 is shifted from red to yellow after being processed by RBAF. The reason of color distortion is that the PCA is not good enough to be taken as a model that can effectively separate the luminance from the chromaticity of images. The explanation is as follows.

According to the pipeline of RBAF, the PCA transformation on input logarithmic RGB image could be expressed as the following linear algebraic equation:

$$\begin{cases} P_1 = t_{11} \log R + t_{21} \log G + t_{31} \log B \\ P_2 = t_{12} \log R + t_{22} \log G + t_{32} \log B \\ P_3 = t_{13} \log R + t_{23} \log G + t_{33} \log B \end{cases} \tag{1}$$

where $[logR, logG, logB]$ are respectively the R, G, B color channels of input image in the logarithmic space. $[t_{ij}]$ indicates the corresponding transform matrix of PCA. $[P_1, P_2, P_3]$ are the first, second, and third principal components respectively. RBAF assumes that the PCA can encode a chromaticity model by treating

P_1 as the luminance of images, which is independent of the chromaticity of image described by P_2 and P_3. In other words, the chromaticity of images defined by P_2 and P_3 will form a pure chromatic space, which is not affected by the change of the luminance of images defined by P_1.

Fig. 3. (a) Original HDR images (b) Ground truth images of (a), which are captured using normal (non-HDR) cameras without any processing. (c) The results obtained by processing the original HDR images (a) with RBAF.

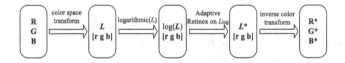

Fig. 4. The proposed pipeline for HDR image rendering.

In order to investigate the independence of chromatic components, we simultaneously fixate P_2 and P_3 in linear algebraic Eq. 1 while P_1 could be arbitrary values. Thus, we derive the Eq. 1 as follows:

$$\begin{cases} P_2 = t_{12} \log R + t_{22} \log G + t_{32} \log B \\ P_3 = t_{13} \log R + t_{23} \log G + t_{33} \log B \end{cases} \tag{2}$$

We solve this linear algebraic equation and get

$$\begin{cases} R = c_1 G^{x_1} \\ B = c_2 G^{x_2} \end{cases} \tag{3}$$

P_2 and P_3 are fixed, thus x_1, x_2, c_1, and c_2 are decided by P_1. If P_2 and P_3 are independent of P_1, the chromatic space defined by P_2 and P_3 should not be changed when P_1 is changing. For example, we probe the color space $[R \ G \ B]$ when setting $c_2 = 0$, $c_1 = 1$, $x_1 = 0.33$. Note all of coefficients in Eq. 3 are decided by P_1 and thus can reflect the changing of P_1. We get

$$\begin{cases} R = G^{0.33} \\ B = 0 \end{cases} \tag{4}$$

If P_2 and P_3 can really represent a luminance independent chromatic space, $[R \ G \ B]$ defined by Eq. 4 should have the same chromaticity. Figure 2(d) shows the relationship between R and G, and the corresponding chromatic space. Obviously, the defined chromatic space $[R \ G \ B]$ with fixed P_2 and P_3 does not have same chromaticity but has undergone the color change that is shifted from red to yellow. This phenomenon is well consistent with the color distortion in Fig. 3, in which the color of objects labeled with the arrow in ground truth image is shifted from red to yellow after being processed by RBAF.

3 Improving RBAF by Using a Luminance-Chromaticity Independent Model

In Sect. 2, we point out that the global logarithmic mapping utilized in RBAF results in desaturation. We also indicate that the PCA based color model is not good enough to separate the luminance from the chromaticity of images, which is the reason that leads to the color distortion in the results of RBAF. In this section, we improve RBAF by using a luminance-chromaticity independent model that can truly process brightness and chromaticity independently. Figure 4 shows the pipeline of the proposed algorithm.

Different from RBAF, here we first use a simple color model to express luminance of image and the chromaticity of image respectively. We will prove that this color model is truly luminance-chromaticity independent. The luminance-chromaticity independent model means that the chromaticity of an image is not affected by the change of luminance of the image.

Our model extracts the luminance component of an image by calculating the maximum value among R,G,B channels for each pixel.

$$L(x,y) = max\{R(x,y), G(x,y), B(x,y)\} \tag{5}$$

Then, the computed luminance of each pixel is used to normalize each of the R, G, and B color values.

$$r(x,y) = \frac{R(x,y)}{L(x,y)}, g(x,y) = \frac{G(x,y)}{L(x,y)}, b(x,y) = \frac{B(x,y)}{L(x,y)} \tag{6}$$

Then a globally logarithmic mapping is only applying on $L(x,y)$ while keeping the normalized chromaticity $[r(x,y), g(x,y), b(x,y)]$ unchanged, which is different from RBAF. After global logarithmic processing on $L(x,y)$, local adaptation is performed on $L(x,y)$ using the surround-based Retinex method designed in RBAF [16], which is expressed as

$$L^*(x,y) = log(L(x,y)) - \beta(x,y) \cdot log(L_{Gauss}(x,y)) \tag{7}$$

where $log(L(x,y))$ indicates the logarithmic mapping on $L(x,y)$, $log(L_{Gauss}(x,y))$ is computed with surround mask. For simplicity, we directly use the technique in RBAF [16] for computing surround mask $log(L_{Gauss}(x,y))$. $\beta(x,y)$ weights the surround mask based on the pixel value at coordinate (x,y)

$$\beta(x,y) = 1 - \frac{1}{1 + e^{-7 \cdot (L(x,y) - 0.5)}} \tag{8}$$

We finaly replace the original luminance $L(x,y)$ with $L^*(x,y)$ and recover a RGB image for displaying by

$$\begin{cases} R^*(x,y) = L^*(x,y) \cdot r(x,y) \\ G^*(x,y) = L^*(x,y) \cdot g(x,y) \\ B^*(x,y) = L^*(x,y) \cdot b(x,y) \end{cases} \tag{9}$$

3.1 Analysis of the Independence of Luminance and Chromaticity

Our algorithm takes the color coefficient $[r(x,y), g(x,y), b(x,y)]$ normalized by luminance $L(x,y)$ as chromatic component, which is truly independent from brightness component. Figure 5 shows the relationship between the color coefficient $[r(x,y), g(x,y), b(x,y)]$ and the luminance $L(x,y)$. This figure demonstrates that the chromaticity keeps almost unchanged when the luminance is changing.

Fig. 5. (a) chromaticity diagram. (b) The ground truth of the scene. (c) Result processed by RBAF. (d) Result processed by our algorithm (Color figure online).

For example, the chromaticity (e.g., the point labeled with arrow) almost keeps same when the luminance on vertical coordinate varies. Moreover, as analyzed above, the PCA-based color model is not good enough to separate the luminance from chromaticity of images and thus results in color distortion (Fig. 5(b) and (c)). In contrast, our algorithm overcomes the color distortion of RBAF and achieve good performance on color fidelity (Fig. 5(d)).

To evaluate our luminance model, another common way for computing the luminance of image is used, i.e., the mean of RGB values of each pixel is adopted as luminance component. Figure 6 shows the reproducing results by taking two different strategies for luminance computation. From this figure we can obviously observe that taking the max of RGB values as luminance obtains better performance of rendering than taking the mean of RGB values as luminance.

In contrast to other linear combination based luminance model (e.g., mean and PCA), the results also indicate that the luminance model $L(x, y)$ is more effective on controlling the dynamic range of images, which can greatly enhance the luminance contrast and visibility of HDR images.

Fig. 6. Top: The rendering results when taking the max of RGB values as luminance. Bottom: The results when taking the mean of RGB values as luminance (Color figure online).

3.2 Experimental Results and Comparison

We compared our method with other typical tone mapping methods, including: the MSRCR of Rahman *et al.* [13,18], the RBAF of Meylan [16]. Figure 7 totally show six commonly used HDR images. From the viewpoint of luminance contrast enhancement, the MSRCR performs very poor on luminance enhancement on all of the HDR images listed here. There are serious halo artifacts in the outputs of MSRCR, mainly because that MSRCR is based on a fixed Retinex filter [16,18].

In contrast, both RBAF and our method employ an adaptive Retinex filter to enhance the luminance and thus avoid the problem of halo artifacts. Although both RBAF and our method achieve very good performance on luminance enhancement, our method performs better than RBAF on local luminance contrast enhancement(e.g., the local contrast between sky and cloud in the third and fourth rows of Fig. 7)

We further investigated the performance of color fidelity between RBAF and our method. Both RBAF and our method perform well on enhancing the luminance visibility of scenes. However, RBAF obtains poor performance on keeping the color fidelity in comparison to our method. The results of RBAF on color enhancement are consistent with the previous analysis that the global logarithmic mapping on all of the R,G,B channels results in desaturation (e.g., the color checker in the first row of Fig. 7). In contrast, by applying our method, the details in both dark areas and bright areas are well enhanced while remaining better hue and saturation. Moreover, there is contrast loss of color information in the results of RBAF, since the PCA is not a good color model for separating the luminance from chromaticity of images. For example, all of the images in Fig. 7 seem grayer and unnatural after be enhanced by RBAF. However, our method provides a good appearance of colors on all of the processed images.

Original HDR image   Proposed   RBAF   MSRCR

Fig. 7. Results of different methods. From left to right: Original HDR images, the results of the proposed algorithm, RBAF [16], and MSRCR [13,18], respectively.

4   Conclusion and Discussion

HDR image rendering has been widely studied and a large number of algorithms exist. They enhance the quality of the rendered images, but still suffer from various problems. One of the common drawbacks is the color distortion when enhancing the local contrast and luminance visibility on both bright and dark areas. We provided in this work a method to render HDR images by improving the Retinex based adaptive filter (RBAF). RBAF was inspired from by the Retinex theory of color vision and has its efficiency in increasing the local luminance contrast while preventing halo artifacts.

However, RBAF performs poor on color enhancement. We proved that global logarithmic mapping on R,G,B channels utilized by RBAF is the main reason of resulting in desaturation. Moreover, we also pointed out that PCA is not a good model for separating the luminance from the chromaticity of images and it is the key reason of serious color shift in the results of RBAF. In order to overcome the problems of RBAF, we improved the RBAF by adopting a model which is truly color and luminance independent. In particular, our method uses the max among R,G,B values of each pixel as luminance component of images and utilizes it to normalize the R,G,B values of each pixel to obtain the chromaticity component of images. In comparison to other models (e.g., PCA-based [16] or HSV-based [1,5,14]), this simple model is really luminance-chromaticity independent and thus it can minimize the chromatic changes induced by the processing of luminance. Moreover, we only apply the global logarithmic mapping on the luminance channel, which can avoid the problem of desaturation introduced by global logarithmic mapping on all of the R,G,B color channels.

We tested our method on various HDR images, compared it with other typical algorithms and showed that it efficiently increases the luminance visibility while preventing color distortion. As an important future direction, we are interested in quantitatively testing the color fidelity of an image after being processed by HDR rendering algorithms. A feasible way is to use the angular error, which is frequently utilized in color constancy literature [7,9], to measure the color difference between the original HDR image and the enhanced image.

Acknowledgments. This work was supported by the 973 Project under Grant 2013 CB329401 and the NSFC under Grant 61375115, 91420105. The work was also supported by the 111 Project of China (B12027).

References

1. Choi, D.H., Jang, I.H., Kim, M.H., Kim, N.C.: Color image enhancement using single-scale retinex based on an improved image formation model. In: Proceedings of the EUSIPCO (2008)
2. Devlin, K.: A review of tone reproduction techniques. Technical report, Computer Science, University of Bristol, CSTR-02-005 (2002)
3. Drago, F., Myszkowski, K., Annen, T., Chiba, N.: Adaptive logarithmic mapping for displaying high contrast scenes. Comput. Graph. Forum. **22**, 419–426 (2003). Wiley Online Library
4. Fattal, R., Lischinski, D., Werman, M.: Gradient domain high dynamic range compression. ACM Trans. Graph. (TOG) **21**, 249–256 (2002)
5. Fu, X., Sun, Y., LiWang, M., Huang, Y., Zhang, X.P., Ding, X.: A novel retinex based approach for image enhancement with illumination adjustment. In: 2014 IEEE International Conference on Acoustics, Speech and Signal Processing (ICASSP), pp. 1190–1194. IEEE (2014)
6. Funt, B., McCann, J., Ciurea, F.: Retinex in matlab. J. Electron. Imag. **13**(1), 48–57 (2004)

7. Gao, S., Han, W., Yang, K., Li, C., Li, Y.: Efficient color constancy with local surface reflectance statistics. In: Fleet, D., Pajdla, T., Schiele, B., Tuytelaars, T. (eds.) ECCV 2014, Part II. LNCS, vol. 8690, pp. 158–173. Springer, Heidelberg (2014)

8. Gao, S., Li, Y.: A retinal mechanism based color constancy model. In: Liu, C.-L., Zhang, C., Wang, L. (eds.) CCPR 2012. CCIS, vol. 321, pp. 422–429. Springer, Heidelberg (2012)

9. Gao, S., Yang, K., Li, C., Li, Y.: A color constancy model with double-opponency mechanisms. In: 2013 IEEE International Conference on Computer Vision (ICCV), pp. 929–936. IEEE (2013)

10. Gao, S., Yang, K., Li, C., Li, Y.: Color constancy using double-opponency. IEEE Trans. Pattern Anal. Mach. Intell. (2015) (in press). doi:10.1109/TPA-MI.2015. 2396053

11. Hood, D.C., Finkelstein, M.A.: Sensitivity to light. In: Boff, K.R., Kaufman, L., Thomas, J.P. (eds.) Handbook of Perception and Human Performance. Sensory Processes and Perception, vol. 1. Wiley, New York (1986)

12. Horn, B.K.: Determining lightness from an image. Comput. Graph. Image Process. **3**(4), 277–299 (1974)

13. Jobson, D.J., Rahman, Z.U., Woodell, G.A.: A multiscale retinex for bridging the gap between color images and the human observation of scenes. IEEE Trans. Image Process. **6**(7), 965–976 (1997)

14. Kimmel, R., Elad, M., Shaked, D., Keshet, R., Sobel, I.: A variational framework for retinex. Int. J. Comput. Vis. **52**(1), 7–23 (2003)

15. Land, E.H., McCann, J.: Lightness and retinex theory. JOSA **61**(1), 1–11 (1971)

16. Meylan, L., Susstrunk, S.: High dynamic range image rendering with a retinex-based adaptive filter. IEEE Trans. Image Process. **15**(9), 2820–2830 (2006)

17. Pattanaik, S.N., Ferwerda, J.A., Fairchild, M.D., Greenberg, D.P.: A multiscale model of adaptation and spatial vision for realistic image display. In: Proceedings of the 25th annual conference on Computer graphics and interactive techniques, pp. 287–298. ACM (1998)

18. Rahman, Z.U., Jobson, D.J., Woodell, G.A.: Retinex processing for automatic image enhancement. J. Electron. Imag. **13**(1), 100–110 (2004)

19. Reinhard, E., Stark, M., Shirley, P., Ferwerda, J.: Photographic tone reproduction for digital images. ACM Trans. Graph. (TOG) **21**, 267–276 (2002)

20. Webster, M.A.: Human colour perception and its adaptation. Netw. Comput. Neural Syst. **7**(4), 587–634 (1996)

21. Yang, K., Gao, S., Li, C., Li, Y.: Efficient illuminant estimation for color constancy using grey pixels. In: 2015 IEEE Conference on CVPR, pp. 1–10 (2015)

Graph Cuts Based Moving Object Detection for Aerial Video

Xiaomin Tong[✉], Yanning Zhang, Tao Yang, and Wenguang Ma

School of Computer Science ShaanXi Provincial Key Laboratory of Speech and Image
Information Processing, Northwestern Polytechnical University, Xi'an, China
xmtongnwpu@gmail.com, ynzhang@nwpu.edu.cn, yangtaonwpu@163.com

Abstract. With the wide development of UAV technology, moving target detection for aerial video has become a hot research topic in the computer vision. Most existing methods are under the registration-detection framework and can only deal with simple background scenes. They tend to go wrong in the complex multi background scenarios, such as viaduct, building and trees. In this paper, we break through the single background constraint and perceive the complex scene accurately by automatic estimation of multiple background models. First, we divide the scene into several color blocks and estimate the dense optical flow. Then, we calculate an affine transformation model for each block with large area and merge the consistent models. Finally, we calculate subordinate degree to multi-background models pixel to pixel for all small area blocks. Moving objects are segmented by energy optimization method solved via Graph Cuts. The extensive experimental results on public aerial videos show that, due to multi background models estimation, analyzing each pixels subordinate relationship to multi models by energy minimization, our method can effectively remove buildings, trees and other false alarms and detect moving objects correctly.

Keywords: Aerial video · Object detection · Multi-model estimation · Graph cuts

1 Introduction

Moving target detection for aerial video is one of the core technologies of UAV surveillance system. It can be widely applied in military domain such as battlefield reconnaissance and surveillance etc. Also, it can support civil purposes such as border patrol, nuclear radiation detection, etc. Due to its wide application, low cost, high cost effectiveness, no casualties risk, strong survival ability, good maneuvering performance and convenience, moving target tracking algorithm for aerial video has become a hot research topic in the computer vision. This study can not only make UAVs eyes more clearly, but also guarantee advanced processing and application such as behavior analysis, importance analysis.

We are faced with core difficulties in moving object detection for aerial video, such as motion mutation caused by UAV fast motion, low resolution noisy

© Springer International Publishing Switzerland 2015
X. He et al. (Eds.): IScIDE 2015, Part I, LNCS 9242, pp. 231–239, 2015.
DOI: 10.1007/978-3-319-23989-7_24

images, small target, low contrast, complex background, scale changes, occlusion, etc. With the UAV development, researchers have proposed many algorithms to solve above problems. However, these methods are all under registration-detection framework, which assumes only single background in scenario and will identify all the regions generating parallax error as targets. As a result, tracking failure usually happens in the complex scenarios with multi background, trees, buildings and so on. Therefore, the current tracking cannot satisfy application need and it is very necessary to break through the technology in complex scene.

Automatic estimation of multiple background models for complex scenario can provide a solution for perceiving the scene accurately. This paper first focuses on automatic estimation of multiple background models for complex scenarios. Then the pixels motion information and subordinate degrees to multi-background models are analyzed by optical flow. Based on the neighborhood information and the subordinate degree, we segment the moving objects via energy minimization. Since we estimate multiple background models and perceive the complex scene correctly, our algorithm can effectively remove the buildings, trees and other false alarms and improve the locating precision. In addition, the adoption of energy minimization which makes both use of the analysis of neighborhood continuity and subordinate degree can significantly improve segmentation precision.

The rest of this paper is organized as follows. Section 2 summarize and analyze the related work in recent years. Section 3 proposes the moving object detection algorithm based on multi-model estimation for aerial video of complex scenario. The experimental results are reported in Sect. 4, which demonstrate the accuracy and effectiveness of our approach. Finally, the conclusions are drawn in Sect. 5.

2   Related Work

Moving object detection for aerial video [1] is widely used in battlefield reconnaissance, border surveillance and traffic monitoring, etc. The existing moving object detection algorithms for aerial video are mainly under bottom-up framework, also named as Data-driven method, which does not rely on prior knowledge and extract the moving information directly from the image sequences. Using bottom-up method to realize moving object detection for aerial video mainly includes three steps: (1)image matching, (2)object detection, (3)object classification. The existing algorithms for moving object detection include classical COCOA system [2]. The procedure of this system contains image stabilization, frame difference and block tracking. However, this algorithm often fails in scenario scaling due to the Harris corner-based image stabilization. Cohen I et al. [3,4] proposed a moving object detection and tracking system. First they aligned the images by estimating the affine transformation model iteratively. Then the normalized optical flow field was applied for the moving detection and the graph model was constructed to resolve and maintain the dynamic template of moving objects. This system run fast but it cannot solve the complex scaling scenarios.

Aryo Ibrahim [5] proposed the MODAT framework. Instead of Harris corner, they adopted SIFT features to fulfill the image matching. However, all above methods can only deal with simple background scenes and assume only the moving objects can cause the parallax error. They tend to go wrong in the complex multi background scenarios, such as viaduct, building and trees. Chad et al. [6] proposed a moving object detection method for aerial video with low frame rate. They constructed an accurate background model to solve object detection and shadow problem. However, the application of this method is restricted because we need to know the camera calibration parameters in advance and start tracking object manually. Shen et al. [7] proposed a moving object detection method for aerial video based on spatiotemporal saliency. However, this method still cannot overcome the parallax error problem and the false alarm rate is high in complex scenarios. As shown in Fig. 1(b), the false alarms occurred at building and trees (labeled by the red circles) by using the traditional method. The real objects may be missed due to the inaccurate model estimation. In this paper, we propose a moving object detection algorithm based on multi-model estimation for aerial video and Graph Cuts [8] is utilized to divide the foreground pixels into different objects. Our method can not only handle the moving object detection in the complex multiple background scenarios, but also remove the buildings, trees and other false alarms effectively.

(a)Original image   (b)Moving detection by traditional method   (c)Moving detection by our method

Fig. 1. The comparison of moving detection results by different methods.

3   Multi-Model Estimation Based Moving Object Detection

In order to overcome influence of the complex multi background scenarios, this paper proposes a moving object detection algorithm for aerial video based on multi-model estimation. First, the scene is divided into several color blocks. Then, the affine transformation model is estimated based on the dense optical flow. Third, subordinate degree is computed between each pixel and multi-background model to judge if the pixel belongs to moving object or not. Finally, moving objects are segmented by energy optimization method solved via Graph Cuts.

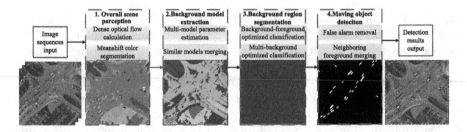

Fig. 2. The flowchart of moving object detection based on multi-model estimation.

3.1 Algorithm Flow

The flowchart of the proposed framework is shown in Fig. 2. Our approach mainly includes four steps: overall perception of scene, background model extraction, background region segmentation and moving object detection. First, the overall perception of scene divides the scene into several color blocks and estimates the dense optical flow. Here, Mean shift pyramid segmentation method from OPENCV is adopted for color blocks segmentation and Gunnar Farneback algorithm [9] is used for computing dense optical flow. Second, to confirm the multiple background models included in the scenario, background model extraction calculates affine transformation model for multiple color block and merges the consistent models. Then in the third step, the background region segmentation will be transformed to background and foreground binary classification, multi background region and multi label classification problem. This problem can be solved by the energy optimization method, which can obtain smooth and continuous global optimal solution. Fourth, after obtaining foreground region, we merge the blocks and remove false objects based on the moving consistency and region proximity. Afterwards, the moving object detection is finished and the accurate detected results are obtained.

3.2 Multi-model Estimation

Estimating the background model parameters of complex scenarios accurately can ensure the correct scene perception, accurate object segmentation and robust object tracking. The current multi-model estimation methods like JLinkage, only adapts to small samples and unable to solve the big samples like multi-model estimation under complex scenarios. In the aerial video, the background blocks with consistent color often belong to the same background and the background area is much lager than that of objects. Therefore, this paper first divides the scenarios into color blocks and selects the blocks with big area as the candidate background blocks. Afterwards, affine transformation model is estimated for each background block.

3.3 Background Segmentation Based on Graph Cuts

We define the set of points which does not belong to the large background region as Ω. Then points of Ω can be judged as background region point or not based on the existing multiple background models and large background regions. This paper proposes an energy minimization based algorithm for optimized classification. First, we define the scenario point belonging to category $l = BNum + 1$, where $BNum$ is the number of background regions, and the rest one is the foreground region. For each category, we need to define and solve a label function $f : \Omega \to L$ where $L = \{0, 1, 2, ..., BNum\}$ are all the possible category labels for all the points in Ω. Label $i > 0$ corresponds to the ith background model m_i. Label 0 corresponds to no background models but corresponds to the category of foreground points. Given an pixel p, if $f(p) > 0$, it belongs to background points. Otherwise if $f(p) = 0$, this pixel belongs to foreground points. Denoting energy function as follows:

$$E(f) = E_d(f) + E_s(f) \tag{1}$$

where data term E_d represents the sum of classification cost of all the classified points. The smooth term is a regularizer that encourages neighboring pixels to share the same label. Therefore, the classification problem is transformed to minimizing $E(f)$ and finding corresponding solution. However, minimizing $E(f)$ directly is very difficult due to the above classification problem is the coupling of foreground and background, and background and background classification. This paper decomposes above problem into two optimized solution modules $f = \{fs, fc\}$: 1) optimizing fs for background segmentation, 2) optimizing fc for classifying different background categories. In the first module, in order to segment the background regions, we transform this optimized classification problem to solve the binary energy minimization. If a pixel belongs to background, its label is 0, otherwise 1. The energy function includes a one variable data term and pairwise smoothing terms, where data term represents the cost of dividing pixels to background. The smoothing term corresponds to the continuous smoothness prior of the background region. The Graph Cuts is adopted for optimizing and solving energy minimization problem. In the second module, the problem of dividing background points into different background models is transformed to a multi-labeling energy minimization problem, which can also be solved via Graph Cuts.

3.4 Moving Object Detection

The pixels divided into foreground may come from true moving object, and also may belong to false alarm of parallax error caused by buildings and others. How to distinguish these two category points is the key of segmenting moving object accurately. As we known, when moving target is compensated by the background model, the parallax error only causes by the object itself, which represents the absolute motion vector of the object. Then the object motion between two neighboring frames is approximately linear motion. As a result, the motion vectors of inliers belonging to one object are similar. According to above analysis, we will first calculate the motion vectors of foreground blocks compensated by the background model, and determine the moving object by analyzing similarity of the motion vectors.

4   Experimental Results and Analysis

In order to evaluate the proposed multi-model estimation based moving object detection algorithm for aerial video, we perform the comparison experiments on the public DAPAR VIVID and KIT AIS Data Set database. In DAPAR VIVID database (http://vision.cse.psu.edu/data/vividEval/datasets/datasets.html), Eg Test01 dataset contains many moving cars but the background is relatively simple. In KIT AIS Data Set (http://www.ipf.kit.edu/downloads.php), shooting frame rate is 1FPS and it includes viaduct, overpasses, buildings, trees and other complex scenarios, which is very challenging for the moving object detection algorithms of aerial video. Shen et al. [7]proposed a moving object detection method for aerial video based on spatiotemporal saliency. This method can accurately handle moving target detection under simple scenarios. However, it has not adopted multiple background analysis for the scenarios, and detection missing and false alarm will happen frequently in complex scenarios.

We compare our algorithm with Shen [7] on KIT AIS Data Set. The results are shown in Fig. 3. StuttgartCrossroad01 dataset (Fig. 3(a)) contains overpasses multiple background complex scenarios as well as complex elements such as trees and shadow which will influence the detection results. All the factors will bring substantial challenge to the detection algorithms. In the 3th column of Fig. 3(a), the objects in blue boxes are the detection results of this paper. The objects in red boxes are stationary targets. The detection results show our approach can segment and detect moving objects accurately in the complex background situation of overpasses. Since Shen [7] approach cannot perceive multiple background of the scenario and cannot obtain accurately background information, the situations such as inaccurate moving segmentation and false alarm target will happen. We can see these situations in the 4th column of Fig. 3(a). The blue bounding boxes show the detected objects. The yellow boxes show the false detection and missing targets. Figure 3(b) shows the comparison results on the MunichCrossroad01 dataset. The characteristic of this dataset is that the false objects of parallax error caused by the trees and other elements occupy a large proportion of the image area. As shown in the 4th column of Fig. 3(b), the detection rate of traditional method is low and the false alarm is high. Figure 3(c) shows the detection results on Munich Crossroad02 dataset with many buildings. The results demonstrate our approach can perceive scenario and detect moving objects correctly due to multiple background models estimation.

In order to quantitatively analyze the detection accuracy of this paper, we calculate recall R, accuracy P and comprehensive evaluation indicators $F1$ on different datasets. Higher $F1$ indicates the experimental method is more effective.

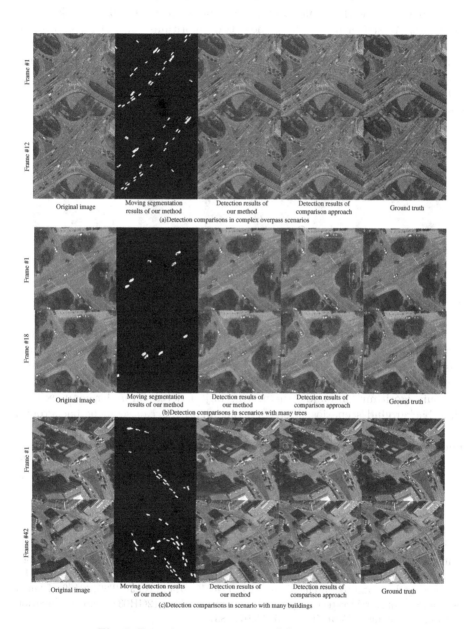

Fig. 3. Detection comparisons in different scenarios.

Figure 4 shows the comparison results of our paper and Shen [7] method. The results from left to right are the statistical results of DAPAR VIVID EgTest01, StuttgartCrossroad01, MunichCrossroad01 and MunichCrossroad02 of KIT AIS Data Set. As shown in Fig. 4(a), these two algorithms can both achieve high detection rate under simple background and their detection precisions are similar. However, $F1$ of our algorithm under complex background is higher than Shen [7], i.e. on StuttgartCrossroad01 dataset, $F1$ of our result is 0.949, which is higher than 0.808 of Shen [7]. These results show the significant superiority of our algorithm, as shown in Figs. 4(b), (c) and (d).

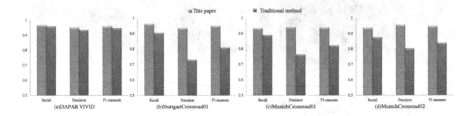

Fig. 4. The statistical result of our method and the traditional method.

5   Conclusions

This paper studies the moving object detection problem under complex scenarios for aerial videos. We propose a novel moving object detection algorithm based on multi-model estimation and optimized classification.Since we break through the single background constraint and adopt multiple background models, our algorithm can handle the moving object detection under complex multiple background scenarios. Moreover, our algorithm can segment background and foreground region accurately due to the using of Graph Cuts, optical flow information and continuous smooth constraints. The experimental results on many aerial videos indicate our algorithm can correctly perceive multiple background information of scenarios and detect moving object accurately in multiple background, buildings and other the complex scenes caused by parallax.

Acknowledgment. This work is supported by the National Natural Science Foundation of China (No.60903126,No.61272288,No.61303123, No.61231016), Shaanxi Provincial Natural Science Foundation (2013JM8027), 2013 New People and New Directions Foundation of School of Computer Science in NPU (No.13GH014604), the NPU Foundation for Fundamental Research (No.JC201120,No.JC201148,No.JC T20130109), Plan of Soaring Star of Northwestern Polytechnical University (No. 12GH0311), and Foundation of China Scholarship Council (No.201303070083).

References

1. Zhu, S., Wang, D.W., Low, C.B.: Ground target tracking using UAV with input constraints. J. Intell. Robot. Sys. **69**(1–4), 417–429 (2013)

2. Ali, S. and Shah, M.: COCOA: Tracking in aerial imagery. In: Airborne Intelligence, Surveillance, Reconnaissance (ISR) Systems and Applications, III, p. 62090D (2006)
3. Cohen, I. and Medioni, G.: Detecting and tracking moving objects for video surveillance. In: Proceedings of the IEEE Computer Society Conference on Computer Vision and Pattern Recognition, **2**, pp. 319–325. IEEE Press (1999)
4. Medioni, G., Cohen, I., Bremond, F., Hong, S., Nevatia, R.: Event detection and analysis from video streams. IEEE Trans. Pattern Anal. Mach. Intell. **23**, 873–889 (2001)
5. Ibrahim, A.W.N., Pang, C.W., Seet, G.L.G., Lau, W.S.M., Czajewski, W.: Moving objects detection and tracking framework for uav-based surveillance. In: Pacific-Rim Symposium on Image and Video Technology (PSIVT), pp. 456–461 (2010)
6. Aeschliman, C., Park, J., Kak, A.C.: Tracking vehicles through shadows and occlusions in wide-area aerial video. IEEE Trans. Aerosp. Electron. Sys. **50**(1), 429–444 (2014)
7. Shen, H., Li, S.X., Zhu, C.F., Chang, H.X., Zhang, J.L.: Moving object detection in aerial video based on spatiotemporal saliency. Chin. J. Aeronaut. **26**(5), 1211C1217 (2013)
8. Boykov, Y., Kolmogorov, V.: An experimental comparison of in-cut/max-flow algorithms for energy minimization in vision. IEEE Trans. Pattern Anal. Mach. Intell. (PAMI) **26**(9), 1124–1137 (2004)
9. Farnebäck, G.: Two-frame motion estimation based on polynomial expansion. In: Bigun, J., Gustavsson, T. (eds.) SCIA 2003. LNCS, vol. 2749, pp. 363–370. Springer, Heidelberg (2003)

Compressed Hashing for Neighborhood Structure Preserving

Lijie Ding, Xiumei Wang$^{(\boxtimes)}$, and Xinbo Gao

School of Electronic Engineering, Xidian University, Xi'an 710071, China
dinglijiedlj@163.com, wangxm@xidian.edu.cn,
xbgao@mail.xidian.edu.cn

Abstract. Hashing methods play an important role in large-scale image retrieval, and have been widely applied due to fast approximate nearest neighbor search and efficient data storage. However, most existing hashing methods are not taken the low-dimensional manifold into account in nearest neighbors search. In this paper, we propose an effective hashing method to preserve the intrinsic structure of high-dimensional data points in the low-dimensional manifold space. In particular, we introduce a compressed algorithm to learn a smaller synthetic data set represent the database in the original space so that the approximated nearest neighbors can be quickly discovered. To this end, we exploit the manifold learning generate appropriate binary code. Experimental results on benchmark data sets show that the proposed approach is effective in comparison with state-of-the-art methods.

Keywords: Approximate nearest neighbor search · Hashing · Manifold · Precision-recall

1 Introduction

In recent years, with the rapid development of the Internet, cloud computing and mobile devices. The amount of global data comes into the era of ZB, and still growthing in the form of exponential each year. How to collect, store, analyse, visualization display effective and make full use of large-scale data to obtain benefits become a core problem. Among large-scale data processing technologies, Approximate Nearest Neighbor (ANN) search [1] is widely used in Content Based Image Retrieval (CBIR), object recognition and classification.

As an effective ANN search technology, hashing methods have attracted more and more attention. Hash can transform an image into compact binary coding sequence so that the high-dimensional data in original space embedded into low-dimension Hamming space. Therefore, hashing methods can save memory space greatly because the dimension of the short hash codes is much lower than that of high-dimensional samples. What's more, hash codes as binary codes in the computer memory, and the Hamming distance among hash codes can be calculated through the XOR operation, thereby hashing can realize effective and fast ANN search. An issue of particular interest in ANN search is to obtain compact representations that can approximate the data distances, which requires the hash algorithm to satisfy the following conditions.

X. He et al. (Eds.): IScIDE 2015, Part I, LNCS 9242, pp. 240–248, 2015.
DOI: 10.1007/978-3-319-23989-7_25

First of all, given a new query point, the algorithm can quickly calculate the hash codes. Second, the hash codes preserve as much information as possible. Finally, hashing method maps similar images to similar binary codes.

Some classical hash methods have been proposed for ANN search. One of the most well-known hashing methods is Locality Sensitive Hashing (LSH) [2], which gets binary hash codes by random projects and quantizes high-dimensional data. But LSH needs long hash codes to maintain data similarity and desired discriminative power, which will lead to a larger storage cost and time consumption. In contrast to the random projections hash scheme, recent much research work has focused on obtaining compact codes based learning scheme. There are a variety of powerful hash techniques have improved the compactness of hash codes, such as Iterative Quantization (ITQ) [3] uses sequential projections to alleviate the problem of unbalanced variances and continuously reduce the quantization error to learn hash function. Optimized product quantization (OPQ) [4] puts forward to decomposing high-dimensional space data to low-dimensional space, and to get the effectively hash codes by minimizing the quantization distortion associated with spatial decomposition and code books. In addition, K-means hashing (KMH) [5] can learn binary compact codes through partition the feature space by k-means quantization and estimates the Euclidean distance of the original space by Hamming distance of each cluster indexes. Although hash algorithm based on product quantization [6] can get effective retrieval performance, the process to optimize objective function is difficult and with restrictions on constructing the data feature. Besides, hash approaches based on Graph were proposed. Spectral hashing (SH) [7] and anchor graph hashing (AGH) [8] construct undirected graph to preserve similarity relationship between data points, then solve the similarity matrix eigenvalue and eigenfunction by spectral graph theory to obtain effective and efficient hash functions.

Manifold learning methods [9, 10] aim to preserve the neighborhood structure of samples by building graph model of high-dimensional data points in the low dimensional manifold space, which is favorable for large-scale hashing. SH and AGH are hashing approaches based on manifold. However, SH makes an impractical assumption that the data should uniformly distribute in the high dimensional space and the manifold structure of original data points is destroyed. Towards to AGH, not all graph eigenvectors generated are suitable for hashing especially when hash bits increases. In this paper, an effective hashing approach is proposed. It introduces a data compressed algorithm to learn a new set of synthetic reference vectors represent data in the original space, which can avoid the infeasibility when find data point's nearest neighbors at the large scale database and speed up the computation when to find the approximated nearest neighbors. This process is free from any restrictive data distribution assumption. Moreover, learning a low-dimensional embedding and reserving the neighborhood distribution for synthetic reference vectors by manifold learning so that improve suitable for hashing through removing the graph eigenvector style.

The rest of the paper is organized as follows: The compressed hashing for neighborhood structure preserving approach (SNCH) and its analysis are presented in Sect. 2. Show the setting of our experiments and experimental results in Sect. 3. Finally, we conclude the paper in Sect. 4.

2 Compressed Hashing for Neighborhood Structure Preserving

Given a data set $X = \{x_1, x_2, \ldots, x_n\} \subset \mathbb{R}^d$, where each sample is a d-dimensional column vector and there are n training samples. The aim of the proposed method is to determine a group of hashing functions H, which can transform the samples into r-bit binary hash codes.

Aiming to quickly find nearest neighbors relationships among the data points and solve the question of high complexity for large-scale image retrieval. Instead of conducting the exact nearest neighbor relationships among the whole database, approximated nearest neighbor relationships can be constructed. To describe our algorithm, we first learn a compressed set $S = [s_1, s_2, \ldots, s_m]$ as synthetic data set [11], where $m \ll n$. Then we approximate data neighborhood structure, similarities between a sample x_i and the compressed data point s_j can be computed using the squared Euclidean distance by:

$$w(x_i, s_j) = \exp\left(-\frac{||x_i - s_j||^2}{\sigma^2}\right) \tag{1}$$

We can convert similarities into a probability via the following formula:

$$p_{ij} = \frac{w(x_i, s_j)}{\sum\limits_{i=1}^{m} w(x_i, s_j)} \tag{2}$$

In information theory, to match two distributions as well as possible can be achieved by minimizing a cost function which is a sum of Kullback-Leibler divergences [12] between distribution P and Q for each object:

$$L_{KL}(Q_i||P_i) = \sum_i \sum_j q_{ij} \log \frac{q_{ij}}{p_{ij}} \tag{3}$$

In order to match compressed set and the whole database, ideally, we want the probability $Q_i = 1$ for all training data set. We define our loss function to sum over the Kullback-Leibler divergences between this "perfect" distribution and distribution p_i for all inputs in whole data set X:

$$L_{KL}(S) = -\sum_{i=1}^{n} \log(p_i) \tag{4}$$

For convenience, we introduce two matrices P and T respectively:

$$P_{ij} = \frac{p_{ij}}{p_i}, \quad T_{ij} = (\delta_{y_i, y_j} - p_i) \tag{5}$$

where, $\delta_{ij} \in \{0,1\}$ and take on value 1 if and only if object i and j belong to same cluster. We minimize the objective with gradient descent with respect to the compressed set S:

$$\frac{\partial L}{\partial S} = 2\big(X(T \circ P) - S\big(diag\big((T \circ P)^T 1_n\big)\big)\big) \tag{6}$$

Here $\circ$ is the Hadamard product, $1_n = [1, 1, \ldots, 1]^T \in \mathbb{R}^n$.

Avoid to graph eigenvectors generated are not suitable for hashing, we introduce a non-spectral embedding method t-distributed stochastic neighborhood embedding (t-SNE) [13]. t-SNE provides an effective technique for dimensionality reduction. Moreover, t-SNE can model similar data points by means of small pairwise distance, so that keep the low-dimensional representations of very similar high-dimensional data points that lies on or near a low-dimensional, non-linear manifold close together and preserve as much of the significant structure of the high-dimensional data as possible in the low-dimensional map. To obtain low-dimensional data representations $Y_s = [y_1, y_2, \ldots, y_m]$ in regard to the high-dimensional compressed data set, we minimizing the Kullback-Leibler divergence (3) with respect to conditional probability p_{ij} in the high-dimensional compressed data set and the joint probability defined using the t-distribution in the low dimensional embedding space.

The goal about our hashing method is to learn binary codes such that neighbors in the input space are mapped to similar codes in the Hamming space. Given a new input x_q, we are approximate this data point neighborhood structure in S by minimize the following question:

$$\min \sum_{i=1}^{n} w(x_q, s_i) \|y_q - y_i\|^2 \tag{7}$$

y is defined as manifold-based embedding. Differentiating objective function with respect to y_q, and making derivative is 0 will lead to the optimal solution

$$y_q = \frac{\sum_{i=1}^{n} w(x_q, s_i) y_i}{\sum_{i=1}^{n} w(x_q, s_i)} \tag{8}$$

Therefore, the low-dimensional embedding for the training data becomes

$$Y = W_{XS} Y_S \tag{9}$$

where $W_{ij} = p_{ij}$, and $Y_S = [y_1, y_2, \ldots, y_m]$ is the embedding for the compressed set. Then, the hash functions for training data are constructed by thresholding Y at zero:

$$H = sgn(Y) = sgn(W_{XS} Y_S) \tag{10}$$

Based on hash functions $H(x) = \{h_k(x)\}_{k=1}^{r}$, given a sample x_i, $i = 1, 2, \ldots, n$, its compact hash codes can be obtained by $B_{x_i} = H(x_i)$. The flowchart of the proposed approach is illustrated as Fig. 1.

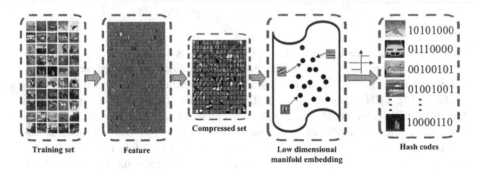

Training set   Feature   Compressed set   Low dimensional manifold embedding   Hash codes

Fig. 1. Flowchart of the proposed SNCH approach

3 Experimental Results

3.1 Experimental Setting

We evaluate the effectiveness of the approximate nearest neighbor search about hashing algorithm is proposed in this paper on CIFAR10 [14] and MNIST [15] two databases. The CIFAR10 database is a subset of 80 million small images. It including 60000 color images and is divided into 10 categories with 6000 samples for each class. Each image is represented by a 512-dimensional GIST feature vector. The MNIST database consists of 70000 images of handwritten digits from '0' to '9', each digit image have $28 \times 28 = 784$ pixels.

We measure the hashing performance mainly through precision-recall curve. The definitions of precision and recall are as follows:

$$precision = \frac{Number\ of\ retrieved\ relevant\ images}{Total\ number\ of\ all\ retrieved\ images} \quad (11)$$

$$recall = \frac{Number\ of\ retrieved\ relevant\ images}{Total\ number\ of\ all\ relevant\ images} \quad (12)$$

Besides, we compare SNCH with the representative and state-of-the-art unsupervised hashing approaches. They are SH, LSH, ITQ, AGH and PCAH [16].

3.2 Experimental Result

All of the evaluation results on MNIST database are shown in Fig. 2. Figure 2(a)–(c) are precision-recall curves at 24 bits, 32 bits and 48 bits respectively. As we can see from these figures, the precision decreases when the recall increases. That is because the value of precision is sensitive to true positive rate while the recall is sensitive to false positive rate. And our method outperforms all competitors on precision-recall curves. Figure 2(d) is the mean average precision(MAP) curves, as shown in the figure, our method has much better MAP curves than other methods. Figure 2(e) and (f) are the precision curves and the recall curves with the number of top retrieved samples at 32 bits, they show our method gain about 7 % in precision curves and about 10 % in recall curves over the best competitor.

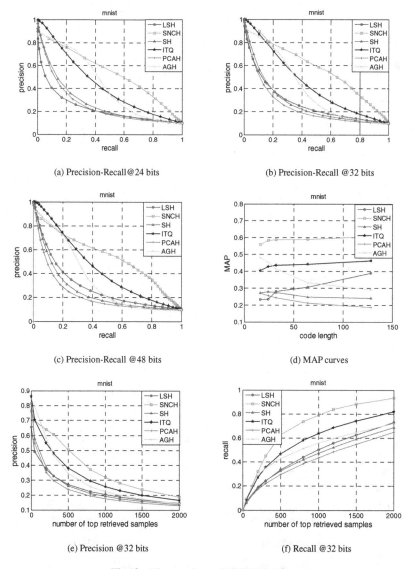

(a) Precision-Recall@24 bits

(b) Precision-Recall @32 bits

(c) Precision-Recall @48 bits

(d) MAP curves

(e) Precision @32 bits

(f) Recall @32 bits

Fig. 2. The results on MNIST database

Figure 3 shows the results on CIFAR10 database. The precision-recall curves at 16 bits, 24 bits and 32 bits as shown in Fig. 3(a)–(c) respectively, our hashing approach performs better than other hashing approaches. Figure 3(d)–(f) show the MAP curves, the precision curves and the recall curves with the number of top retrieved samples. As shown in these figures, our hashing approach achieves the highest search accuracy performance.

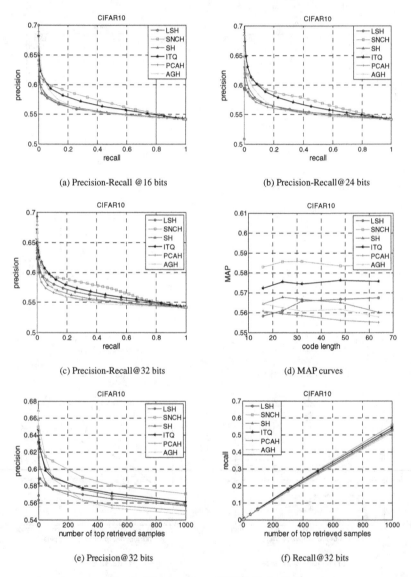

(a) Precision-Recall @16 bits

(b) Precision-Recall@24 bits

(c) Precision-Recall@32 bits

(d) MAP curves

(e) Precision@32 bits

(f) Recall@32 bits

Fig. 3. The results on CIFAR10 database

The training time for all hash methods being compared are given as benchmark in Tables 1 and 2. As shown in these tables, SNCH take long time than other hashing approaches.

Table 1. Training time on MNIST database

Method	Training time (s)				
	16 Bits	24 Bits	32 Bits	48 Bits	64 Bits
SNCH	3.506	3.680	3.906	4.232	4.394
LSH	0.005	0.007	0.011	0.008	0.012
SH	0.118	0.123	0.147	0.242	0.300
ITQ	0.160	0.207	0.239	0.429	0.528
PCAH	0.086	0.097	0.146	0.128	0.133
AGH	0.145	0.146	0.157	0.150	0.151

Table 2. Training time on CIFAR10 database

Method	Training time (s)				
	16 Bits	24 Bits	32 Bits	48 Bits	64 Bits
SNCH	3.786	3.706	3.843	4.160	4.580
LSH	0.003	0.005	0.005	0.008	0.009
SH	0.065	0.082	0.101	0.171	0.257
ITQ	0.130	0.164	0.198	0.327	0.453
PCAH	0.051	0.057	0.062	0.073	0.098
AGH	0.136	0.132	0.131	0.128	0.145

From the experimental results on two database are shown in Figs. 2 and 3, we find the proposed SNCH achieves best performance with shorter binary code outperform ITQ, LSH, SH, PCAH and AGH. SNCH as a manifold learning hashing method, it compared with the state-of-the-art manifold-like hashing method AGH, in most cases SNCH performs better. Therefore, we can conclude that our hashing approach provides an effective solution to solve the approximate nearest neighbor search problem.

4 Conclusion

In this paper, we proposed an effective compress-based unsupervised manifold learning hashing approach. We first learn a smaller synthetic data set with a data compress algorithm and preserve the underlying manifold structure of the data by low-dimensional embedding method. We further showed that this hashing method is feasibility when find data point's nearest neighbors at the large scale database. Moreover, our hashing approach free from any restrictive data distribution assumption and remove the graph eigenvectors style so that obtain high hashing accuracy with short binary codes. Experimental comparison showed that our proposed approach achieved very promising hashing performance over the state-of-the-art hashing methods.

Acknowledgement. This research was supported by the National Natural Science Foundation of China (61472304, 61125204, 61432014 and 61201294), the Fundamental Research Funds for the Central Universities (Grant No. K5051202048, BDZ021403, JB149901), Program for Changjiang Scholars and Innovative Research Team in University (No. IRT13088).

References

1. Indyk, P., Motwani, R.: Approximate nearest neighbors: towards removing the curse of dimensionality. In: Proceedings of ACM Symposium on Theory of Computing, pp. 604–613 (1998)
2. Datar, M., Immorlica, N., Indyk, P.: Locality-sensitive hashing scheme based on p-stable distributions. In: Proceedings of ACM Symposium on Computational Geometry, pp. 253–262 (2004)
3. Gong, Y., Lazebnik, S.: Iterative quantization: a procrustean approach to learning binary codes. In: Proceedings of IEEE Conference on Computer Vision and Pattern Recognition, pp. 817–824 (2011)
4. Ge, T., He, K., Ke, Q.: Optimized product quantization for approximate nearest neighbor search. In: Proceedings of IEEE Conference on Computer Vision and Pattern Recognition, pp. 2946–2953 (2013)
5. He, K., Wen, F., Sun, J.: K-means hashing: an affinity-preserving quantization method for learning binary compact codes. In: Proceedings of IEEE Conference on Computer Vision and Pattern Recognition, pp. 2938–2945 (2013)
6. Jegou, H., Douze, M., Schmid, C.: Product quantization for nearest neighbor search. IEEE Trans. Pattern Anal. Mach. Intell. **33**(1), 117–128 (2011)
7. Weiss, Y., Torralba, A., Fergus, R.: Spectral hashing. In: Proceedings of Advances in Neural Information Processing Systems, pp. 1753–1760 (2009)
8. Liu, W., Wang, J., Kumar, S.: Hashing with graphs. In: Proceedings of International Conference on Machine Learning, pp. 1–8 (2011)
9. He, X., Cai, D., Niyogi, P.: Laplacian score for feature selection. In: Proceedings of Advances in Neural Information Processing Systems, pp. 507–514 (2005)
10. Wang, S.J., Chen, H.L., Peng, X.J.: Exponential locality preserving projections for small sample size problem. J. Neurocomput. **74**(17), 3654–3662 (2011)
11. Kusner, M., Tyree, S., Weinberger, K.Q.: Stochastic neighbor compression. In: Proceedings of International Conference on Machine Learning, pp. 622–630 (2014)
12. Carter, T., Fe, S.: An introduction to information theory and entropy. J. Complex Syst. Summer Sch. 129–134 (2011)
13. Van der Maaten, L., Hinton, G.: Visualizing data using t-SNE. J. Mach. Learn. Res. **9**, 2579–2605 (2008)
14. http://www.cs.toronto.edu/~kriz/cifar.html
15. http://yann.lecun.com/exdb/mnist
16. Wang, X.J., Zhang, L., Jing, F.: Annosearch: image auto-annotation by search. In: Proceedings of IEEE Conference on Computer Vision and Pattern Recognition, pp. 1483–1490 (2006)

Abnormal Crowd Motion Detection Using Double Sparse Representations

Jinzhe Jiang, Ye Tao, Wei Zhao, and Xianglong Tang[✉]

Pattern Recognition Center, Harbin Institute of Technology, Harbin, China
tangxl@hit.edu.cn

Abstract. Sparse representation method has been well used in image analysis, restoration and recognition, and it has also been introduced to analysis of video crowd movements recent years. To improve its accuracy of detecting abnormal events in crowd videos, a double sparse representation method is proposed. The method has two sparse processes, one of them judges whether the region of interest is normal, the other finds out whether the region is abnormal. The two judgments will be processed by fuzzy integral to obtain a final result for this region. Experiments are proceed in different datasets to validate the advantages of our algorithm. The results show that our method achieves higher accuracy than previous methods which are used for analysis of video crowd movements.

Keywords: Abnormal event · Crowd analysis · Double sparse representations

1 Introduction

Due to increasing populations and higher mobility, mass events, such as sports events, festivals, or concerts, attract growing numbers of attendees, and thus security measures are becoming more and more important. Nevertheless, despite adequate precautions even video surveillance are adopted, deadly stampedes and crowd disasters still occur rather frequently [16].

1.1 Relate Work

To understand human behavior, experimental studies are conducted by researches. Parameters such as crowd density, speed, flow, and crowd pressure [18] are determined either manually [17] or by means of digital image processing [12]. They usually do not adopt real data except experimental data. To avoid occlusions and to facilitate automatic video analysis, video-based experiments are typically carried out using top-view cameras. Former researches often detect and track individuals, but holistic approaches that make use of optical flow features are proposed

J. Jiang—The work is supported by the National Natural Science Foundation of China (Projects No. 61171184 and 61201309).

X. He et al. (Eds.): IScIDE 2015, Part I, LNCS 9242, pp. 249–259, 2015.
DOI: 10.1007/978-3-319-23989-7_26

recently. Over the years, various visual tracking approaches have been reported that were specifically developed for tracking pedestrians in crowded scenes [19]. Recently, simulations of pedestrian dynamics are considered to be used into the design of visual tracking systems. Ali and Shah [2] present a tracking framework inspired by the cellular automaton model [11]. They automatically calculate force fields that integrate information on human behavior as well as the locations of obstacles and important regions such as exit doors. In their previous work [1], Ali and Shah propose a flow segmentation framework which enables them to detect changes in flow patterns. Barbara and Christian [10] proposed a crowd disaster automatic video analysis system in 2012. Their system extracts optical flow fields of image sequences before mass disasters happen. Models of walking scenes are needed if the camera viewpoints and walking scenes are changed. This makes the system hardly to be used widely.

1.2   Abnormal Events Detection and Sparse Representation

Abnormal events detection is the most direct way to find and avoid crowd disasters happened in the video surveillance area. Two key issues need to be properly addressed, event representation and abnormal detection.

Abnormal event representation: Social force model is adopted in [15]. Some other methods consider the spatialCtemporal information, such as Optical Flow [4], chaotic invariant [22], mixtures of dynamic textures [13], high-frequency/ local/size-adapted spatio-temporal features [21]. There is also energy model to count the number of people and detect the crowd energy abnormality, which often represents the abnormal events [23].

Abnormal detection: abnormal detection is a one-class learning problem, most conventional algorithms [7] intend to detect testing sample with lower probability as anomaly by fitting a probability model over the training data. There are several statistics models, such as Gaussian model [9], Hidden Markov Model [21], Latent Dirichlet Allocation (LDA)[14].

High dimensional feature represents the events better than low dimensional ones, but requires more training data as the dimension increases. Sparse representation is suitable to represent high dimensional samples, it can find a group of features which can accurately represent the normal or abnormal events with least number of features.

As shown in Fig. 1(a) abnormal events generate a group of huge sparse reconstruction coefficients, most of which are nonzero elements, while the normal events generate a group of little coefficients with less nonzero elements. As shown in Fig. 1(b), normal event generates a small reconstruction costs, while abnormal event is dissimilar to any of the normal sample, thus generates a large reconstruction cost [5]. Based on above theories, sparse can be used to detect abnormal events in the video surveillance.

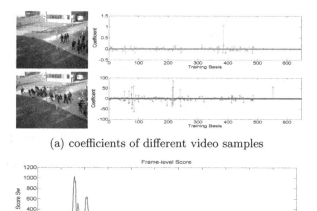

(a) coefficients of different video samples

(b) sparse reconstruction cost of a video sequence

Fig. 1. velocity error curves of four scenes

1.3 Motivation and Contribution

Sparse representation is suitable to deal with high dimension data. It is used as a kind of classifier, solving a one class classification problem in crowd events detection systems. Every sparse process contains a dictionary which represents a class, and the sparse process judges whether the coming sample is in this class. In previous researches, the dictionary is made up of a few typical normal samples. There are many different kinds of normal events. Normal events may be judged to be abnormal if this kind of events are not included into the normal dictionary. To fix this problem we proposed two improvements for previous sparse representation methods

- A dynamic dictionary updating process is add to the sparse representation. Some atypical samples whose sparse reconstruction cost is not as large as the abnormal events are added into the dictionary dynamically after being judged to be a normal events.
- After updating the dictionary, we find it still exists some normal events judged to be abnormal ones. To deal with these samples, we build an abnormal dictionary to fix a new sparse representation classifier, the new dictionary contains obvious abnormal events features. All the coming samples will be judged twice, lots of samples mentioned above are classified correctly at last.

2 Sparse Representation and Multi-scale Histogram of Optical Flow

[6] propose a new feature descriptor called Multi-scale histogram of optical flow, and for event representation, all the types of basis are concatenated by Multi-scale histogram of optical flow with various spatialor temporal structures.

A video frame are usually expressed by a sequence of features like $b = [B_1, B_2, \cdots, B_N]^T$, where N is the number of necessary features. To express video samples with high dimensional features, sparse representation provides a reasonable way. Multiply those features by a group of coefficients which represent the weight of features, we can get

$$y = [B_1 X_1, B_2 X_2, \cdots, B_N X_N]^T, \tag{1}$$

where $X_1, X_2, \cdots, X_N$ are coefficients. To obtain the best group of coefficients, $y - \sum_{n=1}^{N} B_n X_n$ should be small enough. That means we need to establish an optimization problem expressed as

$$x^* = \min \left\| y - \sum_{n=1}^{N} B_n X_n \right\|_2^2, \tag{2}$$

where the L_2 norm $\|\|_2$ denotes the euclidean distance. Express features as a matrix $\varPhi$, Eq. 2 would be

$$x^* = \min_x \| y - \varPhi x \|_2^2. \tag{3}$$

In order to use less features when fix a sample, there must be some 0 elements in coefficients vector x . A L_0 norm $\|x\|_0$ is added into Eq. 3,

$$x^* = \min_x \| y - \varPhi x \|_2^2 + \lambda \|x\|_0. \tag{4}$$

L_0 norm computes the number of nonzero element, so Eq. 4 computes the best group of coefficients with least features. L_1 norm is used here to replace the L_0 norm cause the computing of L_0 norm is a NP problem. [8] proved that the L_1 norm has the same effect as L_0 norm, here

$$x^* = \min_x \| y - \varPhi x \|_2^2 + \lambda \|x\|_1. \tag{5}$$

Equation 5 is the last form for sparse representation, where x^* is the reconstruction coefficients. [6] modified it to Eq. 6, and sparse reconstruction cost is expressed as Eq. 7

$$x^* = \min_x \frac{1}{2} \| y - \varPhi x \|_2^2 + \lambda \|x\|_1, \tag{6}$$

$$SRC = \frac{1}{2} \| y - \varPhi x \|_2^2 + \lambda \|x\|_1. \tag{7}$$

As shown in Fig. 1(b), the normal frame has a small reconstruction cost, while the abnormal frame usually generates a large reconstruction cost. Therefore, the SRC can be adopted as an anomaly measurement for this a one-class classification problem.

3 Dynamic Dictionary Updating and Double Sparse Representations Method

3.1 Dynamic Dictionary Updating

The purpose of dictionary updating is to make the dictionary describe the normal bases more accurate. The initial dictionary only contains the normal bases of training data, we try to add some test samples normal bases into the dictionary, so that the dictionary can deal with the test data better.

We address the problem of how to select comprehensive bases to fix the dictionary. The coming basis should be judged whether it is a normal one after enough sample have been chosen to make up the dictionary $\Phi = [\boldsymbol{b}_1, \boldsymbol{b}_2, \cdots \boldsymbol{b}_N]^T$. All kinds of normal bases should be added into the dictionary while the dimension of dictionary is not changed. Scan the dictionary and Find a basis whose distance to this coming normal basis is least, fuse the frame into dictionary. The distance here is measured by sparse reconstruction cost, all the SRC of elements in dictionary Φ is expressed by $SRC^{dic} = [SRC_1^{dic}, SRC_2^{dic}, \cdots, SRC_N^{dic}]^T$, and the coming samples SRC can be shown as SRC^y. The dynamic dictionary updating is shown as

$$\begin{cases} \boldsymbol{\omega}_{m^*} = \frac{1}{2}(\boldsymbol{b}_{m^*} + \boldsymbol{y}), \\ m^* = \min_m \left\| SRC_m^{dic} - SRC^y \right\|_2, m = 1, \cdots, N \end{cases}, \quad (8)$$

where m^* is index of basis in the dictionary whose distance to coming basis $\boldsymbol{y}$ is least, $\boldsymbol{\omega}_{m^*}$ is a new basis which will replace the m^*th basis of dictionary Φ, $\boldsymbol{b}_{m^*}$ is the m^*th basis in the old dictionary. We summarize the algorithm in Algorithm 1.

Algorithm 1. Purpose-driven lattice Boltzmann model

Input: Φ, SRC^{dic}, SRC^y
Output: $\Phi_{new}, SRC_{new}^{dic}$
  1: **if** SRC^y is normal **then**
  2:     $m^* = \min_m \left\| SRC_m^{dic} - SRC^y \right\|_2$
  3:     $\boldsymbol{\omega}_{m^*} = \frac{1}{2}(\boldsymbol{b}_{m^*} + \boldsymbol{y})$
  4:     $\boldsymbol{b}_{m^*} = \boldsymbol{\omega}_{m^*}, \Phi_{new} = [\boldsymbol{b}_1, \boldsymbol{b}_2, \cdots, \boldsymbol{\omega}_{m^*}, \cdots \boldsymbol{b}_N]$
  5:     $SRC_{m^*}^{dic} = mean(SRC_{m^*}^{dic}, SRC^y)$,
        $SRC_{new}^{dic} = [SRC_1^{dic}, SRC_2^{dic}, \cdots, SRC_{m^*}^{dic}, \cdots SRC_N^{dic}]$
  6: **end if**

3.2 Double Sparse Representations Method

One sparse representation process can be considered as a classifier judging whether the testing sample belongs to the class described by sparse dictionary. It is like a one class problem. The sparse representation with a dictionary whose elements is all normal samples can find out whether the testing sample is a normal sample. In a similar way, a testing sample can be judged whether to be an abnormal sample if the dictionary contains only abnormal samples. There may be several kinds of motion classes in crowd scenes. In theory, lots of classifiers can be established, and these sparse classifiers can be combined to be a strong multiple classifier system which can give a more accurate classify result of a testing sample. The number of classifiers can be very large, but two classifiers is enough for us.

The double sparse representation in our work is such a system, which contains two sparse representation processes, the first is similar to [6], the dictionary of which contains only normal bases, and the other one established with a dictionary contains abnormal bases gained by first sparse representation process. The abnormal dictionary can be made up of bases which have very low probabilities to be normal ones. The equations of second sparse representation is

$$x^* = \min_{x} \frac{1}{2} \|y - \Phi_{ab}x\|_2^2 + \lambda\|x\|_1, \tag{9}$$

$$SRC^{ab} = \frac{1}{2} \|y - \Phi_{ab}x\|_2^2 + \lambda\|x\|_1, \tag{10}$$

where Φ_{ab} is an abnormal dictionary. As we can see, Eqs. 9 and 10 is no different with Eqs. 6 and 7, but a new dictionary Φ_{ab}. Similar to normal dictionary Φ, Φ_{ab} needs a dynamic dictionary updating process, kinds of abnormal bases are added to this dictionary. The first a few test samples of videos is judged whether to be a normal one, Φ will be updated if the sample is normal, while the sample will be added into Φ_{ab} if it is judged to be an abnormal one. Samples comes later (after Φ_{ab} is established) will be judged by two sparse representation classifiers respectively. These two sparse processes are combined, in parallel. The final judgment will be given by using fuzzy integral [24]. As the final result is given, the updating process must be continued. The testing basis will be added into Φ if the final result is normal, otherwise this basis will be added into Φ_{ab}.

We summarize the multiple sparse representation algorithm in Algorithm 2.

4   Experiments

To test the effect of our proposed method, the algorithm is applied in several published datasets. UMN dataset [20] is adopted in this paper to test our algorithms performance in glob abnormal events detection, while UCSN dataset [13] is used to test the algorithms property when dealing with the local abnormal events.

4.1   Glob Abnormal Events Detection

The picture results are shown in Fig. 2, the normal/abnormal results are annotated as red/green color in the indicated bars respectively. In Fig. 3, the ROC

Algorithm 2. multiple sparse representations

Input: test frames, Φ, $\Phi_{ab} = \emptyset$
Output: normal/abnormal
  loop
    if Φ_{ab} is not complete **then**
      $y \rightarrow$ Eq. 6.7
      if y is normal **then**
        update Φ(Algorithm 1)
      else
        add y into Φ_{ab}
      end if
    else
      $y \rightarrow$ Eq. 6.7, $y \rightarrow$ Eq. 10.11
      $SRC, SRC^{ab} \rightarrow$ fuzzy integral and give result
      if normal **then**
        update Φ(Algorithm 1)
      else
        update Φ_{ab} (Algorithm 1)
      end if
    end if
  end loop

curves by are shown to compare our SRC to two other measurements, which are
i. Sparse without a dynamic dictionary selection, the original method brought
by [6]. ii. Sparse with dynamic dictionary selection process. In the ROC curve,
the more each curve hugs the left and top edges of the plot, the better the classification. Fig 6 shows that our double sparse methods curve is nearest to the
left and top edges of the plot, that means it is better than the rest two methods.
The Area under ROC (AUC) is used to assess a classification system, too. The
more a curve hugs the left top edges, the larger AUC value it will get. As the
ROC curve is obtained, it is easy to compute the AUC of those curves.

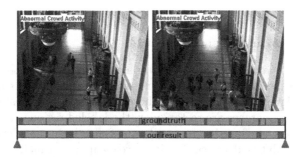

Fig. 2. The qualitative results of the global abnormal event detection for the second
scene from UMN dataset (Color figure online)

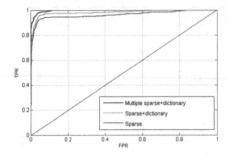

Fig. 3. ROC curve of three methods detecting glob events

Table 1 provides the quantitative comparisons to the state-of-the-art methods. The AUC of our method is 0.9957, which outperforms [14] and is comparable to [22]. As shown in the Table 1, our methods AUC is higher than the previous sparse methods, its multiple sparse process making some of the bases be judged twice, the judgment of some bases among them is changed to be right, this is the reason why our AUC is higher than other sparse methods.

Table 1. AUC of each method

method	Area under ROC
Chaotic Invariants[22]	0.99
Social Force[15]	0.96
Optical flow[3]	0.84
double sparse with dictionary selection	0.9957
sparse with dictionary selection	0.9812
Sparse[6]	0.9624

4.2 Local Abnormal Events Detection

Some image results are shown in Fig. 4. Our algorithm can detect bikers, skaters, small cars, even the frames which contains two abnormal events can be labeled accurately. We label the whole patch whose basis is judged to abnormal with green color. As the patches are rectangular, the labeled regions are rectangular, too. The algorithm can find the abnormal events effectively though the abnormal targets shape couldnt be detected. In Fig. 8, we compare our method with sparse, and sparse with a dictionary updating. It is easy to find that our ROC curve outperforms others. Figure 8 is shows the ROC of three methods dealing with local events by frame level. It is difficult to get pixel level results, because the green areas only represent the event patches not the accurate position of every

Fig. 4. The qualitative results of the local abnormal event detection for the several scenes from UCSD Ped1 dataset

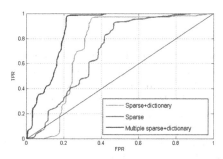

Fig. 5. ROC curve of three methods detecting local events

pixel of those bikers, skaters, small cars, etc. Table 2 provides the quantitative comparisons to the three kinds of sparse methods. The AUC of our method is 0.8784, which outperforms [6]. As shown in the Table 4, our methods AUC is higher than the previous sparse methods, its double sparse process making some of the bases be judged twice, the judgment of some bases among them is changed to be right. Some abnormal targets distribute in two or three bases, the basis which contains little abnormal region become hard to be detected as abnormal. This makes the sparse algorithms performance in local events is not as good as in glob ones, but our algorithm is better than the previous sparse methods obviously and can find the abnormal regions effectively (Fig. 5).

Table 2. AUC of each method

method	Area under ROC
Multiple sparse with dictionary selection	0.8784
sparse with dictionary selection	0.7328
Sparse[6]	0.6568

5 Conclusion

A new sparse method which is used to judge the station of crowd videos is proposed. It is made up of two sparse processes, each of them contains a dynamic dictionary updating process. The dynamic dictionary updating makes the dictionary describes the normal or abnormal station more comprehensive. The two sparse processes give two probabilities. A final result is given after the probabilities are input into a fuzzy integral Algorithm. The proposed Algorithm can make most of the bases which distribute near the edge of normal an abnormal samples be judged more accurate. The experiments results shows that the new method has a higher accuracy when detecting crowd abnormal events in the videos.

References

1. Ali, S., Shah, M.: A lagrangian particle dynamics approach for crowd flow segmentation and stability analysis. In: IEEE Conference on Computer Vision and Pattern Recognition, CVPR 2007, pp. 1–6. IEEE (2007)
2. Ali, S., Shah, M.: Floor fields for tracking in high density crowd scenes. In: Forsyth, D., Torr, P., Zisserman, A. (eds.) ECCV 2008, Part II. LNCS, vol. 5303, pp. 1–14. Springer, Heidelberg (2008)
3. Barron, J.L., Fleet, D.J., Beauchemin, S.S.: Performance of optical flow techniques. Int. J. Comput. Vis. **12**(1), 43–77 (1994)
4. Cao, T., Wu, X., Guo, J., Yu, S., Xu, Y.: Abnormal crowd motion analysis. In: 2009 IEEE International Conference on Robotics and Biomimetics (ROBIO), pp. 1709–1714. IEEE (2009)
5. Cong, Y., Yuan, J., Liu, J.: Sparse reconstruction cost for abnormal event detection. In: 2011 IEEE Conference on Computer Vision and Pattern Recognition (CVPR), pp. 3449–3456. IEEE (2011)
6. Cong, Y., Yuan, J., Liu, J.: Abnormal event detection in crowded scenes using sparse representation. Pattern Recogn. **46**(7), 1851–1864 (2013)
7. Daniyal, F., Cavallaro, A.: Abnormal motion detection in crowded scenes using local spatio-temporal analysis. In: 2011 IEEE International Conference on Acoustics, Speech and Signal Processing (ICASSP), pp. 1944–1947. IEEE (2011)
8. Hale, E.T., Yin, W., Zhang, Y.: A fixed-point continuation method for l1-regularized minimization with applications to compressed sensing. CAAM TR07-07, Rice University (2007)
9. Kim, J., Grauman, K.: Observe locally, infer globally: a space-time mrf for detecting abnormal activities with incremental updates. In: IEEE Conference on Computer Vision and Pattern Recognition, CVPR 2009, pp. 2921–2928. IEEE (2009)
10. Krausz, B., Bauckhage, C.: Loveparade 2010: automatic video analysis of a crowd disaster. Comput. Vis. Image Underst. **116**(3), 307–319 (2012)
11. Lee, B.H., Koo, Y.H., Oh, J.Y., Cheon, J.S., Tahk, Y.W., Sohn, D.S.: Fuel performance code cosmos for analysis of lwr uo2 and mox fuel. Nucl. Eng. Technol. **43**(6), 499–508 (2011)
12. Liu, X., Song, W., Zhang, J.: Extraction and quantitative analysis of microscopic evacuation characteristics based on digital image processing. Phys. A: Statist. Mech. Appl. **388**(13), 2717–2726 (2009)
13. Mahadevan, V., Li, W., Bhalodia, V., Vasconcelos, N.: Anomaly detection in crowded scenes. In: 2010 IEEE Conference on Computer Vision and Pattern Recognition (CVPR), pp. 1975–1981. IEEE (2010)
14. Mehran, R., Oyama, A., Shah, M.: Abnormal crowd behavior detection using social force model. In: IEEE Conference on Computer Vision and Pattern Recognition, CVPR 2009, pp. 935–942. IEEE (2009)

15. Raghavendra, R., Del Bue, A., Cristani, M., Murino, V.: Abnormal crowd behavior detection by social force optimization. In: Salah, A.A., Lepri, B. (eds.) HBU 2011. LNCS, vol. 7065, pp. 134–145. Springer, Heidelberg (2011)
16. Rodriguez, M., Laptev, I., Sivic, J., Audibert, J.Y.: Density-aware person detection and tracking in crowds. In: 2011 IEEE International Conference on Computer Vision (ICCV), pp. 2423–2430. IEEE (2011)
17. Seyfried, A., Steffen, B., Klingsch, W., Boltes, M.: The fundamental diagram of pedestrian movement revisited. J. Stat. Mech: Theory Exp. **2005**(10), P10002 (2005)
18. Steffen, B., Seyfried, A.: Methods for measuring pedestrian density, flow, speed and direction with minimal scatter. Phys. A: Statist. Mech. Appl. **389**(9), 1902–1910 (2010)
19. Sugimura, D., Kitani, K.M., Okabe, T., Sato, Y., Sugimoto, A.: Using individuality to track individuals: clustering individual trajectories in crowds using local appearance and frequency trait. In: 2009 IEEE 12th International Conference on Computer Vision, pp. 1467–1474. IEEE (2009)
20. UnusualcrowdactivitydatasetofUniversityofMinnesota.   http://mha.cs.umn.edu/movies/crowdactivity-all.avi
21. Wang, B., Ye, M., Li, X., Zhao, F., Ding, J.: Abnormal crowd behavior detection using high-frequency and spatio-temporal features. Mach. Vis. Appl. **23**(3), 501–511 (2012)
22. Wu, S., Moore, B.E., Shah, M.: Chaotic invariants of lagrangian particle trajectories for anomaly detection in crowded scenes. In: 2010 IEEE Conference on Computer Vision and Pattern Recognition (CVPR), pp. 2054–2060. IEEE (2010)
23. Xiong, G., Cheng, J., Wu, X., Chen, Y.L., Ou, Y., Xu, Y.: An energy model approach to people counting for abnormal crowd behavior detection. Neurocomputing **83**, 121–135 (2012)
24. Zhu, H., Tang, X.: Classifier geometrical characteristic comparison and its application in classifier selection. Pattern Recogn. Letters **26**(6), 829–842 (2005)

Nonparametric Discriminant Analysis Based on the Trace Ratio Criterion

Jin Liu[1(✉)], Qianqian Ge[1], Yanli Liu[1], and Jing Dai[2]

[1] School of Electronic Engineering, Xidian University, Xi'an 710071, China
jinliu@xidian.edu.cn, geqianqina_ex@163.com,
ylliu@stu.xidian.edu.cn
[2] School of Electromechanical and Information,
Hebei United University, Qian'an College, Tangshan 064400, China
djmagic@163.com

Abstract. Feature extraction is a hot topic in machine learning and pattern recognition. This paper proposes a new nonparametric linear feature extraction method called nonparametric discriminant analysis based on the trace ratio criterion (TRNDA). The motivation comes principally from the nonparametric maximum margin criterion (NMMC). Based on nonparametric extensions of commonly used scatter matrices, an NMMC is one of the effective nonparametric methods of discriminant analysis for linear feature extraction. However, it is sensitive to outliers. By the proposed TRNDA, new scatter matrices are designed for reducing the influence of outliers, and the trace ratio algorithm is used to learn a set of orthogonal projections in succession. We evaluate the proposed method by several benchmark datasets and the results confirm its effectiveness.

Keywords: Feature extraction · Nonparametric discriminant analysis · Trace ratio algorithm · Orthogonal projections

1 Introduction

Linear discriminant analysis (LDA) [1] is an efficient method for feature extraction and has been widely used for face recognition. LDA aims to find an optimal low-dimensional projection of the original high dimensional data, which is obtained by maximizing the ratio of trace of the inter-class scatter matrix to the intra-class scatter matrix. While LDA is effective for classification under the assumption that the distribution of each class is Gaussian, its performance deteriorates when the real world data are not Gaussian. Another drawback is that it can attract at most $c - 1$ features, since the rank of S_b is at most $c - 1$, where S_b is the inter-class scatter matrix and c is the number of class.

Nonparametric discriminant analysis (NDA) [2] and many extensions have therefore been developed to solve the above problems. The new nonparametric inter-class scatter matrix is calculated based on the vectors that can preserve classification structure, but it is designed for the two class classification problems. Then, it is extended for the general case [3]. In [4], a new path for dimensionality called the marginal fisher analysis (MFA) method is proposed, and new scatter matrices are

X. He et al. (Eds.): IScIDE 2015, Part I, LNCS 9242, pp. 260–271, 2015.
DOI: 10.1007/978-3-319-23989-7_27

calculated based on a penalty graph and an intrinsic graph that describe the inter-class separability and intra-class compactness, respectively. In order to uncover the essential low-dimensional structure of data under no assumption on the distribution of data, a nonparametric margin maximum criterion (NMMC) is discussed in [5], where for each data, its furthest intra-class neighbor and nearest inter-class neighbor are found to calculate the intra-class and inter-class scatter matrices. However, an NMMC may suffer from problems in some applications. Firstly, push-pull marginal discriminant analysis (PPMDA) [6] shows that only the samples that are close to the margin and potentially contribute to the increase of the margin are necessary for classification, but PPMDA neither discriminates the vectors around the boundary having different contributions for classification nor effectively tightens the intra-class distribution. Second, Yang et al. [7] pointed that outliers may bias the estimation of data distribution. So the intra-class variance may be biased evaluated. In this paper, we propose a new algorithm called nonparametric discriminant analysis based on the trace ratio criterion (TRNDA). Compared with NMMC, the TRNDA has a number of advantages: (1) The local inter-class scatter matrix is constructed by the local gradients, and therefore the local margin is enlarged. (2) The local intra-class scatter matrix is constructed by the k farthest intra-class neighbors and that are weighted according to their distance, which can reduce the impact of outliers. (3) The trace ratio criterion is used to find the optimal subspace. Just as what is described in [4] that most of the dimension reduction methods can be reformulated within the graph embedding framework, it is important to describe the similarity between data points. The trace ratio algorithm [8, 9] based on Euclidean distance that can directly describe the relation of data points is used to find the low-dimension space. However, there is no close-form solution to it and it is usually transformed into simpler but not equivalent problem. Recently, a series of trace ratio optimization methods have been proposed [10, 11].The optimal projection obtained by the trace ratio algorithm is orthogonal, and then the algorithm can preserve the similarity relation of the projection data [10].

The rest of the paper is organized as follows. In Sect. 2, a brief review of LDA, NMMC and PPMDA is given. Then we describe the new method in Sect. 3. The experimental evaluations are presented in Sect. 4. Finally, Sect. 5 concludes the paper.

2   Related Works

Assume that there is a training set $X=[x_1,\ldots,x_N]$. An element x_i in X with label $\omega_i \in \{1,\ldots,c\}$ is a D-dimensional vector and c is the number of classes. The number of samples in class ω_i is denoted by N_i. Then $N = \sum_{i=1}^{c} N_i$ is the total number of the training set. The goal of discriminant analysis is to learn a projection matrix $W \in R^{D \times d}$ such that $y_i \in R^d$ $(d \ll D)$. The projection of x_i is:

$$y_i = W^T x_i. \tag{1}$$

Moreover, the inter-class variance is maximized and the intra-class variance is minimized in the projection space.

2.1   LDA

LDA assumes that each class meets Gaussian distribution. Its parametric form of the scatter matrix S_b^{LDA} and S_w^{LDA} is defined as:

$$S_b^{LDA} = \sum_{i=1}^{c} N_i(m_i - m)(m_i - m)^T, \tag{2}$$

$$S_w^{LDA} = \sum_{i=1}^{c} \sum_{j=1}^{N_i} (x_j - m_i)(x_j - m_i)^T, \tag{3}$$

where m_i is the mean vector of class i, and the mean vector of the training set is $m = \frac{1}{N} \sum_{i=1}^{N} x_i$.

The optimal projection matrix is obtained by maximizing the ratio of trace between inter-class variance and intra-class variance, as shown below:

$$W = \arg\max_{W} \frac{tr(W^T S_b^{LDA} W)}{tr(W^T S_w^{LDA} W)}, \tag{4}$$

Where $tr(\cdot)$ is the trace of a matrix. However, there is no close-form solution to Eq. (4) [9]. Then Eq. (4) is usually transformed into the trace differential problem, i.e.,

$$W = \arg\max_{W} tr(W^T (S_b^{LDA} - S_w^{LDA}) W). \tag{5}$$

The eigenvectors corresponding to the first d largest eigenvalues construct the transforming matrix W, but it is deviate from the true one.

2.2   Nonparametric Maximum Margin Criterion

For NMMC [5], the scatter matrices S_b^{NMMC} and S_w^{NMMC} are calculated based on the inter-class nearest or intra-class farthest point of each sample. These two kinds of neighbors for a sample $x_i \in \omega_i$ are defined as:

$$x^E = \{x' \notin \omega_i | \|x' - x_i\| \le \|z - x_i\|, \forall z \notin \omega_i\}, \tag{6}$$

$$x^I = \{x' \in \omega_i | \|x' - x_i\| \ge \|z - x_i\|, \forall z \in \omega_i\}, \tag{7}$$

Given a sample x_i, its nonparametric margin is defined as $\|\Delta_i^E\|^2 - \|\Delta_i^I\|^2$, where $\Delta_i^E = x_i - x^E$ and $\Delta_i^I = x_i - x^I$. NMMC aims to find a subspace where the nonparametric margin obtains the maximum value. As a result, the scatter matrix of NMMC has the form:

$$S_b^{NMMC} = \sum_{i=1}^{N} \omega_i \left(\Delta_i^E\right)\left(\Delta_i^E\right)^T, \qquad (8)$$

$$S_w^{NMMC} = \sum_{i=1}^{N} \omega_i \left(\Delta_i^I\right)\left(\Delta_i^I\right)^T, \qquad (9)$$

where ω_i is the weight and $\omega_i = \frac{\left\|\Delta_i^I\right\|^\alpha}{\left\|\Delta_i^I\right\|^\alpha + \left\|\Delta_i^E\right\|^\alpha}$ with $\alpha \in [0, \infty]$.

The criterion of NMMC has the form:

$$W = \arg \max_{W} tr\left(W^T \left(S_b^{NMMC} - S_w^{NMMC}\right) W\right). \qquad (10)$$

It is obvious that the main difference between NMMC and LDA is the scatter matrix. LDA evaluates the scatter matrix under the assumption that each class follows the Gaussian model, while NMMC does not make any assumption on data distribution. Furthermore, the local margin structure is emphasized in NMMC.

2.3   Push-Pull Marginal Discriminant Analysis

For a sample $x \in \omega_i$, we first need to find its inter-class k nearest neighbor in class ω_j with local mean value $\hat{m}_j$. Then with respect to $\hat{m}_j$, we find its inter-class k nearest neighbor in class ω_i with local mean value $\hat{m}_i$. By pushing $\hat{m}_i$ away from $\hat{m}_j$, the local margin is enlarged, and simultaneously by pulling $\hat{m}_i$ close to x, the intra-class density is enhanced.

The optimal projection matrix W can be calculated by

$$J(W) = \frac{W^T S_b^{PPMDA} W}{W^T S_W^{PPMDA} W}, \qquad (11)$$

where

$$S_b^{PPMDA} = \sum_{i=1}^{c} \sum_{\substack{j=1 \\ j \neq i}}^{c} \sum_{n=1}^{N_i} \left(\hat{m}_n^i - m_n^j\right)\left(\hat{m}_n^i - m_n^j\right)^T, \qquad (12)$$

$$S_W^{PPMDA} = \sum_{i=1}^{c} \sum_{\substack{j=1 \\ j \neq i}}^{c} \sum_{n=1}^{N_i} \left(\hat{m}_n^i - x_n^i\right)\left(\hat{m}_n^i - x_n^i\right)^T. \qquad (13)$$

Note that m_n^j is the local mean with $m_n^j = \frac{1}{k}\sum_{p=1}^{k} x_{np}$, x_{np} is the pth nearest neighbor of x_n belonging to class ω_j, and $\hat{m}_n^i$ is the local mean of m_n^j in class ω_i. Finally, a method similar to FLDA or regularized methods is used to avoid the singular problem [6].

We can see that NMMC only use one sample pair between a point and its nearest or farthest point to calculate the scatter matrix that may be deviate from the true ones, if there are outliers. Moreover, the projection vectors are related, which may destroy the relation of data in the input space. In addition, PPMDA needs high computational cost (the complexity for nearest neighbor search in PPMDA is $O(kN^2D)$, while NMMC has a complexity of $O(kND)$) and assumes that all the vectors equally contribute for classification.

3 Nonparametric Discriminant Analysis Based on the Trace Ratio Criterion

Similar to NMMC, TRNDA aims to find a projection matrix such that the data in the same class get close to each other, while the data in different class get apart. The local structure is described by two kinds of neighborhoods.

Heterogeneous Neighborhoods: For a data point $x_i \in \omega_i$, the set of its heterogeneous neighbors N_i^E is the k nearest data which are not in the same class with x_i.

Homogeneous Neighborhoods: For a data point $x_i \in \omega_i$, the set of its homogeneous neighbors N_i^O is the k farthest data which are in the same class with x_i.

The local class margin for x_i is defined as

$$J_i = \frac{\sum_{j=1}^{k} w_{ij} \Delta_{ij}^E (\Delta_{ij}^E)^T}{\sum_{j=1}^{k} \Delta_{ij}^O (\Delta_{ij}^O)^T}, \tag{14}$$

Where

$$\Delta_{ij}^E = x_i - N_{ij}^E, \ j = 1, \ldots, k, \tag{15}$$

$$\Delta_{ij}^O = x_i - N_{ij}^O, \ j = 1, \ldots, k, \tag{16}$$

$$w_{ij} = 1 - \frac{\left\| \Delta_{ij}^E \right\|}{\sum_{p=1}^{k} \left\| \Delta_{ip}^E \right\|}, \tag{17}$$

$\left\| \Delta_{ij}^E \right\|$ is the Euclidean distance from x_i to its j nearest heterogeneous neighbor. The local class margin describes the local structure around the point x_i, the points in the same class with x_i are pulled towards x_i, and the data in different class with x_i are pushed away from x_i by maximizing the local class margin. The weighting function is used to de-emphasize the samples that are far from the point x_i. Compared with the above method, the new scatter matrix has the advantages that: (1) the local structure is effectively captured by putting different weights on local gradient. (2) the intra-class scatter is minimized by making the k far-apart samples in the same class close simultaneously, which can reduce the influence of outliers and make full use of the intra-class samples. Figure 1 depicts a geometric description of TRNDA.

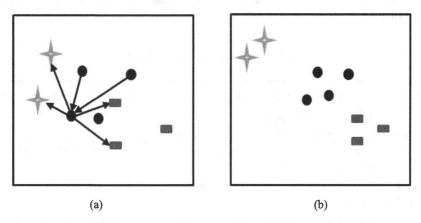

(a)   (b)

Fig. 1. A geometric description of TRNDA: (a) the local structure in the original space, and (b) the local structure in the projected space. Data in the same color and shape belong to the same class.

The nonparametric inter-class and intra-class scatter matrices are defined as

$$S_b^{TRNDA} = \sum_{l=1}^{c} \sum_{i=1}^{N_l} \sum_{j=1}^{k} w_{ij} \left(\Delta_{ij}^E\right) \left(\Delta_{ij}^E\right)^T, \tag{18}$$

$$S_w^{TRNDA} = \sum_{l=1}^{c} \sum_{i=1}^{N_l} \sum_{j=1}^{k} \left(\Delta_{ij}^O\right) \left(\Delta_{ij}^O\right)^T. \tag{19}$$

The objective function of TRNDA is defined as

$$J(W) = \arg\max_{W} \frac{tr\left(W^T S_b^{TRNDA} W\right)}{tr\left(W^T S_w^{TRNDA} W\right)}. \tag{20}$$

where $W^T W = I$, S_b^{TRNDA} and S_w^{TRNDA} are the nonparametric inter-class and intra-class scatter matrices, respectively, as defined in Eqs. (18) and (19). It has been shown that a good feature space in which most samples get well separated is better than the one that maximizes total separation [12, 13]. Thus, we learn the features step by step such that the features learned before can be used to learn the next feature and the new feature is the optimal one that maximizes the trace ratio criterion in the subspace orthogonal to the space spanned by the previously extracted feature [14].

Let $W_i = [w_1, w_2, \ldots, w_d]$ be the feature space and its orthogonal complementary space be $V = [v_1, v_2, \ldots, v_D]$. Then we have $V = I - W_i(W_i^T W_i)^{-1} W_i^T$. By considering that w_k is orthogonal to W_{k-1}, w_k can be written as a linear combination of V, i.e., $w_k = V\alpha$, where α is a $D - k + 1$-dimensional vector. We can see that the first feature w_1 is the generalized eigenvector corresponding to the maximum eigenvalue of matrix S_b^{TRNDA} with respect to S_w^{TRNDA}. For more details, please refer to [14]. The main procedure of TRNDA is shown in Table 1.

Table 1. The main procedure of TRNDA

Input:	Training set $X = \left\{ (x_i, \omega_i) \right\}_{i=1}^N$, testing set $Z = [z_1, \ldots, z_m]$, neighborhood size k, and the number of the feature d.
Output:	$D \times d$ feature matrix W.
Step 1	Construct the heterogeneous neighborhoods and homogeneous neighborhoods for each x_i;
Step 2	Construct the scatter matrix S_b and S_w using Eqs. (18) and (19), respectively;
Step 3	For $n = 1, \ldots, d$

Do the Cholesky decomposition $S_w = Q_i Q_i^T$, where Q_i is a full column rank lower triangular matrix. Let $S = Q_i^{-1}(Q_i^{-1})^T$ and obtain its maximum eigenvalue corresponding eigenvector v_i. Choose $w_i = (Q_i^{-1})^T v_i / \left\| (Q_i^{-1})^T v_i \right\|$ and $W_i = [W_{i-1}, w_i]$.

Update S_b and S_w:

$$S_{w_{n+1}} = \left(I - W_i W_i^\dagger \right) S_w \left(I - W_i W_i^\dagger \right) + c_0 W_i W_i^\dagger S_w W_i W_i^\dagger$$

$$S_{b_{n+1}} = \left(I - W_i W_i^\dagger \right) S_b \left(I - W_i W_i^\dagger \right)$$

$W_i^\dagger = (W_i^T W_i)^{-1} W_i^T$ and c_0 is a small positive number.

4 Experiments

In this section, the performance of the proposed method is evaluated and compared with MMC, NMMC, MFA and PPMDA under the same experiment condition on ORL, XM2VTS, and extended Yale B face datasets. ORL dataset is used to test the performance under condition that pose and expression vary at the same time. XM2VTS is used to test the performance in large pose variation, and the extended Yale B dataset is used to evaluate the performance under the condition that there is a variation in large lighting condition. Figure 2 shows the sample images in the three datasets.

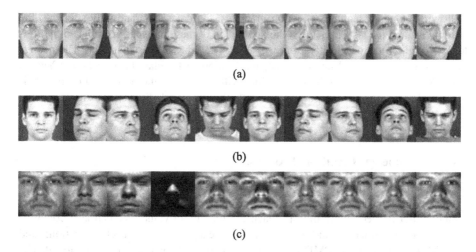

Fig. 2. Samples of the three face dataset: (a) ORL, (b) XM2VTS, and (c) Yale B.

4.1 Datasets

The ORL face image dataset [15] contains 400 images belonging to 40 persons taken at different time. Each person has 10 different images that contain variations in lighting conditions, facial expression, and with or without glasses. The images are taken with a tolerance for some tilting and rotation of the face of up to 20°. In our experiments, all images are down sampled into 32 × 32 pixels with 256 gray levels for computation efficiency. In our experiment, five and seven images per subject are randomly selected for training, and the others are used for testing.

The XM2VTS dataset [16] contains two recordings of 295 subjects taken over a period of four months. The first recording consists of speech while the second consists of rotating head movements. In our experiment, 10 images of each person are selected in the second recording. The original images are down sampled into 46 × 56 pixels with 256 gray levels for computation efficiency. In our experiment, five and seven images per subject are randomly selected for training, and the others are used for testing.

The Yale B [17] contains 38 individuals under nine poses and 64 illumination conditions. 2414 face images under pose 00 are used in the experiment. All the face images are down sampled into 32 × 32 pixels with 256 gray levels for computation efficiency. In our experiment, 10 and 20 images per subject are randomly selected for training, and the others are used for testing.

5 Comparative Methods and Settings

For PPMDA, the neighbor k and the classifier are set according to [6] on the ORL and Yale B datasets. However, the neighbor $k = 1$ is chosen for PPMDA on the XM2VTS dataset. As in MFA method, the heat kernel is used as the weight ($w_{ij} = \exp(-\|x_i - x_j\|/2t^2)$) and t is set to be the mean value of the Euclidean distance between all samples. The number of neighbors k is set to $N_c - 1$ in TRNDA and MFA.

Following standard feature extraction practices, the dataset is randomly split into the training set and the test set. At first the dimension of the input space is reduced by PCA to $N - c$. In the trace ratio iteration, we set $c_0 = 0.9$ based on experience. The nearest neighbor classifier with Euclidean distance is adopted for classification on the Yale B and XM2VTS datasets, while the classifier with the cosine metric is used on the ORL dataset. The average recognition rates with the projected dimensions are calculated over 40 independent experiments.

5.1   Experimental Results and Analysis

Figures 3, 4 and 5 show the recognition rates over the number of features on three face databases. We can see that: (1) the results illustrate that TRNDA outperforms other methods, but it needs more time to reach the maximum value. We can also conclude that the local structure and outliers have great influence on classification; and (2) NMMC outperforms PPMDA on most of the datasets, which is opposite to [6]. The reason is that samples with large illumination variation on the Yale B dataset and the images were not pre-processed using histogram equalization that can reduce the effect of illumination, and hence PPMDA does not work as well as NMMC in these dataset. For XM2VTS dataset, it is difficult for PPMDA to effectively capture the boundary information. Since XM2VTS is one of the most challenging data sets, there are much more types of images with large variations in the dataset, and the points from different classes may overlap and being the outliers. Therefore the methods do not work well on this dataset; and (3) Note that TRNDA performs better than MFA, while their differences lie in the construction of the graph that characterize the intra-class compactness. This demonstrates that marginal samples play an important role for classification and (4) the recognition rates of different methods increase as the number of features increases before they reach their maximum values. However, there is a little decrease after they reach their maximum values, probably due to the fact that the features corresponding to the small values have negative influence on their performance.

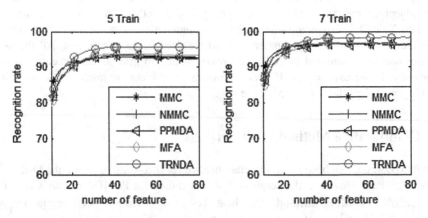

Fig. 3. The average recognition rates on ORL dataset with 5, 7 images per individual randomly selected for training.

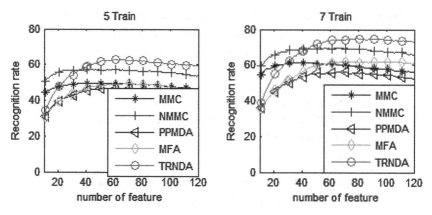

Fig. 4. The average recognition rates on XM2VTS dataset with 5, 7 images per individual randomly selected for training.

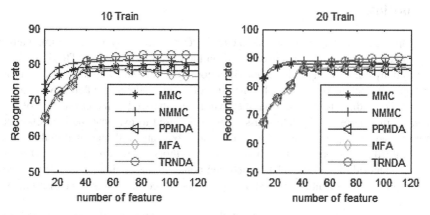

Fig. 5. The average recognition rates on Yale B dataset with 10, 20 images per individual randomly selected for training.

The number of eigenvectors is different in different datasets, since it has to do with the PCA stage and that of classes.

Table 2 tabulates the maximum average experimental results of different methods on three data sets. The "Tn" represents n facial images of each person. They are randomly selected for training. We can see that the recognition rates of different methods increase with the increase of the training samples and that TRNDA has better performance than other methods.

In addition, we study the effect of k on the recognition rate of TRNDA based on the ORL and XM2VTS datasets. In the experiment, seven samples for each object were randomly selected for training. By varying k from 2 to $N_i - 1$. Table 3 lists the maximum recognition rates with respect to k. As can be seen, the recognition rate as a function of k is nearly constant for a wide range of k values. This phenomenon allows us to easily choose an appropriate range of values for k.

Table 2. Face recognition rates (%) of different methods on different datasets

Dataset	MMC	MFA	PPMDA	NMMC	TRNDA
ORL (T5)	92.91	93.89	93.55	92.84	95.78
ORL (T7)	96.48	96.81	96.56	96.88	98.83
XM2VTS (T5)	49.65	49.97	46.88	57.24	62.97
XM2VTS (T7)	61.33	62.08	56.15	69.67	75.02
Yale B (T10)	80.80	79.07	78.73	81.34	82.95
Yale B (T20)	88.11	86.64	86.07	89.10	90.91

Table 3. Face recognition rates (%) of TRNDA versus k

Dataset	k = 2	k = 3	k = 4	k = 5	k = 6
ORL (T7)	98.71	98.79	98.83	98.90	98.83
XM2VTS (T7)	71.92	73.86	74.60	74.93	75.02

6 Conclusion

This paper proposes a new nonparametric discriminant analysis algorithm, namely nonparametric discriminant analysis, based on the trace ratio criterion (TRNDA) that is motivated by the existing nonparametric algorithms. In TRNDA, the local marginal structure information is captured by the sum of local gradients. Different weights are given to the gradient terms such that they will be emphasized in the calculation of local scatter. Furthermore, all the intra-class samples are used to evaluate the intra-class distribution, and a sample and its heterogeneous neighbors are well separated in the feature space learned by the proposed method. A set of orthogonal features is obtained by the trace ratio algorithm, and the relation between points in the original space is kept in the feature space. Experimental results show that the proposed method is more discriminant.

Acknowledgments. We would like to thank the associate editor and all anonymous reviewers for their constructive comments and suggestions. This research was partially supported by the National Science Foundation of China (Grant No. 61101246) and the Fundamental Research Funds for the Central Universities (Grant No. JB150209).

References

1. Fukunaga, K.: Introduction to Statistical Pattern Recognition, 2nd edn. Academic Press, New York (1990)
2. Fukunaga, K., Mantock, J.M.: Nonparametric discriminant analysis. IEEE Trans. Pattern Anal. Mach. Intell. **6**, 671–678 (1983)
3. Li, Z., Lin, D., Tang, X.: Nonparametric discriminant for face recognition. IEEE Trans. Pattern Anal. Mach. Intell. **31**(4), 755–761 (2009)
4. Yan, S., Xu, D., Zhang, B., Zhang, H.J., Yang, Q., Lin, S.: Graph embedding and extensions: a general framework for dimensionality reduction. IEEE Trans. Pattern Anal. Mach. Intell. **29**(1), 40–51 (2007)

5. Qiu, X., Wu, L.: Nonparametric Maximum Margin Criterion for Face Recognition. In: IEEE International Conference on Image Processing. ICIP 2005, 2, 918–921, IEEE press, (2005)

6. Gu, Z., Yang, J., Zhang, L.: Push-pull marginal discriminant analysis for feature extraction. Pattern Recogn. Lett. **31**(15), 2345–2352 (2010)

7. Yang, N., He, R., Zheng, W.S., Wang, X.: Robust large margin discriminant tangent analysis for face recognition. J. Neural Comput. Appl. **21**(2), 269–279 (2012)

8. Wang, H., Yan, S., Xu, D., Tang, X., Huang, T.: Trace ratio vs. ratio trace for dimensionality reduction. In: IEEE Conference on Computer Vision and Pattern Recognition, CVPR 2007, pp. 1–8. IEEE press (2007)

9. Yan, S., Tang, X.: Trace quotient problems revisited. In: Leonardis, A., Bischof, H., Pinz, A. (eds.) ECCV 2006. LNCS, vol. 3952, pp. 232–244. Springer, Heidelberg (2006)

10. Zhao, M., Zhang, Z., Chow, T.W. S.: ITR-score algorithm: an efficient trace ratio criterion based algorithm for supervised dimensionality reduction. In: The 2011 International Joint Conference on Neural Networks, pp. 145–152. IEEE press (2011)

11. Ngo, T.T., Bellalij, M., Saad, Y.: The trace ratio optimization problem. J. SIAM Rev. **54**(3), 545–569 (2012)

12. Wang, L., Sugiyama, M., Yang, C., Zhou, Z.H., Feng, J.: On the margin explanation of boosting algorithms. In: COLT, pp. 479–490 (2008)

13. Schapire, R.E., Freund, Y., Bartlett, P., Lee, W.S.: Boosting the margin: a new explanation for the effectiveness of voting methods. In: Proceedings of the Fourteenth International Conference on Machines Learning, Nashville, Tennessee, USA, pp. 1651–1686 (1998)

14. Li, G., Wen, C., Wei, W., Xu, Y., Ding, J., Zhao, G., Shi, L.: Trace ratio criterion for feature extraction in classification. J. Math. Probl. Eng.**2014,** 1–9 (2014)

15. ORL dataset. http://www.cad.zju.edu.cn/home/dengcai/Data/FaceData.html

16. XM2VTS dataset. http://www.ee.surrey.ac.uk/CVSSP/xm2vtsdb/

17. YALE B dataset. http://www.cad.zju.edu.cn/home/dengcai/Data/FaceData.html

Traffic Sign Recognition Using Deep Convolutional Networks and Extreme Learning Machine

Yujun Zeng, Xin Xu[⊠], Yuqiang Fang, and Kun Zhao

College of Mechatronic Engineering and Automation,
National University of Defense Technology, Changsha, China
{YujunZeng,XinXu,YuqiangFang}@nudt.edu.cn

Abstract. Traffic sign recognition is an important but challenging task, especially for automated driving and driver assistance. Its accuracy depends on two aspects: feature exactor and classifier. Current popular algorithms mainly use convolutional neural networks (CNN) to execute feature extraction and classification. Such methods could achieve impressive results but usually on the basis of an extremely huge and complex network. What's more, since the fully-connected layers in CNN form a classical neural network classifier, which is trained by conventional gradient descent-based implementations, the generalization ability is limited. The performance could be further improved if other favorable classifiers are used instead and extreme learning machine (ELM) is just the candidate. In this paper, a novel CNN-ELM model is proposed, which integrates the CNN's terrific capability of feature learning with the outstanding generalization performance of ELM. Firstly CNN learns deep and robust features and then ELM is used as classifier to conduct a fast and excellent classification. Experiments on German traffic sign recognition benchmark (GTSRB) demonstrate that the proposed method can obtain competitive results with state-of-the-art algorithms with less computation time.

Keywords: Traffic sign recognition · Convolutional neural network · Extreme learning machine

1 Introduction

Traffic sign recognition is a promising subfield of object recognition with various applications, which could provide reliable safety precaution and guiding information to drivers in motorway or urban environment. Nowadays, it is an indispensable opponent of the driver assistance system (DAS) and unmanned ground vehicle (UGV). Even though a lot of algorithms have been put forward, just as the German traffic sign recognition benchmark (GTSRB) [1] shows, there are still problems such as viewpoint changes, color distortion, motion blur, contrast degradation, occlusion, underexposure or overexposure, etc., which make it challenging to achieve a satisfying recognition accuracy.

Traditional methods for traffic sign recognition generally share a baseline that consists of hand-crafted features (e.g. HOG [2]) and regular classifiers (e.g. SVM [3],

X. He et al. (Eds.): IScIDE 2015, Part I, LNCS 9242, pp. 272–280, 2015.
DOI: 10.1007/978-3-319-23989-7_28

Random Forests [4], etc.), but Razavian et al. [5] state that hand-crafted features, which are also called shallow features, are not discriminative enough as databases become larger and larger and generic deep features should push the recognition performance even further. Convolutional neural networks (CNN) [6] is one of the deep neural networks (DNN) models and it works to resemble the perceptual processing of human visual cortex and can learning robust and more discriminative features. The state-of-the-art high-rated traffic sign recognition algorithms are almost CNN-based and it is reported that compared with the results of current best-performing methods using hand-crafted features, CNN works superiorly.

There are a variety of CNN variants having been proposed. Sermanet and LeCun [7] fed both the high-level and low-level features extracted by different convolutional layers to the fully-connected layers. This method combined global invariant features with the local detailed ones and the accuracy record was 99.17 %. Cireşan et al. [8] preprocessed input images by contrast-limited adaptive histogram equalization (CLAHE) and boosted the CNN's performance with multilayer perceptrons (MLPs) trained with HOG feature descriptors. Its recognition accuracy reached 99.15 % and was further increased to 99.46 % through multiple CNNs [9] trained with data pre-processed with different contrast normalization methods.

Recently, more and more proofs indicate that the generalization ability of the fully-connected layers in CNN is limited and just applying CNN-learnt features to simple classifiers will be beneficial. Both [5] and [10] integrated the CNN learnt representations to a simple linear support vector machine (SVM) classifier. After testing on many standard datasets (e.g. Oxford 102 flowers [11], Caltech-UCSD birds [12], MIT indoor scenes 67 [13], PASCAL VOC 2007 [14], etc.), all astonishingly produced the best performance. Nevertheless, cross-validation is usually inevitable so as to decide proper parameters for the training of SVM. The time cost will be unacceptable as the size of dataset grows and even that its generalization performance is not guaranteed to be optimal.

Huang et al. proposed a powerful learning algorithm called extreme learning machine (ELM) [15–17], which was based on the single-hidden layer feedforward neural network (SLFN). It randomly chose the hidden node parameters (i.e. input weights and biases) and determined the output weight by solving the least-squares solution of SLFN, which could be treated as a general linear system. It was quite training-efficient and able to ensure a satisfying classification result at the same time. In [18], ELM is trained with features learnt by stacked denoising autoencoder (SDA, another DNN model) and used to detect ships in remote sensing images. It outperformed SVM not only in detection performance but also time cost. Meanwhile, Ribeiro et al. [19] and Huang et al. [20] resorted to deep belief networks (DBN, also a DNN model) to build up a DBN-ELM architecture to undertake pattern recognition and attained competitive (and often better) results than other successful approaches, DBN-SVM included. Consequently, it is believed that ELM would bring in exciting performance when cooperating with CNN.

In this paper, a novel CNN-ELM model architecture is proposed and applied to traffic sign recognition, which is able to combine the outstanding feature learning capability of CNN with the high generalization performance of ELM. Firstly CNN is trained using raw traffic sign images. Then the fully-connected layers of the trained

CNN are removed, resulting in a feature extractor that learns deep and robust features. ELM is used as classifier to conduct a fast and excellent classification based on the CNN-learnt features. Experimental results clarify that the proposed method could reach comparable performance (99.40 %) of the state-of-the-art approaches with less computation burden.

The rest of this paper is organized as follows. The architecture of the proposed CNN-ELM model is described in details in Sect. 2. Then experimental results and comparisons are shown in Sect. 3. At last, the conclusion and future work are given in Sect. 4.

2 The CNN-ELM Model

For the purpose of making full use of the advantages of both CNN and ELM, a novel CNN-ELM traffic sign recognition model is proposed in this section (see Fig. 1). Before sent to CNN for feature extraction, the average image of the traffic signs is subtracted to ensure illumination invariance to some extent. In contrast to SVM or conventional neural networks that are trained using backpropagation gradient descent, ELM could ensure a prominent classification performance while reducing the computation burden remarkably, so that CNN is trained as a feature extractor, which means that only the convolutional and maxpooling layers are retained when the training of the whole CNN is completed. ELM is fed with features extracted by CNN and trained as the classifier. In the following subsections, more details are described.

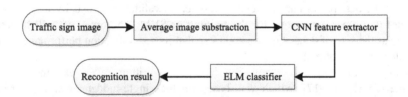

Fig. 1. Flowchart of the CNN-ELM model for traffic sign recognition.

2.1 Feature Extraction with CNN

As Fig. 2 shows, CNN can be considered to be made up of two main parts. The first contains alternating convolutional and maxpooling layers. The input of each layer is just the output of its previous layer. As a result, it forms a hierarchical feature extractor that maps the original input images into feature vectors. Then the extracted features vectors are classified by the second part, that is, the fully-connected layers.

Different from other deep learning models, CNN can be trained using backpropagation because the amount of hyperparameters is diminished greatly due to its combination of three architectural characteristics, i.e. local receptive fields, shared weights and spatial or temporal sub-sampling. Hence, all the hyperparameters are jointly optimized through backpropagation. We refer to [8] to build up the CNN. The difference is that an extra convolutional layer with 200 feature maps of 1×1 neuron is

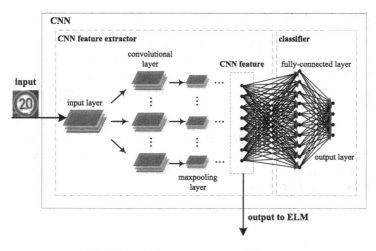

Fig. 2. CNN feature extractor architecture.

added before the fully-connected layer, Table 1 shows the whole setting of the CNN architecture. Note that the max pooling layer is non-overlapping and no rectification and inner-layer normalization operation is used.

Table 1. CNN architecture parameters.

Layer	Type	Number of maps and neurons	Kernel size	Stride
1	input	1 or 3 maps of 48×48 neurons	—	—
2	convolutional	100 maps of 46×46 neurons	3×3	1
3	max-pooling	100 maps of 23×23 neurons	2×2	2
4	convolutional	150 maps of 20×20 neurons	4×4	1
5	max-pooling	150 maps of 10×10 neurons	2×2	2
6	convolutional	250 maps of 8×8 neurons	3×3	1
7	max-pooling	250 maps of 4×4 neurons	2×2	2
8	convolutional	200 maps of 1×1 neuron	4×4	1
9	fully-connected	43 neurons	—	—

Since the traffic signs images are relatively invariable in shape and the size of the samples in GTSRB dataset varies from 15×15 to 250×250, here we assume that the influence coming from cropping and wrapping is considered neglectable. Thus, only images in bounding boxes given by the annotations are cropped and resized to 48×48 uniformly.

The fully-connected layers of CNN are equivalent to a general SLFN classifier. Since CNN is used to extract deep features rather than conduct classification, the first eight layers of the CNN are taken out as a feature extractor while the fully-connected layers are removed when training is done.

2.2   Classification Using ELM

Compared with conventional gradient descent-based implementations, ELM is a more promising learning algorithm for SLFN. Once input weights and hidden layer biases are chosen randomly, together with hidden nodes using infinitely differentiable activation functions, ELM is able to possess both an extremely fast convergence rate and an impressive generalization performance.

Huang et al. have proved that standard SLFNs (see Fig. 3), whose M hidden nodes use infinitely differentiable activation functions, could approximate arbitrary samples with zero error, which means given a training set of instance-label pairs $(\mathbf{x}_i, \mathbf{l}_i)$, $i = 1,2, ...,N$, where $\mathbf{x}_i = [x_{i1}, x_{i2},..., x_{in}]^T \in \mathbf{R}^n$ and $\mathbf{l}_i = [l_{i1}, l_{i2},..., l_{im}]^T \in \mathbf{R}^m$, there exist β_j, $\mathbf{w}_i$ and b_i that make Eq. 1 hold true.

$$\sum_{j=1}^{M} \beta_j g(\mathbf{w}_{ij} \cdot \mathbf{x}_i + b_j) = \mathbf{o}_i = \mathbf{l}_i \quad i = 1, 2, ..., N . \tag{1}$$

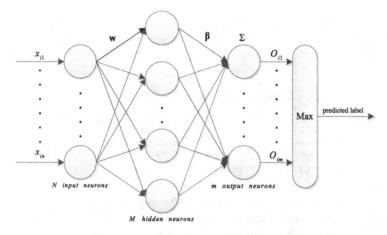

Fig. 3. ELM architecture.

where β_j is the weight vector which connects the j th hidden node with the output nodes, $\mathbf{o}_i$ is the SLFN output vector for the i th sample, $\mathbf{l}_i$ is the label vector of i th sample, and $\mathbf{w}_{ij}$ is the weight vector connecting the i th sample and the j th hidden node, b_j is the bias of the j th hidden node. $g(\bullet)$ is the activation function.

Equation 2 is the compact version of Eq. 1, where $\mathbf{H}_{\mathbf{w},\mathbf{b},\mathbf{x}}$ is named the hidden layer output matrix (see Eq. 3).

$$\mathbf{H}_{\mathbf{w},\mathbf{b},\mathbf{x}}\boldsymbol{\beta} = \mathbf{L} , \quad \boldsymbol{\beta} = \begin{bmatrix} \beta_1^T \\ \vdots \\ \beta_M^T \end{bmatrix}_{M \times m} , \quad \mathbf{L} = \begin{bmatrix} \mathbf{l}_1^T \\ \vdots \\ \mathbf{l}_N^T \end{bmatrix}_{N \times m} . \tag{2}$$

$$\mathbf{H}_{\mathbf{w,b,x}} = \begin{bmatrix} g(\mathbf{w}_1 \cdot \mathbf{x}_1 + \mathbf{b}_1) & \cdots & g(\mathbf{w}_M \cdot \mathbf{x}_1 + \mathbf{b}_M) \\ \vdots & \cdots & \vdots \\ g(\mathbf{w}_1 \cdot \mathbf{x}_N + \mathbf{b}_1) & \cdots & g(\mathbf{w}_M \cdot \mathbf{x}_N + \mathbf{b}_M) \end{bmatrix}_{N \times M} . \tag{3}$$

Fortunately, since Eq. 2 represents a general linear system, the smallest training error could be achieved by computing the corresponding least-squares solution $\boldsymbol{\beta} = \mathbf{H}_{\mathbf{w,b,x}}^{\dagger}\mathbf{L}$, where $\mathbf{H}_{\mathbf{w,b,x}}^{\dagger}$ is the Moore–Penrose generalized inverse of $\mathbf{H}_{\mathbf{w,b,x}}$.

Altogether, the ELM training algorithm consists of the following three steps.

- Randomly assign hidden node parameters $\mathbf{w}_j$ and b_j, $j = 1,2,\ldots, N$.
- Calculate the hidden-layer output matrix and its Moore–Penrose generalized inverse.
- Calculate the output weight $\boldsymbol{\beta}$.

During training ELM, there are only two parameters to be set (i.e. the number of hidden neurons and the type of activation function). By default, sigmoid function is chosen as the activation function. However, determining how many hidden neurons are needed through cross-validation is time-consuming. Zhan-Li et al. [21] have proposed a BW-ELM model for traffic sign recognition. It takes the ratio of feature's between-category to within-category sums of squares (BW) as the criterion to guide the selection of HOG features and gets a comparable classification result with ELM, in which the number of hidden nodes is decided empirically. Here we refer to [21] for the initial value and determine the final number experimentally.

3 Experiments and Analysis

We use the MatConvNet toolbox [22] to train the CNN feature extractor. Matlab 2014a is used to conduct all the operations, running on a system with 8 Intel(R) Xeon(R) E5-2643 CPUs (3.30 GHz), 12 GB DDR4. Note that no translation, scale or rotation is done to the GTSRB training dataset. The CNN training ends once the default training epoch (e.g. 50) is reached. Initial weights of the CNN are drawn from a uniform random distribution in the range [−0.01, 0.01]. Each neuron's activation function is sigmoid.

ELM is trained using the features of training dataset extracted by the CNN feature extractor. Different numbers of hidden nodes are chosen and the corresponding recognition rate is recorded in Table 2. Apparently, the recognition rate generally increases as there are more hidden nodes used. It reaches 99.23 % with 4000 hidden nodes, which is already better than that of many methods. So far the highest recognition rate is 99.40 % with 12000 hidden nodes. In addition, considering that the ELM is trained with the hidden node parameters set randomly, we take five trials and the average recognition rate is adopted as the final result.

Figure 4 shows the comparison of the recognition rate between the proposed CNN-ELM method and other recently reported results. Obviously, compared with Multi-scale CNN [7], CNNaug/MLP [8], CNN-SVM, BW-ELM [21] and plain CNN,

Table 2. ELM performance with different numbers of hidden nodes.

Number of hidden nodes	1000	2000	3000	4000	5000	6000
Recognition rate (avg.) [%]	98.70	99.03	99.15	99.23	99.29	99.32
Number of hidden nodes	7000	8000	9000	10000	11000	12000
Recognition rate (avg.) [%]	99.34	99.37	99.35	99.36	99.38	**99.40**

the proposed method performs superiorly and from this it is believed that the CNN-learnt features are discriminative enough but need a classifier with more powerful generalization ability and ELM is preferable. Even though the proposed method works a little inferior to the method of the best performance MCDNN [9] (99.46 %), its complexity of model structure is much lower for the usage of single CNN and no need of data augmentation, which will relieve the training procedure a lot.

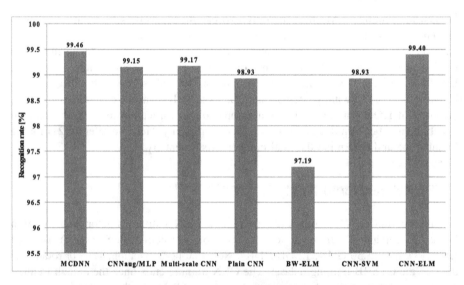

Fig. 4. Recognition accuracy comparison of different methods: MCDNN, CNNaug/MLP, Multi-scale CNN, Plain CNN, CNN-SVM, CNN-ELM, where Plain CNN retains the fully-connected layers to implement classification and CNN-SVM means the combination of CNN feature extractor and SVM classifier.

In terms of training time, the proposed method takes 5–6 h. It is longer than that of methods like BW-ELM, but much faster than MCDNN, which needs to train up to 25 DNNs and costs 37 h even when the GPU parallelization is carried out.

4 Conclusion

In this paper, a novel CNN-ELM model with less computation complexity is proposed and applied to traffic sign recognition, where CNN works as a feature extractor and ELM as the classifier trained on CNN-learnt features. In contrast with state-of-the-art methods, the proposed method could achieve competitive results (99.40 %, without any data augmentation and preprocessing like contrast normalization) with a much simpler architecture that relieves the time-consuming training procedure a lot.

Yet, the fact that the most errors are mainly due to motion blur implies that the performance may be further improved if the CNN is equipped with some layers that could learn blur-invariant features. Once new and larger traffic sign recognition databases are available, the portability of the proposed method is still to be confirmed. Additionally, the recognition accuracy might be further improved if operations like overlapping maxpooling, cross-channel normalization and ReLu [23] are taken and kernel-based ELM [24] is utilized instead of the standard one. The training time could be further cut down if a GPU implementation is accessible.

Acknowledgments. This work is supported by the National Natural Science Foundation of China under Grant 61075072, and 91220301, and the New Century Excellent Talent Plan under Grant NCET-10-0901.

References

1. Stallkamp, J., Schlipsing, M., Salmen, J., Igel, C.: The German traffic sign recognition benchmark: a multi-class classification competition. In: Proceedings of International Joint Conference on Neural Networks, pp. 1453–1460 (2011)
2. Stallkamp, J., Schlipsing, M., Salmen, J., Igel, C.: Man vs. computer: benchmarking machine learning algorithms for traffic sign recognition. Neural Netw. **32**, 323–332 (2012)
3. Ruta, A., Li, Y.M., Liu, X.H.: Robust class similarity measure for traffic sign recognition. IEEE Trans. Intell. Transp. Syst. **11**, 847–855 (2010)
4. Zaklouta, F., Bogdan, S., Hamdoun, O.: Traffic sign classification using kd trees and random forests. In: The 2011 International Joint Conference on Neural Networks (IJCNN). IEEE (2011)
5. Razavian, A.S., et al.: CNN features off-the-shelf: an astounding baseline for recognition. arXiv preprint arXiv:1403.6382 (2014)
6. LeCun, Y., Kavukcuoglu, K., Farabet, C.: Convolutional networks and applications in vision. In: Proceedings of 2010 IEEE International Symposium on Circuits and Systems (ISCAS). IEEE (2010)
7. Sermanet, P., LeCun, Y.: Traffic sign recognition with multi-scale convolutional networks. In: The 2011 International Joint Conference on Neural Networks (IJCNN). IEEE (2011)
8. Cireşan, D., et al.: A committee of neural networks for traffic sign classification. In: The 2011 International Joint Conference on Neural Networks (IJCNN). IEEE (2011)
9. Cireşan, D., et al.: Multi-column deep neural network for traffic sign classification. Neural Netw. **32**, 333–338 (2012)
10. Azizpour, H., et al.: From generic to specific deep representations for visual recognition. arXiv preprint arXiv:1406.5774 (2014)
11. Nilsback, M-E., Zisserman, A.: Automated flower classification over a large number of classes. In: Sixth Indian Conference on Computer Vision, Graphics & Image Processing, 2008. ICVGIP 2008. IEEE (2008)

12. Wah, C., Branson, S., Welinder, P., Perona, P., Belongie, S.: The Caltech-UCSD Birds-200–2011 Dataset. Technical report CNS-TR-2011-001, California Institute of Technology (2011)
13. Quattoni, A., Torralba, A.: Recognizing indoor scenes. In: CVPR (2009)
14. Everingham, M., Van Gool, L., Williams, C.K.I., Winn, J., Zisserman, A.: The PASCAL visual object classes challenge 2012(VOC2012) results. http://www.pascal-network.org/challenges/VOC/voc2012/workshop/index.html
15. Huang, G.-B., Zhu, Q.-Y., Siew, C.-K.: Extreme learning machine: a new learning scheme of feedforward neural networks. In: Proceedings. 2004 IEEE International Joint Conference on Neural Networks, 2004, vol. 2. IEEE (2004)
16. Huang, G.-B., Zhu, Q.-Y., Siew, C.-K.: Extreme learning machine: theory and applications. Neurocomputing 70(1), 489–501 (2006)
17. Huang, G.-B., et al.: Extreme learning machine for regression and multiclass classification. IEEE Trans. Syst. Man Cybern. Part B Cybern. 42(2), 513–529 (2012)
18. Tang, J., Deng, C., Huang, G.-B., Zhao, B.: Compressed-domain ship detection on spaceborne optical image using deep neural network and extreme learning machine. IEEE Trans. Geosci. Remote Sens. 53, 1174–1185 (2014)
19. Ribeiro, B., Lopes, N.: Extreme learning classifier with deep concepts. In: Ruiz-Shulcloper, J., Sanniti di Baja, G. (eds.) CIARP 2013, Part I. LNCS, vol. 8258, pp. 182–189. Springer, Heidelberg (2013)
20. Huang, W., Sun, F.C.: A deep and stable extreme learning approach for classification and regression. In: Proceedings of ELM-2014 Volume 1. Springer International Publishing, pp. 141–150 (2015)
21. Zhan-Li, S., et al.: Application of BW-ELM model on traffic sign recognition. Neurocomputing 128, 153–159 (2014)
22. MatConvNet toolbox. https://github.com/almazan/matconvnet.git
23. Dahl, G.E., Tara, N.S., Geoffrey, E.H.: Improving deep neural networks for LVCSR using rectified linear units and dropout. In: 2013 IEEE International Conference on Acoustics, Speech and Signal Processing (ICASSP). IEEE (2013)
24. Huang, G.-B., Siew, C.-K.: Extreme learning machine with randomly assigned RBF kernels. Int. J. Inf. Technol. 11(1), 16–24 (2005)

Hough Transform-Based Road Detection for Advanced Driver Assistance Systems

JongBae Kim[✉]

Department of Computer Engineering,
Seoul Digital University, Seoul, South Korea
jbkim@sdu.ac.kr

Abstract. This paper proposes a real-time road region detection method for safe driving assistance systems. This is a method on detecting road regions by receiving input images from vehicle black boxes. As such, we propose a method that detects major straight line components through Hough transform in input images, selects a vanishing point from the intersection of detected straight lines using preliminary information from the road environment, and detects road regions using the selected vanishing point and left and right straight lines from the Hough transform. As a result of applying the proposed method to various road environments, approximately 0.37 s were consumed per frame, providing over 81 % detection accuracy.

Keywords: Hough transform · Advanced driver assistance system · Road detection

1 Introduction

With the rapid and recent development of information technology (IT), advanced driver-assistance systems (ADASs) have been developed to provide drivers with more convenience and safety by integrating IT in automobiles [1, 2, 10]. Such IT devices typically include the Lane Departure Warning System that detects traffic in lanes, and the driving vehicle-detection system that detects vehicles that are driving ahead and to the left and right of the driver [3–8]. There are three main reasons to integrate an ADAS in a vehicle: to maintain a safe trailing distance for drivers in the driving lane, to detect nearby obstacles in an effort to avoid collisions, and to control the vehicle speed depending on the traffic and road conditions.

ADASs use sensors (RADAR, LIDAR, etc.) for lane detection and oncoming obstacle detection [3–8]. However, such sensors are, embedded in the front, back, and sides of vehicles, and this creates problems such as high installation costs and frequent damage or malfunctioning from even minor collisions. Although such sensor-based processing systems yield accurate results, the high costs outweigh the advantages. An alternative to these sensor-based systems is a vision-based image-sensor system [10].

Vehicle black boxes are increasingly popular, and they are used not only for recording accidents, but also as an accident-prevention system by including image processing. Black boxes are widely used, owing to the ease of installation, relative low cost, and high-volume storage. Lane detection based on image processing uses an

X. He et al. (Eds.): IScIDE 2015, Part I, LNCS 9242, pp. 281–287, 2015.
DOI: 10.1007/978-3-319-23989-7_29

approximation method that involves lane-pattern matching [5], Kalman filters [6], and wavelet filtering [7, 8]. However, because of inconsistency in the color, shape, and texture of lanes as a result of various road conditions, accurate detection with such methods is not guaranteed. In addition, soiled, snow-covered, rain-covered, and irregularly paved roads also lead to inaccuracies in lane detection. Nevertheless, road-region detection should be prioritized for fast-moving vehicles in order to detect lanes in real-time, given the modest memory available in black boxes. Distinguishing the road region in the input image can help to prevent drivers from diverging off the road and out of the lane. The proposed method can prevent accidents that are caused by distracted driving, and it can be applied to various ADASs by detecting the road region and vanishing point based on the directional information of an edge.

The proposed method can be used to develop a commercializable road-region detection technology by effectively utilizing fast processing speeds and directional information under limited conditions—limitations such as those common to vehicle black boxes (product size, cost, etc.).

First, a video-recording device is installed on the front windshield of the vehicle to record a fixed view of the road in order to process only certain parts of the input image, particularly the edge of the road, rather than the entire image.

Second, the road region is detected with a statistical analysis of the directional information from an edge that tends to be straight, meaning it is less affected by noise and changes in illumination. Two straight lines are extracted, whose directional information for the edge is maximum on the borderline of the road region, and the intersection of the straight lines is selected as the vanishing point.

2 The Proposed Method

Considering that the color and texture information of road lanes, curbs, roads, and non-road regions are different, the proposed method performs a road region segmentation process that separates such information for use as feature information for detecting road regions. A flowchart of the proposed method is shown in Fig. 1.

Gray-level transform is performed in the input image, and size transform of the input image, noise removal, and removal of the upper and lower parts (sky region and road surface region) in the image are performed. In addition, an edge-preserving smoothing step is performed, and after performing image segmentation through a color histogram quantization step, Hough transform is performed at the edge in order to calculate the Hough transform matrix, theta, and rho values. In the Hough transform matrix, six points with peak values are detected, and based on these, two straight lines that satisfy the given condition are found and the road region is distinguished by selecting a pertinent region in the input image.

2.1   Pre-processing and Edge-Preserving Smoothing

For a road region (210×570 pixels) whose image size was reduced 30 % from an input image ($1,920 \times 1,080$ pixels) that excludes the upper and lower parts, an

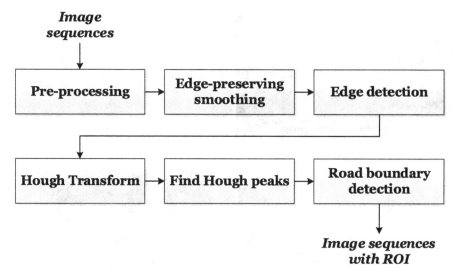

Fig. 1. Flowchart of the proposed method.

edge-preserving smoothing step was performed to make the pixel color values similar to those of the surrounding pixels while maintaining edge strength. In general, road environments contain various noises because of various environmental factors, such as large illumination changes, vehicle movements, and diffused sunlight reflection. Therefore, the simplest method for distinguishing road regions is to detect straight-line segments. In the proposed method, the edge detection included in the image is required for performing Hough transform for straight-line detection. Therefore, in the pre-processing step, to minimize noise effects on the input image, the color differences between the eight-direction neighborhood pixels in an $n \times n$ window size are calculated; if the color variation is less than the threshold, image color simplification is performed by changing the various colors to colors similar to the pixels located in the surrounding eight directions.

Figure 2 is the result of selecting a road ROI region in an input image, and detecting it through the edge-preserving smoothing [11] process for the pertinent region. The edge-preserving smoothing process is performed to every pixel of ROI. The calculations are [11]

$$t_i = \frac{(|R - r_i| + |G - g_i| + |B - b_i|)}{255 \times 3}, \quad i = 1, \ldots, 8 \qquad (1)$$

where R, G, B is the center pixel value in a 8×8 window mask, and r, g, b is the 8-neighborhood pixel value of center pixel. To using the 3×3 convolution mask of the Eq. (1), color values of edge pixel have a small change. Otherwise a pixel value is similar to the surrounding pixel values. In our case, the smoothing process is applied to gray-level image.

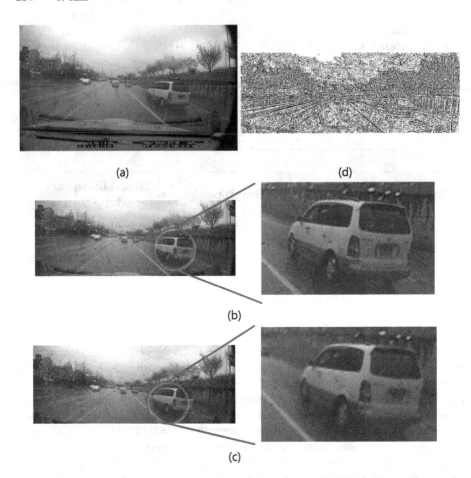

Fig. 2. Results of edge-preserving smoothing, (a) input image, (b) ROI, (c) smoothing results with n = 5, (d) diffidence image of |(b)–(c)|.

2.2   Edge Detecting

In the method proposed for detecting edges in an edge-preserving smoothing image, an edge is detected in the boundary possessed by each color region after performing a color histogram quantization [9]. In the proposed method, straight-line detection in the image is necessary for detecting the road region. To detect the straight line, the image edges are calculated and their straight-line components are extracted.

Figure 3 shows the edge detection image and the canny edge detection image of Fig. 2(b) after performing color histogram quantization. To the color quantization, we have set the color quantized level 6, and quantization window size 3.

In the resulting image, because the canny edge detector detects edge pixels according to variations in the intensity pixels, even a similar object region with identical color is sometimes differentiated as dissimilar regions because of the difference in brightness. However, after performing color simplification through color

(a)   (b)

Fig. 3. Results of color histogram quantization (a) and canny edge detection (b).

histogram quantization, the edge detection result is the outcome of detecting boundaries close to the object region, rather than the difference in brightness.

2.3   Hough Transform and Line Detection

In Fig. 3(a), Hough transform is performed from the edge image. The Hough transform matrix, theta, and rho values are calculated through Hough transform, and the peak points with peak values are determined in the Hough transform matrix. Figure 4(b) is the result of selecting six peak points in the Hough transform matrix of the Fig. 3(a) image, and the result of detecting lines with peak component values through Hough transform in the input image. The six peak points in the Hough transform represent the detection of a total of six straight lines. Therefore, the conditions of the vanishing point location selection are as follows for selecting two straight lines among the detected straight lines. First, the intersection of two straight lines must be located in the center part (50 × 100 pixels) of the input image; if the absolute slope difference of two straight lines is less than or equal to 30 degrees and the slope of one straight line is less than ten degrees, the pertinent straight line is excluded. Figure 4(c) is the result of finding two straight lines through Hough transform in the canny edge result.

3   Results and Future Works

To test the accuracy of the proposed method, the results obtained through manual road region detection and the results obtained using the proposed road region detection method were compared. As a result, approximately 81 % detection accuracy was provided in a total of 137 experiment images, and approximately 0.37 s per frame were consumed on average for processing time. The edge-preserving smoothing and color histogram quantization steps that perform computations based on entire pixels required the most processing time and accounted for most of the consumption. For the remaining image processing, time consumption was minimized by processing the connected pixel regions only in the binary images.

The test results (Fig. 5) indicate that most detection errors occur at the edges of other vehicles, curved roads, and inclined regions. Because of this, in the straight-line detection results of Hough transform, intersections occur in the sky region or are shifted to the left/right side, thus resulting in errors when detecting road regions. The result of testing the proposed method in daytime road images presents a relatively positive

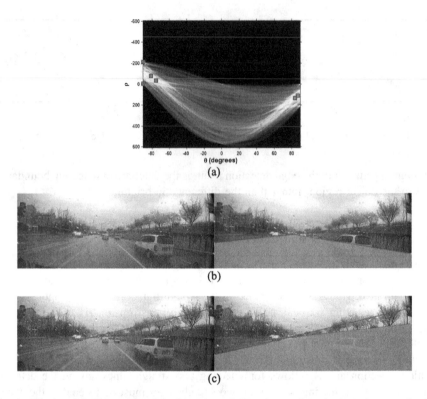

(a)

(b)

(c)

Fig. 4. Results of finding two straight lines through Hough transform in the color quantization (b) and canny edge (c).

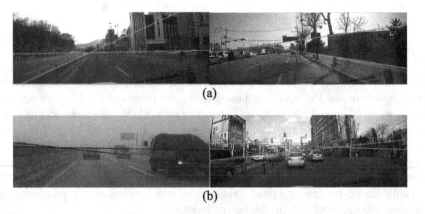

(a)

(b)

Fig. 5. Experimental results of finding two straight lines through Hough transform in the canny edge result, (a) correct results, and (b) incorrect results.

performance; however, in order to apply the proposed method in various road environments, such as during nighttime and in the rain, a more refined implementation is required. However, because the most effective processing is required to minimize heat generation, power consumption, and memory for a vehicle-type black box, additional processing was restricted. In future research, a road region detection study will be performed using intensity pixel texture orientation through Gabor filters.

Acknowledgments. This research is supported by Basic Science Research Program through the National Research Foundation of Korea funded by the Ministry of Education, Science and Technology (2010-0021071) in 2014.

References

1. Bengler, K., Dietmayer, K., Farber, B., Maurer, M., Stiller, C., Winner, H.: Three decades of driver assistance systems: review and future perspectives. IEEE Intell. Transp. Syst. Mag. **6** (4), 6–22 (2014)
2. Wu, C.F., Lin, C.J., Lee, C.Y.: Applying a functional neurofuzzy network to real-time lane detection and front-vehicle distance measurement. IEEE Trans. Syst. Man Cybern. **42**(4), 577–589 (2012)
3. Cualain, D.O., Hughes, C., Glavin, M., Jones, E.: Automotive standards-grade lane departure warning system. IEEE Intell. Transp. Syst. **6**(1), 44–57 (2012)
4. Son, J., Yoo, H., Kim, S., Sohn, K.: Real-time illumination invariant lane detection for lane departure warning system. Expert Syst. Appl. **42**(4), 1816–1824 (2015)
5. Chao, M., Ling, M., YueFei, Z., Mei, X.: Lane detection using heuristic search methods based on color clustering. In: IEEE International Conference on Communications, Circuits and Systems, pp. 368–372 (2010)
6. Cualain, D.O., Hughes, C., Glavin, M., Jones, E.: Automotive standards-grade lane departure warning system. IET Intel. Transp. Syst. **6**(1), 44–57 (2012)
7. Ouyang, D.: Intelligent road control system using advanced image processing techniques. MS theses, University of Toledo, USA (2012)
8. Cheng, X., Chao, M., Chao, C., Zhibang, Y.: Road detection based on vanishing point location. J. Convergence Inf. Technol. **7**(6), 137–145 (2012)
9. Deng, Y., Manjunath, B.S., Shin, H.: Color image segmentation. In: IEEE Conference on CVR, vol. 2, p. 2446 (1999)
10. Kim, J.B.: Detection of traffic signs based on eigen-color model and saliency model in driver assistance systems. Int. J. Automot. Technol. **14**(3), 429–439 (2013)
11. Nikolaou, N., Papamarkos, N.: Color reduction for complex document images. Int. J. Imaging Syst. Technol. **19**(1), 14–26 (2009)

Dictionary Pair Learning with Block-Diagonal Structure for Image Classification

Meng Yang$^{(\boxtimes)}$, Weixin Luo, and Linlin Shen

School of Computer Science and Software Engineering,
Shenzhen University, Shenzhen, China
yangmengpolyu@msn.com

Abstract. Dictionary learning has played an important role in the success of sparse representation. Although several dictionary learning approaches have been developed for image classification, the dictionary pair learning, i.e., jointly learning a synthesis dictionary and an analysis dictionary, is still in its infant stage. In this paper, we proposed a novel model of dictionary pair learning with block-diagonal structure (DPL-BDS), in which a block-diagonal structure of coding coefficient matrix and a block-diagonal structure of analysis dictionary are enforced. With the block-diagonal structures, discrimination of synthesis dictionary representation, coding coefficients and analysis dictionary are introduced into the dictionary pair learning model. An iterative algorithm to efficiently solve the proposed DPL-BDS was presented in this paper. The experiments on face recognition, scene categorization, and action recognition clearly show the advantage of the proposed DPL-BDS.

Keywords: Dictionary pair learning · Block-diagonal structure · Image classification

1 Introduction

With encourage from the success of sparse coding in image processing [1, 2] and inspiration of sparsity mechanism of human vision system [3], sparse representation has been widely applied to many fields, such as computer vision, pattern recognition [4]. As indicated by [5], the dictionary plays an important role in the success of sparse representation and learning the desired dictionary from training data instead of using off-the-shelf bases (e.g., wavelets) has led to state-of-the-art results in many practical applications, such as image denoising [1], face recognition [6, 7], and image classification [8, 9].

According to the way of encoding input signals, the dictionary used in sparse representation could be categorized into synthesis dictionary and analysis dictionary [10]. Therefore dictionary learning approaches could be mainly divided into three categories: synthesis dictionary learning, analysis dictionary learning, and analysis-synthesis dictionary pair learning.

Synthesis Dictionary Learning: Given a training data X, the synthesis dictionary learning wants to learn a synthesis dictionary D, which meets $X \cong DA$ and some additional task-driven regularizations. Most of existing dictionary learning methods

X. He et al. (Eds.): IScIDE 2015, Part I, LNCS 9242, pp. 288–299, 2015.
DOI: 10.1007/978-3-319-23989-7_30

belong to this category. A representative unsupervised dictionary learning method is K-SVD [11], which has shown promising performance in image restoration. For classification tasks, discriminative versions [6, 7] of K-SVD were developed by jointly learning a dictionary and a classifier based on the coding coefficients. The learned dictionary in [6, 7] is a shared dictionary since it could represent all class data. The learned synthesis dictionary can be class-specific, which means that each dictionary atom belongs to a single class. Following this line, class-specific dictionary learning, e.g., dictionary learning with structured incoherence (DLSI) [8] and Fisher discrimination dictionary learning (FDDL) [9], were developed by requiring that a sub-dictionary should represent well for some class but bad for all the other classes. To fully explore the advantage of class-specific dictionaries and shared dictionaries, some hybrid dictionary learning models were proposed. Zhou and Fan [13] learned a hybrid dictionary with a Fisher-like regularizer on the representation coefficients, while Kong and Wang [12] learned a hybrid dictionary by introducing an incoherence penalty term to the class-specific sub-dictionaries. Although promising results have been reported in the synthesis dictionary learning, the coding coefficients need to be computed by the time-consuming sparse coding. In addition, synthesis sparse coding cannot give an intuitive illustration like feature transformation (e.g., wavelet transformation), so analysis sparse coding, which is a dual analysis viewpoint, has started to be studied.

Analysis Dictionary Learning: as indicated in by [10], there is a dual analysis viewpoint to synthesis sparse presentation. Given a training data X, the analysis dictionary learning wants to learn an analysis dictionary Ω, which makes ΩX be a sparse coding coefficient matrix. The representative analysis dictionary learning approach is the analysis K-SVD [10], which is an unsupervised learning approach. Although promising result in image restoration has been reported in [10], analysis K-SVD is still computationally complicated and not designed for classification task due to the lack of enough discrimination.

Analysis-Synthesis Dictionary Pair Learning: To fully explore the ability of learned dictionary and simplify the computation of coding, recently two analysis-synthesis dictionary pair learning approaches have been proposed. Rubinstein and Elad [14] proposed an analysis-synthesis dictionary learning (ASDL) model for the task of image processing. In ASDL, a pair of synthesis dictionary and analysis dictionary is learned from an image patch set by requiring their good representation ability and sparsity of coding coefficients. Different from ASDL, Gu et al. [15] proposed a projective dictionary pair learning (PDPL) for image classification. In PDPL, the synthesis class-specific dictionary and analysis class-specific dictionary are required to well represent some class but bad for all the other classes. For classification tasks, obviously ASDL ignores to introduce discrimination into the learned dictionaries. Although PDPL has introduced the class-specific structure into dictionary pair learning, the discrimination embedded in the training data is not fully exploited since there may be a big correlation between the analysis class-specific dictionaries, and representation coding coefficients may have a big within-class variance.

In this paper we propose a novel model of dictionary pair learning with block-diagonal structures (DPL-BDS). In the proposed model, a block-diagonal structure of coding coefficient is required so that the learned synthesis dictionary is class-specific and the coding coefficient has a small within-class scatter. Meanwhile, a block-diagonal structure of analysis dictionary is also exploited to reduce the disturbance between different analysis class-specific dictionaries. An efficient algorithm was proposed to efficiently solve the proposed model. The DPL-BDS was evaluated on the application of face recognition, scene categorization, and action recognition. Compared with other state-of-the-art dictionary learning methods, DPL-BDS has better or competitive performance in various classification tasks.

The rest of this paper is organized as follows. Section 2 briefly introduces related work. Section 3 presents the proposed DPL-BDS model. Section 4 conducts experiments, and Sect. 5 concludes the paper.

2   Brief Review of Related Work

Recently, Rubinstein and Elad [14] proposed an analysis-synthesis dictionary learning (ASDL) for image deblurring. ASDL explicitly learned a pair of analysis dictionary and synthesis dictionary via

$$\min_{\Omega, D} \|X - DS_\lambda(\Omega X)\|_2^2 \quad s.t. \ \|\omega_i\|_2^2 = 1 \ \forall i \tag{1}$$

where X is a matrix of training samples, D is a synthesis dictionary, $\Omega = [\omega_1; \omega_2; \ldots; \omega_n]$ is an analysis dictionary, and $S_\lambda(\alpha) = \alpha.1(|\alpha| \geq \lambda)$ is a hard thresholding operator. It is easy to see that in ASDL the learned dictionary pair is only required to well represent the training samples, without considering its application to classification task.

Following ASDL, Gu et al. [15] proposed a projective dictionary pair learning (PDPL) for pattern classification. The core idea of PDPL is that both analysis and synthesis dictionaries are class-specific dictionaries. Denote by D_i and Ω_i the synthesis sub-dictionary and analysis sub-dictionary for class i, respectively, the model PDPL can be formulated as

$$\min_{\Omega, D} \sum_{k=1}^{K} \|X_k - D_k \Omega_k X_k\|_F^2 + \lambda \|\Omega_k \bar{X}_k\|_2^2 \quad s.t. \ \|d_i\|_2^2 \leq 1 \ \forall i \tag{2}$$

where the class-specific dictionary pair of class k (e.g., D_k and Ω_k) is required to well represent the k-th training data, and the training data which don't belong to class k (denote by $\bar{X}_k$) should have a small representation on Ω_k. This structured dictionary regularization in PDPL makes the learned dictionary pair suitable for classification task, and some promising results have been presented in [15].

3   Dictionary Pair Learning with Block-Diagonal Structure

Although PDPL has introduced the class-specific structure into dictionary pair learning, the discrimination embedded in the training data is not fully exploited since there may be a big correlation between the analysis class-specific dictionaries, and representation coding coefficients may have a big within-class variance. In this section, we proposed a model of dictionary pair learning with block-diagonal structure (DPL-BDS). Two block-diagonal structures are exploited to introduce more discrimination into the learned dictionary pair in the proposed DPL-BDS.

3.1   Block-Diagonal Structure of Coding Coefficient

When the learned dictionary atoms have labels, the discrimination could be introduced by enforcing coding coefficient of labeled training samples to have a block-diagonal structure. Denote by $X = [X_1, X_2, ..., X_K]$ the labeled training data, where X_i is the i-th class training data. Let $D = [D_1, D_2, ..., D_K]$ be the labeled synthesis dictionary, where D_i is the i-th class synthesis sub-dictionary. The structure of coding matrix, i.e., A, where $X = DA$, would determine the discrimination of the learned dictionary. An intuitive example of a block-diagonal coding coefficient matrix is shown in Fig. 1. The block-diagonal structure of A would make the synthesis sub-dictionary D_i only well represent X_i, but have nearly zero representation for X_j, where $j \neq i$. Thus the learned dictionary would be enforced a strong discrimination.

Labeled Training data   Labeled Synthesis Dictionary   Coding Coefficient

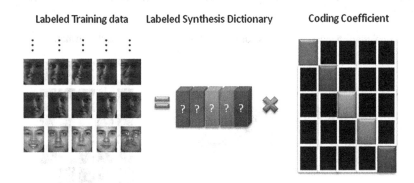

Fig. 1. The discrimination introduced by the block-diagonal structure of coding coefficient.

Let $A=[A_{1,1},\ A_{2,1},...,A_{K,1};A_{1,2},A_{2,2},...,A_{K,2};...;A_{1,K},A_{2,K},...,A_{K,K}]$, where $A_{i,j}$ denotes the coding sub-matrix of X_i over the sub-dictionary D_j. The block-diagonal structured of A here has two properties: (1) $A_{i,j} = 0$ for $i \neq j$ in order to increase the between-class scatter of A and the discrimination of D; (2) the variance of A_i should be small to reduce the within-class scatter. Therefore the dictionary learning model with block-diagonal coding coefficients could be formulated as

$$\min_D \sum_{k=1}^{K} \|X_k - DA_k\|_F^2 + \mu \|A_k - A_k^m\|_F^2 \quad s.t. \ \|d_i\|_2^2 \leq 1 \ \forall i; \ A_{i,j} = 0 \ \forall i \neq j \quad (3)$$

where $A_k = [A_{k,1}; A_{k,2}; ...; A_{k,K}]$ is the coding coefficient of X_k over D, and A_k^m is a mean matrix with the mean vector of A_k as its column vector and the same size to A_k.

3.2   Block-Diagonal Structure of Analysis Dictionary

In DPL-BDS, the analysis dictionary atoms are also class-specific. Denote by $\boldsymbol{\Omega} = [\boldsymbol{\Omega}_1;\boldsymbol{\Omega}_2;\ldots;\boldsymbol{\Omega}_k]$ the analysis dictionary, where $\boldsymbol{\Omega}_i$ is the analysis sub-dictionary associated to class i. An ideal analysis dictionary should also have block-diagonal structure, as shown in Fig. 2. The correlation matrix of analysis dictionary should have a block-diagonal structure. With the block-diagonal structure, the analysis sub-dictionaries associated to different classes would have nearly zero correlation, which can reduce the disturbance between different classes. In our paper, we expect the correlation matrix of the analysis dictionary, i.e., $\boldsymbol{\Omega}\boldsymbol{\Omega}^T$, to have a block-diagonal structure. Therefore the dictionary learning model with block-diagonal analysis dictionary could be formulated as

$$\min_{\boldsymbol{\Omega}} \sum_{k=1}^{K} \left(\sum_{j\neq k} \left\| \boldsymbol{\Omega}_k\boldsymbol{\Omega}_j^T \right\|_F^2 + \|\boldsymbol{\Omega}_k\|_F^2 \right) \quad \text{s.t. } \boldsymbol{A}_k = \boldsymbol{\Omega}\boldsymbol{X}_k \forall k; \tag{4}$$

where the minimization of $\|\boldsymbol{\Omega}_k\|_F^2$, which approximates the minimization of $\|\boldsymbol{A}_k\|_F^2$ because $\|\boldsymbol{A}_k\|_F^2 = \|\boldsymbol{\Omega}\boldsymbol{X}_k\|_F^2 \leq \|\boldsymbol{X}_k\|_F^2 \|\boldsymbol{\Omega}\|_F^2$, would make the representation more stable.

Analysis dictionary   Analysis dictionary transpose   Correlation matrix

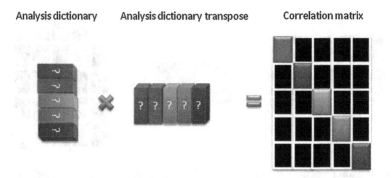

Fig. 2. The discrimination introduced by the block-diagonal structure of analysis dictionary.

3.3   Dictionary Pair Learning with Block-Diagonal Structure

Based on the presented two block-diagonal structures above, we proposed the model of dictionary pair learning with block-diagonal structure (DPL-BDS) as

$$\min_{\boldsymbol{D},\boldsymbol{\Omega}} \sum_{k=1}^{K} \left(\|\boldsymbol{X}_k - \boldsymbol{D}\boldsymbol{A}_k\|_F^2 + \mu\|\boldsymbol{A}_k - \boldsymbol{A}_k^m\|_F^2 + \gamma \sum_{j\neq k} \left\| \boldsymbol{\Omega}_k\boldsymbol{\Omega}_j^T \right\|_F^2 + \gamma\|\boldsymbol{\Omega}_k\|_F^2 \right) \tag{5}$$
$$s.t. \ \|\boldsymbol{d}_i\|_2^2 \leq 1 \ \forall i; \ \boldsymbol{A}_{i,j} = 0 \ \forall i \neq j; \ \boldsymbol{A}_k = \boldsymbol{\Omega}\boldsymbol{X}_k \forall k$$

Based on the constraint that $\boldsymbol{A}_{i,j} = 0$ for $i \neq j$ and $\boldsymbol{A}_k = \boldsymbol{\Omega}\boldsymbol{X}_k$, the proposed DPL-BDS could be rewritten as

$$\min_{D,\Omega} \sum_{k=1}^{K} \left(\left\| X_k - D_k A_{k,k} \right\|_F^2 + \mu \left\| \Omega_k (X_k - X_k^m) \right\|_F^2 + \gamma \sum_{j \neq k} \left\| \Omega_k \Omega_j \right\|_F^2 + \gamma \left\| \Omega_k \right\|_F^2 \right) \tag{6}$$
$$s.t. \ \|d_i\|_2^2 \leq 1 \ \forall i; \ \Omega_i X_j = 0 \ \forall i \neq j; \ A_{k,k} = \Omega_k X_k$$

where μ and γ are two parameters of DPL-BDS and X_k^m is a mean matrix with the mean vector of X_k as its column vector and the same size to X_k.

It can be seen that the proposed DPL-BDS has introduced the discrimination of class-specific representation, discrimination of coding coefficient and discrimination of analysis dictionary via two block-diagonal structures.

3.4   Solving Algorithm of DPL-BDS

In order to learn the pair of synthesis dictionary and analysis dictionary efficiently, we solve a relaxed version of DPL-BDS, which is

$$\min_{D,\Omega} \sum_{k=1}^{K} \left(\begin{array}{c} \left\| X_k - D_k A_{k,k} \right\|_F^2 + \mu \left\| \Omega_k (X_k - X_k^m) \right\|_F^2 + \gamma \sum_{j \neq k} \left\| \Omega_k \Omega_j \right\|_F^2 + \gamma \left\| \Omega_k \right\|_F^2 \\ + \lambda \left\| \Omega_k \bar{X}_k \right\|_F^2 + \tau \left\| A_{k,k} - \Omega_k X_k \right\|_F^2 \end{array} \right) \tag{7}$$
$$s.t. \ \|d_i\|_2^2 \leq 1 \ \forall i;$$

where λ and τ are two scalar constants for relaxing the original model. Equation (7) could be easily solved by alternatively updating D, Ω, and A.

When D and Ω are fixed, A could be updated via

$$\min_A \sum_{k=1}^{K} \left(\left\| X_k - D_k A_{k,k} \right\|_F^2 + \tau \left\| A_{k,k} - \Omega_k X_k \right\|_F^2 \right) \tag{8}$$

There is a closed-form solution for $A_{k,k}$ for every k, which is

$$A_{k,k} = \left(D_k^T D_k + \tau I \right)^{-1} \left(\tau \Omega_k X_k + D_k^T X_k \right) \tag{9}$$

where I is an identity matrix.

When A and D are fixed, Ω could be updated via

$$\min_\Omega \sum_{k=1}^{K} \left(\begin{array}{c} \mu \left\| \Omega_k (X_k - X_k^m) \right\|_F^2 + \gamma \sum_{j \neq k} \left\| \Omega_k \Omega_j \right\|_F^2 + \gamma \left\| \Omega_k \right\|_F^2 \\ + \lambda \left\| \Omega_k \bar{X}_k \right\|_F^2 + \tau \left\| A_{k,k} - \Omega_k X_k \right\|_F^2 \end{array} \right) \tag{10}$$

We solve Ω class by class. When solving Ω_k, we fixed $\Omega_j, j \neq k$. Therefore there is also a closed-form solution for Ω_k, which is

$$\Omega_k = \tau A_{k,k} X_k^T \left(\mu \tilde{X}_k \tilde{X}_k^T + \gamma \sum_{j \neq k} \Omega_j^T \Omega_j + \gamma I + \lambda \bar{X}_k \bar{X}_k^T + \tau X_k X_k^T \right)^{-1} \tag{11}$$

where $\tilde{X}_k = \left(X_k - X_k^m \right)$.

When A and Ω are fixed, D could be updated via

$$\min_D \sum_{k=1}^{K} \left(\left\| X_k - D_k A_{k,k} \right\|_F^2 \right) \quad s.t. \ \|d_i\|_2^2 \le 1 \ \forall i; \tag{12}$$

which could be efficiently solved by the solver presented in [15].

The whole solving algorithm of DPL-BDS is summarized in Table 1. After solving DPL-BDS, the synthesis sub-dictionary for each class, e.g., D_i, and the analysis sub-dictionary, e.g., Ω_i, are also known. When a testing sample, i.e., y, comes, it coding vector on D is

$$\alpha = \Omega y = [\Omega_1; \Omega_2; \cdots; \Omega_K] y = [\Omega_1 y; \Omega_2 y; \cdots; \Omega_K y] \tag{13}$$

Then the class-specific representation residual is used to do classification via

$$identity = \arg \min_i \left\{ \|y - D_i \Omega_i y\|_2 \right\} \tag{14}$$

Table 1. Algorithm of DPL-BDS.

Solving algorithm of DPL-BDS
1. Initialize D and Ω as random matrixes with unit Frobenious norm.
2. While not converge do
3.   Compute A via Eq.(8);
4.   Compute Ω via Eq.(10);
5.   Compute D via Eq.(12)
6. End while
7. Output D and Ω until the condition of convergence is met, or the maximal number of iterations is reached.

4 Experiments

We perform experiments on Extended Yale B [16], AR [17], Scene 15 [18] and UCF sport action dataset [19] to demonstrate the performance of DPL-BDS. In Sect. 4.1, a discussion on the proposed two block-diagonal structures is conducted; then we evaluate DPL-BDS on the applications of face recognition, scene categorization and action recognition in Sects. 4.2, 4.3, and 4.4, respectively.

To clearly illustrate the advantage of DPL-BDS, we compare it with several latest DL methods, such as Discriminative K-SVD (DKSVD) [6], Label Consistent K-SVD (LCKSVD) [7], dictionary learning with structure incoherence (DLSI) [8], Fisher discrimination dictionary learning (FDDL) [9], and the projective dictionary pair learning (PDPL) [15]. Besides, we also report sparse representation based classifier (SRC) [4], collaborative representation based classifier (CRC) [22], linear support vector machine (SVM) and some methods for special tasks.

There are four parameters in the proposed DPL-BDS. As shown in Eq. (6), μ and γ control the regularizations of analysis dictionary disturbance and within-class scatter of coding coefficient, respectively. As shown in Eq. (7), τ and λ are two parameters for relaxing the original model. In our paper, if no specific instruction, the parameters are empirically set as $\tau = \lambda = 0.1$, and $\mu = \gamma = 0.0001$.

4.1   Discussion on Block-Diagonal Structure

In this section, we discuss the roles of two block-diagonal structures in classification. Here we use the extended Yale B dataset (the detailed experimental setting is described in Sect. 4.2) to do this evaluation. In order to make the evaluation more challenging, we add 0.1 % uniform noise to the data. We first test the role of block-diagonal structure on the analysis dictionary by changing the values of γ and fixing the other parameter values as 1e-3. Figure 3 plots the accuracy curve with the values of γ. It can be seen that when there is an appropriate value of γ, the accuracy is the highest. When γ is too small, the disturbance of analysis dictionary would weaken the performance; while a too big γ would make the analysis dictionary have a small energy and harm the representation of data.

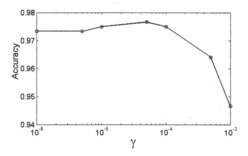

Fig. 3. The recognition rate versus the parameter (i.e., γ) associated to the block-diagonal structure of analysis dictionary.

We then test the role of block-diagonal structure on the coding coefficient. Figure 4 plots the accuracy curves with different μ and λ. λ controls the inter-class distance and the accuracy is the lowest when $\lambda = 0$. μ controls the intra-class distance and the accuracy achieves the highest with $\mu = 1e-2$ when $\lambda = 0.01$. Different combinations of μ and λ would affect the final performance, which show the importance of the block-structure on the coding coefficient.

4.2   Face Recognition

Two face databases, such as Extended Yale B [16] and AR [17], are used to evaluate the performance of the proposed DPL-BDS. The Extended Yale B dataset mainly includes the illumination variation; while the AR dataset contains illumination, expression, and disguise variation.

Following the experimental setting of Extended Yale B in [7], a set of 2,414 face images of 38 persons are used to do the experiment. In this experiment, a half of images per subject are used for training, with the left for testing. The 504-d feature provided by [7] is used as the facial features. The experimental results of all the competing methods are listed in Table 2. We can observe that DPL-BDS and PDPL achieve the best performance, and visibly outperform other methods. All the methods achieved over 95 % accuracy, which is also the reason why the improvement of DPL-BDS is not so big. Meanwhile, we listed the training time and testing time of several methods in Table 2 with the desktop of i5 2.6 GHz CPU and with 8G RAM. We could see that the average running time for recognizing a testing sample of DPL-BDS and PDPL is about 3e-4 s, much faster than FDDL, CRC and SRC. Thus we could see that the dictionary pair learning methods are better than the synthesis dictionary learning approaches in accuracy and efficiency.

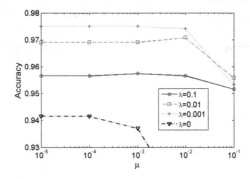

Fig. 4. The recognition rate versus the parameters (i.e., μ and λ) associated to the block-diagonal structure of coding coefficient.

Table 2. Recognition rates and running time (second) on the Extended Yale B.

	SVM	CRC	SRC	DLSI	FDDL	LC-KSVD	PDPL	DPL-BDS
Accuracy	95.6	97.0	96.5	97.0	96.7	96.7	**97.4**	**97.4**
Training	—	0	0	—	5951.2	20.59	6.56	16.95
Testing	—	8.5e-4	207.9	—	2.2	1.6e-4	2.9e-4	3.3e-4

Note: To solve sparse coding, l_1_ls and C code of OMP are applied in SRC and LC-KSVD, respectively

Following the experimental setting of AR in [7], a set of 2,600 images of 50 female and 50 male subjects are extracted. 20 images of each subject are used for training and the remaining 6 images are used for testing. 540-d feature provided by [7] is used as the facial features. The experimental results of all the competing methods are listed in Table 3. In this dataset, DPL-BDS achieves the best performance, e.g., 98.7 % accuracy, and have over 1 % improvements over all the other competing methods except PDPL. It should be noted that both DPL-BDS and PDPL have a very fast speed in testing since only a matrix multiplication is needed for coding.

Table 3. Recognition rates on the AR.

SVM	CRC	SRC	DLSI	FDDL	LC-KSVD	DPDL	DPL-BDS
96.5	98.0	97.5	97.5	97.5	97.8	98.3	**98.7**

4.3   Scene Categorization

The fifteen scene dataset was introduced in [18]. Each category of natural scene has 200 to 400 images, and the average image size is about 250×300 pixels. The fifteen scenes contain bedroom, kitchen, and country scenes. Following the experimental setting in [7], 100 images per category are chosen as the training data, with the rest as the testing data. The image descriptor is generated by extracting SIFT feature in local region, encoding local patch feature, and max pooling in spatial pyramid. In our paper, we use the 3000-d PCA feature of the spatial pyramid features provided by [7] to do the experiment.

The experimental results of all the competing methods are shown in Table 4. Compared to the traditional synthesis dictionary learning methods, such as DK-SVD, LC-KSVD, K-SVD, the proposed DPL-BDS has over 5 % improvements. That indicates that the dictionary pair learning is a very promising direction.

Table 4. Recognition rates on the Scene 15.

SRC	K-SVD	DK-SVD	LC-KSVD1	LC-KSVD2	PDPL	DPL-BDS
91.8	86.7	89.1	90.4	92.9	97.7	**97.8**

4.4   Action Recognition

The benchmark action dataset, UCF sports action dataset [19], is used to conduct the action classification experiment. The dataset collected video clips from various broadcast sports channels (e.g., BBC and ESPN). The action bank features of 140 videos provided by [20] are adopted in the experiment. These videos cover 10 sport action classes: driving, golfing, kicking, lifting, horse riding, running, skateboarding, swinging-(pommel horse and floor), swinging-(high bar) and walking. As the experiment setting in [7], we evaluated the DPL-BDS via five-fold cross validation. Here the dimension of the action bank feature is reduced to 100 via PCA, and the performance of some specific methods for action recognition, such as Qiu 2011 [21], action back feature with SVM classifier [20] are also reported in Table 5. From Table 5, we can see that DPL-BDS still have higher recognition rate compared to the traditional dictionary learning methods and some specific approaches. DPL-BDS outperforms PDPL by 0.7 %, and both DPL-BDS and FDDL all achieve the best performance. However, DPL-BDS has much faster speed than FDDL since there is a time consuming sparse coding in FDDL.

Table 5. Recognition rates on the UCF sports action dataset.

Qiu 2011	SVM	SRC	DLSI	DKSVD	LC-KSVD	FDDL	PDPL	DPL-BDS
83.6	90.7	92.9	92.1	88.1	91.2	**93.6**	92.9	**93.6**

5 Conclusion

This paper presented a novel discriminative analysis-synthesis dictionary learning model for image classification. It has shown that the discrimination of the learned dictionaries could be exploited via two block-diagonal structures of coding coefficient and analysis dictionary. Based on the introduced block-diagonal structures, the discrimination of synthesis dictionary, coding coefficient and analysis is introduced in the proposed dictionary pair learning with block-diagonal structure (DPL-BDS). An efficient solving algorithm is also proposed for the proposed model. The proposed DPL-BDS method was extensively evaluated on several image classification tasks. The experimental results clearly demonstrated that DPL-BDS could outperform/compete with previous state-of-the-art methods, such as SRC, FDDL, LC-KSVD, PDPL and DKSVD.

Acknowledgment. This work is partially supported by the National Natural Science Foundation of China under Grants no. 61402289 and 61272050, National Science Foundation of Guangdong Province (2014A030313558 and 2014A030313556), and Shenzhen Scientific Research and Development Funding Program under Grants JCYJ20140509172609171 and JCYJ20130329115750231.

References

1. Elad, M., Aharon, M.: Image denoising via sparse and redundant representations over learned dictionaries. IEEE TIP **15**(12), 3736–3745 (2006)
2. Yang, J.C., Wright, J., Ma, Y., Huang, T.: Image super-resolution as sparse representation of raw image patches. In: Proceedings of CVPR (2008)
3. Olshausen, B.A., Field, D.J.: Emergence of simple-cell receptive field properties by learning a sparse code for natural images. Nature **381**, 607–609 (1996)
4. Wright, J., Yang, A.Y., Ganesh, A., Sastry, S.S., Ma, Y.: Robust face recognition via sparse representation. IEEE TPAMI **31**(2), 210–227 (2009)
5. Rubinstein, R., Bruckstein, A.M., Elad, M.: Dictionaries for sparse representation modeling. Proc. IEEE **9**(6), 1045–1057 (2010)
6. Zhang, Q., Li, B.X.: Discriminative K-SVD for dictionary learning in face recognition. In: Proceedings of CVPR (2010)
7. Jiang, Z.L., Lin, Z., Davis, L.S.: Label consistent K-SVD: learning a discriminative dictionary for recognition. IEEE TPAMI **35**(11), 2651–2664 (2013)
8. Ramirez, I., Sprechmann, P., Sapiro, G.: Classification and clustering via dictionary learning with structured incoherence and shared features. In: Proceedings of CVPR (2010)
9. Yang, M., Zhang, L., Feng, X.C., Zhang, D.: Fisher discrimination dictionary learning for sparse representation. In: Proceedings of ICCV (2011)
10. Rubinstein, R., Peleg, T., Elad, M.: Analysis K-SVD: a dictionary-learning algorithm for the analysis sparse model. IEEE TSP **61**(3), 661–677 (2013)
11. Aharon, M., Elad, M., Bruckstein, A.: K-SVD: an algorithm for designing overcomplete dictionaries for sparse representation. IEEE TSP **54**(11), 4311–4322 (2006)

12. Kong, S., Wang, D.: A dictionary learning approach for classification: separating the particularity and the commonality. In: Fitzgibbon, A., Lazebnik, S., Perona, P., Sato, Y., Schmid, C. (eds.) ECCV 2012, Part I. LNCS, vol. 7572, pp. 186–199. Springer, Heidelberg (2012)

13. Zhou, N., Fan, J.P.: Learning inter-related visual dictionary for object recognition. In: Proceedings of CVPR (2012)

14. Rubinstein, R., Elad, M.: Dictionary learning for analysis-synthesis thresholding. IEEE TSP **62**(22), 5962–5972 (2014)

15. Gu, S., Zhang, L., Zuo, W., Feng, X.: Projective dictionary pair learning for pattern classification. In: Proceedings of NIPs (2014)

16. Georghiades, A., Belhumeur, P., Kriegman, D.: From few to many: illumination cone models for face recognition under variable lighting and pose. IEEE TPAMI **23**(6), 643–660 (2001)

17. Martinez, A., Benavente, R.: The ar face database. CVC technical report (1998)

18. Lazebnik, S., Schmid, C., Ponce, J.: Beyond bags of features: spatial pyramid matching for recognizing natural scene categories. In: Proceedings of IEEE CVPR (2007)

19. Rodriguez, M., Ahmed, J., Shah, M.: A spatio-temporal maximum average correlation height filter for action recognition. In: Proceedings of CVPR (2008)

20. Sadanand, S., Corso, J.J.: Action bank: a high-level representation of activity in video. In: Proceedings of CVPR (2012)

21. Qiu, Q., Jiang, Z.L., Chellappa, R.: Sparse dictionary-based representation and recognition of action attributes. In: Proceedings of ICCV (2011)

22. Zhang, L., Yang, M., Feng, X.C.: Sparse representation or collaborative representation: which helps face recognition?. In: Proceedings of ICCV (2011)

A Gradient Weighted Thresholding Method for Image Segmentation

Bo Lei[✉] and Jiu-lun Fan

School of Communication and Information Engineering,
Xi'an University of Posts and Telecommunications,
Xi'an 710121, People's Republic of China
leileibo2015@163.com

Abstract. Otsu method is widely used for image thresholding, which only considers the gray level information of the pixels. Otsu method can provide satisfactory result for thresholding an image with a histogram of clear bimodal distribution. This method, however, fails if the variance or the class probability of the object is much smaller than that of the background. In order to introduce more information of the image, a gradient weighted threholding method is presented, which weighs the objective function of the Otsu method with the gray level and gradient mapping (GGM) function. It makes the between-class variance of the thresholded image maximize and the threshold locate as close to the boundary of the object and the background as possible. The experimental results on optical images as well as infrared images show the effectiveness of the proposed method.

Keywords: Image segmentation · Otsu method · GGM function · Gradient weighted

1 Introduction

Image segmentation is the key step of the image processing to image analysis. Thresholding is one of the most commonly used method and is the hot off the press in image segmentation with the characteristics of simple, intuitive and easy to be realized. Many different thresholding techniques have been proposed over the years [1]. Among the global thresholding techniques, Otsu method was one of the better threshold selection methods for general real world images with respect to uniformity and shape measures [2]. By selecting threshold values of the maximum between-class variances or minimum within-class variances for image histogram, this method is optimal in many cases.

Otsu method assumes that the gray levels of the object and the background in an image are Gaussian distribution with equal variances [3, 4]. Generally speaking, the real world images do not have such character. Therefore, Otsu method fails to select the optimal thresholds in some cases, especially where the variance or the class probability of the object is much smaller than that of the background. It has been proved that the threshold of the Otsu method was biased towards the component with larger class variance or larger class probability [5]. In addition, Otsu method only considers

© Springer International Publishing Switzerland 2015
X. He et al. (Eds.): IScIDE 2015, Part I, LNCS 9242, pp. 300–309, 2015.
DOI: 10.1007/978-3-319-23989-7_31

the information of the gray level of the pixels and without considering the space location and the boundary information. Many revised Otsu methods have been proposed [5–8].

In the existing improved Otsu method, they also only used the gray information of the pixels and without considering other information of the image. To thresholding, the best threshold should locate at the boundary of the object and the background. Analysis shows that the gray level changes slowly within the object and the background of an image, and changes fast on the boundary. Generally, the gradient information of an image can reflects the characteristics of the grayscale change. In order to make the threshold close to the boundary of the object and the background, a novel gradient weighted Otsu method is prevented. The new method weighs the objective function of the Otsu method with the gray level and gradient mapping (GGM) function [9]. It makes the between-class·variance of the thresholded image maximize and the threshold locate as close to the boundary of the object and the background as possible. In the experiment, it proved the performance of the proposed method by the optical images and the infrared images with small object.

The paper is organized as follows. In Sect. 2, we first prevent a brief description of the Otsu method. In Sect. 3, we describe the gray level and gradient mapping (GGM) function and the novel gradient weighted method. Section 4 describes the performance investigation over optical and infrared images. Finally, the paper is concluded in Sect. 5.

2 Otsu Method

Let $F = \{f_1, f_2, \ldots, f_{M \times N}\}$ be a digital image of size $M \times N$, where f_i is the gray value of the ith pixel and $f_i \in [0, 1, \ldots, L - 1]$, L is the largest gray level of the image. The amount of the pixels with the gray value g is denoted as $f(g)$. The frequency $h(g)$ of the occurrence of gray-level g is defined as:

$$h(g) = \frac{f(g)}{M \times N}, \quad g = 0, 1, \cdots, L - 1 \tag{1}$$

In the case of single threshold t, the pixels of an image are divided into two classes c_0 and c_1 (object and background, or background and object), i.e., $c_0 = [0, \ldots, t]$, $c_1 = [t + 1, \ldots, L - 1]$. For the threshold t, using discriminate analysis, Otsu showed that the between-class variance $\sigma_B^2(t)$ of c_0 and c_1 is,

$$\sigma_B^2(t) = p_0(t)\mu_0^2(t) + p_1(t)\mu_1^2(t) \tag{2}$$

Where, $p_0(t) = \sum_{g=0}^{t} h(g), \quad p_1(t) = \sum_{g=t+1}^{L-1} h(g), \quad \mu_0(t) = \sum_{g=0}^{t} gh(g)/p_0(t)$ and

$\mu_1(t) = \sum_{g=t+1}^{L-1} gh(g)/p_1(t).$

The optimal threshold t^* can be determined as:

$$t^* = Arg \max_{0 < t < L-1} \sigma_B^2(t) \tag{3}$$

The Otsu method considers the gray level information of the pixels only. It works well when the images to be thresholded have bimodal distribution with similar size. In other words, it works for image whose object is as large as the background.

3   The Gradient Weighted Method

Otsu method can provide satisfactory result for thresholding an image with a histogram of similar size bimodal distribution, as shown in Fig. 1. However, the histogram of the practical image is unimodal or multimodal distribution. So, the Otsu method fails in these conditions. Figure 2(a) shows an infrared image of the Terravic Research Motion IR database [10]. In this image, the area of the object is much smaller than that of the background. As shown in Fig. 2(b), the grayscale of the object is the small peak near the 255. The Otsu method can't find the optimal threshold in this case. The threshold of the Otsu method is near by 50, and the thresholding result is shown in Fig. 2(c) which can not extract the object from the background. The infrared image is based on the thermo radiation which depends closely on the temperature distribution of the object. As it is shown in Fig. 2(a), the temperature distribution of the object (a person) is very different from that of the background. The boundary of the object and the background is clearly. Therefore, we introduce the gradient information of an image and proposed a gradient weighted Otsu thresholding method. The improved method combines the gray level with the gradient information, such that the segmented threshold locates at the maximum of the between class variance and with large gradient. Then, the threshold is much close to the boundary of the object and the background in the image.

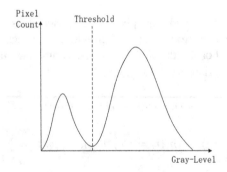

Fig. 1. Optimal threshold in gray-level histogram

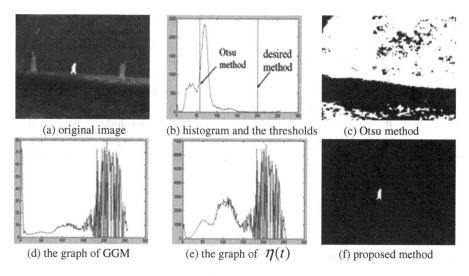

(a) original image   (b) histogram and the thresholds   (c) Otsu method

(d) the graph of GGM   (e) the graph of $\eta(t)$ (f) proposed method

Fig. 2. Segmentation results of an infrared image

3.1 Gray Level and Gradient Mapping (GGM) Function

The gradient of the pixel shows the gray level differences between the pixel and the neighborhood pixels. In the inner of the object and the background, the differences of the gray levels are small. The gradient is large on the boundary of the object and the background. The gradients could be different to the same gray level in the image. The gray level and gradient mapping function give the mapping relation of the gray level and the gradient, such that each gray level is corresponding to one gradient.

Let $f(x, y)$ represents the gray level of the pixel (x, y) in the image. To $f(x, y)$, the gradient of (x, y) is defined by a two-dimensional vector:

$$\nabla f = \left| \frac{\partial f}{\partial x}, \frac{\partial f}{\partial y} \right| \tag{4}$$

The modulus of ∇f is,

$$mag(\nabla f) = \left| G_x^2 + G_y^2 \right|^{1/2} = [(\frac{\partial f}{\partial x})^2 + (\frac{\partial f}{\partial y})^2] \tag{5}$$

Let $G(x, y)$ be the gradient function of the image, then $G(x, y) = \nabla f$. The pixels with gray level $i(i \in [0, 1, \ldots, L-1])$ are represented in set $R_i = \{(x, y) | f(x, y) = i\}$, the cardinality of R_i is n_i. The gray level and gradient mapping (GGM) function is defined as:

$$T(i) = \frac{\sum\limits_{(x,y) \in R_i} G(x, y)}{n_i}, \ i \in [0, 1, \ldots, L-1] \tag{6}$$

where, $T(i)$ is the average gradient of the gray level i.

3.2   Weighted Otsu Method Based on GGM Function

The basic thought of the Otsu method is to select a threshold by maximize the between-class variances or minimize the within-class variances of an image. The method only considers the gray level information in the image, but not considers the location information of the gray level. Therefore, a gradient weighted Otsu method is presented in this paper. The new method weighs the objective function of the Otsu method with the GGM function, which makes the optimal threshold not only with a maximum between-class variance but also with a large probability at the boundary of the object and the background. Specifically, the objective function of the gradient weighted Otsu method is,

$$
\begin{aligned}
\eta(t) &= T(t) * \sigma_B^2(t) \\
&= T(t) * [p_0(t)\mu_0^2(t) + p_1(t)\mu_1^2(t)]
\end{aligned}
\tag{7}
$$

where $T(\bullet)$ is the weight. The best threshold is selected by the following formula.

$$
\begin{aligned}
t^* &= Arg \max_{0<t<L-1} [\eta(t)] \\
&= Arg \max_{0<t<L-1} \{T(t) * [p_0(t)\mu_0^2(t) + p_1(t)\mu_1^2(t)]\}
\end{aligned}
\tag{8}
$$

In formula (7), the first term $T(t)$ is the weight and the second term $p_0(t)\mu_0^2(t) + p_1(t)\mu_1^2(t)$ is the between-class variance of the image. The optimal threshold is selected to maximize the multiplication of the two terms, that is, each of the term should be the largest at the same time. For the weight factor $T(t)$, the larger the $T(t)$, the larger the average gradient (large probability t at the boundary of the object in the image).

To the infrared image shown in Fig. 2(a), the histogram has a small peak near to 255 (as shown in Fig. 2(b)). As analysis before, the Otsu method can not find the optimal threshold in this condition. The maximum value of the objective function in the Otsu method is near by 50, which can be seen from the Fig. 2(b). Observing the infrared image, the grayscale changes fast in the light values. That's to say, the gradient value will be large in this area. Figure 2(d) shows the graph of the GGM which is computed by formula (6). In Fig. 2(d), the average gradient values are large and change fast when the gray level over 150. Therefore, for the effect of the average gradient weight, the maximum value of $\eta(t)$ is near to 200 and the optimal threshold is close to this value also. Figure 2(e) shows the graph of $\eta(t)$ and Fig. 2(f) shows the segmentation result of the gradient weighted Otsu method. The object is well extracted.

4   Experiment Results

Experiments are simulated on PC with Matlab7, Intel Core 2.33 GHz CPU and 2G memory. In order to evaluate the performances of the proposed method, the experiments include two parts. One part is to compare the proposed method with the Otsu method using standard optical images. It's demonstrated that the proposed method can get better segmentation results for the images which the variance or the class probability of the object is much smaller than that of the background. For the reason of paper length, in the following experiments, there lists two standard optical images, number image and SAR image with size of 730 × 203 and 1024 × 1026. The segmentation results are shown as Figs. 3 and 4.

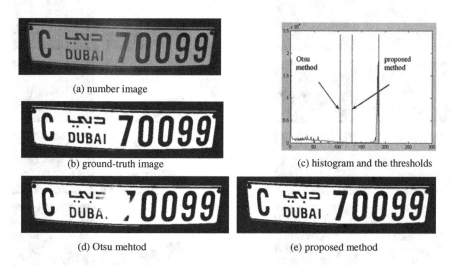

(a) number image

(b) ground-truth image

(c) histogram and the thresholds

(d) Otsu mehtod

(e) proposed method

Fig. 3. Segmentation results for the number image

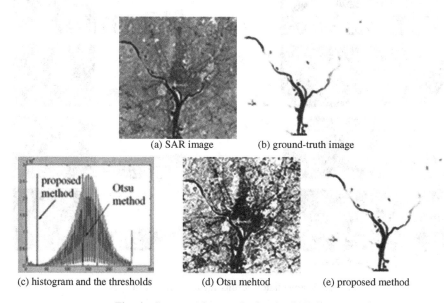

(a) SAR image

(b) ground-truth image

(c) histogram and the thresholds

(d) Otsu mehtod

(e) proposed method

Fig. 4. Segmentation results for the SAR image

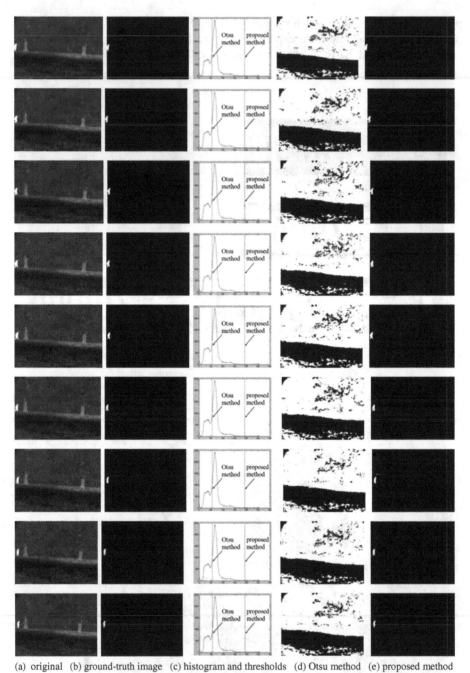

(a) original   (b) ground-truth image   (c) histogram and thresholds   (d) Otsu method   (e) proposed method

Fig. 5. Segmentation results for the infrared images

The second part shows the segmentation results for the infrared images in OTCBVS Benchmark Dataset by the Otsu method and the proposed method. The infrared image sequences are selected from Terravic Research Motion IR Database irw10. The sequences have 500 images. It shows the object (a person) enters the FOV from the left and continues walking right until exiting the FOV. The object enters the FOV in frame 209 and exists in frame 480. The proposed method can extract the objects from all the sequences effectively. Figure 5 shows the segmentation results for the 9 images from frame 217 to frame 225.

In order to evaluate the segmentation result, we used the misclassification error (*ME*) measure for illustration. *ME* [1] was defined as,

$$ME = 1 - \frac{|B_O \cap B_T| + |F_O \cap F_T|}{|B_O| + |F_O|} \tag{9}$$

where B_O and F_O denote the background and foreground area pixels of the manually thresholded images, B_T and F_T denote the background and foreground area pixels in the image that are thresholded using either the proposed method or the Otsu method, and $|\cdot|$ is the cardinality of the set. This error reflects the percentage of wrongly assigned pixels, which ranges from zero for no error, to one for completely wrong. In the experiment, each image was first thresholded manually, and then by the Otsu method and the proposed method, and their misclassification errors computed. Table 1 shows the thresholds and the *ME* values of the two methods for the optical images. Table 2 shows the thresholds and the *ME* values of the two methods for the infrared images of the OTCBVS database.

Figure 3 shows the segmentation results for the number image. The histogram is similar to bimodal distribution with large difference in the class variances as shown in Fig. 3(c). The variance of the class near 0 is much larger than that of the class near 255. In this case, the threshold of the Otsu would bias to the component with larger class variance as shown in Fig. 3(c). The threshold of the proposed method is close to the boundary between the object and the background for considering the gradient information of the pixels. Figure 3(d) is the thresholded image by the Otsu method. It can be seen that the Otsu method can segment the image basically, but lost some information in the middle area. The proposed gradient weighted Otsu method can extract the object from the background well (Fig. 3(e)).

Table 1. The optimal thresholds and *ME* values for the optical images

	Otsu method		Proposed method	
	Threshold	*ME*	Threshold	*ME*
number	105	0.0357	130	0.0107
SAR	137	0.3385	27	5.0541e-004

Table 2. The optimal thresholds and *ME* values for the infrared images from OTCBVS database

	Otsu method		Proposed method	
	Threshold	*ME*	Threshold	*ME*
000217.jpg	55	0.6575	193	5.2083e-005
000218.jpg	56	0.6432	198	0
000219.jpg	55	0.6462	216	1.4323e-004
000220.jpg	55	0.6521	197	1.3021e-005
000221.jpg	55	0.6493	179	2.4740e-004
000222.jpg	56	0.6418	190	1.1719e-004
000223.jpg	56	0.6645	203	2.6042e-005
000224.jpg	56	0.6527	184	1.4323e-004
000225.jpg	56	0.6607	192	2.6042e-005

Figure 4 shows the segmentation results for a SAR image. It can be seen that the object has small proportion of the image from the image and its histogram. In the histogram, the object is the small peak near by 0. The grayscale distribution is bimodal with large difference in the size. The Otsu method fails in this case. It can not get the optimal threshold. Figure 4(d) shows the segmentation result by the Otsu method. It can be seen that the object was not extracted from the background. The threshold of the Otsu method is 137 which close to the maximum value of the bigger peak in the histogram. The proposed gradient weighted method introduces the gradient information and its threshold is 27 which at the valley of the object and the background gray level distribution. The proposed method provides satisfactory results shown as Fig. 4(e). The misclassification error is 5.0541e-004 for the proposed method. That's mean the segmented results by the proposed method is close to the results by manually.

Figure 5 shows the segmented results for the infrared images from frame 217 to frame 225 of the Terravic Research Motion IR Database irw10. The object (a person) of the infrared image is much smaller than the background. The gray scale distribution of the object is the small peak near 255 in the histogram as shown in Fig. 5(c). The distribution of the background gray level is bimodal. As the analysis before, the threshold of the Otsu method is near by the average value of the background gray levels and it can not extract the object effectively. The optimal threshold of the proposed gradient weighted method is close to the boundary between the object and the background. Figure 5(e) shows the segmented results by the proposed method, which extracts the object from the background well. Table 2 shows the thresholds and the misclassification error for the infrared series by the two methods in this paper. The *ME* values are close to 0 for the proposed gradient weighted method.

5 Conclusions

In this paper, a revised Otsu method named gradient weighted thresholding method is proposed. This novel method weighs the objective function defined in Otsu method with the average gradient information of the threshold, and selects a threshold value that has large average gradient and also maximizes the between-classes variance in the gray-level histogram. Therefore, the optimal threshold of the proposed method will close to the boundary between the object and the background. The experimental results show the effectiveness of the proposed method for the images which the variance or the class probability of the object is much smaller than that of the background. Experiment results of optical images as well as infrared images show the effectiveness of the proposed method.

Acknowledgements. This work is supported by National Science Foundation of China (Grant No. 61102095,61202183,61340040). And thanks for Dr. Liu Ying on the help of the language.

References

1. Sezgin, M., Sankur, B.: Survey over image thresholding techniques and quantitative performance evaluation. J. Electron. Imaging 13(1), 146–165 (2004)
2. Otsu, N.: A thresholding selection method from gray-level histograms. IEEE Trans. Syst. Man Cybern. **9**(1), 62–66 (1979)
3. Kittler, J., Illingwotth, J.: On threshold selection using clustering criteria. IEEE Trans. Syst. Man Cybern. **15**, 652–655 (1985)
4. Kurita, T., et al.: Maximum likelihood thresholding based on population mixture models. Pattern Recogn. **25**, 1231–1240 (1992)
5. Hou, Z., et al.: On minimum variance thresholding. Pattern Recogn. Lett. **27**, 1732–1743 (2006)
6. Fujiki, M.: An image thresholding method using a minimum weighted squared-distortion criterion. Pattern Recogn. **28**, 1063–1071 (1994)
7. Ng, H.-F.: Automatic thresholding for defect detection. Pattern Recogn. Lett. **27**, 1644–1649 (2006)
8. Fo, J.-L., Lei, B.: A modified valley-emphasis method for automatic thresholding. Pattern Recogn. Lett. **33**, 703–708 (2012)
9. Songtao, L., Dongming, Z.: Gradient-based polyhedral segmentation for range images. Pattern Recogn. Lett. **24**, 2069–2077 (2003)
10. http://www.cse.ohio-state.edu/otcbvs-bench

Study on Image Classification
with Convolution Neural Networks

Lei Wang[1,2(✉)], Yanning Zhang[1], and Runping Xi[1]

[1] School of Computer, Shan Xi Provincial Key Laboratory of Speech and Image
Information Processing, Northwestern Polytechnical University,
Xi'an 710072, People's Republic of China
Wlei598@163.com
[2] Faculty of Information Engineering, East China Institute of Technology,
Nanchang 330013, People's Republic of China

Abstract. Classification is one of the most popular topics in image process. In this paper, it provides the specific process of convolutional neural network in deep learning. Here builds typical convolution neural network which parameters and the connection mode can be adjusted. In addition, we present our preliminary classification results. Through the experiments of a convolutional neural network on the Mixed National Institute of Standards and Technology database (MNIST), we compared with the classification results and analyzed in the experimental results with the parameters. The experimental results show that image classification effect is very good used by convolutional neural network.

Keywords: Image classification · Handwritten digits · Deep learning · Convolutional Neural Networks (CCN)

1 Introduction

With the rapid development of digital photography, video and digital storage technology, the digital image data are acquired more and more by people. The data of these images contains vast amounts of information, it is very hard for our human to handle these information in real time. Therefore, let the computer automatically recognize and understand the images become more and more urgently. The classification of image is a very important research content for image understanding, a large number of scholars have do a more extensive and research deeply for a long time. Image classification and recognition has great significance for the application of some advanced technology, such as automatic acquisition, human-computer interaction, intelligent context-aware, and so on [1].

The basic idea of the deep convolution neural network is a way to simulate a multi-layer structure of the human brain. It will extract incrementally the feature of the input data from low to high layer, and eventually form an ideal feature for pattern classification so as to enhance the accuracy of classification or prediction. LeCun and Cortes [2] and LeCun et al. [3] proposed a convolution neural networks (CNNs) which is the first true learning algorithm for multiple hidden layers, it can extract a spatial relationships of data, and improve the training performance of back-propagation

© Springer International Publishing Switzerland 2015
X. He et al. (Eds.): IScIDE 2015, Part I, LNCS 9242, pp. 310–319, 2015.
DOI: 10.1007/978-3-319-23989-7_32

(BP) algorithm by reducing the training parameters. Hinton and Salakhutdinov [4] built the deep neural network had achieved amazing results in 2012. It focused on the algorithm improvement and introduced a concept of weight attenuation in the training of the network so as to effectively reduce the amplitude of the weight and prevent the network over-fitting. With this improvement, it overcame somewhat the problem of efficiency and effectiveness in traditional network training. Before conducting a global multi-simulation study, neural network is first divided into several sub neural networks which consists of two layers, and trains these sub neural networks layer by layer, and then the initial value of the multilayer neural networks will be achieved by accumulating that of these sub neural network, and finally using global optimization algorithm such as BP algorithm to fine-tune. Depth study as an emerging research in the field of machine learning has been widely concerned by academia and industry. It also has achieved good results in image recognition, speech recognition, natural language process, information retrieval and other fields [5–12].

In this paper, it research the method of image classification which based on Convolution Neural Networks (CNN). There build typical convolution neural network which parameters and the connection mode can be adjusted. Aiming at the layer of network, the number of iterations, the rate of learning, adjusting different parameters to construct the neuron number etc. The experiments of a convolutional neural network on the Mixed National Institute of Standards and Technology database (MNIST), and we compared with the classification results and analyzed in the experimental results with the parameters. The experimental results show that image classification effect is very good used by convolutional neural network.

2 Deep Learning

In recent year, deep learning is a concept which is put forward by Hinton [13] in 2006, it establishes an abstract expression for the data of image, voice and text by using a multilayer Neural Networks, and constructing NN model which contains the multilayer nodes and having unsupervised training. When in the face of complex intelligence problems, Deep Neural Networks can handle it better. The method of information processing of network model will be more similar with the imitation of the human brain, so it can be better applied to a variety of areas such as image segmentation, scene classification, speech recognition, etc.

2.1 Introduction of Convolution Neural Network (CNN)

In 1980, Kunihiko Fukushima proposed Neocognitron first introduced the concept of CNN [14], which is the first model after deep learning in practice and theoretical analysis. Many scholars has made a significant contribution for the development of CNN. In 1988, LeCun et al. [15] introduced the BP algorithm into CNN. In 2003, Behnke [16] wrote a book about the CNN on CNN summarized. In the same year, Simard et al. [17] on CNN simplified. In 2011, Cirean et al. [18], who has been improved achieve of CNN and its

GPU version in further. After that they do the experiment in more image databases use by frame of CNN, it obtained the best results than ever.

Convolution neural network is one of artificial neural networks, it has become a hot topic in speech analysis and image recognition field. Convolution neural network, also name as CNN, is a model of deep layer of the neural networks contains layer of convolution. Usually, the architecture of convolution neural network consists of two layers that can be trained by the nonlinear convolution generated, two fixed subsample layer and a full connection layer, the number of the hidden layer are generally at least 5 or more. The model of deep convolution neural networks not only significantly improves the accuracy of image recognition, but also avoids the time consuming by manual feature extraction work, and making online operation efficiency is greatly improved. Deep learning become mainstream technology in identification and classification of image, it will replace the combination of artificial and machine learning method.

2.2   The Structure of Convolution Neural Networks

Convolution neural networks is a multilayer supervised learning neural network, the convolution-like layer and sub-layers of hidden layers are realized preclude convolution neural network feature extraction capabilities of the core module. The model of network can implement reverse adjustment of weight parameter layer by layer in the network by using a gradient descent method and minimum loss function and improve the accuracy of the network through frequent iterative training [19]. Low hidden layer in convolution neural network is made of alternating layers and sub-sampled convolution layers. The top layer is fully connected corresponding traditional multilayer perception hidden layer and a logistic regression classifier. The first layer of fully connected input feature images obtained to feature extraction by roll laminated and sub sampling layer. The last output layer is a classifier that can preclude the use of logistic regression, regression of softbacks or support vector machine even classified the input image.

Convolution neural network include comprising convolution and subsample layer is structure of these two special layers. It is a non-fully connected neural network structure. Convolution layer defines the corresponding accepted domain, for each neuron, it receives signal which is transferred only from accepted domain. Each feature plane constituted by neurons, characterized by a plurality of planar layers together constitute convolution, complete tasks extracted features, and all neurons on the same plane with consistent characteristics connection weights. The connection matrixes which have the same characteristics of planar neurons are coherent, because of the same size of neuronal receptive fields. Convolution and sub-sampling process is: convolution process includes training input data and the filter f_x convolution operation, together with the training with the offset vector b_x, resulting convolution layer $C_x = f_x(x) + b_x$; subsampling process includes adding the neighborhood, the weighted scalar $W_x + 1$, plus the paranoid vector $b_x + 1$, the results generated feature maps $S_x + 1$ through sigmoid activation function as shown in Fig. 1.

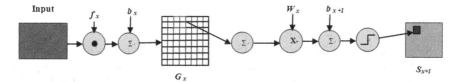

Fig. 1. Process of convolution and subsampling

3 Method of Image Classification Based on Convolution Neural Network (CCN)

To better understand the classification method processes used here in CNN shown in the flowchart presented in Fig. 2. The frame of CNN consists of two convolution layers and two sampling sub-layers alternating. Each neuron of input layer connected with the local receptive field in the layer of before and extracts the local feature. After the local feature is extracted, characterized in that the positional relationship between the others and also determined. When we used the same kernel of convolution at the same time extracting feature, so the weight value of having the same feature graphs are shared. Subsample layer is responsible for the characteristic convolution layer obtained sub-sample, so that the extracted feature with the zoom invariant. At the last layer of CNN is generally connected to the last layer of a few fully connected layers, the number nodes of final output is the numbers of classified in target. The purpose of the training is to make the output of CNNs close to the original classification labels, to achieve the correct classification.

First, the size of handwritten digital image in MNIST is 28×28, on this basis, the frame of CNN classify in handwritten image shown in Fig. 2. And the details as follows:

(1) Layer of C1 is a feature extraction layer, it obtain six two-dimensional feature maps which size is 24×24. It used by a 5×5 kernel of convolution size. The size of the convolution kernel determines size of a neuron's receptive field. The convolution kernel with a characteristic diagram using 5×5 are the same. The results of obtained convolution is not directly stored in the C1 layer, but the first is calculated by an activation function. And it used as a characteristic value of the C1 layer neurons, Sigmoid activation function and generally function in the actual operation, when the convolution offset term for the image block x, using the convolution kernel w convoluted, the offset term is b, the output of the convolution for y is:

$$y = Sigmoid(wx + b) \qquad (1)$$

(2) Layer of S1 is a layer of sub-sampling, which takes six characteristic images which size is 12×12 by C1 in all sum of non-overlapping x sub-blocks which size of 2×2, multiplied by a weight w, plus calculating a sub-sampling process offset term b is obtained:

$$y = Sigmoid[w * sum(x_i) + b] \qquad (2)$$

Since characterized image size of 24 × 24 in C1 layer, the results of sub-sampling on 12 × 12 characterized sub-graphs obtained finally. Now we should control the speed of scale. Because the zoom is exponential scaling, reduced too fast also means rougher image feature extraction, image will lose more in detail features in CNN. So we choose each sub-sampling the each layer are 2 used in scaling factor.

(3) Layer of C2 is also a feature extraction layer. It similar with C1 but also have some differences. The kernel size of convolution is 5 × 5 in C2. Through the processing of C1 and S1, actually every neuron of S1 receptive fields corresponds to covering 10 × 10 (C1 convolution kernel size is 5 × 5, S2 of the sample sub-block size of 2 × 2, 5 × 5 × 2 × 2 = 10 × 10)in the original image. After size of 5 × 5 convolution kernel to get feature extraction of S1 in C2, which further expansion of receptive fields. Each feature of C2 when convoluted is characterized by S1 in FIG several or all feature maps are combined into an input and do convolution.

(4) The remaining convolution layer and sub-sampled layer, the working principle of these layers are the same as the previous layer. With incensement of the depth s, the features of extract are more abstract and more expressive ability. After C1, S1, C2, S2 convolution layers and sampling several layers, the extracted features are very expressive capability.

(5) The output layer is a layer of the subsample layer fully connected, there are 12 × 5 × 5 = 300 neurons, and each neuron is connected to the output of a neuron. It has 10 neurons of output (the number of handwritten species) in total. The representation from subsample layer to the output layer corresponds to classify by the mapping of the vector.

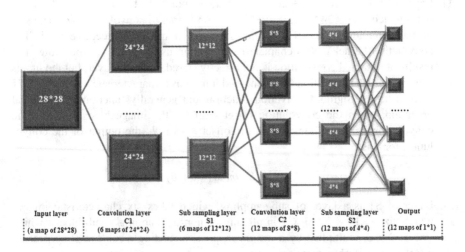

Fig. 2. The process of image classification based on CCN

4 Results and Analysis of the Experiment

4.1 The Brief Introduction of MNIST Database

The MNIST database (handwritten digit) has a training sample included 60,000 samples with the size of 28 * 28, and a test sample included 10,000 samples as shown in Fig. 3. These images in these training sample have been normalized and concentrated to a fixed-size image.

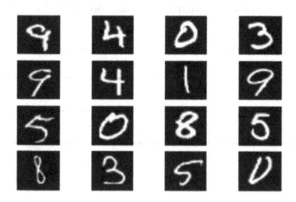

Fig. 3. A few samples from MNIST database

4.2 Assessments of Classification Accuracy

The analysis of classification accuracy is an important part in the evaluation of image classification. Through the accurate analysis, we can quantitatively judge the accuracy of the classification model. The confusion matrix almost used as the common method. The confusion matrix is also called the error matrix. It reflects the relevant information between classification results and ground-truth category. And it is a foundation of the analysis of the overall accuracy. M represents a confusion matrix. The element of m_{ij} indicates that the total number of pixels should actually belong to the category j but is assigned to the category i in the test samples. The value of the diagonal element, m_{ii} is the larger, and the classification accuracy is the higher.

The overall accuracy (OA) is the overall evaluation of the quality of the classification result. It is equal to the total sum of pixels divided by the number of pixels correctly classified. The pixels of correctly classified distribute along the diagonal of the confusion matrix, which shows the number of correctly classified pixels to the real classification. According to the confusion matrix, OA is calculated as:

$$p = \sum_{i=1}^{c} m_{ii}/N \qquad (3)$$

In the above formula, c represents the number of categories, m_{ii} represents the elements on the diagonal of the confusion matrix, $\sum_{i=1}^{c} m_{ii}$ indicates the total number of test samples.

4.3   Analysis of Results

In the experiment, 10,000 training samples of MNIST dataset were used to train convolution neural network (CNN).The learning rate of convolution neural network is defined as 1. Each batch of 50 samples goes on training to adjust the weight. The iterative training conducts five times. 1000 test samples were used to test for the classification results. Due to the training and testing samples are randomly selected, so each type will not necessarily be exactly equal, but may be more accessible. The Class with the highest true positive rate was the "0" class and the lowest rate is the "9" class. The data of the confusion matrix after the test, are showed in Table 1.

Table 1. Confusion matrix of CNN classification results

Class	Ground-truth										Total
	0	1	2	3	4	5	6	7	8	9	
0	91	0	0	0	0	0	0	0	1	0	92
1	0	110	0	0	0	0	0	0	2	0	112
2	2	0	85	0	2	0	3	7	3	0	102
3	1	0	4	87	0	2	0	0	2	0	96
4	1	1	0	0	75	0	4	2	0	21	104
5	2	3	0	5	3	66	2	0	6	1	88
6	3	1	0	0	3	0	98	2	0	0	107
7	0	3	8	0	0	0	0	88	1	5	105
8	1	2	1	3	1	0	3	0	79	1	91
9	0	0	1	1	1	0	1	3	1	95	103
Total	101	120	99	96	85	68	111	102	95	123	1000
Accuracy%	90.1	92	85.8	90.6	88.2	97.1	88.3	86.2	83.2	77.2	100

In the training, it found that if convolution kernel is too small, we can't extracted from local characteristics effective. And if the convolution kernel is too large, the complexity of the extracted features may far exceed the ability to represent the kernel of convolution. So choose the appropriate kernel of convolution is 5 * 5, which is crucial to improve the learning ability of CNN. Of course, how to choose the appropriate the times of iterations training and parameter of the learning rate also are difficulties for tuning parameters for the model of CNN. For analysis of the data in Table 1, during the test with 1,000 test samples, it was found that the rate of classification in class "0", class "1", class "3", class "5" is more than 90 %, and the rate of classification in class "2", class "4", class "6", class "7", class "8" is more than 80 %. This illustrates the effect of the test is well.

In the experiment, we found the different number of iterative training that is the number of filters in each layer of convolution neural network structure, which has a relatively larger impact for the training speed and the final results of training. Selecting the appropriate number of filter in each layer, can appropriately shorten the training time and maintain a certain classification rate. But too many filters in each layer, it may cause network training cannot converge when the sample size is not very abundant in the case. This can't achieve purposes of effective identification. Therefore, we think about the situation of test in this paper, the times of iterations in training is different. In addition to the computation time, they obtained different result in the correct classification accuracy. When the times of training iterations rose from 2 to 3, correct classification accuracy increased significantly. When the times of training iterations from 4 to 10, the correct classification accuracy is also a continuous slight increase. It indicates that the number of iterations has a significant impact on the performance of convolution neural network. And the inadequate training iterations will weaken capabilities of extracting features in convolution neural network. In training process of convolution neural network, the curve of the misclassification rate (MCR) is as shown showed in Fig. 4. The abscissa represents the number of iterations of training, the ordinate represents the MCR, In the drawing triangular mark represents misclassification rate (training MCR) of the network during training in the training set, a circular mark represents misclassification rate (test MCR) of the network in the test set.

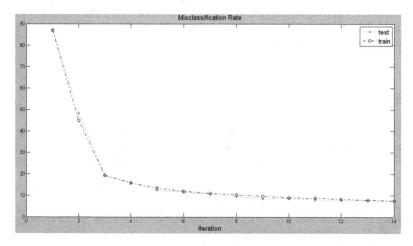

Fig. 4. The curve of misclassification rate during the iterative training process

Figure 4 shows that MCR of test data reaches 9.6 % in the lowest point, after the ninth times iteration. Although in the subsequent training process the training MCR is slightly lower, but the MCR of test data is instead of increasing after the 10th iterations and then remains stable. So that it can be considered after the completion of the first 12 times iterations the network achieves the convergence of the current best network training, the testing MCR is 7.8 %. Although the bigger training data samples are better ensure training effect in the convolution neural network. But there are no sufficient data in many classification problems, so that these large-scale network structure may not be well trained. In this paper we only selects 1000 training examples randomly, it is mainly used to compare with the learning ability and the effect by convolution neural network in the case that the samples data are relatively rare.

5 Conclusion

As a new method, the deep learning has been widely research in classification and recognition of image and speech recognition etc. The convolutional neural network as one of model in deep learning, it collect advantages of unsupervised learning and supervised learning. And it can automatically extract image features. In this paper, it builds typical convolution neural network which parameters and the connection mode can be adjusted based on the research of CNN. The experiments of a convolutional neural network on the (MNIST), and we compared with the classification results and analyzed in the experimental results with the parameters. In future work, we will do some further study in deep learning model to improve the accuracy of classification in scene image, Aerial image and remote sensing image applications.

Acknowledgment. This work is supported by the National Natural Science Foundation of China (No. 61272288, No. 61231016, and No. 61301192), Foundation of China Scholarship Council (No. 201303070083), Foundation of NPU New Soaring Star, NPU New People and New Directions (No. 13GH014604), and NPU Soaring Star (No. 12GH0311). The National High-Tech Research and Development Program of China (863 Program SS2015AA010502). Shaan xi Provincial Natural Science Foundation (2013JM8027).

References

1. Zhou, L.: Research on key technologies of scene classification and object recognition. National University of Defense Technology, Changsha (2012)
2. LeCun, Y., Cortes, C.: MNIST handwritten digit database. AT&T Labs (1998). http://yann.lecun.com/exdb/mnist/
3. LeCun, Y., Bottou, L., Bengio, Y., et al.: Gradient-based learning applied to document recognition. Proc. IEEE **86**(11), 2278–2324 (1998)
4. Hinton, G.E., Salakhutdinov, R.R.: Reducing the dimensionality of data with neural network. Science **313**(5786), 504–507 (2006)
5. Yu, D., Deng, L.: Deep learning and its applications to signal and information processing. IEEE Sig. Process. Mag. **28**(1), 145–154 (2011)
6. Arel, I., Rose, C., Karnowski, T.: Deep machine learning-a new frontier in artificial intelligence. IEEE Comput. Intell. Mag. **5**(4), 13–18 (2010)
7. Bengio, Y.: Learning deep architectures for AI. Found. Trends Mach. Learn. **2**(1), 1–71 (2009)
8. Bengio, Y., Courville, A., Vincent, P.: Representation learning: a review and new perspectives. IEEE Trans. Pattern Anal. Mach. Intell. **35**(8), 1798–1828 (2013)
9. Anthes, G.: Deep learning comes of age. Commun. ACM **56**(6), 13–15 (2013)
10. Jones, N.: The learning machines. Nature **505**(7482), 146–148 (2014)

11. Hu, X., Zhu, J.: Deep learning: new hotspot in machine learning. Commun. CCF **9**(7), 64–69 (2013). (in Chinese)
12. Yu, K., Jia, Y., Chen, Y., et al.: Deep learning: yesterday, today, and tomorrow. J. Comput. Res. Dev. **50**(9), 1799–1804 (2013). (in Chinese)
13. Krizhevsl, A., Sutskever, I., Hinton, G.: Image net classification wittier deep convolutional neural networks. In: Advances in Neural Information Processing Systems, vol. 25, pp. 1106–1114 (2012)
14. Fukushima, K.: Neocognitron: a self-organizing neural network model for a mechanism of pattern recognition unaffected by shift in position. Biol. Cybern. **36**(4), 193–202 (1980)
15. LeCun, Y., Boser, B., Denker, J.S., et al.: Backpropagation applied to handwritten zip code recognition. Neural Comput. **1**(4), 541–551 (1989)
16. Behnke, S.: Hierarchical Neural Networks for Image Interpretation. LNCS, vol. 2766. Springer, Heidelberg (2003)
17. Simard, P., Steinkraus, D., Platt, J.C.: Best practices for convolutional neural networks applied to visual document analysis. In: Proceedings of the 7th International Conference on Document Analysis and Recognition (ICDAR), pp. 958–962 (2003)
18. Ciresan, D.C., Meier, U., Masci, J., et al.: Flexible, high performance convolutional neural networks for image classification. In: Proceedings of the 22nd International Joint Conference on Artificial Intelligence, vol. 2, pp. 1237–1242. AAAI Press (2011)
19. Guyon, I., Albrecht, P., LeCun, Y., et al.: Design of a neural network character recognizer for a touch terminal. Pattern Recogn. **24**(2), 105–119 (1991)

Multiscale Change Detection Method
for Remote Sensing Images Based
on Online Learning Framework

Jianlong Zhang[(⊠)] and Jianfeng Zhai

School of Electronic Engineering, Xidian University, Xi'an, Shaanxi, China
Jlzhang@mail.xidian.edu.cn

Abstract. Change detection for remote sensing images is very important for urban planning, disaster evaluation etc. Traditional detection methods include supervised and unsupervised learning algorithm. A novel semi-supervised multiscale change detection method based on online learning framework is presented in this paper. Firstly, mean-variance classifier and SVM classifier are trained at the different scales of 2*2 pixels block and original pixel respectively. Initial training set is extracted from the ground truth. Secondly, the difference image is obtained according to two phase remote sensing images, and arranged by the unit of 16*16 pixel block. Image blocks are input into the mean-variance classifier and SVM classier to be detected one by one, it is cascade connection between two classifiers. The error correction rules are used to choose the misclassified instances to retrain the classifiers. Experiment results show that the method in this paper can efficiently decrease the FN (false negative numbers) to improve the performance of change detection algorithm.

Keywords: Change detection · Online leaning · Semi-supervised learning · Multi-scale analysis · SVM

1 Introduction

The human's ability to transform the earth becomes stronger and stronger with the development of technology and society. Surface of earth is being changed rapidly because much infrastructure such as highroad, railway, airport and dam is being constructed. It's very important to timely and exactly monitor this change information, analyze the change reason and the results, this work can help us to understand the interaction and relationship between nature and human. Also change detection is very meaningful for assessing the effect of human activity and protecting the nature balance. Change detection of remote sensing image has been applied into many fields for urban layout, land monitoring, resources investigation and assessment for natural disaster.

According to the information processing level, change detection methods of remote sensing image can be classified into three categories: (1) Change detection based pixel [1, 2]. These methods detect the change using the original time series images and look for the change area in the difference image produced by input images. These methods have been successfully and widely applied in the fields of low resolution remote

© Springer International Publishing Switzerland 2015
X. He et al. (Eds.): IScIDE 2015, Part I, LNCS 9242, pp. 320–330, 2015.
DOI: 10.1007/978-3-319-23989-7_33

sensing image. Their most important characters are simple and direct without much priori knowledge. But it's difficult to analyze and describe the change detection of high resolution remote sensing images because the detection result has too much pieces. (2) Change detection based feature [3, 4]. These methods extract the features from original images to analyze and detect. Their major application fields are the terrestrial objects with special edge and local feature. They have evident advantages in the high resolution image or application with typical geometry shape but they are not suitable for SAR (Synthetic Aperture Radar) image change detection. (3) Change detection based objective. These algorithms belong to high level analysis method based objective model and they are easily limited by the factor of algorithm accuracy for objective detection, especially the complex background application. Objective detection methods [5, 6] have the high level semantic information by cooperating with the methods of objective segmentation and classification.

Machine learning and pattern recognition are the hot points over the past few decades. Also they are one of the core content for Artificial Intelligent field. Change detection for remote sensing images can be as two classification problem, namely change areas and unchanged areas, so supervised learning or unsupervised learning methods are applied to detect the change regions. Celik [7] presents an unsupervised learning algorithm to use the dual-tree complex wavelet transform to individually decompose each input composition into on low-pass sub-band and six directional high-pass sub-bands at each scale. To avoid illumination variation issue possibly incurred in the low-pass sub-band, only the DT-CWT coefficient difference resulted from the six high-pass sub-bands of the two satellite images under comparison is analyzed in order to decide whether change has incurred in each subband pixel intensity.

Gong [8] presents an unsupervised distribution-free change detection approach for SAR images based on an image fusion strategy and a fuzzy clustering algorithm. The image fusion technique is introduced to generate a difference image by using complementary information from a mean-ratio image and a log-ratio image. Wavelet fusion rules based on an average operator and minimum local area energy are chosen to fuse the wavelet coefficients for a low-frequency band and a high-frequency band, respectively. A reformulated fuzzy local-information C-means clustering algorithm is proposed for classifying changed and unchanged regions in the fused difference image. This algorithm incorporates the novel fuzzy way and wavelet fusion to enhance the changed information and reduce the effect of speckle noise. Xin [9] uses a improved dynamic Fisher classifier to identify the changes in the joint intensity histogram to detect the change regions. By considering the relationship between the pixel and its neighborhood, local mean dynamic Fisher discriminant analysis is proposed to introduce the neighborhood's information. Meanwhile,the parameters of the classifier are adjusted according to the current detection results, which avoids the influences of initial conditions. These algorithms mentioned above use the unsupervised learning to cluster the difference image into two classes, which are change regions and unchanged regions, they detect the change regions without priori knowledge and can be applied into many fields. However it's difficult to obtain the higher accuracy of detection and performance for them. They only adapt some special conditions for remote sensing images.

A novel change detection framework for remote sensing image is present in this paper, which incorporate online learning, semi-supervised learning and multi-scale analysis to improve detection performance with little priori knowledge. Two kinds of classifier are employed in this method, and connection is cascade between classifiers. Only few training instances are obtained from ground truth to train the initial classifier. Difference image is arranged by the unit of image blocks to be input into classifiers one by one, then expert rules are introduced to select the misclassified instances from detection result and again be input into training set for training. The rest of this paper is organized as follows. Section 2 describes the proposed change detection framework with online learning and multi-scale analysis. Section 3 presents the experimental study. Conclusion is given in Sect. 4.

2   Paper Preparation

The framework of this paper is shown as Fig. 1. It includes online learning, semi-supervised learning, classifier cascade and multi-scale analysis. This method has a novel construction to detect the change of remote sensing images and it can improve the adaptive learning ability of finding the change regions. The contributions of this paper are as follow: (1) Thinking of online learning is introduced to this method. The difference image generated by time series images is divided into several blocks of 16*16 pixels to be detected one by one. Then misclassified instances are chosen by using expert rules and the adaptive ability for the incorrect instances is improved by updating the training data set to retrain the classifiers. (2) Semi-supervised learning is utilized. Only few classical instances in the ground truth is select to be as initial training data set for classifiers, so we use priori information but need much more, it just is the core of semi-supervised learning method. (3) Two classifiers with cascade connection are used to classify the change regions and unchanged regions according to input image

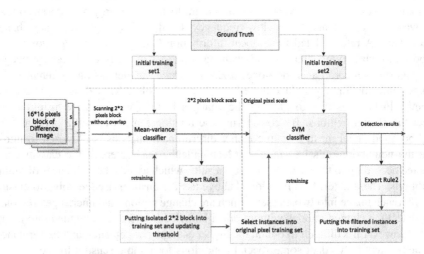

Fig. 1. The construction of framework for this paper.

blocks, mean-variance classifier is the first one and SVM classifier is second one, output of the former is the input of the latter. (4) The framework of multi-scale analysis and detection is employed for this paper, here coarse scale is 2*2 pixels block using by mean-variance classifier and the fine scale is original pixel level using by SVM classifier. The final detection regions fuse the classified results from two scales.

The work process of this paper can be divided into two parts according to Fig. 1. One part is initial training process, in which few classical training instances are chosen to train the mean-variance classifier and SVM classifier at the scale of 2*2 pixels block and original pixel respectively. The other part responses the major detection process and it can be summarized into four steps. First step, the difference image is generated by input images, the difference image can be obtained by algorithm of log-ratio, mean ratio or direct subtraction. Second step, the difference image is arranged by the unit of 16*16 pixel block. These image blocks are not overlapped each other and they are input into the two classifiers in turn. Third step, each image block is introduced into the mean-variance classifier and SVM classifier one by one, classified result of mean-variance classifier is the input of SVM classifier, the former works at the coarse scale of 2*2 pixel block and the latter works at the fine scale of original pixel. Fourth step, misclassified instances are chosen from classified result by using expert rules. Some misclassified instances at the scale of 2*2 pixel block are chosen to retrain SVM classifier at the scale of original pixel by returning the 2*2 block into four pixels. Then these misclassified instances are placed into training data set again to retrain classifiers to improve detection performance of total framework, that's all the process of change detection.

2.1 Difference Image

The necessary step is obtaining the difference image because the change information is hidden in it. Not only change information is extracted but also random noise can be removed by achieving the difference image. Suppose the difference image is D, remote sensing image are X1 and X2, they can be expressed as formula (1).

$$X1 = \sum_{i=1}^{M}\sum_{j=1}^{N} x_{ij}^1, X2 = \sum_{i=1}^{M}\sum_{j=1}^{N} x_{ij}^2, D = \sum_{i=1}^{M}\sum_{j=1}^{N} d_{ij} \tag{1}$$

Here coefficients of M and N are image resolution in formula (1).

Common methods of obtaining the difference image include subtraction, log-ratio and mean ratio, they are shown as formula (2), (3), and (4) respectively.

$$D = \sum_{i=1}^{M}\sum_{j=1}^{N} \left| x_{ij}^1 - x_{ij}^2 \right| \tag{2}$$

$$D = \sum_{i=1}^{M}\sum_{j=1}^{N} \left| \log \frac{x_{ij}^1}{x_{ij}^2} \right| = \sum_{i=1}^{M}\sum_{j=1}^{N} \left| \log x_{ij}^1 - \log x_{ij}^2 \right| \tag{3}$$

$$D = \sum_{i=1}^{M} \sum_{j=1}^{N} \left| \frac{m_{ij}^1}{m_{ij}^2} \right|, m_{ij} = \frac{1}{K^2} \sum_{i=1}^{K} \sum_{j=1}^{K} x_{ij} \tag{4}$$

Here, K is the neighborhood size of difference image pixel.

Especially denote that we should choose the proper method for difference image according to the real condition.

2.2 Classifiers

Mean and variance are two kinds of statistic information used in image processing. There exists some grey distribution difference between object and background, in theories, object and background can be separated easily by image's mean and variance, but in fact it is not true. The difference between object and background is not obvious for the effect of random noise, so separation mistake usually takes place. Threshold selection for segmentation becomes a key factor. For the same reason it's difficult to confirm the segmentation threshold in change detection by using mean and variance. In this paper classifier cascade mode is employed and mean-variance classifier is used as the former classifier to detect coarse change region, formula (5) shows how to confirm the threshold T for mean-variance classifier.

$$T = \frac{f(x, y) - \alpha \mu_n}{\beta \sigma} \tag{5}$$

Here, α and β are the adjustable weighted factors for mean and variance respectively.

The SVM is a statistically robust learning method based on the structural risk minimization [10]. It trains a classifier by finding an optimal separating hyperplane which maximizes the margin between two classes of data in the kernel induced feature space. Also SVM has still excellent data generalization ability to be trained by relative few training instances so it can effectively resolve the problem of over leaning and dimension disaster existed in traditional classifiers. SVM has been applied to the fields of pattern recognition, detection, tracking and artificial intelligent.

Without loss of generality, suppose that there are m instances of training data. Each instance consists of an (x_i, y_i) pair where $x_i \in IR^{IN}$ is a vector containing attributes of the ith instance, and $y_i \in \{+1, -1\}$ is the class label for the instance. The objective of the SVM is to find the optimal separating hyperplane $x^T \cdot w + b = 0$ between the two classes of data. To classify a testing instance, the decision function is

$$f(x) = x^T \cdot w + b \tag{6}$$

The corresponding classifier is sgn($f(x)$). The SVM finds the optimal separating hyper-lane by solving the following quadratic programming optimization problem:

$$\arg\min_{w,b,\xi} \frac{1}{2}\|w\|^2 + C\sum_{i=1}^{m}\xi_i \qquad (7)$$

$$\text{s.t. } y_i \cdot (w^T \cdot x_i + b) \geq 1 - \xi_i, \xi_i \geq 0, \text{ for } i = 1, 2, \ldots, m$$

In the objective function, minimizing corresponds to maximizing the margin between $W \cdot X + b = 1$ and $W \cdot X + b = -1$. The constraints aim to put the instances with positive labels at one side of the margin $W \cdot X + b \geq 1$, and the ones with negative labels at the other side $W \cdot X + b \leq -1$. The variables $\xi_i, i = 1, \ldots, m$ are called slacks. Each ξ_i denotes the extent of x_i falling outside its corresponding region. C is called cost parameter, which is a positive constant specified by the user. The cost parameter denotes the penalty of slacks. The objective function of the optimization problem is a tradeoff between maximizing the margin and minimizing the slacks. A larger C corresponds to assigning higher penalty to slacks, which will result in less slacks but a smaller margin. The value of the cost parameter C is usually determined by cross-validation.

2.3   Initial Training of Classifiers

The characteristic of semi-supervised learning is that only few instances is used for training, in this paper two classifiers of mean-variance and SVM both need initial training. But these two classifiers work at the different scale of 2*2 pixel block and original pixel respectively, all of training instances are chosen from grey image mapped by ground truth. There are 100 positive instances and 500 negative instances for mean-variance classifier, and 400 positive instances and 2000 negative instances for SVM classifier.

2.4   Online Learning

Once the training process of supervised or semi-supervised learning is finished, their classified characteristic could not be changed because training process is offline, so adaptive ability of classifiers is limited. Online learning can help the classifiers to improve their adaptive learning ability. In this paper online learning framework is introduced to make better the data generalization ability of classifiers. This paper employed the initial classified ability of semi-supervised learning, then chosen misclassified instances by given expert rules to retrain the classifiers to adapt to these hard-classified instances. Robustness and adaptation of classifiers are improved gradually by adding the misclassified instances into training data set again and again. Such framework incorporates the process of training and testing to decrease the need for number and diversity of training data, so the data generalization ability of classifiers has been improved. Fig. 2 shows the online learning construction for this paper, classifiers include mean-variance classifier and SVM classifier.

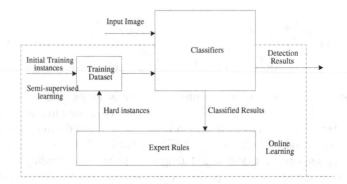

Fig. 2. Construction of online learning in this paper

2.5   Expert Rules

Expert rules are the key module of online learning framework, misclassified instances can be chosen from classified results by setting the proper expert rules to improve the robustness for these instances. Principles of setting expert rules are as followed: (1) They must satisfied the need of framework to choose the misclassified instances from classified results; (2) They must conform the fundamental law of physics; (3) They are best to be structural data for remote sensing image.

Expert rules in this paper belong to the method of connected domain labeling. Suppose R is connected domain in detection results, $N = Num[f_i \in R]$ is number of all instances in R, f_i denotes the instances in R. We give two expert rules for mean-variance classifier and SVM classifier respectively.

1. Isolated rule: isolated instances at the scale the 2*2 pixels block should be as the isolated noise to be filtered.
2. Area rule: *Discard   conneted   region,   when* $N_{CA} \leq T$

When number of connected area instances N_{CA} is less than threshold T at the scale original pixel this connected domain should be as the unchanged region to discard because change regions should be stable and continuous.

3   Simulation Experiments

The proposed method is tested in the conditions that PC specifications are CPU pentium-i5 3.2GHz and RAM 4G. All the algorithms in the experiments are implemented with Matlab2013b. There are two sets of SAR image to be used for simulation experiments. Data set 1 represents a section (301*301 pixels) of two images acquired by European Remote Sending 2 satellite SAR sensor over an area near the city of Bern, Switzerland, in April and May 1999, respectively. Data set 2 is a section (290*350 pixels) of two SAR images acquired by a RADARSAT SAR sensor and provide by Defense Research and Development Canada Ottawa.

We used the same framework of this paper to detect the change region but different algorithms to acquire the difference image, one is log-ratio and the other is mean ratio. We called the former as LROL and the latter as MROL. LROL, MROL, Gong's method [9] and CS_KSVD [11] are compared in simulation experiments. Fig. 3(a)–(d) shows the two input images, ground truth and LROL detection result for Bern. Fig. 4 (a)–(d) shows two input images, ground truth and MROL detection result for Ottawa. On the other hand we calculated some performance coefficient include FN (number of false Negatives), FP (number of false positives), OE (FN+FP), PCC = ((TN+TP)/(TN +FN+TP+FP)) and KAPPA for each algorithm. Tables 1 and 2 give performance coefficients of compared algorithms for Bern and Ottawa respectively.

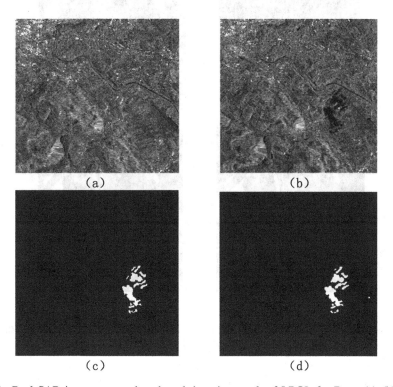

(a)   (b)

(c)   (d)

Fig. 3. Real SAR images, ground truth and detection result of LROL for Bern. (a)–(b) time phases input images. (c) ground truth. (d) detection result.

According to the Figs. 3 and 4, Tables 1 and 2 we can know that LROL and MROL in this paper obtained the best experiment results in the four algorithms for Bern and Ottawa respectively, and FN coefficient of these two algorithms is minimum among the compared algorithms, meanwhile difference between FN and FP is also the least. Experiment results show that our framework of change detection has obtained better performance, because online learning and semi-supervised learning are introduced to our algorithm by using priori knowledge to decrease the FN (false negative numbers), meantime FP (false positive numbers) is still very low. PCC and KAPPA check the comprehensive performance of detection algorithm, so the closer FP is to FN, the higher PCC and KAPPA are.

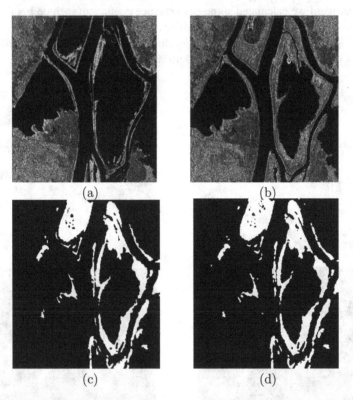

Fig. 4. Real SAR images, ground truth and detection result of MROL for Ottawa. (a)–(b) time phases input images. (c) ground truth. (d) detection result.

Table 1. Detection results comparison for Bern in Fig. 3

Method	FN	FP	OE	PCC	KAPPA
CS_KSVD	159	148	307	99.66 %	0.866
Gong's method	133	159	292	99.68 %	0.871
MROL	314	80	394	99.57 %	0.843
LROL	165	122	287	99.68 %	0.876

Table 2. Detection results comparison for Ottawa in Fig. 4

Method	FN	FP	OE	PCC	KAPPA
CS_KSVD	585	1894	2879	97.56 %	0.9025
Gong's method	270	994	1264	98.75 %	0.9490
MROL	756	530	1286	98.73 %	0.9527
LROL	695	1560	2255	97.78 %	0.9147

4 Conclusion

In this paper we present a novel change detection framework based on online learning and multiscale for remote sensing images. Semi-supervised learning is introduced to obtain better performance by using the priori knowledge. Multiscale analysis and classifier cascade can decrease the misclassified negative instances because we think that change is not alone and independent but continuous and nonlocal. Online learning can help our method to get the better adaptive learning ability. Experiment results showed that our framework can obviously improve the performance of change detection. In the future we hope that initial training for classifiers can be automatic not artificial selection, better classifiers and expert rules can be found to improve the change detection performance further.

Acknowledgments. This paper was supported by the Fundamental Research Funds for the Central Universities (K5051202048) and the National Natural Science Foundation Project (61172146).

References

1. Radke, R.J., Andra, S., Al-Kofahi, O., Roysam, B.: Image change detection algorithms: a systematic survey. IEEE Trans. Image Process. **14**(3), 294–307 (2005)
2. Celik, T.: Change detection in satellite images using a genetic algorithm approach. IEEE Geosci. Remote Sens. Lett. **7**(2), 386–390 (2010)
3. Thonfeld, F., Menz, G.: Coherence and multitemporal intensity metrics of high resolution SAR images for urban change detection. DLR, Oberpfaffenhofen (2011)
4. Gamba, P., Dell, F.A., Lisini, G.: Change detection of multitemporal SAR data in urban areas combining feature-based and pixel-based techniques. IEEE Trans. Geo-sci. Remote Sens. **44**, 2820–2827 (2006)
5. Oe, Z., Zhu, W.C.: Object-based cloud and cloud shadow detection in landsat imagery. Remote Sens. Environ. **118**, 83–94 (2012)
6. Lu, P., Stumpf, A., Kerle, N., Casagli, N.: Object-oriented change detection for landslide rapid mapping. IEEE Geo-sci. Remote Sense Lett. **8**(4), 701–705 (2011)
7. Celik, T., Ma, K.K.: Unsupervised change detection for satellite images using dual-tree complex wavelet transform. IEEE Trans. Geosci. Remote Sens. **48**(3), 1199–1210 (2010)
8. Gong, M., Zhou, Z., Ma, J.: Change detection in synthetic aperture radar images based on image fusion and fuzzy clustering. IEEE Trans. Image Process. **21**(4), 2141–2151 (2012)

9. Xin, F.F., Jiao, L.C., Wang, G.T., Wan, H.L.: Change detection of SAR images based on wavelet domain Fisher classifier. J. Infrared Millimeter Waves **30**(2), 173–178 (2011)
10. Vapnik, V.N.: Statistical Learning Theory. Wiley, New York (1998)
11. Fang, L., Li, S., Hu, J: Multitemporal image change detection with compressed sparse representation. In: ICIP, pp. 2673–2676 (2011)

Spatiotemporal Saliency Detection Using Slow Feature Analysis and Spatial Information for Dynamic Scenes

Yang Wu$^{(\boxtimes)}$, Zhaohui Wang, Xin Xu, Shengrong Gong, Quan Liu, and Chunping Liu

School of Computer Science and Technolgoy, Soochow University, Suzhou, China
krukey@foxmail.com, cpliu@suda.edu.cn

Abstract. Slow feature analysis (SFA) can extract slowly varying signals from quickly varying input data. Inspired by the temporal slowness principle, we propose a novel spatiotemporal saliency algorithm for dynamic scenes analysis. In the training phase, slow feature functions are learned from different video patches using SFA. At the stage of saliency computation, we first exploit two-layer slow feature functions to extract pixel-level high-level motion features, which represent temporal slowness of every local space-time cuboid. Temporal saliency of each location is measured by the average of the corresponding feature vector. Finally, a saliency map is generated by combining the proposed temporal saliency and existing spatial saliency. The algorithm is qualitatively and quantitatively evaluated on challenging video sequences, and achieves competitive performance in contrast to the state-of-the-art algorithm.

Keywords: Slow feature analysis · Spatiotemporal saliency · Dynamic scenes

1 Introduction

Saliency detection is commonly used as a preprocess technique for many areas, such as video quality assessment [1], video compression [2], and tracking [3], etc. In this paper, we aimed to investigate spatiotemporal saliency detection for dynamic scenes, such as swaying trees, a crowd, a flock of birds, waves, snow, rain, smoke, or egomotion of a camera.

One of the important problems is motion feature description. Several saliency models exploit frame difference to extract the simplest motion features, such as [2]. Other researchers consider motion feature by optical flow estimation, such as [17]. Besides, Mahadevan et al. [9] utilize linear dynamic system to model spatiotemporal stimulus statistics, which is an autoregressive model to describe dynamic textures. In this paper, we intend to use an unsupervised feature learning method (e.g., Slow Feature Analysis, Deep Belief Nets, and other methods) to model spatiotemporal saliency instead of previous motion characterization.

© Springer International Publishing Switzerland 2015
X. He et al. (Eds.): IScIDE 2015, Part I, LNCS 9242, pp. 331–340, 2015.
DOI: 10.1007/978-3-319-23989-7_34

As SFA has been widely applied to many fields, such as invariant object recognition [13], human action recognition [14,18], dynamic scene classification [15] and change detection [16], these work demonstrates that SFA has good potential to describe motion feature. Thus, we try out to apply SFA to saliency detection for dynamic scenes.

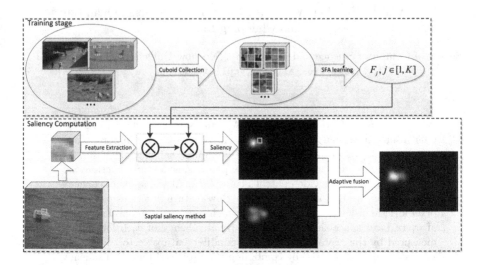

Fig. 1. The diagram of the SFA-based method. At the training stage, a large number of cuboids are extracted from different sequences to learn slow feature functions. At the stage of saliency computation, a spatial saliency map and a temporal saliency map are fused to generate the final saliency map.

In this paper, we propose a SFA-based spatiotemporal saliency algorithm for dynamic scenes, inspired by temporal slowness principle. Our framework is shown in Fig. 1. So what is the temporal slowness principle? Berkes and Wiskott [12] points out that sensor signals and surrounding environment changes on different time scales, and they propose slow feature analysis to extract slowly varying features from a quickly varying signals. Thus, through slow feature analysis, we intend to learn out high-level motion representations of surrounding environment changes directly from input data in an unsupervised manner. Our method is aimed to suppress slowly varying regions to make quickly varying objects (or regions) "pop out" by the extracted motion features.

The main contributions of this paper are summarized as follows. Firstly, Slow Feature Analysis is used in unsupervised learning of invariances [11]. In this paper, we introduce SFA to spatiotemporal saliency for dynamic scenes. To the best of our knowledge, this is the first work that uses the temporal slowness or SFA to detect salient objects. Secondly, we exploit two-layer SFA kernels to extract high-level motion characterization for temporal saliency, which is robust to complicated background movement.

The remainder of this paper is organized as follows. Section 2 reviews recent works on spatiotemporal saliency detection. Section 3 details the proposed approach for dynamic scenes analysis. Video sequences used in experiments and experimental results are presented in Sect. 4. Section 5 concludes this paper.

2   Previous Work

In recent years, the research on spatiotemporal saliency for videos has increasingly received attention. Several models have been proposed for salient object detection. Given a simple assumption that one of important goals of the visual system is to find potential targets, Zhang et al. [7] builds up a Bayesian probabilistic framework of what the visual system should calculate to optimally achieve this goal. Guo et al. [2] proposed a novel quaternion representation of an image and developed a multiresolution spatiotemporal saliency detection model called phase spectrum of quaternion Fourier transform (PQFT) to compute the spatiotemporal saliency map. Seo et al. [8] measure the likeness of a voxel to its surroundings by computing local regression kernels to generate the saliency map. However, the fore-mentioned models are not robust to dynamic scenes. To overcome complexity of scenes, Mahadevan and Vasconcelos [9] proposed a spatiotemporal saliency algorithm based a center-surround framework, motivated by biological mechanisms of motion-based perceptual grouping. This saliency detection is robust to complicated dynamic scenes. However, it requires high computation cost. Then, Zhou and Shi [10] proposed a saliency model using bio-inspired features for dynamic scenes. Their bottom-up saliency model requires relatively low computation cost and achieves competitive performance. Nevertheless, this method produces poor performance for some special scenes (e.g., smoky environment).

3   The SFA and Spatial Based Spatiotemporal Saliency

There are four main steps in our algorithm: collection of training cuboids, unsupervised SFA learning, saliency computation and proto-objects detection in a saliency map.

3.1   Collection of Training Cuboids

In this section, a large number of cuboids are required to perform subsequent SFA learning. Before cuboids are extracted, we perform normalization for each video sequences, so that the input signals are of zero mean with unit variance. Then, spatiotemporal cuboids are randomly sampled from video patches, and this procedure is divided into two steps.

Generation of Initial Points: Given a video patch, first frame is selected for edge detection to generate an edge image. Fixed initial points are randomly sampled from the edge image to extract cuboids.

Extraction of Local Cuboids: Centered at the selected point, a cuboid with the size of $h \times w \times d$ is extracted. In this paper, the size of a cuboid is set to $10 \times 10 \times 5$. Figure 2 shows an example of the collected cuboids.

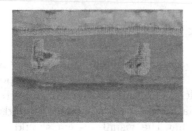

Fig. 2. Example of training cuboids.

In the first step, since the edge image contains plentiful motion information implicitly, "canny" detector is selected for edge detection to obtain rich edge information from video frames. Alternatively, we can also utilize other detectors for edge detection. At the step of cuboids extraction, in order to count the temporal information in the neighbor frames, each extracted cuboid is reconstructed by Δt successive frames. This process is presented in Fig. 3.

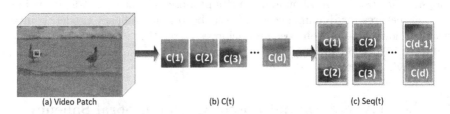

Fig. 3. The reformation process of a cuboid. The middle time sequence $c(t)$ is a cuboid from video patch. Then, the cuboid is transformed to the right time sequence $Seq(t)$ by Δt successive frames. Here, $\Delta t = 2$.

3.2 Unsupervised SFA Learning

SFA can learn slowly varying features from input sensor signals [11,12]. According to Sect. 3.1, a large number of local space-time cuboids are extracted for training. The training procedure is detailed as follow. For each single cuboid, the input signals are expanded to a nonlinear space using a quadratic function. For example, the quadratic expansion increases the dimensionality of input vector from n to $n + n(n + 1)/2$. Since the dimensionality increases greatly via the quadratic expansion, Principal Component Analysis (PCA) is performed to reduce the dimensionality of the input vector to 50, which is sufficient for subsequent experiments. For all reduced vector sequences, the solution of SFA is equivalent to the generalized eigenvalue problem [12], denoted by (1)

$$AW = BWA, A = < \dot{x}\dot{x}^T >_t, B = < xx^T >_t \tag{1}$$

where x is a vector sequence, A is the covariance matrix of the first-order derivative of the input and B is the covariance matrix of the input signal.

Based on the above-mentioned SFA algorithm, all cuboids collected from different video sequences are mixed to learn slow feature functions for saliency detection. The training stage is shown in Fig. 1.

3.3   Saliency Computation

In our approach, saliency is measured by change rate of local space-time information. Given a video frame, each location corresponds to a cuboid containing local space-time information. Hence, a saliency map is generated by computing internal slowness of every cuboid. In one video patch, the saliency map is computed as follows.

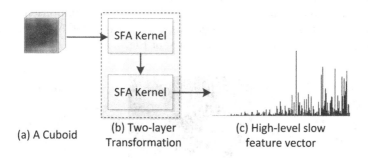

(a) A Cuboid   (b) Two-layer Transformation   (c) High-level slow feature vector

Fig. 4. Two-layer SFA feature learning.

Given a video patch with d successive frames, the middle frame is selected as the current frame, corresponding to the final saliency map. According to Sect. 3.1, each location (denoted by $l_{x,y}$) in the current frame corresponds to a cuboid with the size of $h \times w \times d$. After the reformation shown in Fig. 3, each cuboid is represented as a vector sequence (denoted by $Seq_l(t)$) with the time length of $d - \Delta t + 1$, wherein the vector at each time point is produced by concatenating Δt successive patches. Then, with the learned slow feature functions from Sect. 3.2, each sequence is transformed to a new time sequence with the size of $K \times (d - \Delta t + 1)$ in a cascaded way, wherein K represents the number of slow feature functions (In this paper, $K = 200$). Afterwards, the jth high-level slowness at the location l is calculated by (2)

$$Slow_l(j) = \frac{1}{d-1} \sum_{t=1}^{d-\Delta t+1} [Seq_l(t+1) \otimes \mathcal{F}_j - Seq_l(t) \otimes \mathcal{F}_j]^2, j \in [1, K] \tag{2}$$

where l is the position of the current frame, $\mathcal{F}_j$ is the jth slow feature function and $\otimes$ is the two-layer slow feature transformation. The feature extraction is shown in Fig. 4. Followed by (2), the K-dimensional slow feature vector is calculated to represent local space-time information at the location $l_{x,y}$. Finally, we compute the average of the K-dimensional feature vector to generate the final temporal saliency map using (3)

$$Sal(l) = g * \frac{1}{K} \sum_{j}^{K} Slow_l(j), l \in CurrF \tag{3}$$

where g is a 2-dimensional gaussian filter with variance as σ ($\sigma = 1$).

Afterwards, we exploit a boolean map approach in [19] to generate a spatial saliency. Finally, a temporal saliency map and a spatial saliency map are combined using the nonlinear fusion scheme in [20] to generate the final saliency map. An example is shown in the middle image of Fig. 5.

3.4   Proto-Objects Detection in a Saliency Map

A saliency map provides the location of the proto-objects in the original image. We employ the algorithm in [2] to find salient proto-objects from a saliency map. The approach searched out the first K focus of attention by "inhibition of return" (IoR), where K is the number of object candidate area (OCA). A example of search result is presented in the right image of Fig. 5.

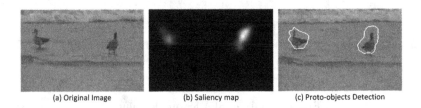

(a) Original Image   (b) Saliency map   (c) Proto-objects Detection

Fig. 5. An example of saliency detection.

4   Experiments

In this section, our method is tested on the standard datasets from [9] qualitatively and quantitatively. There are 18 video sequences with ground-truth in this datasets, including a static camera that monitors a distant pedestrian walkway or a crowded highway, highly varying dynamic backgrounds (e.g., water, smoke, rain and snow) and egomotion, and so on. All sequences are available in [21].

4.1   Two-Layer SFA Transformation

Here, we compare two-layer SFA transformation with one-layer, presented in Fig. 6. The results of Fig. 6 demonstrate that two-layer SFA kernels can improve the performance of salient object detection in dynamic scenes. Through the extracted high-level motion features using two-layer transformation, varying backgrounds are well suppressed.

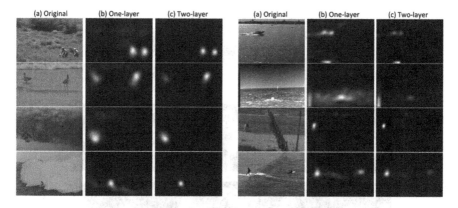

Fig. 6. Comparison of saliency maps based on one-layer and two-layer SFA feature learning.

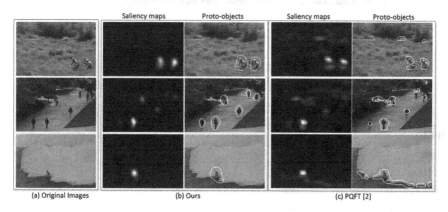

Fig. 7. Saliency detection in complex scenes.

4.2 Qualitative and Quantitative Analysis

Before saliency computation, the sequences are resized to a small scale. This operation improves the performance of the approach and reduces the computation cost. We compare our algorithm with PQFT [2], presented in Fig. 7.

In order to obtain more intuitive performance, the proposed algorithm is compared with some previous methods, shown in Fig. 8. Our method can separate regions efficiently that change on different time scales, and make quickly varying object candidate areas "pop out". The proposed algorithm is superior to DiscSal [9] slightly. PQFT [2] and the method of Monnet et al. [5] show inferior performance. GMM [6] and KDE [4] almost make no sense, and both produce worst results.

To make sure that we perform a quantitative analysis, 50 frames of each sequence are computed for saliency map. We measure the performance of the proposed algorithm on different video sequences by equal error rate (EER) [9].

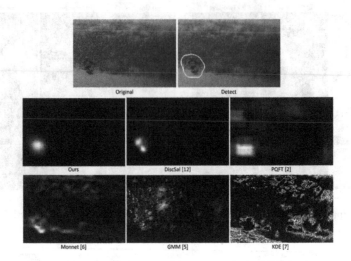

Fig. 8. Saliency maps of different methods.

Table 1. Equal Error Rates of different algorithms.

%	Ours	DiscSal [9]	Zhou [10]	PQFT [2]	Monnet [5]	KDE [4]	GMM [6]
Birds	**4.5**	5	8.5	5.7	7	20	23
Boats	**4.8**	9	5.2	8.2	15	13	15
Bottle	2.5	2	3.5	**1.7**	17	38	25
Chopper	5	5	8.3	6.4	8	43	35
Cyclists	**3.9**	8	–	10.1	28	44	36
Flock	22.9	**15**	–	38.9	31	33	34
Freeway	8.3	**6**	–	20.8	31	21	25
Hockey	24	24	**21.1**	25	29	35	39
Jump	**7.5**	15	17.2	16.2	31	33	34
Land	8.5	**3**	6.7	10.2	16	54	40
Ocean	**9.7**	11	–	18.1	11	19	30
Peds	**6.7**	7	12.1	12.8	11	17	11
Rain	6	**3**	7	4.2	17	23	15
Skiing	3	3	5.6	5.9	11	46	36
Surf	2	4	7.9	3.6	10	36	23
Surfers	**4.5**	7	7	8.4	10	25	35
Traffic	7.2	**3**	–	10.8	9	39	34
Zodiac	1	1	5.5	5.2	3	20	40
Average	**7.3**	7.6	8.9	11.8	16	33.1	29.7

Table 1 shows equal error rates of the various methods (DiscSal [9], Zhou [10], PQFT [2], Monnet [5], KDE [4] and GMM [6]) for each video sequence, wherein experimental results of Monnet, KDE and GMM are presented in [9]. The results show that the proposed method outperforms all others, achieving an average ERR of 7.3 percent (Ours). Although DiscSal has the similar performance as ours, its computation cost (32fps) is 8 times greater than ours (4fps). In addition, two saliency methods (Zhou and PQFT) are not robust to some dynamic scenes, achieving inferior performance in contrast to ours method and DiscSal. Those traditional approaches (Monnet, KDE and GMM) obtain low performance relatively.

According to Table 1, our method has lower EER than DiscSal for some special scenes, such as "flock" and "hockey", where the foreground objects cover a substantial portion of the image area. The proposed approach is difficult to handle these cases. Besides, due to the foggy environment ("freeway", "traffic"), the method is not capable to capture accurate motion characterizations for temporal saliency. Nevertheless, our method is still superior to others except for DiscSal in the above-mentioned scenes.

5   Conclusion

In this paper, we propose a SFA-based spatiotemporal saliency model for background subtraction. The new algorithm is inspired by the temporal slowness, and applies slow feature analysis to saliency computation for the first time. The learned slow feature functions contain internal slowness representation of local space-time information. SFA is successfully applied to spatiotemporal saliency, and our experiments demonstrate that SFA has good potential to describe motion information.

Acknowledgments. This work is supported by National Natural Science Foundation of China (NSFC Grant No. 61272258, 61170124, 61301299, 61272005), and a prospective joint research projects from joint innovation and research foundation of Jiangsu Province (BY2014059-14). The corresponding author is Chunping Liu.

References

1. Ćulibrk, D., Mirković, M., Zlokolica, V., Pokrić, M., Crnojević, V., Kukolj, D.: Salient motion features for video quality assessment. IEEE Trans. Image Process. **20**(4), 948–958 (2011)
2. Guo, C., Zhang, L.: A novel multiresolution spatiotemporal saliency detection model and its applications in image and video compression. IEEE Trans. Image Process. **19**(1), 185–198 (2010)
3. Mahadevan, V., Vasconcelos, N.: Saliency-based discriminant tracking. In: IEEE Conference on Computer Vision and Pattern Recognition, CVPR 2009, pp. 1007–1013 (2009)

4. Elgammal, A., Duraiswami, R., Harwood, D., Davis, S.L.: Background and foreground modeling using nonparametric kernel density estimation for visual surveillance. Proc. IEEE **90**(7), 1151–1163 (2002)

5. Monnet, A., Mittal, A., Paragios, N., Ramesh, V.: Background modeling and subtraction of dynamic scenes. In: Proceedings of the Ninth IEEE International Conference on Computer Vision, pp. 1305–1312 (2003)

6. Gaber, M.M., Stahl, F., Gomes, J.B.: Background. In: Gaber, M.M., Stahl, F., Gomes, J.B. (eds.) Big Data on Small Devices. SBD, vol. 2, 1st edn, pp. 7–22. Springer, Heidelberg (2014)

7. Zhang, L., Tong, M.H., Marks, T.K., Shan, H., Cottrell, G.W.: SUN: a bayesian framework for saliency using natural statistics. J. Vis. **8**(7), 32.1–32.20 (2008)

8. Seo, H.J., Milanfar, P.: Static and space-time visual saliency detection by self-resemblance. J. Vis. **9**(12), 15.1–15.27 (2009)

9. Mahadevan, V., Vasconcelos, N.: Spatiotemporal saliency in dynamic scenes. IEEE Trans. Pattern Anal. Mach. Intell. **32**(1), 171–177 (2010)

10. Zhou, Y., Shi, K.: Spatiotemporal saliency based on distributed opponent oriented energy. In: 2012 21st International Conference on Pattern Recognition (ICPR), pp. 2021–2024 (2012)

11. Wiskott, L., Sejnowski, T.J.: Slow feature analysis: unsupervised learning of invariances. Neural Comput. **14**(4), 715–770 (2002)

12. Berkes, P., Wiskott, L.: Slow feature analysis yields a rich repertoire of complex cell properties. J. Vis. **5**(6), 579–602 (2005)

13. Franzius, M., Wolbert, N., Wiskott, L.: Invariant object recognition and pose estimation with slow feature analysis. Neural Comput. **23**(8), 1–35 (2011)

14. Zhang, Z., Tao, D.: Slow feature analysis for human action recognition. IEEE Trans. Pattern Anal. Mach. Intell. **34**(3), 436–450 (2012)

15. Theriault, C., Thome, N., Cord, M.: Dynamic scene classification: learning motion descriptors with slow features analysis. In: 2013 IEEE Conference on Computer Vision and Pattern Recognition (CVPR), pp. 2603–2610 (2013)

16. Wu, C., Du, B., Zhang, L.: Slow feature analysis for change detection in multispectral imagery. IEEE Trans. Geosci. Remote Sens. **52**(5), 2858–2874 (2014)

17. Zhou, F., Kang, S.B., Cohen, M.F.: Time-mapping using space-time saliency. In: 2014 IEEE Conference on Computer Vision and Pattern Recognition (CVPR), pp. 3358–3365 (2014)

18. Sun, L., Jia, K., Chan, T.H., Fang, Y., Wang, G., Yan, S.: DL-SFA: deeply-learned slow feature analysis for action recognition. In: 2014 IEEE Conference on Computer Vision and Pattern Recognition (CVPR), pp. 2625–2632 (2014)

19. Zhang, J., Sclaroff, S.: Saliency detection: a boolean map approach. In: 2013 IEEE International Conference on Computer Vision (ICCV), pp. 153–160 (2013)

20. Liu, Z., Zhang, X., Luo, S., Le Meur, O.: Superpixel-based spatiotemporal saliency detection. IEEE Trans. Circ. Syst. Video Technol. **24**(9), 1522–1540 (2014)

21. http://www.svcl.ucsd.edu/projects/background_subtraction/

Face Verification Across Aging Based on Deep Convolutional Networks and Local Binary Patterns

Huanhuan Zhai[✉], Chunping Liu, Husheng Dong, Yi Ji, Yun Guo,
and Shengrong Gong[✉]

School of Computer Science and Technolgoy, Soochow University, Suzhou, China
JessieZhh@foxmail.com, shrgong@suda.edu.cn

Abstract. This paper proposes a novel method to learn a set of high-level feature representations for face verification across aging. Conventional hand-crafted features are not capable to overcome aging effects. In order to obtain an accurate face representation, we apply the combination of a nine-layer deep convolutional neural network and Local Binary Pattern(LBP) histograms, both of which are essential to face recognition. On account of the need of large quantity data in deep learning methods, we train the model on the publicly available cross-age face dataset CACD (Cross-Age Celebrity Dataset), which contains more than 160000 face images of 2000 different celebrities. Experiments on the CACD and LFW (Labeled Faces in the Wild) dataset demonstrate that the proposed approach outperforms the state-of-the-art methods. In addition, hairstyle, facial expression, changes of background and occlusion provide discriminative cues to the system of face verification.

Keywords: Face verification · Aging · Deep convolutional networks

1 Introduction

Face verification has been an active research area with the increasing need of law enforcement and surveillance. According to Zhao et al. [1], face verification across aging could be more challenging because as age crossing, great changes would appear in faces which indeed increased difficulty. In recent years, several researchers have tried a variety of methods in this area. Most of the methods focus on age estimation [2,3] and age simulation [4,5]. They can be roughly categorized as generative methods since aging needs to be modeled.

Traditional face verification pipeline can be divided into four parts: detect, align, represent, and verification. Face recognition is to classify a face image into the right class. However, face verification is the task of judging whether two input instances belong to the same class or not. It is not hard to find that face verification is far more challenging than recognition. In the laws of nature driven, with age crossing, subtle change will occur in the face. Some researchers study the effect of LBP (Local Binary Pattern) features [6,7] in face verification

© Springer International Publishing Switzerland 2015
X. He et al. (Eds.): IScIDE 2015, Part I, LNCS 9242, pp. 341–350, 2015.
DOI: 10.1007/978-3-319-23989-7_35

and have achieved significant improvements. Therefore, we consider texture is rather suitable for the description of this kind of change. However, hand-crafted descriptors are not completely capable to represent the appearance of face, so we take deep learning methods into account. The primary advantage of deep neural network is to learn discriminative features through autonomous learning without supervision. Thus, we take it into consideration that if it can be applied into our face verification across aging. The answer is in the affirmative.

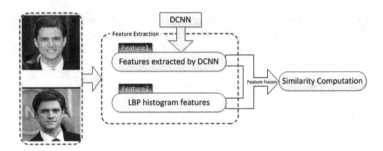

Fig. 1. The pipeline of our method for face verification. Given two images, DCNN features (featrue1) and LBP histogram features (feature2) are integrated to calculate matrix similarity. Then the approach judge whether the two instances come from the same individual or not by the value of similarity.

Deep learning is especially suitable for dealing with large training sets, with many recent successes [8,9] in diverse domains. Although deep learning methods have achieved remarkable performance in face recognition, age-related facial image analysis has not been extensively studied. We propose a novel method to combine automatically extracted features and hand-crafted features, as the framework shown in Fig. 1. The deep convolutional neural network (DCNN) is learned to classify all the face images available for training by their identity which represents the category of celebrities. The last hidden layer values of neuron activations are used as our rough features. With the carefully designed nine-layer deep convolutional neural network, combined with texture descriptor LBP histogram [7], we get a perfect feedback. The proposed system differs from the majority of contributions in the field in that it uses deep learning framework and takes aging effects into consideration. In general, the proposed method can be summarized as follow: (1) set up a nine-layer deep model to automatically learn discriminative face representations across aging, with the big dataset and well-designed layers. (2) Integrate LBP histogram features [7] into the features extracted by DCNN, which make face representations much more convincing.

2   Previous Work

Face Verification and Age Progression: The goal of face verification is to judge whether two input instances belong to the same individual or not. When joined with age constraint, the difficulty of verification would be dramatically increased. As a result, how to obtain a discriminative feature is particularly important.

In previous work, parts of the existing methods adapt the idea of age estimation. Ramanathan et al. [10] designed a discriminative approach for face verification across age progression. In order to reduce the influence of illumination, they exploited probabilistic eigenspace technique and Bayesian model. They later designed a facial growth model for face verification task [11]. However, these approaches usually require prior information like the actual age of the individual. Then, age simulation techniques involved generating age models which use the shape and texture information from the face images were presented. Tiddeman et al. [12] proposed a model by studying the texture change of face with age progression. Their model simulated wrinkles in face images by applying locally weighted wavelet functions at different scales and orientations to extract features. Lanitis et al. [13] proposed a method for simulating aging effects in face images. They modeled age as a quadratic function of the PCA coefficients extracted from parameters to estimate age. Ling et al. [14] also used a discriminative method for face verification across age progression. They put forward a face representation called gradient orientation pyramid, for each face image. This model generated a Gaussian pyramid, then computed gradient orientation pixel by pixel in each layer. While the above methods explicitly address the aging effects, they usually require additional information about the images being compared. Nevertheless, their methods considering about aging, which itself is a challenging problem.

Deep Learning and Big Data: In recent years, with the development of Internet technology, we are moving towards the era of Big Data. Such a large amount of data along with the growth of internet resources make us probe more powerful models which can effectively deal with problems occur in computer vision, such as illumination, occlusion and change of attitudes. Since 2006, with the boom of interests in deep learning, more and more researchers participate in the area of new deep models. A lot of deep learning algorithms [8,9,15] are proposed for face recognition which indeed achieve fantastic performance. There are a few methods that also use deep convolutional neural network for face verification. Taigman et al. [9] proposed a deep model in their project called deepface, which closed the gap to human-level performance in face verification. They exploited 3D face model along with a nine-layer deep convolutional neural network to obtain a face representation that generalized well to other datasets. Besides, their method advanced the state of the art significantly in the Labeled Faces in the Wild benchmark (LFW). Yi Sun et al. [8] learned a face representation by joint identification and verification to conduct a supervised training. They also learned a set of high-level feature representation through deep learning [16]. They changed the mode of face images to enhance the robustness of verification. Considering about other deep models, Gary B.Huang et al. [17] proposed deep learning as a natural source for obtaining additional, complementary representations. They made use of convolutional deep belief networks to learn features in high-resolution images. However, aging effects were not considered by the methods above. In contrast to existing methods, we take age progression into consideration and demonstrate that the method proposed achieves inspiring performance in face verification across aging.

3   Face Representation and Verification

3.1   DCNN Architecture and Training

We train our DCNN on a multi-class face recognition task with our carefully designed deep convolutional network. The convolution and pooling operation are used to extract visual features hierarchically, from local low-level features to global high-level features. We stack a nine-layer network which contains 4 convolutional layers and 3 max-pooling layers. Each of the first three convolutional layer is followed by a max-pooling layer. In each convolutional layer, weights are shared. The last two layers are fully connected. The last hidden layer neuron activations are treated as our rough features. Finally, the output layer predicts the identity of each image by a probability distribution. The predicted id is corresponding to the maximum value of the distribution. An illustration of the deep convolutional network structure is shown in Fig. 2. When the size of the input instance changes, the map size of the following layers will change accordingly. We train a multi-layer convolutional network with the goal to classify the face image into the right category. As shown in Fig. 2, a face image of size 67×59 pixels is given to a convolutional layer with 60 filters of size 4×4. The resulting 60 feature maps are then fed to a max-pooling layer with a stride of 2. This is followed by another convolutional layer which has 40 filters of size 3×3. These three layers are mainly used to extract low-level features, such as simple edges and texture feature. The max-pooling makes our features extracted from convolutional layer more robust to subtle changes in images. Before training the whole network, we use 2D face alignment to ensure that for some minor adjustments in faces, so our deep net would be more adaptive. Thus, we obtain a set of face features after extraction layer by layer. The feature extraction process is defined as $f = Conv(x, \theta)$, in which $Conv(.)$ means the function of convolutional feature extraction while θ represents the parameters to be learned, x is the input image and f becomes the feature vector.

Finally the output of the last fully connected layer is fed to a N-way softmax (N means the number of classes) which propose a probability distribution over n different identities. Given an input, the probability assigned to the $n-th$ class is the output of the softmax activation function:

$$p_n = \frac{\exp(o_n)}{\sum_h \exp(o_h)} \tag{1}$$

The goal of training is to maximize the probability of the correct class. We achieve this by minimizing the cross-entropy loss [9] for each training sample. If n is the index of the true label for a given input, the loss is: $L = -\log p_n$. We can

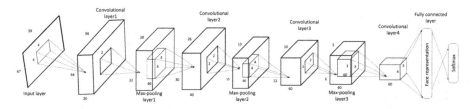

Fig. 2. The deep convolutional neural network in this paper. The map numbers and dimensions of all the layers are illustrated as the length, width, height of the cuboids.

use stochastic gradient descent to update parameters of L. In each convolutional layer, feature maps of the previous layer are convoluted by a convolution kernel that can be learned, then the output feature maps will be obtained through an activation function. Each output map may be the value of multiple input maps.

For each input image, we also compute the LBP histogram features [7] at the same time. Given an image $I(x, y)$, define the pyramid of I as: $G_k(I) = I(x, y, k)$, $k = 0, \ldots, s$ with $G_k(I) = [I(x, y, k-1) \otimes \Phi(x, y)]$, where $\Phi(x, y)$ is the Gaussian kernel and s is the number of pyramid layers. For each layer of the pyramid, we divide it into blocks with the size 8×8 and compute LBP histogram for each block. Finally, the LBP histograms of each block are connected in a vector with the dimension of 3304. After extracted by the proposed model, features are concatenated together to form the final face representation for verification.

3.2 Similarity Metric

Verifying whether two input instances belong to the same individual or not has been researched in face recognition. In effect, supervised methods show an obvious advantage over unsupervised methods. In this paper, we utilize matrix cosine similarity to calculate the distance between two feature vectors extracted from two input instances.

Cosine similarity is the common similarity calculation of information retrieval. The object should be expressed as vector before calculating. For example, suppose a set of feature maps belong to one image, we should pull them into a column vector, define object i as $D_i = (w_{i1}, w_{i2}, w_{i3}, \ldots, w_{in})$, object j as $D_j = (w_{j1}, w_{j2}, w_{j3}, \ldots, w_{jn})$, the cosine similarity between them is defined as: the similarity between them is: $cos(D_i, D_j) = \dfrac{\sum_{k=1}^{n} w_{ik} w_{jk}}{\sqrt{\sum_{k=1}^{n} w_{ik}^2} \sqrt{\sum_{k=1}^{n} w_{jk}^2}}$.

4 Experiments

4.1 Preprocessing

Face Alignment. As mentioned before, in unconstraint environment, face images may be influenced by face expression, illumination and occlusion, the verification results will be affected by these constraints especially when two faces are not symmetry properly. Facial feature location is to determine the location of facial feature points (such as eyes, mouth) on the basis of the region obtained by face detection. The basic idea of facial feature point positioning is to handle the constraint of the texture feature of local organs and the feature points of organs.

Early facial feature points positioning mainly focus on the localization of several key points such as center of eyes and mouth. Later, it is widely accepted that applied with more key points along with mutual restraint will improve the accuracy and stability. In this paper, we adapt the method called flandmark. Flandmark is an open source code library for the detection of facial key points that can accurately detect facial position, such as the corner of eyes and mouth. Besides, it can meet real-time processing.

Fig. 3. Samples of the CACD dataset. Numbers on the top are the age of the celebrities.

4.2 Experiment Setup

The experiments are performed on CACD aging dataset [18] (Fig. 3) and LFW dataset. The CACD aging dataset contains more than 160000 face images of 2000 celebrities with age ranging from 16 to 62. To the best of our knowledge, it is by far the largest publicly available cross-age face dataset. With the big data fed to our deep net, the need of training is easily met. Training on the LFW dataset are processed in the same pipeline that was used to train on the CACD dataset. To evaluate the performance, we use True Positive Rate(TPR) and False Positive Rate(FPR) to measure the effectiveness of the experiments.

4.3 Experiments on the CACD Dataset

To investigate the effects of aging in face verification, we perform several experiments using various subsets of the CACD dataset. Before experiments are performed, we re-number the face images according to their names to form the identity label which can facilitate our subsequent operations. Then for each celebrity, we take one half of them for training to learn parameters of the deep model. The size of the training set is more than 80000 images, taking our mini-batch size which is set to 50 into consideration, we selectively drop out some and retain 80000 images for training. Also, experiments are performed to measure the similarity metric methods. According to the results shown in Table 1, matrix cosine similarity outperforms other measurements.

Table 1. Comparison of different similarity metric measurements

Measurement	Accurate(%)
χ^2 Distance	86.3
Euclidean Distance	81.4
Bhattacharyya Distancee	82.1
Matrix Cosine Similarity	89.5

Table 2. Face verification performance on CACD-VS (While the CACD-VS dataset has not been released, we manually selected 2,000 positive cross-age image pairs and 2,000 negative pairs to form a subset for face verification as the website mentioned [24].)

Method	Verification Accuracy(%)
High-Dimensional LBP	81.6
Hidden Factor Analysis	84.4
Cross-age Reference Coding	87.6
DCNN + LBPH	**89.5**
Human, Average	85.7
Human, Voting	94.2

With more than 80000 images left, we manually selected 2000 positive cross-age image pairs and 2000 negative pairs to form a subset for face verification task similar to the one used in LFW. Afterwards, we take the first image of each pair to form a big batch to extract features using the trained deep model, and the left half is processed in the same way. To evaluate the performance of our approach named as DCNN+LBPH, we compare the verification accuracy with several results before as shown in Table 2, our method outperforms others except human participation approaches. The performance of algorithms are also evaluated using the ROC curves, we test the performance of other discriminative approaches. The approach proposed in [14] used Gradient Orientation Pyramid

(GOP) for representation and SVM for classification. l_2 is a baseline approach that uses the l_2 norm to compare two normalized images. Bayesian + PFF [10] is the approach combining Bayesian framework and PointFive Face (PFF) for face verification purposes. Results are shown in Fig. 4. A separate validation subset is needed during training to avoid overfitting. After each training epoch, we observe the errors on the validation subset and select the model that provide the lowest error. It is easy to find that with the increasing class number, the validation error rates drop.

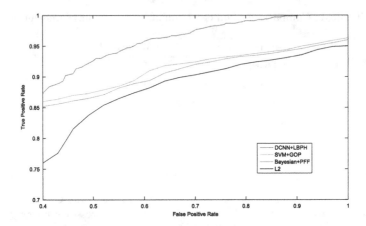

Fig. 4. TPR-FPR curve for the experiment on the CACD subset including 2000 positive pairs and 2000 negative ones.

4.4 Experiments on the LFW Dataset

The vision community has made significant progress on face verification in unconstrained environments. LFW dataset is a standard face dataset for face recognition. Given some very hard cases due to aging effects, large lighting and face pose variations in LFW, any improvement over the state-of-the-art is very remarkable.

Table 3. Accuracy comparison of the proposed model and the-state-of-the-art methods on LFW.

Method	Accuracy(%)
PLDA [19]	90.07
Joint Bayesian [20]	90.90
Linear rectified units [21]	80.73
GSML [22]	84.18
OSS, TSS, full [23]	86.83
DCNN+LBPH	**91.40**

After training on the whole dataset,our tests are taken on the view 2 subset that has lists all the intra-personal pairs and extra-personal pairs which can facilitate our calculation. With the same pipeline as experiments on the CACD dataset, the first step is to obtain face representation, then calculate the similarity of each pair to verify whether they are belong to the same individual. Table 3 gives the final accuracy of our approach using the deep learning framework, in comparison with other systems on the LFW dataset. Empirical studies demonstrate the effectiveness of the proposed approach.

5   Conclusion

In this paper, we propose a deep learning based method for face verification across age progression. According to previous studies, an ideal face representation needs to be invariant to pose, illumination, expression, and sometimes, image quality. Therefore, hand-crafted features can easily be limited. We obtain novel representations through deep convolutional network. By combining such deep learning representations with hand-crafted descriptors, we obtain a powerful and effective representation and achieve competitive performance in face verification across aging. In future, our primary pursuit would be to apply this approach for face verification tasks on videos.

Acknowledgments. This work is supported by National Natural Science Foundation of China (NSFC Grant No. 61272258, 61170124, 61301299, 61272005), and a prospective joint research projects from joint innovation and research foundation of Jiangsu Province (BY2014059-14). The corresponding author is Shengrong Gong.

References

1. Zhao, W., Chellappa, R., Phillips, P.J., Rosenfeld, A.: Face recognition: a literature survey. ACM Comput. Surv. (CSUR) **35**(4), 399–459 (2003)
2. Fu, Y., Huang, T.S.: Human age estimation with regression on discriminative aging manifold. IEEE Trans. Multimedia **10**(4), 578–584 (2008)
3. Guo, G., Fu, Y., Dyer, C.R., Huang, T.S.: Image-based human age estimation by manifold learning and locally adjusted robust regression. IEEE Trans. Image Process. **17**(7), 1178–1188 (2008)
4. Suo, J., Chen, X., Shan, S., Gao, W.: Learning long term face aging patterns from partially dense aging databases. In: 2009 IEEE 12th International Conference on Computer Vision, pp. 622–629 (2009)
5. Suo, J., Zhu, S.C., Shan, S., Chen, X.: A compositional and dynamic model for face aging. IEEE Trans. Pattern Anal. Mach. Intell. **32**(3), 385–401 (2010)
6. Ojala, T., Pietikainen, M., Maenpaa, T.: Multiresolution gray-scale and rotation invariant texture classification with local binary patterns. IEEE Trans. Pattern Anal. Mach. Intell. **24**(7), 971–987 (2002)
7. Mahalingam, G., Kambhamettu, C.: Face verification with aging using AdaBoost and local binary patterns. In: Proceedings of the Seventh Indian Conference on Computer Vision, Graphics and Image Processing, **30**(12), pp. 101–108. ACM (2010)

8. Sun, Y., Chen, Y., Wang, X., Tang, X.: Deep learning face representation by joint identification-verification. In: Advances in Neural Information Processing Systems, pp. 1988–1996 (2014)
9. Taigman, Y., Yang, M., Ranzato, M.A., Wolf, L.: Deepface: closing the gap to human-level performance in face verification. In: 2014 IEEE Conference on Computer Vision and Pattern Recognition (CVPR), pp. 1701–1708 (2014)
10. Ramanathan, N., Chellappa, R.: Face verification across age progression. IEEE Trans. Image Process. **15**(11), 3349–3361 (2006)
11. Ramanathan, N., Chellappa, R.: Modeling age progression in young faces. In: 2006 IEEE Computer Society Conference on Computer Vision and Pattern Recognition, pp. 387–394 (2006)
12. Tiddeman, B., Burt, M., Perrett, D.: Prototyping and transforming facial textures for perception research. IEEE Comput. Graph. Appl. **21**(5), 42–50 (2001)
13. Lanitis, A., Taylor, C.J., Cootes, T.F.: Toward automatic simulation of aging effects on face images. IEEE Trans. Pattern Anal. Mach. Intell. **24**(4), 442–455 (2002)
14. Ling, H., Soatto, S., Ramanathan, N., Jacobs, D.W.: Face verification across age progression using discriminative methods. IEEE Trans. Inf. Forensics Secur. **5**(1), 82–91 (2010)
15. Sun, Y., Wang, X., Tang, X.: Deep learning face representation from predicting 10,000 classes. In: 2014 IEEE Conference on Computer Vision and Pattern Recognition (CVPR), pp. 1891–1898 (2014)
16. Sun, Y., Wang, X., Tang, X.: Hybrid deep learning for face verification. In: 2013 IEEE International Conference on Computer Vision (ICCV), pp. 1489–1496 (2013)
17. Huang, G.B., Lee, H., Learned-Miller, E.: Learning hierarchical representations for face verification with convolutional deep belief networks. In: 2012 IEEE Conference on Computer Vision and Pattern Recognition (CVPR), pp. 2518–2525 (2012)
18. Chen, B.-C., Chen, C.-S., Hsu, W.H.: Cross-age reference coding for age-invariant face recognition and retrieval. In: Fleet, D., Pajdla, T., Schiele, B., Tuytelaars, T. (eds.) ECCV 2014, Part VI. LNCS, vol. 8694, pp. 768–783. Springer, Heidelberg (2014)
19. Simon, P., Li, P., Fu, Y., Mohammed, U.: Probabilistic models for inference about identity. IEEE Trans. Pattern Anal. Mach. Intell. **34**(1), 144–157 (2012)
20. Chen, D., Cao, X., Wang, L., Wen, F., Sun, J.: Bayesian face revisited: a joint formulation. In: Fitzgibbon, A., Lazebnik, S., Perona, P., Sato, Y., Schmid, C. (eds.) ECCV 2012, Part III. LNCS, vol. 7574, pp. 566–579. Springer, Heidelberg (2012)
21. Nair, V., Hinton, G.E.: Rectified linear units improve restricted boltzmann machines. In: Proceedings of the 27th International Conference on Machine Learning (ICML-10), pp. 807–814 (2010)
22. Nguyen, H.V., Bai, L.: Cosine similarity metric learning for face verification. In: Kimmel, R., Klette, R., Sugimoto, A. (eds.) ACCV 2010, Part II. LNCS, vol. 6493, pp. 709–720. Springer, Heidelberg (2011)
23. Wolf, L., Hassner, T., Taigman, Y.: Similarity scores based on background samples. In: Zha, H., Taniguchi, R., Maybank, S. (eds.) ACCV 2009, Part II. LNCS, vol. 5995, pp. 88–97. Springer, Heidelberg (2010)
24. http://bcsiriuschen.github.io/CARC/

Radar Model Correction of Asteroid 4179 Toutatis via Optical Images Observed by Chang'E-2 Probe

Duzhou Zhang[1], Lei Sun[2], Ting Xiao[2], and Wei Zhao[2(✉)]

[1] Beijing Institute of Control Engineering, Beijing 100190, China
DZ_zhang@hit.edu.cn
[2] Harbin Institute of Technology, Harbin 150006, China
{sunlei,zhaowei}@hit.edu.cn, hitxiaoting@163.com

Abstract. Asteroid 4179 Toutatis has been modeling by ground-based radar observations until December 13th, 2012, when distinct optical images of Toutatis was captured by Chang'E-2's flew by at the shortest range for the first time. The surface details of optical images are abundant enough to reinforce the radar model descriptions of Toutatis's surface structure. Under this context, we customized a method of frequency domain data fusion, which combines the depth information of radar model and the depth values estimated from optical image by shape from shading algorithm, in correcting Toutatis' radar model. A more accurate model with abundant surface characteristics had been resulted.

Keywords: Chang'E-2 probe · Asteroid 4179 · Radar model correction · Multi-source data fusion · Shape from shading

1 Introduction

Asteroids are remnant building blocks of the formation of the solar system [1] and the clues asteroids hold are considered as important keys to understand the origin and evolution of solar system. Besides, the fact that Near-Earth Asteroids (NEA) periodically impact the Earth, with consequences possibly as extreme as global catastrophe, underscores the importance of knowing as much as possible about these objects [2].

Mankind discovered over 5000 asteroids so far, and has been discovering new ones rapidly and continuously. Asteroid 4179 could be regarded as one of them that is found by accident and of necessity. Multi-way observation of Asteroid 4179, Toutatis, has been carried out since its discovery in 1989, including photometry, spectroscopy, radar imaging and other Earth-based observations [3]. Asteroid 4179, as an Apollo near-Earth asteroid in an eccentric orbit that originates from the main belt, has a large eccentricity of 0.63 and the lowest orbital inclination of 0.46°. It is currently in a 3:1 mean-motion resonance with Jupiter, which leading to a series of close approaches 4 years apart [4]. During each of the close approaches of Toutatis from 1992 to 2012, astronomers have accumulated a lot of ground-based radar-derived model.

This research is supported by National Science Foundation of China (Grant No. 61201309, No. 61171184) and Key Innovation Fund of China Aerospace Science and Technology Corporation.

© Springer International Publishing Switzerland 2015
X. He et al. (Eds.): IScIDE 2015, Part I, LNCS 9242, pp. 351–361, 2015.
DOI: 10.1007/978-3-319-23989-7_36

In 1992, Hudson and Ostro used a comprehensive physical model to figure out a three-dimensional shape, spin state and the ratios of the principal moments of inertia of Toutatis. They reported that Toutatis is rotating in a long-axis mode characterized by periods of 5.41 days (rotation about the long axis) and 7.35 days (average for long-axis precession), and its dimensions along the principal axes are $(1.92, 2.40, 4.60) \pm 0.10$ km [5, 6]. Moreover, high-resolution radar data have been obtained in 1996, which show that the asteroid's surface properties are strikingly uniform. Furthermore, detailed geological features, such as complex linear structures and concavities, have been revealed at an average spatial resolution of 34 m [4].

The optical images of Toutatis were obtained on December 13th, 2012, 08:30 (UTC time), when Chang'E-2 probe flew by Toutatis in deep space 7 million kilometers away from the Earth [3]. The close observation from Chang'E-2's flyby has revealed several new remarkable discoveries. The shape of Toutatis resembles that of a ginger root with sharper shoulder and longer neck than radar-derived model intuitively [1]. The maximum measuring size via frames of optical images is statistically estimated at 4.7549 ± 0.4695 km long and 1.9485 ± 0.1923 km width, respectively, which shows a good agreement with that of radar model [7].

Our recent task is carried out based on the three-dimensional data of Toutatis recorded by over 20,000 vertices [4], which was generated by delay-Doppler data observed by ground-based Goldstone and Arecibo radar [4], and optical images taken by CMOS camera on Chang'E-2 probe. Due to the principles of radar observation, the depth information in radar model is much reliable, while the details on the surface are quite sparse. Relative to radar model, optical images provide us with higher resolution. Consequently, it's reasonable to obtain a more accurate model with abundant surface characteristics by adding optical images to radar model.

In this paper, a method, special for Asteroid 4179, on depth information fusion to correct its radar model is proposed. The method is inspired by James Edwin Cryer [8] who combined the depth information achieving from Shape From Stereo and Shape From Shading in frequency domain.

The rest of this paper is organized as follows: Sect. 2 describes the detailed method we employed. Section 3 presents the fused model. The discussion and conclusion give out in Sect. 4.

2   Depth Information Fusion Method

2.1   Image Pixels Mapping to Model Vertices

Since only one side of Toutatis surface was captured by CMOS camera [1], it's necessary to choose a specific 3D points sub-set from 20,000 vertices set on radar model, which takes on as similar as the side it appears in optical image. A ray tracing module, Persistence of Vision Raytracer (POV-Ray), is used here to give out radar model output image, when provided radar data as input.

We defined a function $c : P_I \rightarrow P_R$ to map the coordinates of pixels in optical image to the coordinates of vertices in radar model, where P_I represents the pixels in optical image and P_R the pixels in radar output images. Function c could be deduced by the procedure showed in Fig. 1, which illustrates the process we confirm the correspondence between vertices of radar model and pixels of optical image. Step ① computes the mapping relation between vertices of radar model and pixels of output image of ray tracing module. Step ② determines the specific posture of radar model by comparing optical image with output images and step ③ establishes a pixel level mapping between

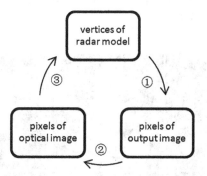

Fig. 1. The process of calculating the correspondence between vertices of radar model and pixels o optical image.

pixels of optical image and pixels of output image using Voronoi. Obviously, output image of ray tracing module serves as a bridge between radar model and optical image.

Vertices of Radar Model to Its Output Image. Formation of radar output image in POV-Ray is shown in Fig. 2. Camera locates at point $P_C(x_C, y_C, z_C)$. Length and width of view plane are L_v and W_v, respectively, L_I and W_I are for those of output image. Output image of ray tracing module is rendered with projection transformation. Consider a vertex on radar model $P_{R_i}(x_{R_i}, y_{R_i}, z_{R_i})$, its projection on the view plane is $P_{V_i}(x_{V_i}, y_{V_i}, z_{V_i})$, where $x_{V_i} = x_{R_i} \cdot (z_C - z_{V_i})/z_C$ and $y_{V_i} = y_{R_i} \cdot (z_C - z_{V_i})/z_C$. Pixel $P_{RO_i}(x_{RO_i}, y_{RO_i})$ in the output image correspond to P_{R_i} is computed by $x_{RO_i} = (W_I - 1)(x_{V_i}/W_v + 0.5)$ and $y_{RO_i} = (L_I - 1)\left(\frac{y_{V_i}}{L_v} + 0.5\right)$. In this way, we get the mapping relation between P_R and P_{RO}.

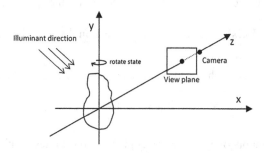

Fig. 2. Image-forming condition in ray tracing module.

Posture Matching Radar Output Images with Optical Image. Output images of radar model appear to be very different in diverse rotate states of radar model and diverse illuminant directions (Fig. 2). After rendering the radar model in different rotate states and illuminant directions, POV-Ray outputs a series of images of radar model, shown in Fig. 3. Then, we extract the contour feature of radar output images and the optical image, respectively. Feature extraction methods include invariant moment, chain code and boundary distance [9]. A particular posture is afterwards obtained by matching these contour features.

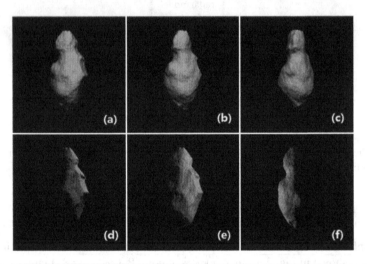

Fig. 3. (a)∼(c)Output images of POV-Ray in different rotate states. (d)∼(f) Output images of in different illuminant directions.

Map Between Vertices of Radar Model and Pixels of Optical Image. It was nearest neighbor fusion that Hanning et al. [10] adopted to set up a pixel level map between different data sources describe the same object. The Voronoi diagram [11] is used to set up a pixel level map between vertices of radar model and pixels of optical image here.

Let $S \subseteq \mathbb{R}^2$ be a finite set with $|S| \geq 3$. A Voronoi-cell $V_S(s)$ of $s \in S$ is the set of points with minimal Euclidean distance to s among all other points in S

$$V_S(s) = \{x \in \mathbb{R}^2 | \parallel x - s \parallel = \min_{y \in S} \parallel x - y \parallel\} \tag{1}$$

It can be shown that each $V_S(s)$ is a convex set. For $x \in \mathbb{R}^2$ the points

$$N_S(x) = \{s \in S | \parallel x - s \parallel = \min_{y \in S} \parallel x - y \parallel\} \tag{2}$$

are the nearest neighbors of x in S. Here, N_S is the Voronoi-tessellation of $\mathbb{R}^2$ w.r.t. S, and x is an inner point w.r.t. S if $|N_S(x)| = 1$; otherwise, x is a border point.

Let $\mathbb{R}^2$ enumerate all the pixels in P_I, and P_{RO_i} refers to all the pixels on the output image projected by P_{R_i}. We assume that all the pixels in $V_{P_{RO}}(P_{RO_i})$ are connected to P_{RO_i}, and $N_{P_{RO}}(P_{I_i})$ locates P_{RO_i} to which P_{I_i} is related. As shown in Fig. 4, $V_{P_{RO}}(P_{R_i})$ and $N_{P_{RO}}(P_{I_i})$ establishes a pixel level connection between P_I and P_{RO}. With the addition of mapping relationship between P_R and P_{RO}, function $c : P_I \rightarrow P_R$ is worked out.

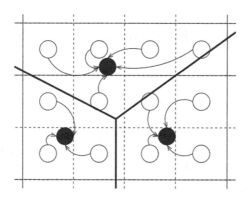

Fig. 4. Nearest neighbor assignment: Black circles represent pixels in $\mathbf{P_{RO}}$. White circles represent pixels in $\mathbf{P_I}$. And Bold lines represent edges of Voronoi-graph of $\mathbf{P_{RO}}$. Arrows point from a pixel $\mathbf{P_{I_i}}$ in $\mathbf{P_I}$ to pixels in $\mathbf{N_{P_{RO}}(P_{I_i})}$ providing a connection between $\mathbf{P_I}$ and $\mathbf{P_{RO}}$.

2.2 Depth Map Extraction from Optical Image

The 20,000 three-dimensional vertices [4] of radar-derived model provide the depth information directly. So the depth map extraction implements the 3rd dimensional value of depth from optical image. Shape from shading (SFS) algorithm is used to do so.

There has been extensive work on methods of SFS [12–14]. Due to Pentland's work [12], the direction of illumination is required to be known first, in obtaining accurate three-dimensional surface shape from two-dimensional shading, because the changes in image intensity are primarily a function of those in surface orientation relative to the illuminant.

Estimation of Illuminant Direction. Two classical solutions, in which tilt angle and slant angle involved, are commonly used to find the light source direction [12, 13]. Meanwhile, Lambertian reflectance and isotropical orientation assumptions are usually made before using them.

Tilt is defined as the angle between the projection of the surface normal on the image plane and the horizontal axis, and slant is defined as the angle between the surface normal and optical axis. Lambertian reflectance function is appropriate for the surface so that the image intensity I can be expressed as $I = \rho f(N \cdot L)$, where N is the surface normal, L is the illumination direction, V is the viewer's direction. If f is the flux emitted toward the surface, ρ is the albedo of the surface. And changes in surface orientation are usually assumed as isotropically distributed. Therefore, the change in

N is also isotropically distributed. That is, the sum of dN over all image points and orientations about those points has an depth map expected value of zero.

Prentland's method is used in this paper. When compute the expected value of dI along a particular image direction given n measurements of dI along the image direction, is therefore

$$E(dI) = \rho f \frac{1}{n-1} \sum_{}^{n} (dx_N dx_L + dy_N dy_L + dz_N dz_L)$$

$$= \rho f \frac{1}{n-1} \left(x_L \sum_{}^{n} dx_N + y_L \sum_{}^{n} dy_N + z_L \sum_{}^{n} dz_N \right) \qquad (3)$$

$$= \rho f \frac{1}{n-1} (x_L E(dx_N) + y_L E(dy_N) + z_L E(dz_N))$$

$$= \rho f (x_L d\bar{x} + y_L d\bar{y} + 0)$$

A maximum-likelihood estimation of the illuminant tilt can be expressed as follows: Let dI_i be the average of dI over the region along image direction (dx_N, dy_N).

Then $(d\bar{I}_1, d\bar{I}_2, \cdots, d\bar{I}_n)^T = \begin{pmatrix} dx_1 & dy_1 \\ dx_2 & dy_2 \\ \cdots & \cdots \\ dx_n & dy_n \end{pmatrix} \begin{pmatrix} \hat{x}L \\ \hat{y}L \end{pmatrix}$, where the tilt of the illuminant

direction is $actan(\hat{y}L/\hat{x}L)$. At the same time, let $k = \left[E(dI^2) - E(dI)^2 \right]^{1/2}$, then the illuminant direction is estimated as: $x_L = \hat{x}L/k, y_L = \hat{y}L/k, z_L = [1 - (\hat{x}L^2 + yL^2)/k^2]^{1/2}$.

Shape from Shading. The goal of SFS is to drive a 3D scene description from one image. In our task, a method of computing SFS based on gradient direction [13] is employed.

In the imaging process, the illumination coordinate system (short for S-image plane) can be derived from the viewer's coordinate system (short for V-image plane), the transformation between the two coordinate systems is

$$(x_L, y_L, z_L)^T = \begin{pmatrix} \cos \sigma_L \cdot \cos \tau_L & \cos \sigma_L \cdot \sin \tau_L & -\sin \sigma_L \\ -\sin \tau_L & \cos \tau_L & 0 \\ \cos \tau_L \cdot \cos \tau_L & \sin \sigma_L \cdot \sin \tau_L & \cos \sigma_L \end{pmatrix} (x, y, z)^T \qquad (4)$$

where σ, τ are the slant and tilt angle of illumination respectively and (x, y, z) is the V-image plane and T means transpose. Consider the surface of a sphere of radius R: $Z(x, y) = \sqrt{R^2 - x^2 - y^2}$ where $R > 0; -R \leq x \leq R; -R \leq x \leq R$. So, $Z_x = -y/\sqrt{R^2 - x^2 - y^2}, Z_y = -y/\sqrt{R^2 - x^2 - y^2}$. According to Theorem 2.1 of [13], the tilt of the surface normal viewed from the illumination coordinate system is $actanII_y/II_x$, where $(II_x, II_y)^T = \begin{pmatrix} \cos \sigma_L \cdot \cos \tau_L & \cos \sigma_L \cdot \sin \tau_L \\ -\sin \tau_L & \cos \tau_L \end{pmatrix} (I_x, I_y)^T$ and (I_x, I_y) are the first derivatives of image intensity. Then use a coordinate transformation to get the tilt of surface normal in viewer's system. From the above method, we have the illuminant direction; similarly, in the support of Theorem 2 in [13], we can get the slant angle of surface normal. What is more, In the Cartesian coordinate system, slant and tilt are the same with (x, y, z).

Hereby, we obtain the depth map of optical image. Figure 5(a) manifests the fluctuation in the depth map of optical image. Depth values in Fig. 5(a) are projected onto six planes with different height in Fig. 5(b). It can be perceived that most depth values are on the top of three planes, which constitute an ocean of detail information.

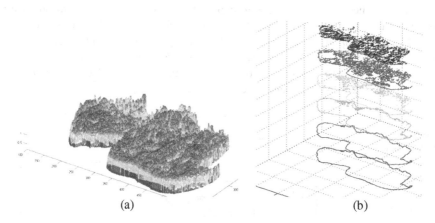

(a)   (b)

Fig. 5. (a) Depth map of optical image. (b) Planes depth values projected onto.

2.3   Data Fusion

The depth information obtained from optical image using SFS algorithm provides affluent detail feature information. In the frequency domain the high spatial frequencies are related to the details of the shape, low spatial frequencies are related to the coarse shape information [15]. Our method for combining radar model and optical image is pretty straightforward: keeping all the original frequency information of radar model and adding with the high frequency information of depth information from the optical image.

Hall and Hall [16] describes a model, which use the high/low frequency emphasis technique, to simulate the characteristics of human visual system. Hall and Hall's filters in the frequency domain are shown as follows:

$$\text{Low}(\omega) = \frac{2\alpha}{\alpha^2 + \omega^2} \tag{5}$$

and

$$\text{High}(\omega) = \frac{\alpha^2 + \omega^2}{2\alpha_0\alpha + (1 - \alpha_0)(\alpha^2 + \omega^2)} \tag{6}$$

where $\omega = \sqrt{\mu^2 + \nu^2}$. μ and ν represent the two dimensional frequencies in Fourier domain.

The flowchart of our method is shown in Fig. 6. What the difference between James's [8] method and ours is that we try to keep all the information of the depth map of radar model, while James's method reserve the low frequency information of depth map of shape from stereo.

The low frequency information of the optical image depth map is calculated by inverting Hall and Hall's high pass filter. By subtracting the low frequency information from the Fourier Transform of the optical image depth map, the high frequency information of the optical image depth map is worked out. Meanwhile, to produce a fused depth map we expected, the high frequency information of optical image depth map, which is multiplied by a parameter $\lambda(0 \leq \lambda \leq 1)$, is added with the depth map of radar model, see Fig. 6. Afterwards, the inverse Fourier Transform is applied to transform the combined frequency information into time domain. The combined depth map is normalized to preserve energy conservation finally.

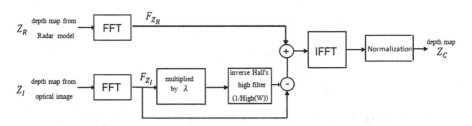

Fig. 6. Flowchart of proposed method for combining the depth values. Z_R, Z_I and Z_C stands for depth maps of radar model, optical image and the combined depth map.

Mathematically, the integration method can be described as:

$$F_{Z_C} = F_{Z_R} + \lambda \times F_{Z_I} \times \left(1 - High(\omega)^{-1}\right) \qquad (7)$$

where $High(\omega)^{-1}$ is the inverse of Hall and Hall's high pass filter, and F_{Z_C}, F_{Z_R}, and F_{Z_I} are the Fourier Transform of the combined depth map, radar model depth map and optical image depth map, respectively. λ acts as a role of adjusting the perception of the high frequency information of the optical image in the combined depth map.

3   Fused Model

The proposed method is tested on real data, including optical image and radar model. The combined depth information is taken under assumption of various λ according to Eq. (7). Optical image, radar model and combined model are shown in Fig. 7. Figure 7 (c)–(h) present the combined models.

Apparently, the detail information on the optical image has been added onto the combined models successfully, which reserve the original shape of the radar model. In Fig. 7(b) and (h) look the same with each other since $\lambda = 0$ means nothing has been added to the radar model, based on Eq. (7). The roughness of the fused model appears to be increasingly evident with the value of λ grows. It verifies the effect of λ.

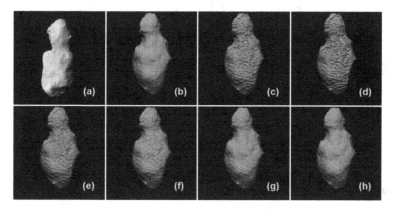

Fig. 7. (a) Optical image. (b) ~ (h) Are output images of ray tracing module. (b) Radar model. (c) ~ (h) Are combined model with different λ in Eq. (7). (c) $\lambda = \mathbf{1.0}$ (d) $\lambda = \mathbf{0.8}$ (e) $\lambda = \mathbf{0.6}$ (f) $\lambda = \mathbf{0.4}$ (g) $\lambda = \mathbf{0.2}$ (h) $\lambda = \mathbf{0}$.

Usually, point set deviation(PSD) statistical approach is used as a portrait of distribution uniformity for a sequence of points in value space. In analyzing the depth information of radar model or combined model in its depth value range, PSD takes a nice look on the value distribution.

PSD is defined in Eq. (8). Given a real sequence $\varsigma = \{a_1, a_2, \cdots, a_i, \cdots, a_n\}$ in interval $[0, 1)$, for $\forall \alpha, \beta \in [0, 1], \alpha < \beta$, $A([\alpha, \beta); n) = A([\alpha, \beta); n; \varsigma)$ indicate number of elements in the interval $[\alpha, \beta)$.

$$D_n = D_n(\varsigma) = \sup_{0 \le \alpha < \beta < 1} \left| \frac{A([\alpha, \beta); n; \varsigma)}{n} - (\beta - \alpha) \right| \tag{8}$$

PSD curve of the combined depth values and depth information of radar model are presented in Fig. 8. The lines of fused models largely move in tandem with the line of radar model. Moreover, deviations of the radar model and the fused models are nearly the same at some depth intervals, which indicate that the fused models maintain the low frequency component of radar model. What's more, several flat areas exist on the line of radar model, while no such flat area on that of combined models, which justifies, again, that the detail characteristics of the optical image gives betterness onto the final model.

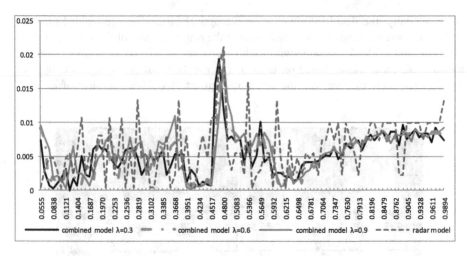

Fig. 8. Statistics of point set deviation of combined models and radar model. The horizontal coordinate is the depth intervals, the vertical one is the point set deviations.

4   Conclusion and Discussion

In this paper, we concentrate our mind in the integration of two data source to get a more accurate 3D-model, with abundant surface characteristics as we expected. According to the experimental results and the statistic of PSD, all the information of the radar model and high frequency from the optical image are both passed on to the fused model. And with the increasing of λ, the surface characteristics of the fused model are growing, while the dissimilarity to the original radar model is raising.

The method we used in finding the illuminant direction produces a good result in some degree. However, two assumptions were involved, which may lead to the inaccuracy of the fused model. In the future, we would like to use engineering parameters of imaging, under which the optical image of asteroid 4179 was taken, to get a better illuminant direction. One of the parameters is the angle formed by the Sun, Asteroid 4179 and the Probe Chang'E-2 (SAP). Also, the appropriate value of λ is uncertain at present. In other words, it remains unclear that how effective the optical image acting on the final model could satisfy the planetary scientists' expectation.

References

1. Huang, J., Ji, J., Ye, P., Wang, X., Yan, J., Meng, L., Wang, S., Li, C., Li, Y., Qiao, D., Zhao, W., Zhao, Y., Zhang, T., Liu, P., Jiang, Y., Rao, W., Li, S., Huang, C., Ip, W-H., Hu, S., Zhu, M., Yu, L., Zou, Y., Tang, X., Li, J., Zhao, H., Huang, H., Jiang, X., Bai, J.: The ginger-shaped asteroid 4179 Toutatis: new observations from a successful flyby of Chang'e-2. Nature-Scientific reports, 3 (2013)
2. Hudson, S.: Three-dimensional reconstruction of asteroids from radar observations. Remote Sens. Rev. **8**(1–3), 195–203 (1994)

3. Zou, X., Li, C., Liu, J., Wang, W., Li, H., Ping, J.: The preliminary analysis of the 4179 Toutatis snapshots of the Chang'E-2 flyby. Icarus **229**, 348–354 (2014)
4. Hudson, R.S., Ostro, S.J., Scheeres, D.J.: High-resolution model of asteroid 4179 Toutatis. Icarus **161**(2), 346–355 (2003)
5. Hudson, R.S., Ostro, S.J.: Shape and non-principal axis spin state of asteroid 4179 Toutatis. Sci. NY. Wash. **270**, 84 (1995)
6. Ostro, S.J., Hudson, R.S., Jurgens, R.F., Rosema, K.D., Campbell, D.B., Yeomans, D.K., Chandler, J.F., Giorgini, J.D., Winkler, R., Rose, R., Howard, S.D., Slade, M.A., Perillat, P., Shapiro, I.I.: Radar images of asteroid 4179 Toutatis. Sci. NY. Wash. 80 (1995)
7. Liu, P., Huang, J., Zhao, W., et al.: In: International Conference on "Asteroids, Comets, Meteors", Helsinki (2014)
8. Cryer, J.E., Tsai, P.S., Shah, M.: Integration of shape from shading and stereo. Pattern Recogn. **28**(7), 1033–1043 (1995)
9. Gonzalez, R.C., Woods, R.E.: Digital Image Processing. Prentice Hall, Upper Saddle River (2002)
10. Hanning, T., Lasaruk, A., Tatschke, T.: Calibration and low-level data fusion algorithms for a parallel 2D/3D-camera. Inf. Fusion **12**(1), 37–47 (2011)
11. Preparata, F.P., Shamos, M.I.: Computational Geometry: An Introduction. Springer, New York (1985)
12. Pentland, A.P.: Finding the illuminant direction. JOSA **72**(4), 448–455 (1982)
13. Lee, C.H., Rosenfeld, A.: Improved methods of estimating shape from shading using the light source coordinate system. Artif. Intell. **26**(2), 125–143 (1985)
14. Zhang, R., Tsai, P.S., Cryer, J.E., Shah, M.: Shape-from-shading: a survey. IEEE Trans. Pattern Anal. Mach. Intell. **21**(8), 690–706 (1999)
15. Ping-Sing, T., Shah, M.: Shape from shading using linear approximation. Image Vis. Comput. **12**(8), 487–498 (1994)
16. Hall, C.F., Hall, E.L.: A nonlinear model for the spatial characteristics of the human visual system. IEEE Trans. Syst. Man Cybern. **7**(3), 161–170 (1977)

Coherence Enhancing Diffusion for Discontinuous Fringe Patterns with Oriented Boundary Padding

Haixia Wang[1], Li Shi[1], Ronghua Liang[1(✉)], Yi-Peng Liu[1], and Xiao-Xin Li[2]

[1] College of Information Engineering, Zhejiang University of Technology, Hangzhou 310014, China
{hxwang, rhliang, liuyipeng}@zjut.edu.cn,
1187839694@qq.com
[2] College of Computer Science and Technology, Zhejiang University of Technology, Hangzhou 310023, China
mordekai@zjut.edu.cn

Abstract. Optical interferometric techniques are very attractive in various research and application fields for the non-contact, high accuracy and full filed measurements they offer. Fringe patterns, as the recorded results of these techniques, often require denoising at the pre-processing step to increase the accuracy and robustness of information retrieval process. Coherence enhancing diffusion (CED) has been an effective iterative and oriented denoising technique for continuous fringe patterns. However, when applied to discontinuous fringe patterns, CED tends to blur the discontinuous regions. This paper proposes an orientation based segmentation method, together with the oriented padding method, to adapt CED for discontinuous fringe patterns. Simulated fringe patterns are tested and quantitative results are given to demonstrate the performance of the proposed method. Experimental results are also given for verification.

Keywords: Optical interferometry · Image denoising · Oriented padding · Discontinuous fringe patterns

1 Introduction

Accurate and efficient measurement is widely required in the modern industry. Optical interferometric techniques [1] have been proven to be attractive in both research and engineering fields for non-contact, highly sensitive and full-field measurements. Various optical interferometric techniques have been designed for different application fields such as mechanical engineering and material engineering. Common results of these optical interferometric techniques are fringe patterns with the following discrete numerical presentation

$$f(x, y) = a(x, y) + b(x, y) \cos[\varphi(x, y)] + n(x, y) \tag{1}$$

© Springer International Publishing Switzerland 2015
X. He et al. (Eds.): IScIDE 2015, Part I, LNCS 9242, pp. 362–369, 2015.
DOI: 10.1007/978-3-319-23989-7_37

where $a(x,y)$, $b(x,y)$, $\varphi(x,y)$ and $n(x,y)$ are the background intensity, fringe amplitude, phase distribution and additive noise, respectively. Usually the phase $\varphi(x,y)$ is directly related to the measurement information and needs to be retrieved. The phase retrieval from a single closed fringe pattern is of interest but difficult, especially with the presence of noise, either additive noise or speckle noise.

Fringe pattern denoising is therefore often required as pre-processing [2, 3]. The cosine function in (1) with a continuously varying phase gives a flow-like structure to fringe patterns, which can be well measured by the fringe orientation [4–8]. Coherence enhancing diffusion [6, 8] is such an iterative and oriented denoising technique based on the partial differential equations. It is carried out through an iterative operation as follows

$$f(x,y;t+1) = f(x,y;t) + \beta \times f_t(x,y;t), f(x,y;0) = f_0(x,y), \qquad (2)$$

where $f(x,y;t)$ denotes the image intensity at pixel (x,y) and iteration t; $f_t(x,y;t)$ is the first order derivative of $f(x,y;t)$ with respect to time t; $f_0(x,y)$ is the original noisy fringe pattern; β is a positive constant; the symbol $\times$ denotes multiplication. A large iteration number ranging from 100 to 200 is quite common.

In (2), the term $f_t(x,y;t)$ controls the evolvement of the fringe pattern and plays an essential role during the denoising. This term can be generally represented as [8],

$$\begin{aligned}
f_t &= \lambda_1 (f_{xx} \sin^2 \theta + f_{yy} \cos^2 \theta - 2f_{xy} \sin \theta \cos \theta) \\
&+ \lambda_2 (f_{xx} \cos^2 \theta + f_{yy} \sin^2 \theta + 2f_{xy} \sin \theta \cos \theta) + r(f_x, f_y, \theta, \theta_x, \theta_y)
\end{aligned} \qquad (3)$$

where f_{xx}, f_{xy} and f_{yy} are second order derivatives of f; θ is the fringe orientation; λ_1 and λ_2 control the diffusion strength in the direction perpendicular and parallel to the fringe orientation, respectively; $r(f_x, f_y, \theta, \theta_x, \theta_y)$ is a complementary term to make the diffusion accurately follow the designed orientation, which involves first order derivatives f_x, f_y, θ_x, and θ_y [8].

The implementation of f_t involves the calculation of the fringe derivatives, which is estimated with the information of neighboring pixels. Oriented boundary padding [9] has provided a supplementary for CED to solve the boundary problem. However, when comes to the discontinuous fringe pattern, the CED tends to blur the discontinuous region due to the miscalculated orientation and derivatives of pixels in discontinuous region. To make CED applicable to discontinuous fringe patterns, there are two tasks needs to be performed. First, the discontinuity needs to be identified. Since the orientation in the CED is estimated by averaging neighboring pixels, the consistency of the orientation within the estimation region is a reasonable measure of the continuity of the fringe pattern. Second, special cares needs to be taken in the discontinuous regions, which is in fact the problem of irregular boundary padding.

Thus, an orientation based fringe pattern segmentation method is proposed to identify the discontinuous region of the fringe pattern. The oriented boundary padding is also adapted to irregular boundaries to help the CED to deal with discontinuous fringe patterns. Together a CED for discontinuity technique (DCED) is proposed. The rest of the paper is organized as follows. The orientation based segmentation is presented in Sect. 2; the oriented irregular boundary padding is presented in Sect. 3; results are given and discussed in Sect. 4; a conclusion is drawn in Sect. 5.

2   Orientation Based Segmentation

The orientation estimation method used in the CED is the gradient based orientation [10], which is expressed as

$$\bar{\theta} = \frac{1}{2}\arctan 2\left\{ -\sum_{u=x-\varepsilon}^{x+\varepsilon}\sum_{v=y-\varepsilon}^{y+\varepsilon} 2f_{x\sigma}(u,v)f_{y\sigma}(u,v), \sum_{u=x-\varepsilon}^{x+\varepsilon}\sum_{v=y-\varepsilon}^{y+\varepsilon}\left[f_{y\sigma}^2(u,v) - f_{x\sigma}^2(u,v)\right]\right\} \quad (4)$$
$$\in [-\Pi/2, \Pi/2]$$

where $f_{x\sigma}$ and $f_{y\sigma}$ are the fringe derivatives after Gaussian filtering with kernel size σ. Averaging of neighboring orientations is performed to increase the robustness in this method. In discontinuous region, the average will cancel out each other and bring a result that is not similar to either side of the discontinuous region. The orientation consistency with the avenging region can be used to identify the discontinuous regions.

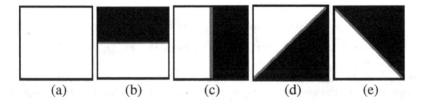

Fig. 1. Averaging areas

The averaging region initially is set as Fig. 1(a). For the same region, eight more averaging regions are set as Fig. 1(b–e), where the red region is the line pass center pixel and not included into estimation in this stage. The orientation difference between these eight orientations ($\theta_i, i = 1 : 8$) and the initial orientation is estimated. The regions with differences above threshold are identified as discontinues regions, which require further processing.

After the discontinuous regions are identified, the actual boundary of the discontinuity needs to be estimated. The identification of boundaries of discontinuity is performed in four steps.

(1) As Fig. 1(b–e) shows, the red region is now included into orientation calculation and eight new orientations are obtained ($\theta_{ci}, i = 1 : 8$).
(2) For every pixel in the discontinuous region, estimate the eight orientation differences as $d = |\theta_{ci} - \theta_i|$.
(3) Identify the pair that has largest orientation differences by estimating $|\theta_{ci=1,3,5,7} - \theta_{ci=2,4,6,8}|$. Since the orientations in the two sides of boundaries are supposed to be most different from each other.

(4) For the selected pair, the side with smaller d is regarded as the side the center pixel belongs.

(5) Repeat step 2 to step 4 until all pixels in the discontinuity region are identified.

3   Oriented Padding Method for Irregular Boundaries

Oriented boundary padding is a technique that increases the image size by appending pixels in the boundary regions along fringe orientation of existing boundary pixels [9]. The original method is for fringe patterns with regular boundary regions. After the segmentation, the fringe pattern is generally divided into more than one fringe region with irregular boundaries. To make sure the discontinuous regions are not affected by the CED, the oriented padding is applied to extend the new boundaries, which is irregular. Two modifications are required to make the oriented padding applicable for irregular fringe patterns. First, the connected boundaries to form the padding matrix need to be identified. Second, the padding matrix is performed for closed boundary instead of one side at a time in the original method. The identification of the connected boundaries is easy by Morphological Operators [11].

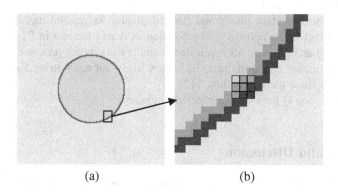

(a)   (b)

Fig. 2. Irregular boundary padding (Color figure online)

As Fig. 2(a) shows, the blue area is the known fringe pattern, the green line indicates the boundary pixels while the red lines indicates the padded pixels. As enlarged and shown in Fig. 1(b), for every boundary pixel, depends on the shape of the boundary line, there are a number of known pixels (6 in the example including the boundary pixel itself) and a number of padded pixels (3 in the example). Based on the orientation of the boundary pixel, the following equation can be established as

$$
\begin{aligned}
&(1 - |\gamma|)(1 - |\eta|)f(x, y) + |\gamma|(1 - |\eta|)f[x + sign(\gamma), y] \\
&+ (1 - |\gamma|)|\eta|f[x, y + sign(\eta)] + |\gamma||\eta|f[x + sign(\gamma), y + sign(\eta)] \approx \\
&(1 - |\gamma|)(1 - |\eta|)f(x, y) + |\gamma|(1 - |\eta|)f[x - sign(\gamma), y] \\
&+ (1 - |\gamma|)|\eta|f[x, y - sign(\eta)] + |\gamma||\eta|f[x - sign(\gamma), y - sign(\eta)]
\end{aligned}
\tag{5}
$$

which can be simplified as

$$|\gamma|(1 - |\eta|)f[x + sign(\gamma), y] + (1 - |\gamma|)|\eta|f[x, y + sign(\eta)] + |\gamma||\eta|f[x + sign(\gamma), y + sign(\eta)] \approx$$
$$|\gamma|(1 - |\eta|)f[x - sign(\gamma), y] + (1 - |\gamma|)|\eta|f[x, y - sign(\eta)] + |\gamma||\eta|f[x - sign(\gamma), y - sign(\eta)]$$

(6)

where

$$\gamma = \pm\alpha\cos\theta(x, y), \eta = \pm\alpha\sin\theta(x, y) \tag{7}$$

Six pixels are involved on Eq. (6). Some of them are padded pixels. For every boundary pixels, such an equation can be established. Unknown padded pixels are shared between neighboring pixels. Totally n_1 (the number of boundary pixels) equations are established with total n_2 (the number of padded pixels) unknowns. Solving this series of equations can gives the values of the padded pixels, which can forms a matrix equation AB = C with B as the unknowns. Therefore, A is of size $n_1 \times n_2$ and B is of size $n_2 \times 1$. Every entry of B is a padded pixel. The total process can be presented as

(1) For a boundary pixel (x, y), establish Eq. (6);
(2) Identify the boundary pixels and padded pixels. Assign the multiplier of each padded pixel to the corresponding position in A and the rest in C;
(3) Repeat (1) and (2) until all connected boundary pixels are processed;
(4) Solve the matrix equation. Since n_1 is very likely not equal to n_2, Pseudo-inverse is used to solve these equations [12].
(5) Repeat (1) to (4) for every connected boundary.

4   Results and Discussion

The proposed DCED technique for discontinuous fringe patterns is applied to simulated and experimental fringe patterns, where the proposed technique is compared with CED using only zero or symmetric padding. The mean square errors (MSEs) between the denoised results and true value are calculated to evaluate the performance. A simulated fringe pattern (200×200) with unit amplitude, orientation and density is shown in Fig. 3(a). Half of the phase is quadratic and half is linear. Additive noise with mean of zero and standard derivations (STD) of 0.5, and speckle noise [13] are simulated into the fringe pattern, which are shown in Fig. 3(b), (c), respectively. The CED is applied with $\lambda_1 = 0.005$ and $\lambda_2 = 1$ for 100 iterations for all simulated examples.

Figure 4(a–c) show the denoising results of the fringe pattern in Fig. 3(c) using the proposed DCED, CED with zero padding and symmetric padding, respectively. The MSEs are shown in Table 1. It can be seen that the DCED has preserved the original structure of the discontinuous regions even severe noise is presented. The performance is better than CED with zero or symmetric padding.

The next simulated fringe pattern (200×200) has circular discontinuity as shown in Fig. 5(a). The same levels of additive noise and speckle noise are simulated into the

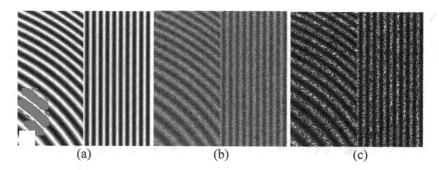

Fig. 3. Simulated fringe patterns. (a) Noiseless fringe pattern; (b) fringe pattern with additive noise; (c) fringe pattern with speckle noise.

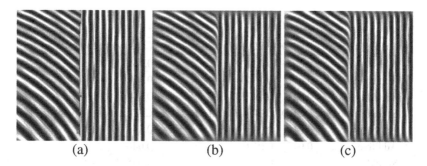

Fig. 4. Denoising results of Fig. 3(c). (a) DCED result; (b) CED with zero padding; (c) CED with symmetric padding.

fringe pattern, which are shown in Fig. 5(b), (c), respectively. The CED is applied with the same parameter setting. The denoising results of the fringe pattern with speckle noise in Fig. 5(c) using DCED, CED with zero padding and symmetric padding are shown in Fig. 6(a–c), respectively. The MSEs are given in Table 1 as well. The

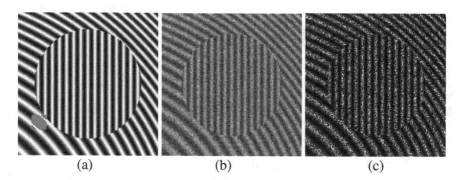

Fig. 5. Simulated fringe patterns. (a) Noiseless fringe pattern; (b) fringe pattern with additive noise; (c) fringe pattern with speckle noise.

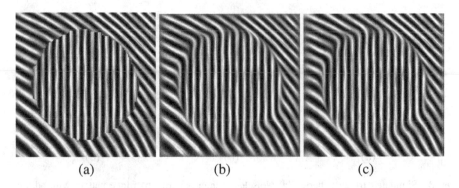

Fig. 6. Denoising results of Fig. 5(c). (a) DCED result; (b) CED with zero padding; (c) CED with symmetric padding.

Table 1. Mean square errors

First example	DCED	CED with zero padding	CED with zero padding
Noiseless	0.0053	0.029	0.031
Additive noise	0.012	0.034	0.036
Speckle noise	0.021	0.042	0.044
Second example	DCED	CED with zero padding	CED with zero padding
Noiseless	0.010	0.043	0.041
Additive noise	0.018	0.049	0.046
Speckle noise	0.029	0.057	0.055

performances of the proposed method are again seen to be better than with zero padding and symmetric padding.

At last, an experimental fringe pattern (240×240) from fringe pattern projection technique is shown in Fig. 7(a). The CED is applied to it with $\lambda_1 = 0.005$ and $\lambda_2 = 1$ for 80 iterations. The denoising results using the DCED and CED with zero padding

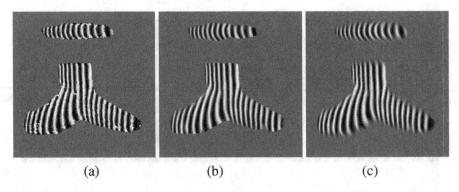

Fig. 7. Experimental fringe pattern. (a) Original fringe pattern; (b) DCED result; (d) CED with zero padding.

are shown in Fig. 7(b–c), respectively. In this case, using zero or symmetric padding produces identical result. The DCED result seems better. The experimental results are considered as consistent with the simulation examples.

5 Conclusion

A coherence enhancing diffusion for discontinuous fringe patterns (DCED) is proposed with orientation based segmentation and oriented padding. The orientation based segmentation identifies the boundaries of discontinuous regions by examining the consistency of orientation within the averaging window. The oriented padding is adapted to irregular boundaries with rearrangement of the matrix equation. The simulation and experimental results show that the DCED can actually solving the noise problem of discontinuous fringe patterns.

Acknowledgment. This research is partially supported by Natural Science Foundation of China (61402411) and Zhejiang Provincial Natural Science Foundation (LY14F020014, LY14F020015).

References

1. Robinson, D.W.: Interferogram Analysis: Digital Fringe Pattern Measurement Techniques. Institute of Physics, Bristol (1993)
2. Wang, H., Kemao, Q.: Frequency-guided demodulation for a single fringe pattern. Opt. Express **17**, 15118–15127 (2009)
3. Li, K., Kemao, Q.: Fast frequency-guided sequential demodulation of a single fringe pattern. Opt. Lett. **35**, 3718–3720 (2010)
4. Yu, Q., Sun, X., Liu, X., Qiu, Z.: Spin filtering with curve windows for interferometric fringes. Appl. Opt. **41**(14), 2650–2654 (2002)
5. Tang, C., Han, L., Ren, H., Zhou, D., Chang, Y., Wang, X., Cui, X.: Second-order oriented partial-differential equations for denoising in electronic-speckle-pattern interferometry fringes. Opt. Lett. **33**(19), 2179–2181 (2008)
6. Wang, H., Kemao, Q., Gao, W., Soon, S.H., Lin, F.: Fringe pattern denoising using coherence-enhancing diffusion. Opt. Lett. **34**(8), 1141–1143 (2009)
7. Villa, J., Quiroga, J.A., Rosa, I.: Regularized quadratic cost function for oriented fringe-pattern filtering. Opt. Lett. **34**(11), 1741–1743 (2009)
8. Wang, H., Kemao, Q.: Comparative analysis on some spatial-domain filters for fringe pattern denoising. Appl. Opt. **50**(12), 1687–1696 (2011)
9. Wang, H., Kemao, Q., Liang*, R., Wang, H.: Oriented boundary padding for iterative and oriented fringe pattern denoising techniques. Sig. Process. **102**, 112–121 (2014)
10. Zhou, X., Baird, J.P., Amold, J.F.: Fringe-orientation estimation by use of a Gaussian gradient filter and neighboring-direction averaging. Appl. Opt. **38**(5), 795–804 (1999)
11. Serra, J.: Image Analysis and Mathematical Morphology. Academic Press, San Diego (1982)
12. Golub, G.H., Van Loan, C.F.: Matrix Computations, 3rd edn, pp. 486–497. Johns Hopkins, Maryland (1996)
13. Kaufmann, G.H., Galizzi, G.E.: Speckle noise reduction in TV holography fringes using wavelet thresholding. Opt. Eng. **35**(1), 9–14 (1996)

A Cooperative Fusion Method of Multi-sensor Image Based on Grey Relational Analysis

Runping Xi[1($\boxtimes$)], Liangshui Jin[1], Yanning Zhang[1], Fujun Zhang[1],
Gen Yang[1], and Miao Ma[1,2]

[1] School of Computer Science and Engineering, Shaanxi Key Laboratory
of Speech and Image Information Processing,
Northwestern Polytechnical University,
Xi'an 710072, People's Republic of China
xrp@nwpu.edu.cn
[2] College of Computer Science, Shaanxi Normal University, Xi'an 710062,
People's Republic of China

Abstract. Current multi-sensor image fusion algorithms have difficulties working at many different scenes. In this paper, a cooperative fusion algorithm based on the grey correlation theory and automatic evaluation feedback is proposed in order to achieve better fusion results. A synergetic mechanism is introduced to adjust the algorithm parameters for better fusion results. Experiments show that the collaborative integration of the results is consistent with the subjective evaluation of human eyes.

Keywords: Cooperative fusion · Automatic evaluation · Grey relational analysis

1 Introduction

Nowadays an increasing number of researchers tend to put an eye on image fusion, which is a state-of-the-art image enhancement method, since image sensor technology evolved from a single sensor model into multi-sensor models. Multi-sensor means more information are recorded and extracted. Image fusion is to generate a new image, which contains richer visual information than source image via the complementary information of different source image from the same scene.

Infrared and visible cameras are the common devices for image acquisition. Infrared cameras have optimum environmental adaptability, which are operable at night or in bad weather, and good for hiding, strong anti-jamming, etc. However, due to the constraints of the imaging mechanism, infrared image is blurry, noisy and low in contrast. Although the visible images always have high spatial resolution, they might be strongly influenced by the weather and illumination. To take the full use of the complementary information of the infrared and visible images, the two are integrated to obtain a higher confidence.

Over the past few years, various fusion methods have been developed [1–5]. According to the image fusion process, the fusion method can be divided into three classes including pixel level, feature level and decision level in principle [6]. Pixel level

X. He et al. (Eds.): IScIDE 2015, Part I, LNCS 9242, pp. 370–379, 2015.
DOI: 10.1007/978-3-319-23989-7_38

fusion can obtain better visual information by combining multiple sources. Feature level fusion must extract some features to process source images [7]. Decision level fusion is firstly to identify, and then correlate the results of the various decisions [8]. However, these methods need to be further improved to obtain a higher fusion quality and to be adaptive to different scenes. The most appropriate methods are rarely publicly reported due to its great significance in military application.

In this paper, we propose a novel cooperative fusion scheme, as well as automatic evaluation method to supervise algorithm to improve fusion results. The outline of the paper is as follow: in Sect. 2, we present comparative results of some fusion methods. In Sect. 3, we present general framework for our algorithm. In Sect. 4, the auto-evaluation algorithm is proposed. Section 5 provides the result of our fusion algorithm and we compared the experiment results with some other algorithms Sect. 6 presents the concluding remarks for this paper.

2 General Approaches to Fusion Image

In this section, we will provide and compare the results of the common fusion approaches. The data set comes from OSU Database [9]. Figure 1 is the experiment results.

By a subjective comparison of the fusion results shown in Fig. 1, we choose PCA [10, 11] and HSI [12] as the primary fusion algorithm of our principal fusion algorithm for their quick and high quality fusion effect.

HSI method is commonly used in image fusion field. Its principle idea is as follow: Firstly, the three-band image information in RGB color space must be transformed to the three parameters–hue, saturation and intensity in HSI space, in which H represents spectral information and I represents spatial information. Secondly, the infrared image contrast must be stretched to obtain the same mean and variance of intensity parameter with the visible image. Finally, instead of the original gray value, the stretched grey value is selected to implement HSI inverse transformation with hue and saturation for fusion image.

Specifically to this application environment, we find that the structure of intensity information is clear in the infrared image. If we continue to stretch the contrast for the same mean and variance of intensity parameter with the visible image, the high intensity information in the fusion image will be compressed. As shown in the left of Fig. 1(m), the contrast of intensity information decreases between pedestrians and the roof. If intensity information is directly replaced, the high intensity information can be retained as well as the algorithm's time cost is reduced. The results are shown in Fig. 1(o).

PCA method has similar principle with HSI method. In PCA, we must first compute correlation matrix between images with 3 or more than 3 bands' data. And principal component images are obtained by computing eigenvalues and eigenvectors with correlation matrix. Then we must stretch the visible image contrast to obtain the same mean and variance with the infrared image. Finally, instead of the first principal component, the stretched visible image is selected to implement PCA inverse transformation with other principal components for fusion image.

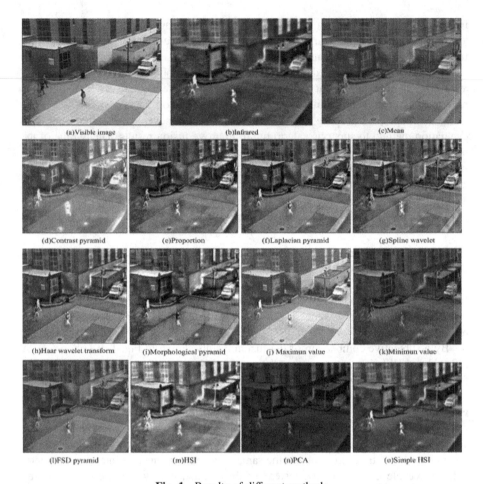

Fig. 1. Results of different methods

Compared with the HSI method, PCA method can lead to slighter distortions of spectral features and retain more spectral features of source images. Besides, PCA method can use more than 3 bands' image for fusion, but HSI method can use only 3 bands.

3 Overview of Our Approach

In order to obtain better fusion data, the cooperative method cannot simply be applied to algorithms or formulas. It needs to design cooperative processing mechanism via proceeding from the actual scene.

Our algorithm is a machine for clients. We use different fusion algorithms (assuming the number of the algorithms is N), to compute different fusion images, and then to evaluate fusion images based on grey correlation analysis [12, 13] through performance parameters. During the design process, the synergetic mechanism is introduced between different fusion algorithms for intelligently coordinating image fusion process, and automatically and timely evaluating algorithm performance.

Detailed algorithm process is shown in Fig. 2. First, images from two sensors must be fused with different fusion strategies. Then, we must evaluate fusion quality of different fusion results, and adjust the fusion image with better performance according to the evaluation results, and rebuild the fusion images and evaluate the fusion quality again until the fusion images with better performance in the two adjacent fusion processes are stable. At that time, we consider the final fusion image as optimal performance fusion result. Finally, cooperative fusion image is outputted.

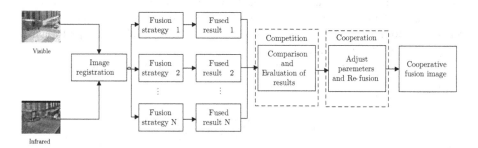

Fig. 2. The collaborative algorithm flow chart

In the algorithm, the competition module is responsible for the calculation of performance indices of every fusion image. The performance indices will be shown in Sect. 4.2. It selects and calculates indices, with the weights fixed (in our experiment, the balanced weights were used) according to the specific application environment. If there is prior knowledge such as expert experience as a guide, we can address the above process by adjusting relevance weights.

Automatic Evaluation of Fusion Results. Automatic evaluation method is based on grey correlation analysis. First, we determine the evaluation performance indices and the corresponding weights according to prior knowledge such as the fusion purpose, expert experience and research interest, etc. Then, we form a comparative sequence using performance indices which come from evaluating fusion algorithms, and select the optimal feature as a reference sequence according to the types of indicators. Finally, through the weighting correlation evaluate fusion algorithms between the two kinds of sequences.

Keep information Feedback and Cooperative. This step is to achieve stable fusion image quality. The feedback information by oscillation adjustment passes to different fusion algorithms to obtain new fusion image. Its purpose is to prevent the probable instability of the first optimal algorithm performance. Then, new indices of fusion images are re-computed. If the best image is the same as the first optimal image, it can be outputted as the final cooperative fusion image with ideal fusion effect and stable fusion performance. Otherwise, it will be not outputted.

The two steps mentioned above are the key steps of cooperative fusion algorithm. Compared with previous methods, the algorithm adds interactive superiority information, and obtains better fusion results based on this information to adjust integration parameters.

4 Automatic Evaluation of Fuse Image

4.1 Fusion Image Performance Index

In general, performance evaluation can be divided into subjective evaluation and objective evaluation [13]. Subjective evaluation refers to human visual perception of an fusion image. Objective evaluation measures the performance of fusion image via quantitative formula. In this paper, we need to automatically evaluate the performance of algorithm, so, the objective evaluation is chosen.

Here, we will consider various performance evaluation indices as follow:

- Mean

$$\overline{F} = \frac{1}{M \times N} \sum_{i=1}^{M} \sum_{j=1}^{N} F(i,j) \tag{1}$$

M and N are respectively row and column of the fusion image. $F(i,j)$ is a gray value at point (i,j). Generally, mean value is difficult to be used to judge the merits of fusion image.

- Standard deviation

$$SD = \sqrt{\frac{\sum_{i=1}^{M} \sum_{j=1}^{N} [F(i,j) - \overline{F}]^2}{M \times N}} \tag{2}$$

The standard denotes discrete degree of image.

- Entropy

$$E = - \sum_{l=0}^{L-1} p_i \log p_i \tag{3}$$

p_i is gray distribution. The greater image entropy is, the better the performance of the fusion will be.

- Cross Entropy

CE_{AF} and CE_{BF} are the cross entropy between source image A, Band fusion image F.

$$CE_{AF} = \sum_{l=0}^{L-1} p_{Ai} \log \frac{p_{Ai}}{p_{Fi}} \tag{4}$$

$$CE_{BF} = \sum_{l=0}^{L-1} p_{Bi} \log \frac{p_{Bi}}{p_{Fi}} \tag{5}$$

The formula of accumulated difference between the source image and image fusion is as (6):

$$CE_{F-AB} = \sqrt{\frac{CE_{AF}^2 + CE_{BF}^2}{2}} \tag{6}$$

The smaller the cross entropy is, the better fusion performance will be.
- Degree deviation
  Degree deviation measures matching degree in the spectral information between source images and fused images.

$$DDR = \frac{1}{M \times N} \sum_{i=1}^{M} \sum_{j=1}^{N} \frac{|F(i,j) - R(i,j)|}{F(i,j)} \tag{7}$$

$R(i,j)$ is gray value of source image R at point (i,j). The lower deviation degree the more spectral information of the fusion image.
- Spectral distortion
  Spectral distortion measures difference in spectral information.

$$D = \frac{1}{M \times N} \sum_{i=1}^{M} \sum_{j=1}^{N} |F(i,j) - R(i,j)| \tag{8}$$

The lower spectral distortion indicates the lower difference and the better performance.

4.2   Grey Relational Analysis Evaluation

In Sect. 3, we propose to determine the optimal fusion method via grey relational analysis [14–16] which uses performance indices analyzing objects. Because the performance indices are not always in the same order of magnitude, it's necessary to apply dimensionless method to them. We choose Deng's correlation degree as correlation model and fix the weights based on the importance degree, determined by prior knowledge such as fusion purpose, expert experience and research interest.

Select the Performance Index and Weights. Assuming the number of performance indices are N. According to the prior knowledge, we set the index's weight $w(k)$. At a specific application, $w(k)$ is affirmatory, and the sum of the index's weight is 1, that is $\sum_{k=1}^{N} w(k) = 1$.

Choose the Reference Sequence. Suppose the number of fusion method is I, and the comparative sequence is $\{x_j | j = 1, \ldots, I\}$, and the performance index is $\{x_0(k) | k = 1, \ldots, N\}$. Selecting the best performance features form reference sequence $x_0(k)$ after calculating fusion images.

Compute the Grey Correlation Degree. In this step, we need to compute the weighting correlation coefficient between comparative sequence $x_j(k)$ and reference sequence $x_0(k)$.

$$\xi_{0j} = \frac{\Delta\min + \varsigma\Delta\max}{\Delta0j(k) + \varsigma\Delta\max} \tag{10}$$

And

$$\begin{cases} \Delta\min = \min_j \min_k |x_0(k) - x_j(k)| \\ \Delta\max = \max_j \max_k |x_0(k) - x_j(k)| \\ \Delta0j(k) = |x_0(k) - x_j(k)| \end{cases} \tag{11}$$

$\zeta = 0.5$, then the correlation degree is as follows:

$$R_{0j} = \sum_{k=1}^{N} \xi_{0j}(k) \times w(k) \tag{12}$$

Assess the Algorithm. Ordering all the grey correlation degree and confirm the performance level of all methods.

5 Experiments and Analysis

Here is the process of our cooperative fusion algorithm:

Step 1: Use PCA and HSI fusing algorithm to fuse the visible pictures and infrared pictures respectively.

Step 2: Calculate the assessment vector, which consists of the mean, the standard deviation, entropy, cross entropy, deviation degree and spectral distortion of the fusion images, and then get the first time fusion gray association degree (GAD).

Step 3: Mark the optimal algorithm by their association degree calculated in Step 2.

Step 4: Adjust the original images' parameters by increasing or decreasing its intensity, contrast, hue and saturation.

Step 5: Redo Step 1 and Step 2.

Step 6: Readjust the images' parameters according to last step's fusion results. The adjustment strategies are as follows: if the fusion image's gray association degree decreases, then adjust the fusion image to the right about by a smaller step. Else, continue its original direction.

Step 7: Stop the algorithm when two optimal algorithms are same and the fusion image's gray association degree is undulating in a reasonable range.

A group of fusion image is shown in Fig. 3.

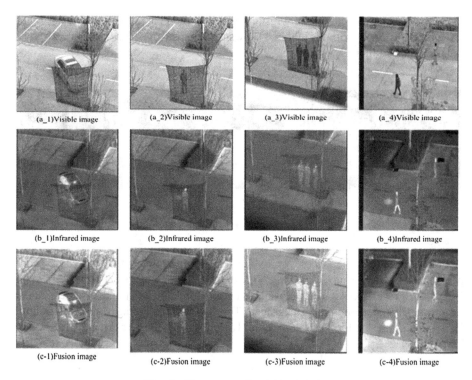

Fig. 3. The cooperative fusion results

According to the algorithm experiment steps, we get the fusion process data and put them in Table 1. In our experiments we use PCA fusion algorithm and HSI fusion algorithm to fuse the visible image and infrared image. The fusion results are PCA_0 and HSI_0. After calculating the gray association degree, we find HSI_0 is better. Then we adjust the input images and get the cooperative fusion image PCA_1−, HSI_1−, PCA_1+ and HSI_1+. The former two images are the results of decreasing the parameters while the latter are the results of increasing the fusing parameters. This time we get HSI_1+, whose gray association is the biggest. HSI_1+ is better, indicating that to enhance the input images' intensity, contrast, hue and saturation may lead to better fusion results. So we continue the fusion process and get HSI_2+ and PCA_2+. Unfortunately the fusion images' gray association degree has decreased but the fluctuation is under the threshold,so the fusion process reaches the end.

Table 1. Gray association degree of cooperative fusion process

GAD	PCA_0	HSI_0	PCA_1−	HSI_1−	PCA_1+	HSI_1+	PCA_2+	HSI_2+	Optimal
First	0.4722	0.8056	——	——	——	——	——	——	HSI_0
Second	0.5994	0.5841	0.5502	0.5228	0.6113	0.8056	——	——	HSI_1+
Third	0.5654	0.5235	0.6660	0.6830	0.7049	0.6479	0.6227	0.7984	HSI_2+

In order to test our algorithm's fusion results, we compared the fusion results with some other fusion algorithms. Figure 4 shows the comparison result and Table 2 shows the performance indices.

| PCA | HSI | shearlet | NSCT | ours |

Fig. 4. Fusion results of different algorithms

We can see that our fusion results are reasonably close to HSI but show better performance in details and some targets are enhanced. In comparison with shearlet and NSCT, their results look better in details but with more dark color and in some area the targets are almost close to the background. To show more advantages of our algorithm, let's examine the performance indices in Table 2 and the gray association degree in Table 3.

Table 2. Performance indices

Performance indices	PCA	HSI	Shearlet	NSCT	Ours
mean	66.9752	67.1986	91.7904	91.7893	67.2891
Standard deviation	61.8327	61.2606	49.0550	48.8684	61.2136
entropy	7.1419	7.2555	7.4658	7.4553	7.2823
Cross entropy	61984.4561	65119.0710	81841.6200	231178.1705	64433.8352
Deviation degree	0.3052	0.2992	0.1526	0.1512	0.2982
Spectral distortion	13.5832	13.2495	7.0740	7.0021	13.2180
Time cost	1.7155 s	0.5071 s	150.9796 s	229.5389 s	4.5312 s

Table 3. Gray association degree of different fusion methods

Fusion method	PCA	HSI	Shearlet	NSCT	Ours
GAD	0.7353	0.7490	0.6493	0.5607	0.7360

From Table 3 we can see that our fusion results' weighted gray association degree is larger than most of the rest. It's close to HSI and PCA because we choose PCA and HSI as the cooperative fusion algorithms, and the optimal fusion image is the result of HSI with adjusted input images. The process of the adjustment of the input images includes PCA fusion and HSI fusion, so the fusion time is much longer. Overall, our algorithm can produce a better fusion image with the gray association theory.

6 Conclusions

A fusion algorithm is proposed in this paper based on grey correlation theory in conjunction with auto-evaluation method. The algorithm changes some adjustable parameters to get better results according to the quality of general picture and detail information of the algorithm to auto-evaluation results. The collaborative fusion experiment shows that our algorithm is able to achieve better consistency with subjective evaluation of the human eye.

Acknowledgments. This work is supported by the National High-Tech Research and Development Program of China (863 Program) (SS2015AA010502), Shaanxi Provincial Natural Science Foundation (2013JM8027), the NPU Foundation for Fundamental Research (JC201148).

References

1. Li, H., Manjunath, B.S., Mitra, S.K.: Multi sensor image fusion using the wavelet transform. Graph. Models Image Process. **57**(3), 35–245 (1995)
2. Piella, G.: A general framework for multiresolution image fusion: from pixels to regions. Inform. Fusion **4**, 259–280 (2003)
3. Sun, J., Zhu, H., Xu, Z., et al.: Poisson image fusion based on Markov random field fusion model. Inf. Fusion **14**(3), 241–254 (2013)
4. Cvejic, N., Bull, D., Canagarajah, N.: Region-based multimodal image fusion using ica bases. IEEE Sens. J. **7**(5), 743–751 (2007)
5. Amolins, K., Zhang, Y., Dare, P.: Wavelet based image fusion techniques—an introduction, review and comparison. ISPRS J. Photogrammetry Remote Sens. **62**(4), 249–263 (2007)
6. Yang, B., Li, S.: Pixel-level image fusion with simultaneous orthogonal matching pursuit. Inf. Fusion **13**(1), 10–19 (2012)
7. Piella, G.: A general framework for multi resolution image fusion: from pixels toregions. Inf. Fusion **4**(4), 259–280 (2003)
8. Khaleghi, B., Khamis, A., Karray, F.O., et al.: Multisensor data fusion: a review of the state-of-the-art. Inf. Fusion **14**(1), 28–44 (2013)
9. OTCBVS Benchmark Dataset Collection. http://www.cse.ohio-state.edu/otcbvs-bench/
10. Patil, U., Mudengudi, U.: Image fusion using hierarchical PCA. In: Image Information Processing (ICIIP), pp. 1–6 (2011)
11. He, C., Liu, Q., Li, H., et al.: Multimodal medical image fusion based on IHS and PCA. Procedia Eng. **7**, 280–285 (2010)
12. Daneshvar, S., Ghassemian, H.: MRI and PET image fusion by combining IHS and retina-inspired models. Inf. Fusion **11**(2), 114–123 (2010)
13. Toet, A., Frankenb, E.M.: Perceptual evaluation of different image fusion schemes. Displays **24**, 25–37 (2003)
14. Xi, Runping, Zhang, Yanning, Yang, Gen: Automatic evaluation method based on the detection results of the evaluation factors and gray correlation analysis. J. Northwest. Polytechnical Univ. **3**, 421–424 (2009)
15. Deng, J.: Grey System Method. Huazhong University of Science Press, Wuhan (1996)
16. Julong, Deng: Control problems of grey system. Syst. Control Lett. **1**(5), 288–294 (1982)

A Self-adaption Fusion Algorithm of PET/CT Based on DTCWT and Combination Membership Function

Tao Zhou[1](✉), Xingyu Wei[1], Huiling Lu[1], and Pengfei Yang[2]

[1] School of Science, Ningxia Medical University, Yinchuan 750004, China
zhoutaonxmu@126.com
[2] Department of Nuclear Medicine, General Hospital of Ningxia Medical University, Yinchuan 750004, China

Abstract. Multi-modality medical image fusion have great value for image analysis and clinical diagnosis, it can enrich medical image information and improve information accuracy by fusing PET/CT medical images. A self-adaption fusion algorithm of PET/CT based on DTCWT and combination membership function is proposed by this paper. Firstly, using DTCWT to decompose registered PET and CT image, and get low-frequency and high-frequency components; Secondly, According to these characters, such as concentrating most energy in low frequency sub-band of the source image and determining image contour, thinking carefully lesions area are smaller in the whole image, How to deal with background of medical image is becoming more critical for highlighting lesions. So the low-frequency components are fused by self-adaption combination membership function. According to the characteristics of high-frequency sub-bands can reflect detail and edge information about medical image, regional energy fusion rule is adopted in high-frequency sub-bands. This paper did two experiments in PET-CT fusion image of lung cancer. (1) Comparison experiment of the algorithm and other pixel-level fusion algorithms; (2) Fusion effect evaluation experiment by objective indicators. The experimental results shown that the algorithm can better retain edge and texture information of lesions.

Keywords: PET/CT · Image fusion · Dual-tree complex wavelet · Combination membership function · Self-adaption

1 Introduction

Medical image fusion is the process of space registering, superposing, transforming and combining multiple images from single or multiple imaging to form a new comprehensive images which is combined anatomical and functional information [1]. Bazelaire [2] antidiastoled inflammatory breast cancer and acute mastitis by MRI and PET/CT images, the results showed that PET/CT could provide richer information and more accurate diagnosis than MRI and also improve the reliability of clinical diagnosis. Because of medical images with high resolution, rich details, border crisscross between organs and tissues and fuzzy and so on, the medical image fusion must take full

© Springer International Publishing Switzerland 2015
X. He et al. (Eds.): IScIDE 2015, Part I, LNCS 9242, pp. 380–391, 2015.
DOI: 10.1007/978-3-319-23989-7_39

account of these factors. In order to get more details information, wavelet transform is an effective means which is multi-resolution, multi-scale and multi-directional method. Such as Tao [3], Lin tao [4] applied wavelet transform to medical image fusion. But wavelet transform the following disadvantages:

(1) For high-frequency components only have three directions which are 0°, 45° and 90°.
(2) Lack of shift invariance.

According to the defects of wavelet transform, Kingsbury [5] proposed the dual-tree complex wavelet transform (DTCWT). He designed a pair of wavelet functions $\psi_h(x)$ and $\psi_g(x)$ which meet Hilbert transform, and constructed complex wavelet function $\psi(x) = \psi_h(x) + i\psi_g(x)$ with the two functions. Then reconstructed the original images through inverse dual-tree complex wavelet transform using complex filter. DTCWT which is a kind of multi-resolution, multi-scale, multi-directional image representation method. It can operate and describe the directions of the image more accurately, because it not only inherits the multi-resolution characteristics and good time localization analysis ability of traditional wavelet transform, but also has the shift invariance, good directional selectivity and limited redundancy. Song [6] took multi-resolution decomposition and reconstruction of multi-focus images by dual-tree complex wavelet, the results of experiment proved that the effect of fusion image can be better by this method. Zhong [7] fused medical images by dual-tree complex wavelet transform. David and Santika [8] proposed a computer aided diagnosis system based on neural network and dual-tree complex wavelet transform for auxiliary diagnosis of breast cancer, the results of the study showed that, the accuracy of the method for the diagnosis of breast cancer was 93.33 %, significantly higher than that of the discrete wavelet transform. Naga Prudhvi Raj [9] used dual-tree complex wavelet transform to denoise the medical images, the results showed that the algorithm can obtain higher quality denoising images. In addition, the medical images can influence the effect of fusion images in a certain extent because of its border crisscross between organs and tissues and fuzzy. Fuzzy mathematics is an effective way to deal with the fuzzy phenomenon. Yang [10] applied fuzzy mathematics to medical image fusion, the results showed that the effect of fusion image can be better by this method and it has good feasibility. Literature [11] used fuzzy logic as the high frequency fusion rule. As fusion rules, the current pixels fusion rules are used. On the selection of threshold and fusion coefficient, they depend on the experience or many trial and error of researchers. Therefore it increases the workload and lengthen the operating time, in the mean time, the scope of application is narrow. In recent years, some scholars try to fuse images adaptively. Liu [12] regarded noise ratio as the objective function to realize the adaptive selection of the decomposition layers. Yang [13] put the information entropy as the objective function to guide the low frequency fusion.

From the above, taking full account of the characteristics of medical images, a self-adaption fusion algorithm of PET/CT based on DTCWT and combination membership function is proposed in this paper. Firstly, using DTCWT to decompose the registration PET and CT image to get two low-frequency components which are X (i + 1,1) and X(i + 1,2) and six directions high-frequency components which are D (i + 1,1), D(i + 1,2), D(i + 1,3), D(i + 1,4), D(i + 1,5) and D(i + 1,6). And six directions are +15°, +45°, +75°, −15°, −45° and −75°. Secondly, according to the characteristics of low frequency sub-bands concentrate most energy of the source image and determine the image contour, fully considering the area of lesions position is smaller in the whole image and it is vital for highlighting the lesions by dealing with the background of medical image reasonably. So the low-frequency components are fused by self-adaption combination membership function. According to the characteristics of high-frequency sub-bands reflect the detail of images and edge information, and they have a great influence on the degree of image sharpness and edge distortion, so the high-frequency sub-bands adopt regional energy fusion rule. Finally, we did two experiments in PET and CT image of lung cancer by the algorithm is proposed in this paper. (1) Comparison experiment of the algorithm and other pixel-level fusion algorithms; (2) Fusion effect evaluation experiment by objective indicators. The experimental results shown that the algorithm can better retain and show the edge and texture information of lesions.

2   Basic Theories

2.1   Dual-Tree Complex Wavelet Transform

Dual-tree complex wavelet transform (DTCWT) is proposed by Kingsbury. The transformation has the following features: (1) approximate shift invariance; (2) good selectivity and directionality; (3) limited redundancy, for 2-D signal, the redundant is 4:1, for n-D signal, the redundant is 2^n:1. DTCWT using a two-way DWT of binary tree structure, one way generates the real part, the other way forms the imaginary part, as shown in Fig. 1, and the main idea is [5] 12 as follows:

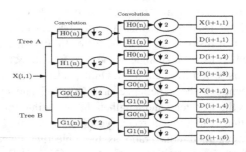

Fig. 1. 2-D dual-tree complex wavelet transform

(1)   For the first layer wavelet decomposition, between Tree A and Tree B filters delay should be kept in a sampling interval, so it can be sure that the samples of Tree B in the two extraction of the first layer are from the missing samples of Tree A;

(2) For the later layers wavelet decomposition except the first layer, the phase frequency of Tree A filter and the phase frequency of Tree B filter should have half a sampling period of the group delay, and the amplitude frequency of two filters should be equal;

(3) In order to ensure linear phase, Kingsbury required Tree A and Tree B odd-length filters in one tree and even-length filters in the other.

DTCWT produces 2 low-frequency component and 6 high-frequency components of complex coefficients at each level, which orient to angles of $\pm 15°$, $\pm 45°$ and $\pm 75°$. However DTCWT is redundant, for n-dimensional signal the redundant is $2^n:1$. So in this paper we use filter is Q-shift dual-tree filter constructed by Kingsbury, which is a pair of orthogonal discrete filter coefficients, and the number of decomposition is two layers (Fig. 2).

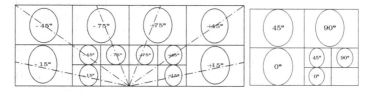

Fig. 2. Direction selectivity compared between DT-CWT and DWT

2.2   Fuzzy Mathematics

Fuzzy mathematics is a mathematical method and the theory of fuzzy phenomena research and treatment, and the basic concept is fuzzy set. Using classical sets to express the intension and extension are clear, but in many expressions, there have many unclear extension concepts, that is the concept of fuzzy [15]. Compared with the classical sets, an element of fuzzy set can be belong to or not belong to a fuzzy set, which has fuzzy boundaries.

Briefly description the above contents, suppose A is an ordinary set, the domain is U, arbitrary mapping uA of U to the [0,1] meet U $\rightarrow$ [0,1], u $\rightarrow$ uA (U), called A is a fuzzy subset of domain U, the mapping uA (U) is the membership function of the fuzzy subsets A, u as the membership of fuzzy set A, and the range of the values is a closed interval [0,1], which reflects the membership degree of u for the fuzzy subsets A. The value is close to 1 indicating u higher degree belongs to A, conversely the lower. At present, there has not a mature and effective method to determine the membership function, and most determination methods still remain in the experience and experiments.

2.3   Evaluation Indicators

See Table 1.

Table 1. Evaluation indicators

Evaluation indicators	Formula	Meaning
Entropy	$E = -\sum_{i=0}^{L-1} P_i \log_2 P_i$	Image entropy is used to indicate how much average information contained in image. The entropy of fusion image is larger, which means the fusion image has more information and better fusion effect
Mean	$U = \frac{1}{MN} \sum_{x=1}^{N} \sum_{y=1}^{N} F(x, y)$	The mean is the arithmetic mean value of all pixels gray-values. The reflection for human eyes is average brightness
Standard deviation	$S = \sqrt{\frac{1}{MN} \sum_{x=1}^{M} \sum_{y=1}^{M} (F(x, y) - u)^2}$	Standard deviation describes the discrete degree of image gray-values and mean. The standard deviation is larger, the distribution of gray levels is more dispersed, and the image contrast is greater. Contrary, the smaller standard deviation, the smaller image contrast
Mutual information	$MI_{GH} = E(G) + E(H) - E(GH)$ $E(G) = -\sum_i p(G_i) \log p(G_i)$ $E(H) = -\sum_i p(H_i) \log p(H_i)$ $E(GH) = -\sum_i \sum_J p(G_i H_j) \log p(G_i H_j)$	Mutual information can be used as a correlation measure of two variables or a variable containing another variable. Here it is used to measure the correlation degree between fusion image and original image, so as to evaluate the fusion effect. The larger mutual information is, the more information and better fusion effect fusion image has
Average gradient	$G = \frac{1}{MN} \sum_{x=1}^{M} \sum_{y=1}^{N} \sqrt{\frac{\nabla F_x^2 + \nabla F_y^2}{2}}$	The average gradient is also called image definition, which reflects the expression ability of minute detail changes in the image. With the average gradient increasing, the image becomes more clear
Signal noise ratio	$SNR = 10 \log_{10} \frac{\sum_{i=1}^{M} \sum_{j=1}^{N} F^2(i,j)}{\sum_{i=1}^{M} \sum_{j=1}^{N} [R(i,j) - F(i,j)]^2}$	The signal-to-noise ratio is used to reflect whether the noise is effectively controlled. The larger the value is, the better the noise be controlled

3 Fusion Algorithm

3.1 Algorithm Idea

The selection of fusion algorithm has a great influence on the quality of fusion image, but every fusion algorithm has its value. The structures of components which get from DTCWT decomposing are similar with the original image, so that the components can be better compared. In this paper, the idea of self-adaption fusion algorithm of PET/CT based on DTCWT and combination membership function is shown in the Fig. 3.

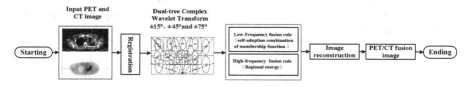

Fig. 3. The fusion algorithm flowchart

Firstly, using dual-tree complex wavelet transform to decompose non-small cell lung cancer PET image and CT image to get the low-frequency and high-frequency components, and the number of decomposition is two layer. Secondly, considering the characteristics of each component to adopt different fusion rules. The low-frequency components are fused by self-adaption combination membership function. While for the high frequency components, regional energy fusion rule is selected. Finally, validated the effectiveness and feasibility of the algorithm by experiments.

3.2 Key Technology

1. Low-Frequency Fusion Rule

Low-frequency component contains the main information and energy, reflects the approximation and average characteristics of the image, and determines the image contour. The medical images have some characteristics, such as huge amounts of data, gray fuzziness, structural complexity, noise significantly and so on, because they are often involved in human tissues and organs. In addition, as a general rule, the lesions in the whole medical image have very low proportion, so in the clinical diagnosis the requirements of medical image background information are higher than other images. The traditional low-frequency fusion method is weighted average, which is a simple operation fusion rule and has low time complexity. However, immobilized fusion coefficient is like to process the image as mean filter, reducing the contrast of the image, but also lead to the edge and contour parts of the fusion image become relatively vague. But choosing different membership functions will cause discrepancy of the evaluation results, and impact on the credibility of the evaluation results. In order to avoid the subjectivity of membership functions selection, as well as the immobilized fusion coefficient, in this paper from the view of self-adaption, the low-frequency components are fused by self-adaption combination membership functions.

Set CT image as A(i,j), PET image as B(i,j), and the sizes are M*N. Using DTCWT to decompose A(i,j) and B(i,j), the layer is 2.

(1) Calculating formulation of combination membership functions is as follows:

$$F(i,j) = \sum_{n=1}^{5} w_n f_n(i,j) \quad (n = 1, 2, \ldots, 5)$$

Where, w_n are weights of f_n, f_n means five membership functions which are gaussian function, generalized bell function, sigmoid function, triangular function and trapezoidal function. As shown in Table 2.

Table 2. Membership functions

Function Name	Formula	Chart
Gaussian function	$f_1(i,j) = G(x,c,\sigma) = e^{-(x-c)^2/2\sigma^2}$	
Generalized bell function	$f_2(i,j) = f(x,a,b,c) = \dfrac{1}{1+\|\dfrac{x-c}{a}\|^{2b}}$	
Sigmoid function	$f_3(i,j) = f(x,a,c) = \dfrac{1}{1+e^{-a(x-c)}}$	
Triangular function	$f_4(i,j) = f(x,a,b,c) = \begin{cases} 0 & x \leq a \\ \dfrac{x-a}{b-a} & a \leq x \leq b \\ \dfrac{c-x}{c-b} & b \leq x \leq c \\ 0 & c \leq x \end{cases}$	
Trapezoidal function	$f_5(i,j) = f(x,a,b,c,d) = \begin{cases} 0 & x \leq a \\ \dfrac{x-a}{b-a} & a \leq x \leq b \\ 1 & b \leq x \leq c \\ \dfrac{d-x}{d-c} & c \leq x \leq d \\ 0 & d \leq x \end{cases}$	

The analytic hierarchy process (AHP) is used to determine the weight w_n of each membership function in this paper. Firstly, constructing judgment matrix A.

$$A = \begin{bmatrix} f_{11} & f_{12} & f_{13} & f_{14} & f_{15} \\ f_{21} & f_{22} & f_{23} & f_{24} & f_{25} \\ f_{31} & f_{32} & f_{33} & f_{34} & f_{35} \\ f_{41} & f_{42} & f_{43} & f_{44} & f_{45} \\ f_{51} & f_{52} & f_{53} & f_{54} & f_{55} \end{bmatrix} = \begin{bmatrix} 1 & 2 & 2 & 2 & 2 \\ \frac{1}{2} & 1 & 1 & 1 & \frac{1}{2} \\ \frac{1}{2} & 1 & 1 & 1 & \frac{1}{2} \\ \frac{1}{2} & 1 & 1 & 1 & \frac{1}{2} \\ \frac{1}{2} & 2 & 2 & 2 & 1 \end{bmatrix}$$

Then calculation the weight vector of 5 functions $w_n = [0.3270 \quad 0.1413 \quad 0.1413 \quad 0.1413 \quad 0.2492]^T$. Next carry out the consistency check. The consistency index is used to measure the judgment quality of judgment matrix. The calculation equation of consistency index is $CI = \frac{\lambda - n}{n - 1}$. By calculating, we get $\lambda = 5.0586$ and $CI = 0.0147$. When n = 5, checking Table 3, we could know $RI = 1.12$ and get $CR = CI/RI = 0.0131 < 0.1$, which means the matrix has good consistency. So the weight vector of 5 functions is $[0.3270 \quad 0.1413 \quad 0.1413 \quad 0.1413 \quad 0.2492]^T$.

Table 3. The value of random consistency index

n	1	2	3	4	5	6	7	8	9	10	11
RI	0	0	0.58	0.90	1.12	1.24	1.32	1.41	1.45	1.49	1.51

According to the AHP we can obtain the combination membership function:

$$F(i,j) = 0.3270 f_1(i,j) + 0.1413 f_2(i,j) + 0.1413 f_3(i,j) + 0.1413 f_4(i,j) + 0.2492 f_5(i,j)$$

(2) Construction the low-frequency fusion rule:

$$C_l(i,j) = \begin{cases} B_l(i,j), & B_l(i,j) < 50 \\ \omega_A^l A_l(i,j) + \omega_B^l B_l(i,j), & others \end{cases}$$

Where, $C_l(i,j)$ is the low-frequency component of fusion image, $A_l(i,j)$ and $B_l(i,j)$ are the low-frequency components of image A and image B. w_A^l and w_B^l are the weight coefficients of $A_l(i,j)$ and $B_l(i,j)$, which $w_A^l + w_B^l = 1$.

Finally, we get w_A^l and w_B^l by the combination membership function, the calculation equations as follows:

$$\omega_A^l = \frac{F_A(i,j)}{F_A(i,j) + F_B(i,j)}, \quad \omega_B^l = \frac{F_B(i,j)}{F_A(i,j) + F_B(i,j)}$$

2. High-Frequency Fusion Rule

High-frequency component reflects the details characteristics and edge information of the image, dealing with the high-frequency coefficients, which directly relates to the image clear or not, and edge distortion serious or not. DTCWT has its own characteristics, so through analysis each sub-band we find that there has correlation of each sub-band coefficient. Therefore, we focus on the above characteristics, choosing regional energy to fuse the high-frequency components. Compared with the pixels fusion rules, the high-frequency fusion rule in this paper considers the correlation of adjacent pixels, as well as the fuzzy problems. Many features of the image can not be reflected by a single pixel which need a window made by some pixels to be shown. So using the region which regards the pixels as the center to process the image that will be made the image features more complete expression. Regional energy is used to reflect the change degree of the grays in the region. Therefore, the fusion rule based on the regional energy can better keep sensitive information of the image, but also improve the image resolution. Set (i,j) as the centre of regional energy, and the calculation equation of regional energy as follows: $E_o^\xi(i,j) = \sum_{m \in S} \sum_{n \in T} w(m,n)[D_o^\xi(i+m)(j+n)]^2$

Where, $w = \frac{1}{16}[1 \quad 2 \quad 1; 2 \quad 4 \quad 2; 1 \quad 2 \quad 1]$, $D_o^\xi(i,j)$ is the high-frequency component, O means image A and image B, $\xi = 1, 2, 3 \ldots 6$ stands for six directions which are $\pm 15°$, $\pm 45°$ and $\pm 75°$.

Defined T_1 as the difference of regional energy between the high-frequency component of image A and the high-frequency component of image B.

$$T_1(i,j) = \frac{E_A^\xi(i,j)}{E_B^\xi(i,j)}$$

The threshold is defined as T which belongs to (0,0.5). If T_1 less than T or greater than or equal $1/T$ means that the difference of regional energy is significant, so choose the larger energy as the high-frequency coefficients of fusion image. Otherwise considered the high-frequency component of image A and the high-frequency component of image B are higher similarity. Then using regional energy to calculate the weights so that we can be adaptive acquisition the high-frequency coefficients of fusion image.

The calculation equations of w_A^ξ, w_B^ξ as follows:

$$\omega_A^\xi = \frac{E_A^\xi(i,j)}{E_A^\xi(i,j) + E_B^\xi(i,j)}, \quad \omega_B^\xi = \frac{E_B^\xi(i,j)}{E_A^\xi(i,j) + E_B^\xi(i,j)}$$

4 Experiments and Results Analysis

Experiment data: A pair of non-small cell lung cancer CT image and PET image with good registration], which sizes are 356 × 356 pixels.

In order to verify the validity and feasibility of the proposed algorithm, the paper did two experiments, the first is comparison experiment of the algorithm and other pixel-level fusion algorithms; the second is fusion effect evaluation experiment by information entropy, mean, standard deviation, mutual information, signal noise ratio and average gradient.

4.1   Experiment 1

In the experiment, the proposed algorithm and other pixel-level fusion algorithms, such as maximum method, minimum method, weighted average method, IHS transform and wavelet transform are compared, the experimental results as shown in Fig. 4.

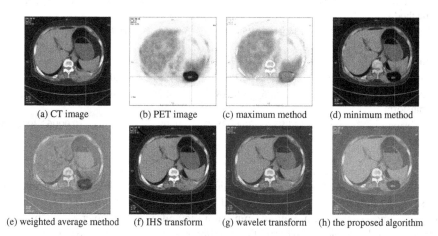

(a) CT image   (b) PET image   (c) maximum method   (d) minimum method

(e) weighted average method   (f) IHS transform   (g) wavelet transform   (h) the proposed algorithm

Fig. 4. Different image fusion algorithms

In Fig. 4, the final fusion images results are different because of the different fusion methods. We can see that, the fusion results of the paper proposed fusion algorithm and weighted average method are the best, which nicely combine the advantages of CT image and PET image in one image. However, the weighted average method is not as good as the paper proposed fusion algorithm in the details, for example, in the location of the lesion, the edge of bone and soft tissues are vague. The fusion image based on maximum method is the worst, so that the fusion image has low-contrast of soft tissues and difficult to distinguish the details. For minimum method, some minimal fluctuations are appeared which do not exist in the original image. The fusion results of wavelet transform and IHS transform are in the middle, but IHS transform is not good as wavelet transform in the visual effect of the edge.

4.2   Experiment 2

For quantitative evaluation the fusion effects of six fusion algorithms in the first experiment, we use six indicators which are entropy, mean, standard deviation, mutual information, signal noise ratio (SNR) and average gradient to evaluate, as shown in Table 4.

From Table 4, it is easy to see that information entropy, standard deviation, mutual information with CT and mutual information with PET of the image which is get from the paper proposed algorithm are respectively 5.9076, 36.9925, 30.0006 and 30.8580. The average gradient of the paper proposed algorithm is less than IHS transform, but it

Table 4. Calculation results of evaluation indicators

	Entropy	Mean	Standard deviation	Average gradient	SNR (PET)	MI (CT)	MI (PET)
Maximum method	4.8989	218.0577	44.5822	5.1917	71.735	20.1092	9.7748
Minimum method	4.6606	51.1303	60.7983	7.0802	43.4406	9.8774	30.3747
Weighted average method	5.4021	134.9165	25.4180	5.2054	51.1955	19.4917	30.2039
IHS transform	5.0404	55.2777	65.1509	8.5035	44.0282	6.6754	22.2091
Wavelet transform	5.8155	81.7506	48.9218	5.7058	46.4746	28.1415	30.1691
Our fusion algorithm	5.9076	124.9586	36.9925	5.9298	54.9741	30.0006	30.8580

has little effect on the lesions. SNR of which is 54.9741 that effectively controlled the noise. Regarding information entropy and mutual information with CT as evaluation indexes, among maximum method, minimum method, weighted average method, IHS transform and wavelet transform, the maximum entropy and the maximum mutual information is the wavelet transform, which are 5.8155 and 28.1415. Compared with weighted average method, information entropy and mutual information with CT of the paper proposed algorithm are respectively increasing 1.58 % and 6.61 %. For other indicators, the algorithm is also superior to other methods, so here does not enumerate.

5 Summary

Images can not be distorted in medical image fusion, retaining original image information, removing image noise and redundant information as much as possible, and making fused medical image can be suitable for human visual perception. Aiming at these requirements, a self-adaption fusion algorithm of PET/CT based on DTCWT and combination membership function is proposed in this paper. Firstly, using DTCWT to decompose the registered PET and CT image, and get the low-frequency and high-frequency components; Secondly, according to the characteristics of low frequency sub-bands concentrating most energy of the source image and determining image contour, fully considering the area of lesions position is smaller in the whole image and it is vital for highlighting the lesions by dealing with the background of medical image reasonably. So the low-frequency components are fused by self-adaption combination membership function. According to the characteristics of high-frequency sub-bands reflecting the detail and edge information in a image, and they have a great influence on the degree of image sharpness and edge distortion, so the high-frequency sub-bands adopt regional energy fusion rule. Finally, validated the effectiveness and feasibility of the algorithm by experiments. The experiment results shown that the algorithm is an efficient fusion method of multimode medical image.

Acknowledgment. The paper supported by national Natural Science Foundation (No: 81160183), Natural Science Foundation of Ningxia (NZ12179, NZ14085), Science research project of Ningxia education Branch (No. NGY2013062), the project of Shaanxi Provincial Key Laboratory of Speech & image Information Processing (SJ2013003) and the special talent project of Ningxia Medical University (XT2011004).

References

1. Hu, C., Huang, Z., Luo, L.H.: Development of medical image fusion. Chin. Med. Equip. J. **31**(4), 157–160 (2010). (in Chinese)
2. de Bazelaire, C., Groheux, D., Chapellier, M., Sabatier, F., et al.: Breast inflammation: indications for MRI and PET-CT. Diagn. Intervent. Imaging **93**(2), 104–115 (2012)
3. Tao, G., Li, D., Lu, G.: Application of wavelet analysis in medical image fusion. J. Xi'an Electron. Sci. Univ. (Nat. Sci.) **31**(1), 82–86 (2004)
4. Tao, G.Q., Li, D.P., Lu, G.H.: Application of wavelet analysis in medical image fusion. J. Xidian Univ. **31**(1), 82–86 (2004). (in Chinese)
5. Kingsbury, N.G.: The dual-tree complex wavelet transform: a new technique for shift invariance and directional filters. In: Proceedings of 8th IEEE Digital Signal Processing Workshop, pp. 86–89. IEEE, Bryce Canyon, Utah, USA (1998)
6. Song, J., Shi, F.: Multi-focus image fusion based on dual-tree complex wavelet transform. Mod. Electron. Technol. **2**(313), 104–107 (2010)
7. Zhong, Q., Xia, L.: Image fusion method based on dual- tree complex wavelet transform. Comput. Eng. Appl. **44**(24), 184–187 (2008)
8. David, D., Santika, D.D.: Computer aided diagnosis system for digital mammogram based on neural network and dual tree complex wavelet transform. Procedia Eng. **50**, 864–870 (2012)
9. Raj, V.N.P., Venkateswarlu, T.: Denoising of medical images using dual tree complex wavelet transform. Procedia Technol. **4**, 238–244 (2012)
10. Yang, F., Wei, H.: Fusion of infrared polarization and intensity images using support value transform and fuzzy combination rules. Infrared Phys. Technol. **60**, 235–243 (2013)
11. Saeedi, J., Faez, K.: Infrared and visible image fusion using fuzzy logic and population-based optimization. Appl. Soft Comput. **12**, 1041–1054 (2012)
12. Liu, F., Yang, B., Gang, K.: Image fusion using adaptive dual-tree discrete wavelet packets based on the noise distribution estimation. In: Processing of 2012 International Conference on Audio, Language and image, pp. 475–479. IEEE Computer Society, United States (2012)
13. Y, Xiaohui, Jia, J., Jiao, L.: Image fusion algorithm in nonsubsampled contourlet domain based on activity measure and closed loop feedback. J. Electron. Inf. Technol. **32**(2), 422–426 (2010)
14. Chakraborty, S., Bhattacharya, I., Chatterjee, A.: A palmprint based biometric authentication system using dual tree complex wavelet transform. Measurement **10**(46), 4179–4188 (2013)
15. Bai, Y.: Medical image fusion research based on fuzzy logic and neural network. Master's thesis of Xidian University (2010)

Automatic Methods for Screening and Assessing Scoliosis by 2-D Digital Images

Weishen Pan[✉], Guangdong Hou, and Changshui Zhang

State Key Laboratory on Intelligent Technology and Systems,
Tsinghua National Laboratory for Information Science and Technology (TNList),
Department of Automation, Tsinghua University,
Beijing 100084, People's Republic of China
pws11@mails.tsinghua.edu.cn

Abstract. The paper proposes an automatic method to screen and assess scoliosis, especially for the early-stage cases. Compared to state-of-art methods based on professional medical images such as radiographs and 3-D surface images, this method only requires 2-D digital images containing human back. The method reaches comparable results with manual results marked on the same images by clinicians, so it provides a feasible and convenient way for potential scoliosis patients to detect scoliosis and for patients to monitor the dynamic changes on scoliosis at home.

Keywords: Automatic · Scoliosis · Screening and assessment · 2-D digital images

1 Introduction

Scoliosis is a common disease related to torso deformity like tilt, bend and twist of body. It causes severely physical and mental hurts to patients. Therefore, to detect scoliosis on time and to assess its severity accurately is important for follow-up diagnosis and treatment.

1.1 Previous Work

The screening and assessment of scoliosis has been studied in the past few decades and some medical methods have been raised. The Cobb angle [3] is the gold standard for assessment of scoliosis. Radiographs provide the direct visual information of spine and become popular for measurement of Cobb angle. The measurement was once finished by manual plotting and calculation on radiographs [3,7]. Since the techniques of image processing and computer vision

---

C. Zhang—This work is supported by Tsinghua University Initiative Scientific Research Program(founded by Tsinghua University Scientific Research Development, No. 20141081231).

© Springer International Publishing Switzerland 2015
X. He et al. (Eds.): IScIDE 2015, Part I, LNCS 9242, pp. 392–400, 2015.
DOI: 10.1007/978-3-319-23989-7_40

develop, semi-automatic and automatic methods have been created to calculate Cobb angle on radiographs [2]. A limitation of methods based on radiographs is that scoliosis is the 3-D deformity but radiographs only provide 2-D information and the measurement of Cobb angle may vary by up to 5 degrees [7]. Another concern is the radiation during taking the radiographs.

Surface images provide the torso depth information for assessment of the torso deformity. And spine deformity is predicted by torso deformity. Methods on laser images [1,8] and moiré topographic images [5] are widely accepted in the domain of scoliosis. The methods on surface images prevent people from exposure to radiation. However, torso shape is not only determined by spine but also muscles, body fat, etc. So the relationship between torso shape and spine shape is necessary for the method. Other indices related to torso deformity have been created as supplements of Cobb angle.

Scoliosis assessment can be carried out on 2-D digital images, which are much easier to achieve than radiographs and surface images. There are a few research results on this topic [9,10]. These methods are all semi-automatic with clinicians' help to manually draw and label on every image.

Although the methods mentioned above perform well, they all rely on professional medical imaging devices or well-trained clinicians. In this paper, our method provides a feasible solution for this new application scenario: the patients can finished scoliosis assessment and detection with 2-D digital images taken at home by digital cameras or smartphone. So our method is significantly convenient, less money-consuming and time-consuming.

1.2 Outline

In this article, we propose a method to detect scoliosis by measuring the torso asymmetry on 2-D digital images. The method is automatic, so patients/potential patients without any medical knowledge are able to finish the measurement at home. We firstly detect human-back contour with the edge information from the image. Then we reconstruct the spine as the symmetric axis of the contour. And further indices can be calculated based on the result of our method.

The remainder of this article is organized as follows. In Sect. 2, we introduce our model and procedure of our algorithm. The implementation and results on images are shown in Sect. 3, which are compared with the ones marked by clinicians. This article is concluded in Sect. 4.

2 Model and Algorithm

To measure the scoliosis, one intuitive idea is to automatically recognize the spine directly from the image by texture on the back, as clinicians do. But the texture varies due to the observed people and shooting situations and not robust for computers to capture.

Another potential idea is to detect the feature points on human back, then to reconstruct the spine as the symmetric axis of pair-wised feature points. Method of [6] is an example to detect curved reflection symmetry and find symmetric axis with SIFT features. But in the scene of scoliosis assessment with 2-D images, human back is smooth with 2-D image. Few SIFT points can be captured.

To be summarized, methods based on features points do not work in this application scenario. Texture on human back is weak and complicate so it's difficult for computers to recognize and produce useful information. But the edge of human back is easy to detect. So we find another way to assess asymmetry of human back:

2.1   Edge-Based Model for Scoliosis Screening

The core element of our model is the edge-based contour of human back. We use the edges to represent the contour and then calculate the asymmetry of human back with help of the contour. Finally the asymmetry is used to decide whether the object has scoliosis and how severe it is.

The contour consists of two relatively symmetry pairs of continuous and smooth curves: waist and shoulder. The waist curve extends from hip to armpit, while the shoulder curve extends from bottom of neck to upper arm. This is shown in Fig. 1(c). And the spine will be reconstructed as a curved symmetry axis of the given contour.

According to the design, our algorithm can be divided into two main steps: back contour detection and spine reconstruction.

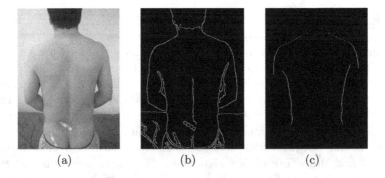

(a)   (b)   (c)

Fig. 1. Introductive image of our model. (a) is the original image collected from a scoliosis patient. (b) is the result of canny edge detector of the grey-scale image of (a). (c) is contour extracted from (a), we clearly see the different parts of the contour: left waist, right waist, left shoulder and right shoulder.

2.2   Procedure of the Algorithm

Back Contour Detection: To get the candidates of the edges, the image is preprocessed with Canny edge detector. As mentioned in Subsect. 2.1, the edges of the contour is the continuous, smooth and longest curves to be observed from the image. A algorithm based on Active Contour Model [4] detects the target curves. The algorithm uses energy to represent the information of the image. But this algorithm needs a closed contour and as described in Subsect. 2.1, the contour of human back can not be represented by a closed contour but four separated curves. Another question is the energy function in [4] where the detection is from outside in and under our situation the algorithm may miss the waist curve. So we modify Active Contour Model into a new version which detects targets curves and change the form of energy function (Fig. 2).

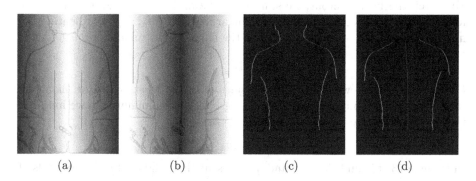

(a)	(b)	(c)	(d)

Fig. 2. Images to explain our algorithm. (a) is the image representing the energy distribution of $E_{external}$ while to find the curves of the back, red ones are the initialized curve and green ones are the target curves we find. (b) is the same kind of image to find the curves of the shoulder. (c) represents the candidate contour. And (d) is the result we get after pruning and smoothing on (c) and the blue lines represent the spine we reconstruct from the image (Color figure online)

In our model, the energy of a given curve is defined as:

$$E_{curve} = E_{internal} + E_{edge} + E_{potential} \tag{1}$$

where E_{curve} is the total energy of the curve consisting of three parts: $E_{internal}$ is the internal energy of the curve, E_{edge} represents the energy from the image, $E_{potential}$ is a predefined energy by priori knowledge to guide the curve to the target position. And let $E_{external}$ as the sum of E_{edge} and $E_{potential}$. The curve is represented by some labelled points on it and we use $\mathbf{v}_i = (x_i, y_i)$ to represent the ith labelled point of the curve.

We use the length of the curve to measure $E_{internal}$. In order to take derivatives, we use the square form:

$$E_{internal} = \sum_{i=1}^{n-1} \alpha |\mathbf{v}_{i+1} - \mathbf{v}_i|^2 \tag{2}$$

We use the result(0 for detected edge pixel and 255 for others) of the image processed by Canny edge detector as a measurement of E_{edge}. A Gaussian filter is also implemented to get the final energy distribution I_{canny}.

$$E_{edge} = \sum_{i=1}^{n} \beta I_{canny}(\mathbf{v}_i) \tag{3}$$

$E_{potential}$ is decided in a potential field on our priori knowledge about the curve of the back and the shoulder.

$$E_{potential} = \sum_{i=1}^{n} \gamma e(\mathbf{v}_i) \tag{4}$$

To search the back contour, we start from the middle of image. Otherwise we may find the outer contour consisting the arm. With (m_x, m_y) representing the middle point of the image, we initialize the curve in the middle of image define $e(\mathbf{v}_i)$ as:

$$e(\mathbf{v}_i) = |m_x - x_i| \tag{5}$$

On the other hand, we start from the sides of image when searching shoulders to avoid meeting body contour and getting into local minimum. So we initialize the curve in the sides of image define $e(\mathbf{v}_i)$ as:

$$e(\mathbf{v}_i) = -|m_x - x_i^?| \tag{6}$$

In formula (3) (4) and (5), α, β and γ represent the weights of different kinds of energy.

After we initialize the position of the curve and the potential field according to its type(left waist, right waist, left shoulder, right shoulder), we can find the target curve by minimizing the curve energy.

Spine Reconstruction: From observation on scoliosis patients, we find that for early-stage scoliosis, deformity of the spine is mild and the spine can be represented by two straight lines. One is the symmetry axis of the torso containing the chest, the other is of the waist. The first one is calculated by the two shoulder curves and the second one by the two waist ones. The symmetry axis of two curves is decided by matching the two curves with each other as much as possible. For the two shoulder curves, let $\mathbf{LS} = \{\mathbf{x}_i^{ls}\}(i = 1, ..., n^{ls})$ to left shoulder curve represented by landmarks and $\mathbf{RS} = \{\mathbf{x}_i^{rs}\}(i = 1, ..., n^{rs})$ to be the right shoulder curve. With a given line L, the function $\mathbf{s}(x, L)$ returns the point which is symmetrical to x about a straight line L. $d(\mathbf{x}, \mathbf{C})$ means the distance from a point $\mathbf{x}$ to a curve $\mathbf{C}$. Then the degree of symmetry between $\mathbf{LS}$ and $\mathbf{RS}$ about $\mathbf{L}$ is define as:

$$S(\mathbf{LS}, \mathbf{RS}, \mathbf{L}) = \frac{\sum_{i=1}^{n^{ls}} d(\mathbf{s}(\mathbf{x}_i^{ls}, \mathbf{L}), \mathbf{RS})}{n^{ls}} + \frac{\sum_{i=1}^{n^{rs}} d(\mathbf{s}(\mathbf{x}_i^{rs}, \mathbf{L}), \mathbf{LS})}{n^{rs}} \tag{7}$$

The symmetry axis of them is decided by:

$$\mathbf{L}^* = \underset{\mathbf{L}}{argmin}\ S(\mathbf{LS}, \mathbf{RS}, \mathbf{L}) \tag{8}$$

The advantage of this spine reconstruction method is that the 2-D edge-based contour is a rough approximation of human back surface. Then a simpler model is more robust.

3   Results

There are few datasets on the problem to assess scoliosis by 2-D digital images. We have created a dataset with images from real world and implemented our method on it. The first part of our dataset contains 7 images collected from scoliosis patients. In the second part, 54 images are from 18 normal people. For each person, one image with upright posture, two with curved-spine postures. The image size is 300 × 400 pixels, the same aspect ratio with images taken by most smartphones and cameras. The original images are 24-bit RGB images.

As mentioned above, there is no benchmark for the problem of assessing scoliosis with 2-D images. What's more, the radiographs and surface images contain much more information than 2-D images. For example, the patient's spine bends forward, but the 2-D image fails to represent this. So the comparison between methods with different kinds of images makes no sense. Whether our algorithm can resume most information from 2-D human back images is what we concern.

We evaluate our algorithm on our dataset consisting of 61 images and invite professional clinicians to plot landmarks to label the spine. We regard the labelling as the true position of the spine and test our algorithm by comparing with the labelling results.

A representative selection of our results is on Figs. 3 and 4. In each image, original image with our predicted spine and the landmarks labelled by clinicians is presented. We can see that our results fit the landmarks well under different situations of real or simulated scoliosis. An index is calculated to measure the fitness between our results and the clinicians'. Within n images, the ith one is with m landmarks $\mathbf{l}_j^i (j = 1, 2, ..., m)$ (clinicians' result) and a predict spine $\mathbf{C}_i$(our result).We define the error function for each image with $d(\mathbf{l}_j^i, \mathbf{C}_i)$ meaning the distance from a point $\mathbf{l}_j^i$ to a curve $\mathbf{C}_i$ with a unit of pixel:

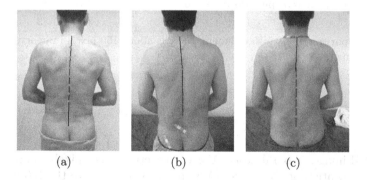

|(a)|(b)|(c)|

Fig. 3. Selected experiment results on scoliosis patients. Blue lines to be the spine reconstructed by our method, green dots to be the landmarks label by clinicians (Color figure online).

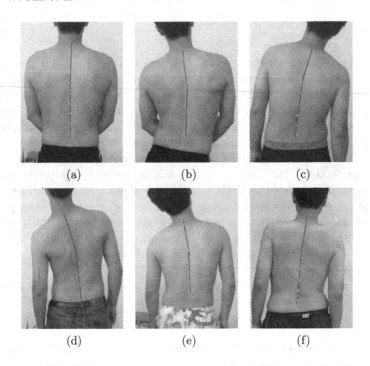

Fig. 4. Selected experiment results on people imitating different kinds of morphology on scoliosis. (a) is taken with a upright posture for comparison.

Table 1. Error analysis

Situation	Number of images	Error Results / pixels
Patients	7	3.89 ± 1.93
Normals with Upright Posture	18	3.36 ± 1.84
Normals with Curved-spine Postures	36	3.76 ± 1.85
Total	61	3.66 ± 1.83

As a comparison, the width of human body is about 150 pixels in images of the experiments.

$$e_i = \frac{1}{m} \sum_{j=1}^{m} d(\mathbf{l}_j^i, \mathbf{C}_i) \tag{9}$$

With all 61 images in our datasets, the average error is 3.66 pixels and the sample standard deviation of the errors is 1.83 pixels. Compared to the 150-pixel-wide body, the error is small and our result is close to the ground truth. And results in Table 1 and Fig. 5 show our method is robust under different situations.

We can conclude that our automatic method obtains results similar with the manual one. So the spine reconstructed provides materials for further assessment

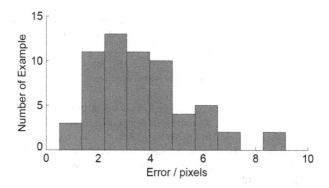

Fig. 5. The Distribution of Errors

of scoliosis. Potential methods include assessing scoliosis with the angle of the two symmetry axes of waist and shoulder, etc.

4 Conclusion

An automatic method was proposed to screen and assess scoliosis on 2-D digital back-torso images. An edge-based contour model is designed to reconstruct the contour of human back and the spine. And experiments on images from scoliosis patients and normal persons verify the algorithm to be reliable and robust.

Future work is to represent the spine with a smooth curve rather than a combination of two straight lines. This can provide more information on human-back asymmetry and help to achieve higher precision when assessing scoliosis.

And designing a method to assess scoliosis with images taken from different angles will be considered in future work.

References

1. Ajemba, P.O., Durdle, N.G., Raso, V.J.: Characterizing torso shape deformity in scoliosis using structured splines models. IEEE Trans. Biomed. Eng. **56**(6), 1652–1662 (2009)
2. Allen, S., Parent, E., Khorasani, M., Hill, D.L., Raso, J.V.: Validity and reliability of active shape models for the estimation of cobb angle in patients with adolescent idiopathic scoliosis. J. Digit. Imaging **21**(2), 208–218 (2008)
3. Cobb, J.R.: Outline for the study of scoliosis. Instr. Course Lect. **5**, 261–275 (1948)
4. Kass, M., Witkin, A., Terzopoulos, D.: Snakes: active contour models. Int. J. Comput. Vis. **1**(4), 321–331 (1988)
5. Kim, H.S., Ishikawa, S., Ohtsuka, Y., Shimizu, H., Shinomiya, T., Viergever, M.A.: Automatic scoliosis detection based on local centroids evaluation on moire topographic images of human backs. IEEE Trans. Med. Imaging **20**(12), 1314–1320 (2001)

6. Liu, J., Liu, Y.: Curved reflection symmetry detection with self-validation. In: Kimmel, R., Klette, R., Sugimoto, A. (eds.) ACCV 2010, Part IV. LNCS, vol. 6495, pp. 102–114. Springer, Heidelberg (2011)
7. Morrissy, R.T., Goldsmith, G.S., Hall, E.C., Kehl, D., Cowie, G.H.: Measurement of the cobb angle on radiographs of patients who have. J. Bone Joint. Surg. Am. **72**, 320–327 (1990)
8. Ramirez, L., Durdle, N.G., Raso, V.J., Hill, D.L.: A support vector machines classifier to assess the severity of idiopathic scoliosis from surface topography. IEEE Trans. Inf. Technol. Biomed. **10**(1), 84–91 (2006)
9. Colombo, A.S., Saad, K.R., João, S.M.A.: Reliability and validity of the photogrammetry for scoliosis evaluation: a crosssectional prospective study. J. Manipulative Physiol. Ther. **32**(6), 423–430 (2009)
10. Stolinski, L., Czaprowski, D., Kozinoga, M., Korbel, K., Janusz, P., Tyrakowski, M., Kono, K., Suzuki, N., Kotwicki, T.: Analysis of anterior trunk symmetry index (atsi) in healthy school children based on 2D digital photography: normal limits for age 7–10 years. Scoliosis **8**(Suppl 2), P10 (2013)

Exemplar-Based Image Inpainting Using Structure Consistent Patch Matching

Haixia Wang[1], Yifei Cai[2], Ronghua Liang[1(✉)], Xiao-Xin Li[2], and Li Jiang[2]

[1] College of Information Engineering,
Zhejiang University of Technology, Hangzhou 310023, China
{hxwang, rhliang}@zjut.edu.cn
[2] College of Computer Science and Technology,
Zhejiang University of Technology, Hangzhou 310023, China
catlwwy@gmail.com, {mordekai, jl}@zjut.edu.cn

Abstract. Image inpainting reconstructs lost or deteriorated parts of images according to the information of surrounding regions. Criminisi has proposed an effective exemplar-based inpainting algorithm, which has the advantages of both texture synthesis and diffusion-based inpainting. Yet, it has its own flaws of fast priority dropping and visual inconsistency. In this paper, we propose a space varying updating strategy for the confidence term to improve the filling priority estimation and a structure consistent patch matching to take the difference distribution of source and target patches into account. Experimental results have demonstrated the improvement of our proposed method.

Keywords: Image inpainting · Exemplar-based · Patch matching · Fast Fourier transform

1 Introduction

The word image inpainting can be traced back to Renaissance when artist repaired missing or damaged rare artworks by hands. With the development of modern computers and digital image processing, algorithms for image inpainting have attracted attentions. Without the requirements of artificial knowledge and professional skills, the capability of image inpainting has been extended to repairing old photographs and removing occlusion.

Image inpainting is an ill-pose problem since there is no well-defined unique solution. To restore the missing information, pixels in the known and unknown regions of the same image are assumed to share same statistical properties or geometrical structures so that the unknown pixels can be estimated based on the known pixels. There are two categories of methods to perform image inpainting. The first category uses the diffusion-based inpainting, where parametric models are established by partial differential equations to propagate the local structures from known regions to unknown regions. This idea was first introduced by Bertalmio et al. [1] who proposed the famous Bertalmio-Sapiro-Caselles-Ballester algorithm. This algorithm simulates the repairing process of professional artists by propagating image information along the isophote.

© Springer International Publishing Switzerland 2015
X. He et al. (Eds.): IScIDE 2015, Part I, LNCS 9242, pp. 401–410, 2015.
DOI: 10.1007/978-3-319-23989-7_41

Inspired by this idea, Chan et al. [2] came up with Total Variation (TV) inpainting model which can maintain edge and perform denoising at the same time. It is improved by replacing the TV model with a curvature driven diffusions model [3]. The diffusion-based inpainting can achieve superior performance when applied to images composed of structure information or when the target region is small. However, when comes to large target region or natural images composed of both texture and structure information, diffusion-based inpainting introduces blur to the restored image.

The other category of image inpainting algorithm is regarded as the exemplar-based inpainting. The idea is inspired by the seminal work on texture synthesis [4, 5]. The original texture synthesis algorithm aims to grow a larger texture image from a given small seed image so that the produced texture image has a similar visual appearance as the seed image [4]. Similar to region-growing, exemplar-based inpainting restores the missing pixel or missing patch one at a time while maintaining coherence with nearby pixels. The first such work was proposed by Ashikhmin [6], which looks for the best match for an unknown pixel only among the set of shifted candidates from the correspondents of the neighbors. It further imposes a certain coherence in the mapping function to improve the visual quality of the synthesis results [7]. Efros and Leung proposed an inpainting method that fills the unknown patches instead of unknown pixels one at a time that greatly improves the inpainting speed [4]. Criminisi et al. [8] presented an exemplar-based image inpainting algorithm dealing both structure and texture information. It adopts the non-parametric sampling concept and set the known regions of the image as source exemplars. Patches with more edge information are assigned with higher priorities [8].

Though the Criminisi algorithm has achieved great success, it has its own flaws. The criterion used for patch matching cannot fully reflect the similarity of two patches. A weighted Bhattacharya distance has been proposed to cope with this problem which however only value the overall intensity distribution but not the spatial distribution of pixels (denoted as CBD algorithm) [9]. The confidence term in the priority estimation has fast dropping effect which decreases the inpainting performance [10].

We thus propose a space varying updating strategy for the confidence term to improve the priority estimation. We also propose a structure consistent patch matching to take the patch difference distribution into account. The remainder of this paper is organized as follows. In Sect. 2, we briefly introduce the Criminisi algorithm. In Sect. 3, we discuss the limitations of Criminisi algorithm and propose our exemplar-based Image Inpainting using structure consistent patch matching. Experimental results are given in Sect. 4 to demonstrate the performance of the proposed method. A conclusion is drawn in Sect. 5.

2 Criminisi Algorithm

Criminisi algorithm is an exemplar-based texture synthesis approach [8]. It has the advantages of both diffusion based inpainting and texture synthesis so that it can restore structure and texture information simultaneously. The keys of this algorithm are the priorities indicating when the points are matched and the criterion indicating how the patches are matched. It directly copies the most similar patch from source exemplars into the target region so that it will not introduce blur. What's more, it takes the structure information into consideration, giving higher priority to the points on the boundary of target region where structure information located. As a result, this algorithm can achieve good performance on both images with large lost regions and images composed of texture and structure information.

For an image I as shown in Fig. 1(a), Φ indicates the source region which is known, Ω indicates the target region which needs to be filled, and $\partial\Omega$ is the boundary target region near Φ where inpainting starts. In the first step, Criminisi algorithm selects a single point p in $\partial\Omega$. A patch surrounding the selected points p is used as the template, which is shown in Fig. 1(b) and indicated by ψ_p. Part of the points in ψ_p that belongs to Φ is used for matching while the rest of points belong to Ω need to be filled. For a wide range of $\partial\Omega$ points, the point filled earlier can affect the points filled afterwards. The processing priorities of these points are thus important to final results. Criminisi algorithm considers a point has higher priority when the known points in ψ_p have higher confidence and present more structure information than the others. For a patch ψ_p centered at pixel p with $p \in \partial\Omega$, its priority is defined as

$$P(p) = C(p) * D(p). \tag{1}$$

The confidence term $C(p)$ measuring the amount of reliable information surrounding the pixel p, which is estimated as the percentage of known points in ψ_p,

$$C(p) = \frac{\sum_{t \in \psi_p \cap \Phi} C(t)}{|\psi_p|}, \tag{2}$$

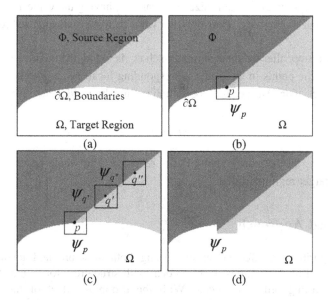

Fig. 1. Structure propagation by exemplar-based synthesis. (a) Source and target region; (b) target patch; (c) target patch and source patches; (d) filling result of one patch.

where t indicates the point coordinates belong to ψ_p and known; $C(t) = 1$ is initialized for points in the source region Φ; $|\psi_p|$ is the total number of points in the target patch ψ_p. The data term $D(p)$ representing the strength of the isophote hitting the boundary $\partial\Omega$ as

$$D(p) = \frac{\left| \nabla I_p^{\perp} \cdot \vec{n}_p \right|}{\alpha}, \tag{3}$$

where α is a normalization factor ($\alpha = 255$ for gray-level image), $\vec{n}_p$ is a unit vector orthogonal to the boundary $\partial\Omega$ at point p, and $\perp$ denotes the orthogonal operator.

In the second step, after a pixel $\hat{p}$ with highest priority is selected, the source patches are compared with the target patch and a most similar one is selected. As Fig. 1 (c) shows, both patches $\psi_{q'}$ and $\psi_{q''}$ are seen to be similar to the target patch $\psi_{\hat{p}}$. How to select the correct source patch, or how to perform the patch matching, is most important to the image inpainting quality. Criminisi algorithm uses the sum of squared differences (SSD) as the criterion of patch matching, which can be written as

$$\psi_{\hat{q}} = \arg \min_{\psi_q \in \Phi} d\left(\psi_{\hat{p}}, \psi_q \right) \tag{4}$$

With

$$d\left(\psi_{\hat{p}}, \psi_q \right) = d_{SSD}\left(\psi_{\hat{p}}, \psi_q \right) = \sum_{t \in \psi_{\hat{p}}} \left[\psi_{\hat{p}}(t) M_{\hat{p}}(t) - \psi_q(t) M_{\hat{p}}(t) \right]^2 \tag{5}$$

where $M_{\hat{p}}$ is a mask with the same size as $\psi_{\hat{p}}$ and ψ_q having the value 1 if the point in $\psi_{\hat{p}}$ is known and 0 otherwise. Thus only the valid known points are used for the patch matching.

In the third step, after the most similar patch $\psi_{\hat{q}}$ is found, the unknown points in $\psi_{\hat{p}}$ are filled with the points in $\psi_{\hat{q}}$ at the corresponding locations, as shown in Fig. 1(d). The confidence value also needs to be initialized for the newly filled pixels, as

$$C(t) = C\left(\hat{p} \right), \forall t \in \psi_{\hat{p}} \cap \Omega, \tag{6}$$

After filling $\psi_{\hat{p}}$, the boundary points $\partial\Omega$ is renewed. The whole process is repeated until whole target region is filled.

3   Proposed Approach

Criminisi algorithm is effective in processing both structure and texture images. However, it still has limitations. First, to deal with structure information, the priority defining the filling order is crucial. With the dropping effect of the confidence term [10], the data term representing structure information soon become insignificant, which may cause problem when filling images rich in structures. Second, the SSD criterion only considers the intensity difference between target patch and source patch and neglects the difference of structure variation within these two patches. We proposed to improve the Criminisi algorithm by (1) modifying the confidence updating strategy to decrease the dropping effect and (2) adding a structure consistent patch matching to evaluate the structure differences between two patches.

3.1   The Space Varying Updating Strategy for the Confidence Term

The priority function has critical defect due to the dropping effect of the confidence term [10]. The filling order is crucial to maintain structure information. However, with the fast dropping of the confidence term, the priority advance provided by the structure information soon becomes insignificant. A method has been proposed to solve the problem by reforming the priority function [10], which changes the multiplication operator to addition operator. However, when one term is generally larger than the other in the priority estimation, it becomes the only factor affecting the priority. We propose a simple but effective updating strategy for the confidence term with two variations to solve the dropping problem.

First, as the center of the patch $\psi_{\hat{p}}$, the matching process is performed with information surrounding the center pixel. The closer to the center, the more reliable the matching is. Instead of initializing the confidences of the newly filled pixels to the same value as in the original updating strategy, we consider the priority of the center pixels $\hat{p}$ higher than its surroundings. The Euclidean distances to the center pixel $\hat{p}$ are used to differentiate pixels in different locations in the patch.

Second, a upper bound and a lower bound are defined to set the space varying confidence to the points in $\psi_{\hat{p}} \cap \Omega$. To decrease the dropping of the confidence value but maintain the idea that the newly filled pixels have smaller confidence than the already existing pixels, the upper bound is set to

$$C_{upper} = \frac{\sum_{t \in \psi_{\hat{p}} \cap \Phi} C(t)}{\left| \psi_{\hat{p}} \cap \Phi \right|}, \tag{7}$$

where $\left| \psi_{\hat{p}} \cap \Phi \right|$ is the number of known pixels in that patch $\psi_{\hat{p}}$. Naturally, the lower bound is set to $C\left(\hat{p} \right)$.

The updating strategy can be summarized as

$$C(t) = \max \left(-\lambda * dis\left(\hat{p}, t \right)^2 + C_{upper}, C\left(\hat{p} \right) \right), t \in \psi_{\hat{p}} \cap \Omega, \tag{8}$$

where $dis\left(\hat{p}, t\right)$ is the Euclidean distance between pixel $\hat{p}$ and pixel t, and λ is a decreasing factor which controls the decreasing rate of the confidence value and set to 0.03 empirically. Addition is used instead of multiplication between $dis\left(\hat{p}, t\right)$ and C_{upper} because we want to fix the decreasing rate for all patches. Figure 2 shows the space varying updating strategy for a single line. The center pixels has the largest confidence the upper bound C_{upper}, the confidence values of its neighbors are dropped quadratically until meets the lower bound.

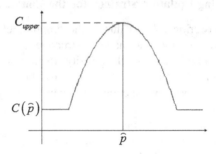

Fig. 2. Space varying updating strategy.

3.2   Structure Consistent Patch Matching

We aim to find a simple but effective criterion to measure the structure consistency of the source patch and target patch, which can also use the information estimated in the SSD calculation so that the computation burden will not be increased. To evaluate whether two patches have similar structure, the most direct way is to evaluate the patch difference. However, after taking the sum of squared differences, the information of how these differences distribute is ignored. A simple and effective way to evaluate the distribution is through the standard deviation of patch differences. The following criterion is thus proposed as

$$d\left(\psi_{\hat{p}}, \psi_q\right) = d_{SSD}\left(\psi_{\hat{p}}, \psi_q\right) * \left\{\left[d_{STD}\left(\psi_{\hat{p}} - \psi_q\right)\right]^{\beta} + 1\right\} \tag{9}$$

With

$$d_{STD}(\psi_{\hat{p}} - \psi_q) = \sqrt{\frac{1}{m}\sum_t \left[\psi_{\hat{p}}(t)M_{\hat{p}}(t) - \psi_q(t)M_{\hat{p}}(t) - \left(\bar{\psi}_{\hat{p}} - \bar{\psi}_q\right)\right]^2} \tag{10}$$

where STD is the standard deviation operator; $m = \sum_t M_{\hat{p}}(t)$; β is an constant value adjusting the weights of d_{STD} over final quality; $\bar{\psi}_{\hat{p}} = \sum_t \psi_{\hat{p}}(t)M_{\hat{p}}(t)/m$ and $\bar{\psi}_q = \sum_t \psi_q(t)M_{\hat{p}}(t)/m$ are the mean values. Without loss of generality, in our work, β is empirically set to 0.8. Thus, the smaller the value of d_{STD} gets, the more similar the structure of these two patches can be. When the value of d_{STD} becomes zero, it means that the structure variations of these two patches are consistent with each other completely. d_{STD} cannot be directly used since when the two patches have same structure but under different illumination condition, $d_{STD} = 0$ causes errors. Thus, the additional 1 is added to d_{STD} before multiplying with d_{SSD}. By adding the standard deviation in the patch matching criterion, our method can achieve more accurate patch matching by taking the structure distribution into account.

4 Experimental Results and Discussion

To verify the effectiveness of our proposed inpainting method, we conduct a set of experiments and compare the results with the Criminisi algorithm [8] and the CBD algorithm [9].

Figure 3 is presented to demonstrate the influence of the β value setting in Eq. (9). For Fig. 3(a) with the target region shown in green, the inpainting results using proposed method with β equaling to 0.5, 0.8 and 1.2 are shown in Fig. 3(b)–(d), respectively. As highlighted in the red rectangle, clear visual discontinuity is presented in Fig. 3(b) and (d) while the result in Fig. 3(c) is more continuous. The same conclusion can be drawn after more images are tested. Therefore, for the rest of our experiments, the β value is set to 0.8.

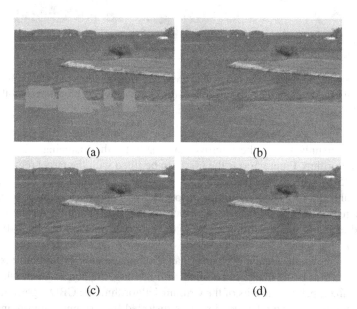

(a)   (b)

(c)   (d)

Fig. 3. Inpainting results of our proposed approach with different β values. (a) The target image. (b) Inpainting result with $\beta = 0.5$. (c) Inpainting result with $\beta = 0.8$. (d) Inpainting result with $\beta = 1.2$

Figure 4 mainly shows the improvement by the filling order of our proposed algorithm. Figure 4(a) is the image to be filled with the target region in green while Fig. 4(b)–(d) are the inpainting results of the Criminisi algorithm, the CBD algorithm and the proposed algorithm. Clear error can be observed in Fig. 4(b)–(c) when the filling priority of the sky is higher than the pyramid. Meanwhile, the proposed algorithm values the structure presented by the pyramid and obtains better performance as shown in Fig. 4(d).

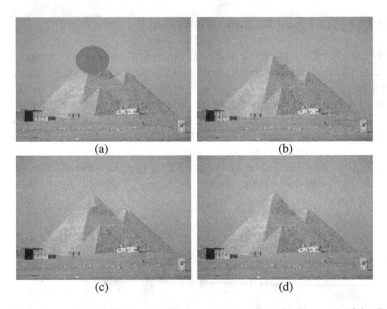

(a)   (b)

(c)   (d)

Fig. 4. Inpainting results. (a) The image with green target region. (b) The result of the Criminisi algorithm. (d) The result of the CBD algorithm. (d) The result of the proposed algorithm (Color figure online).

Figure 5 mainly shows the improvement made by the matching criterion of our proposed algorithm. Figure 5(a) is the image to be filled with the target region in green while Fig. 5(b)–(d) are the inpainting results of the Criminisi algorithm, the CBD algorithm and the proposed algorithm, respectively. Unnecessary object is filled into the target region in Fig. 5(b) and (c), which makes the inpainting results unnatural. Compared to them, the inpainting result of the proposed method in Fig. 5(d) presents more reasonable scene.

Figure 6 shows the inpainting performance of image containing both texture and edges. Figure 6(a) is the image to be filled with the target region in green while Fig. 6 (b)–(d) are the inpainting results of the Criminisi algorithm, the CBD algorithm and the proposed algorithm. The edge continuity is distorted by Criminisi algorithm and the CBD algorithm as in Fig. 6(b) and (c), and is preserved by our proposed method as shown in Fig. 6(d).

(a)   (b)

(c)   (d)

Fig. 5. Inpainting results. (a) The image with green target region. (b) The result of the Criminisi algorithm. (d) The result of the CBD algorithm. (d) The result of the proposed algorithm.

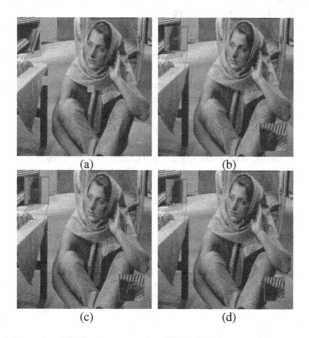

(a)   (b)

(c)   (d)

Fig. 6. Inpainting results. (a) The image to be filled with the target region marked in green. (b) The inpainting result using the Criminisi algorithm. (d) The inpainting result using the CBD algorithm. (d) The inpainting result using the proposed algorithm.

5   Conclusion

In this paper, we propose a space varying updating strategy for the confidence term to improve the filling priority estimation. We also propose a structure consistent patch matching method to take the patch difference distribution into account so that not only the overall similarity but the structure distributions of the target and source patches are evaluated. Experimental results prove the superiority of the proposed algorithm.

Acknowledgment. This research is partially supported by Natural Science Foundation of China (61402411) and Zhejiang Provincial Natural Science Foundation (LY14F020014, LY14F020015)

References

1. Bertalmio, M., Sapiro, G., Caselles, V., Ballester, C.: Image inpainting. In: Proceedings of ACM SIGGRAPH, pp. 417–424. New Orleans, Louisiana (2000)
2. Chan, T.F., Shen, J.: Mathematical models for local nontexture inpaintings. SIAM J. Appl. Math. **62**, 1019–l043 (2002)
3. Chan, T.F., Shen, J.: Non-texture inpainting by curvature-driven diffusion (CCD). J. Vis. Commun. Image Represent. **12**, 436–449 (2001)
4. Efros, A., Leung, T.: Texture synthesis by non-parametric sampling. In: Proceedings of ICCV, pp. 1033–1038. Kerkyra, Greece (1999)
5. Wei, L., Levoy, M.: Fast texture synthesis using tree-structured vector quantization. In: Proceedings of ACM SIGGRAPH, pp. 479–488 (2000)
6. Ashikhmin, M.: Synthesizing natural textures. In: Proceedings of ACM Symposium Interactive 3D Graphics, pp. 217–226 (2001)
7. Wexler, Y., Shechtman, E., Irani, M.: Space-time video completion. In: Proceedings IEEE Computer Society Conference on Computer Vision and Pattern Recognition, vol. 1, pp. 120–127 (2004)
8. Criminisi, A., Pérez, P., Toyama, K.: Region filling and object removal by exemplar-based inpainting. IEEE Trans. Image Process. **13**(9), 1200–1212 (2004)
9. Bugeau, A., Bertalmio, M., Caselles, V., Sapiro, G.: A comprehensive framework for image inpainting. IEEE Trans. Image Process. **19**(10), 2634–2645 (2010)
10. Cheng, W.H., Hsieh, C.W., Lin, S.K.: Robust algorithm for exemplar-based image inpainting. In: The International Conference on Computer Graphics, Imaging and Vision (CGIV), pp. 65–69. Beijing, China (2002)

Content-Based Image Retrieval Using Hybrid Micro-Structure Descriptor

Ying Li[(✉)] and Pei Zhang

School of Computer Science, Northwestern Polytechnical University,
Xi'an 710129, Shaanxi, China
lybyp@nwpu.edu.cn, cszhangpei@mail.nwpu.edu.cn

Abstract. Along with the speedy increase in the size of digital image collections, the content-based image retrieval (CBIR) has already become one of the hot topics in both image processing and computer vision. And it has been widely used in browsing, searching, and retrieving certain interested information from a huge amount of data. In this paper, we present a hybrid micro-structure descriptor (HMSD) to describe the image feature, which is used for image retrieval. This method makes a color quantization and edge orientation detection of the image in both HSV and RGB color space to extract four kinds of micro-structure and then hybrid these four via an effective way. The experimental results show that the information of images such as color, texture, shape, and color layout can be described more effectively by using this method and the accuracy of image retrieval is improved greatly than several well-known methods.

Keywords: Image retrieval · Micro-structure · Descriptor · CBIR

1 Introduction

Image retrieval is one of the most popular topics in the field of pattern recognition and artificial intelligence. In early 1990s, content-based image retrieval (CBIR), a technique, which uses visual contents to search images from large scale image databases according to users' interests, was presented and it has become an active and fast advancing research subject in many fields [1, 2]. The primary goal of the CBIR system is to extract primitive visual features such as color, shape and texture automatically from the images and to retrieve images on the basis of these features [3].

Some algorithms that combine color and texture features together have achieved very good image retrieval results, such as the color edge co-occurrence histogram [4], integrative co-occurrence matrix [5], micro-structure descriptor (MSD) [6], etc. MSD is a combined image feature, which integrate color, texture, shape features and image color layout information, in order to simulate processing in the human visual system. Through this method we can represent image local features in an effectively way, however, it has a weakness in describing the information of color and edge orientation of the image. Furthermore, the complex process of extracting microstructure is also needed to be improved. In order to overcome the lack of MSD, this paper proposed a method called Hybrid MSD (HMSD), that is going to get a color quantization and edge

© Springer International Publishing Switzerland 2015
X. He et al. (Eds.): IScIDE 2015, Part I, LNCS 9242, pp. 411–419, 2015.
DOI: 10.1007/978-3-319-23989-7_42

orientation detection of the image in both HSV and RGB color space to extract four kinds of micro-structure and then hybrid these four via an effective way. Our experimental results show that the proposed method has a better performance than other image retrieval methods such as Gabor feature [7], MTH [8] and MSD [6].

2 Review of Micro-Structure Descriptor

Liu et al. proposed MSD in [6], and their paper holds that the natural images can be regarded as composed by many universal micro-structures and they can be used as the basis of comparison and analysis of different images.

2.1 Color Quantization

Images are usually restored in RGB color model, they can be transformed to HSV color model which is considered more similar to human vision. Given a $W \times N$ color image $g(x, y)$, in MSD, the color will be quantized into 72 colors where the H color channel is quantized into 8 bins, S and V color channels are quantized into 3 and 3 bins, so that we get $8 \times 3 \times 3 = 72$ colors.

2.2 Edge Orientation Detection

Edge orientation detection is one of the most commonly used operations in image analysis, there are many edge detection techniques such as Sobel operator, Robert's cross operator and Prewitt's operator [9]. MSD has proposed an edge orientation detection method for color image in HSV color space as follows:

Given (H, S, V) as a point in the cylinder coordinate system, transforms it to a corresponding point in Cartesian coordinate system as (H', S', V'), where $H' = S \cdot \cos(H)$, $V' = V$, and then for each H', S' and V' channel using Sobel operator. Define $\mathbf{a}(H'_x, S'_x, V'_x)$ as the gradient along x direction and $\mathbf{b}(H'_y, S'_y, V'_y)$ as the gradient along y direction. Let θ be the angle between $\mathbf{a}$ and $\mathbf{b}$. Compute the edge orientation θ of each pixel, and then the orientation will be quantized into m bins uniformly, where $m \in \{6, 12, \ldots, 36\}$. Define $\theta(x, y)$ as the edge orientation map, $\theta(x, y) = \varphi$, $\varphi \in \{0, 1, \ldots, m\}$. According to the experimental results, quantized the orientations into six bins with an interval of $30°$ performs well, that means the value of $\theta(x, y)$ is from 0 to 5.

In MSD, move a 3×3 block from left to right and top to bottom throughout the image to find the micro-structures with similar attributes. In the 3×3 block if the neighboring pixels' values are equal to the value of the center pixel, we may mark them and keep them unchanged. For an edge orientation image $\theta(x, y)$, starting from the origin $(0, 0)$, move the 3×3 block from left to right and top to bottom throughout the image with a step-length of three pixels for micro-structure extracting. Then we can get a micro-structure map $M_1(x, y)$, where $0 \leq x \leq W - 1$, $0 \leq x \leq N - 1$. In a similarly way we can get the extraction micro-structure map $M_2(x, y)$, $M_3(x, y)$ and $M_4(x, y)$ by starting from location $(1, 0)$, $(0, 1)$ and $(1, 1)$ respectively. Finally, these four are merged into a final micro-structure map $M(x, y)$. After that, use the micro-structure map $M(x, y)$ as a mask to extract the underlying color information from the quantized image $C(x, y)$.

3   The Proposed Method HMSD

Though MSD describes image quite effectively, we found that it has three shortages as follows: (1) the color quantization method in MSD is ineffective; (2) the HSV cylindrical coordinate projection is not obvious in the edge detection; (3) the micro-structure extraction process can be rendered more efficient. In order to overcome the lack of MSD, we proposed a method called Hybrid MSD (HMSD) in this paper.

3.1   Edge Orientation Detection

In the three-dimensional color space, we apply Sobel operator to each channel in both RGB and HSV color model. Let $\mathbf{a}(X_h, Y_h, Z_h)$ and $\mathbf{b}(X_v, Y_v, Z_v)$ be the gradients along x and y directions, and let θ be the angle between $\mathbf{a}$ and $\mathbf{b}$ where X_h denotes the gradient in X channel along the horizontal direction. Their norm and dot product can be defined as

$$|\mathbf{a}| = \sqrt{(X_h)^2 + (Y_h)^2 + (Z_h)^2} \tag{1}$$

$$|\mathbf{b}| = \sqrt{(X_v)^2 + (Y_v)^2 + (Z_v)^2} \tag{2}$$

$$\mathbf{a} \cdot \mathbf{b} = X_h X_v + Y_h Y_v + Z_h Z_v \tag{3}$$

$$\theta = \arccos\left(\widehat{a, b}\right) = \arccos\left[\frac{ab}{|a| \cdot |b|}\right] \tag{4}$$

Use the formula (4) to calculate the edge orientation θ of each pixel and quantize the orientation into m bins, where $m \in \{6, 12, \ldots, 36\}$. $\theta(x, y)$ is defined as the edge orientation map, $\theta(x, y) = \varphi, \varphi \in \{0, 1, \ldots, m\}$. In Sect. 4, the experiments show us that using m = 24 with an interval of 7.5° HMSD works the best.

3.2   Color Quantization

In this paper, the color quantization method is determined through a lot of experiments and analysis. In HMSD, we quantize the color image into 120 colors in both RGB and HSV color space. The H, S and V channels are quantized into 6, 4 and 5 bins, the R, G and B channels are quantized into 6, 4 and 5 bins, respectively. So that $6 \times 4 \times 5 = 120$ colors are obtained in both color space. The quantized color images are denoted by $C_1(x, y)$ and $C_2(x, y)$. So that colors are obtained in both color space. Specifically, for the HSV color model, it is first quantized the H from [0, 360) into [0, 179], S and V are quantized from $[0, 1)$ into $[0, 255]$, then quantized the H into $[0, 1, ..., 5]$, S into $[0, 1, 2, 3]$ and V into $[0, 1, 2, 3, 4]$. For the RGB color model, the quantitative method is very similar to HSV, besides the three channels are all quantized into $[0, 255]$ in the first time.

3.3   Micro-Structure Extraction

In this paper we do not intend to use block to detect the whole image. Given an image sized $W \times N$, from left to right and top to bottom, for each pixel its neighboring pixels are on the $0°$, $45°$ and $90°$ directions. If one of the three neighbors has the same value to this pixel, then it is kept unchanged, otherwise set the value to 0. After each pixel in the quantized image has been applied this operation, the micro-structure extraction will be complete. Figure 1 shows the extraction process as below.

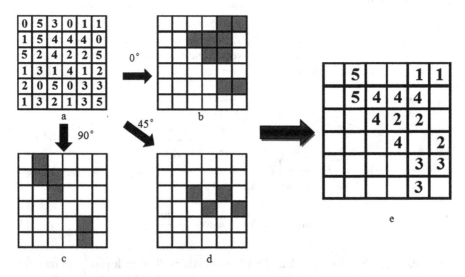

Fig. 1. The micro-structure extraction process in HMSD

Since there are color quantized color image $C(x, y)$ and edge orientation image $\theta(x, y)$ in both RGB and HSV color model, we will get four micro-structure images denoted by MC_1, MC_2, $M\theta_1$ and $M\theta_2$. On one hand, this method traverses a image for only once and is detected only in three directions, so the speed of extraction is quite fast. On the other hand, there have merged two color models of feature information of an image in this method, one is in RGB, and the other is in HSV. This is the reason why it describes the image feature better than MSD.

The next problem is how to describe the hybrid micro-structure features. All the quantized images are denoted by $f(x, y) = w$. For each point $P_0 = (x_0, y_0)$ in the image, let $f(P_0) = w_0$. Define $P_i = (x_i, y_i)$ as the three neighbors on the $0°$, $45°$ and $90°$ orientation of P_0 and $f(P_i) = w_i$, where $i = 1, 2, 3$. Let N be the co-occurring number of w_0 and w_i and $\bar{N}$ be the occurring number of w_0. Traverse every point in the quantized image from left to right and top to bottom, then use formula (5) to express the spatial correlation of neighboring and the distributes of colors.

$$H(w_0) = \begin{cases} \dfrac{N\{f(P_0) = w_0 \wedge f(P_i) = w_i \big| |P_0 - P_i| = D\}}{3\bar{N}\{f(p_0) = w_0\} + \alpha} \\ where \quad w_0 = w_i, i \in \{1, 2, 3\} \end{cases} \tag{5}$$

There are some differences between the quantized images in HSV color model and RGB color model, and this problem exists also in the edge orientation image and common image, so we bring in a parameter α to modify the histograms of the four micro-structure images, the value of α can be estimated approximately through a lot of experiments. The histograms of edge orientation images are 24-dimensional and the histograms of quantized color images are 120-dimensional. If we combine the four histograms directly, the dimensions of the final histogram will be 288. It is very difficult for extracting features or comparing the similarity in such high dimensions. In formula (5), we can see that a histogram is a representation that tells us of how the points with the micro-structural properties distribute in all the points of a color. Hence, we can use formulas (6) and (7) to merge the two histograms of edge orientation micro-structure images and quantized color micro-structure images, respectively.

$$H(i) = Max\{H_a(i), H_b(i)\} \tag{6}$$

In formula (6), $H_a(i)$ describes the micro-structure features in RGB color model $(MC_1, M\theta_1)$, and $H_b(i)$ describes the micro-structure features in HSV color model $(MC_2, M\theta_2)$, and $H(i)$ is the result which integrates them.

Let H_θ be the histogram of hybrid edge orientation micro-structure images, and let H_C be the histogram of hybrid quantized color micro-structure images, we may use formula (7) to show how H_θ and H_C come to form the final H, where β is a correlation coefficient between H_θ and H_C, n is the dimension of H_C.

$$H(i) = \begin{cases} H_C(i) & i < n \\ \beta \cdot H_\theta(i - n) & i \geq n \end{cases} \tag{7}$$

4   Experimental Results

Corel image dataset is the most commonly used dataset in image retrieval [10]. In this paper, our experiments are carried out by using two Corel image datasets, Corel-5000 and Corel-10000. In our experiment, we use the L_1 distance for similarity measure. For each image in the dataset, an M-dimensional feature vector $\mathbf{T} = \{T_1, T_2 \ldots T_M\}$ will be extracted and stored in a file, in HMSD, we set M = 144.

In order to evaluate the effectiveness of the proposed approach HMSD, the precision and recall indices are used, which are the most common measurements for evaluating image retrieval performance. Precision P and recall R are defined in the Eqs. (8) and (9), where I_N is the number of similar images retrieved and N is the total number of images retrieved, M is the total number of similar images. In this paper we set $N = 12$, $M = 100$:

$$P = I_N/N \qquad (8)$$

$$R = I_N/M \qquad (9)$$

Tables 1 and 2 display the average retrieval precisions and recalls of MSD in RGB and HSV color space respectively, Table 3 displays the average retrieval results of HMSD. This is a synthetic result which comes from both RGB and HSV. The retrieval

Table 1. The average retrieval precision and recall of the MSD under different color and orientation quantization levels on Corel-5000 dataset in RGB color space.

Color quantization levels	Texture orientation quantization levels									
	Precision (%)					Recall (%)				
	6	12	18	24	36	6	12	18	24	36
128	50.46	50.24	50.03	49.71	50.01	6.06	6.03	6.00	5.97	6.00
64	49.42	49.32	48.77	48.87	49.14	5.93	5.93	5.85	5.86	5.90
32	46.70	46.89	46.31	46.28	46.52	5.61	5.61	5.56	5.56	5.58
16	39.08	39.53	39.31	39.47	39.26	4.69	4.69	4.72	4.74	4.71

Table 2. The average retrieval precision and recall of the MSD under different color and orientation quantization levels on Corel-5000 dataset in HSV color space.

Color quantization levels	Texture orientation quantization levels									
	Precision (%)					Recall (%)				
	6	12	18	24	36	6	12	18	24	36
192	58.04	57.9	57.74	57.45	57.56	6.96	6.95	6.93	6.89	6.91
128	58.06	57.58	57.84	57.74	57.49	6.97	6.91	6.94	6.93	6.90
108	56.89	56.54	56.54	56.73	56.17	6.83	6.78	6.77	6.81	6.74
72	55.92	55.95	56.02	56.15	55.79	6.71	6.71	6.72	6.74	6.70

Table 3. The average retrieval precision and recall of the HMSD under different color and orientation quantization levels on Corel-5000 dataset.

Color quantization levels	Texture orientation quantization levels									
	Precision (%)					Recall (%)				
	6	12	18	24	36	6	12	18	24	36
192	63.08	64.03	64.80	65.34	65.19	7.57	7.68	7.78	7.84	7.82
144	63.84	64.71	65.35	65.80	65.39	7.66	7.77	7.84	7.90	7.85
128	63.74	64.75	65.54	65.75	65.00	7.65	7.77	7.86	7.89	7.80
120	63.97	65.01	65.64	66.05	65.27	7.68	7.80	7.88	7.93	7.83
108	63.96	64.81	65.58	65.78	64.90	7.68	7.78	7.88	7.89	7.79
72	62.72	63.67	64.29	64.24	62.67	7.53	7.64	7.72	7.71	7.52

Table 4. The average retrieval precision and recall ratios by various methods on Corel-5000 and Corel-10,000.

Dataset	Peformance	TCM	EOAC	Gabor	MTH	MSD	HMSD
Corel-5000	Precision (%)	27.36	31.23	36.22	49.84	55.92	66.05
	Recall (%)	3.28	3.74	4.35	5.98	6.71	7.93
Corel-10,000	Precision (%)	20.42	23.36	29.15	41.44	45.62	55.64
	Recall (%)	2.45	2.81	3.50	4.97	5.48	6.68

results show that the proposed method has a better performance than MSD in a same quantization level.

Table 4 lists the precision and recall results by TCM [11], EOAC [12], Gabor feature, MTH, MSD and the proposed method HMSD. The results show that the performance of HMSD is higher than other methods in both Corel-5000 and Corel-10,000 datasets.

Fig. 2. An example of image retrieval by the MSD on Corel-10,000 dataset.

Fig. 3. The image retrieval result of the same image by the HMSD on Corel-10,000 dataset.

Figures 2 and 3 are the retrieval results of MSD and HMSD on the Corel-10000 for the same image. From Fig. 2 we can see that there are three error results while from Fig. 3 we can see that there is only one error in the retrieval results.

5   Conclusion

This paper has proposed a novel descriptor HMSD for image retrieval. HMSD is an effective descriptor for describing the image. HMSD actuates a color quantization and edge orientation detection of the image in both HSV and RGB color space to extract four kinds of micro-structure, and then to hybridize these four via an effective way. HMSD can effectively represent image local features, and it can extract and describe color, texture and shape features. The experimental results show that the proposed algorithm has a better performance than other retrieval methods such as Gabor feature, MTH and MSD.

Acknowledgments. This work was supported by the Research Fund for the Doctoral Program of Higher Education, China under Grant 20126102110041, the Research Fund for the Key Project of Technology Research Plan of Ministry of Public Security, China under Grant 2014JSYJA018, the Natural Science Research Project of Education Department of Shaanxi Province, China under Grant 12JK0731, and the Royal Academy of Engineering, UK, under Grant 1314RECI025.

References

1. Huang, T., Rui, Y., Chang, S.F.: Image retrieval: past, present, and future. In: International Symposium on Multimedia Information Processing, vol. 108 (1997)
2. Liu, Y., Zhang, D., Lu, G., Ma, W.Y.: A survey of content-based image retrieval with high-level semantics. PR **40**, 262–282 (2007)
3. Jain, A.K., Vailaya, A.: Image retrieval using color and shape. PR **29**, 1233–1244 (1996)
4. Luo, J., Crandall, D.: Color object detection using spatial-color joint probability functions. TIP **15**, 1443–1453 (2006)
5. Palm, C.: Color texture classification by integrative co-occurrence matrices. PR **37**, 965–976 (2004)
6. Liu, G.H., Li, Z.Y., Zhang, L., et al.: Image retrieval based on micro-structure descriptor. PR **4**, 2123–2133 (2011)
7. Manjunath, B.S., Ma, W.Y.: Texture features for browsing and retrieval of image data. IEEE Trans. PAMI **18**, 837–842 (1996)
8. Liu, G.H., Zhang, L., Hou, Y.K., et al.: Image retrieval based on multi-texton histogram. PR **43**, 2380–2389 (2010)
9. Maini, R., Aggarwal, H.: Study and comparison of various image edge detection techniques. IJIP **3**, 1–11 (2009)
10. Müller, H., Marchand-Maillet, S., Pun, T.: The truth about corel - evaluation in image retrieval. In: Lew, M., Sebe, N., Eakins, J.P. (eds.) CIVR 2002. LNCS, vol. 2383, pp. 38–49. Springer, Heidelberg (2002)

11. Liu, G.H., Yang, J.Y.: Image retrieval based on the texton co-occurrence matrix. PR **41**, 3521–3527 (2008)
12. Mahmoudi, F., Shanbehzadeh, J., Eftekhari-Moghadam, A.M., et al.: Image retrieval based on shape similarity by edge orientation autocorrelogram. PR **36**, 1725–1736 (2003)

Improvement on Gabor Texture Feature Based Biometric Analysis Using Image Blurring

Da Huang[1], Kunai Zhang[2], and David Zhang[2(✉)]

[1] Department of Automation, Tsinghua University,
Beijing 100084, People's Republic of China
huangd12@mails.tsinghua.edu.cn
[2] Department of Computing, The Hong Kong Polytechnic University,
Kowloon, Hong Kong
{cskzhang, csdzhang}@comp.polyu.edu.hk

Abstract. Images without blurring are usually considered as high quality samples in biometric recognition. Image acquisition systems are carefully designed in order to capture clear images. However, experimental results show that the performance of Gabor texture feature based biometric recognition methods can be improved by image blurring. The experiments were conducted on the PolyU Palmprint Database using CompCode as well as on the CASIA Iris Database using IrisCode. The blurring method is to adopt a Gaussian filter to the images during pre-processing. Results indicate that there is an optimal range of each dataset and if all the images are blurred to this range, the performance of the whole dataset will reach optimal. A scheme is also proposed to find the optimal range and to blur an image to this range.

Keywords: Image blurring · Gaussian filter · Gabor filter · Palmprint recognition · Iris recognition

1 Introduction

Texture analysis is an important and useful area of research in computer vision that has a long history. Most natural surfaces appear in texture and a successful vision system must be able to analyze texture feature [1]. A large number of texture analysis algorithms have been developed which range from using random field models to multi-channel filtering techniques such as Gabor filters. The use of Gabor filters in extracting textured image features has been shown to be optimal in the sense of minimizing the joint two-dimensional uncertainty in space and frequency [2]. These filters can be considered as orientation and scale tunable edge and line detectors, and the statistics of these micro features in a given range are often used to characterize the underlying texture information. Gabor features have been used in several image analysis applications including texture classification and segmentation [3, 4], image recognition [5–7], image registration and motion tracking [8].

2-D Gabor filters based features are widely utilized in biometrics recognition. Among biometrics, face, fingerprint, iris and palmprint are the most popular biometric features. Face recognition has been researched for decades but it is still a big challenge

© Springer International Publishing Switzerland 2015
X. He et al. (Eds.): IScIDE 2015, Part I, LNCS 9242, pp. 420–430, 2015.
DOI: 10.1007/978-3-319-23989-7_43

to distinguish identical twins. Fingerprint recognition has a long history and has been the most widely used technology in the market [9, 10]. However, it has been reported that fingerprint is easy to be faked. Iris is the most reliable feature in biometrics [11, 12] which was first developed in 1987 [13], while palmprint recognition is relatively novel which developed in 2003 [14] but has comparable accuracy and reliability with iris [15]. Among these biometric methods, the most popular algorithms of both Iris and Palmprint recognition are based on 2-D Gabor filters. The 2-D multichannel Gabor filter is a windowed signal processing methods. It can be used for texture analysis for several reasons: they have tunable orientation and radial frequency bandwidths, tunable center frequencies and optimally achieve joint resolution in spatial and frequency domain.

Usually, people use clear images without blurring for biometric identification. However, it was found that a low-pass filtering stage before feature extraction can increase the recognition performance. Some low-past filters were applied to the sample images (Fig. 1(a)) and the result was that the blurring degree of the ROI images increases. Such images can even performance better than those clear images. Such an experiment was also conducted on iris images and the same result (Fig. 1(b)) was obtained. So far, no other existing method in literature has taken advantage of image blurring to achieve better performance. In the following sections of this paper, the relationship between recognition rate and image blurring will be established using our proposed model.

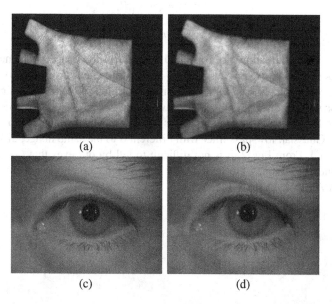

Fig. 1. (a): original clear palmprint image. (b): palmprint image after blurring. (c): original clear iris image. (d): iris image after blurring

To make the experimental results more convincing, appropriate iris recognition and palmprint recognition algorithms should be carefully selected. In the past ten years, a lot of algorithms have been developed for iris recognition and palmprint recognition, most of which utilize texture feature for matching. Among them, the most famous one for iris recognition is called IrisCode developed in 1993 and continuously developed by Daugman [16–19]. After image preprocessing, 2-D Gabor filters are applied to the ROIs for feature extraction. IrisCode is now widely used in commercial iris recognition systems and more than 50 million persons have been enrolled using IrisCode [19]. CompCode was developed by Zhang in 2003 [14], which also uses 2-D Gabor filters for feature extraction, and has been the most accurate, reliable and efficient algorithm for palmprint recognition [15]. In this paper, a group of 2-D Gaussian filters with different parameters will be applied to ROI images to change the blur degree. Experiments on palmprint ROI images are based on CompCode and experiments on iris ROI images are based on IrisCode. The results show that ROI images with some blur can achieve better performance in terms of recognition accuracy than those without blur. From these results, an optimal range can be determined where all the images should be blurred in order to make the performance better.

2 Image Blurring Approach

In our experiment, the Edge Acutance Value (EAV) was used to measure the clear degree of an image. EAV is to calculate the point sharpness value of an image.

$$EAV = \frac{\sum_{i=1}^{m \times n} \sum_{a=1}^{8} \left| df/dx \right|}{m \times n}. \tag{1}$$

where m, n denotes the number of rows and columns of the image, df is the difference in gray values between two pixels and dx is the difference in distance between two pixels. Here df/dx is applied to the eight-neighborhood. When calculating df/dx in horizontal or vertical direction, the weight we use is 1; while in 45° or 135° direction, the weight is $1/\sqrt{2}$. If the EAV of an image is large, it means it contains a lot of blur. By applying the Guassian low-pass filter with different σ to the same image, different EAV can be obtained. From Fig. 2, it is not difficult to tell by our eyes that (a) contains the most blur and (c) contains the least. After calculation, the above conclusion can be verified: EAV(a) > EAV(b) > EAV(c).

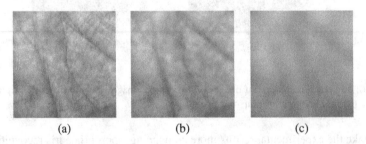

(a)   (b)   (c)

Fig. 2. Palmprint ROIs after blurring. (a): σ = 0.662, (b): σ = 1.42, (c): σ = 12.0.

Assume that the original images of a dataset are all clear images. We define the EAV mean to measure the average EAV of this dataset.

$$EAV\ mean = \frac{\sum_{i=1}^{N} EAV(I_i)}{N}.\tag{2}$$

where I_i is the ith image in the dataset which contains N images.

When a Gaussian low-pass filter with a specific σ is applied to every single image in the dataset, the EAV mean will decrease. It can be assumed that there is an optimal range [E1, E2] on the EAV axis. When the EAV mean is within this range, the recognition accuracy of this dataset will reach the best (Fig. 3).

Fig. 3. Different ranges on the EAV axis.

In order to find the optimal range [E1, E2], the Gaussian filter with a group of fixed σ was applied to the dataset and got a group of corresponding EAV mean and EER. Such an experiment has been done on the PolyU Palmprint Database Session 1 and Session 2 and CASIA Iris Database. The experimental results have verified the assumption that there is an optimal range on the EAV axis.

In order to find the optimal range [E1, E2], the Gaussian filter with a group of fixed σ was applied to the dataset and got a group of corresponding EAV mean and EER. Such an experiment has been done on the PolyU Palmprint Database Session 1 and Session 2 and CASIA Iris Database. The experimental results have verified our assumption that there is an optimal range on the EAV axis.

With such a conclusion, we propose a pre-processing stage which can be utilized in any Gabor texture feature based analysis method (Fig. 4).

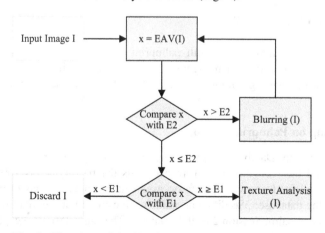

Fig. 4. Flowchart of the blurring stage before texture analysis.

Suppose that all the original sample images are clear images. Before an image goes to texture analysis, the EAV of the image is computed. If the EAV is larger than E2, it should be blurred with a specific low-pass filter to make the EAV lower. Then the EAV of this blurred image will be computed and compared to E2 again. Repeat the previous steps until the EAV is within [E1, E2]. If the blurring step, for example the step of σ in the Gaussian low-pass filter, is reasonably small enough to avoid making the image over blurred, after a few times of blurring the EAV will certainly goes to the optimal range [E1, E2]. But if the EAV is less than E1, the image will be discarded. The algorithm is described below.

```
Input: The clear image I; σ is initialized to a very
small value; ε is the step and initialized to a very
small value.
Output: The properly blurred image I.
LOOP
  x ← EAV(I)
  if x > E2 then
    Blurring(I, σ)
    σ ← σ + ε
    go to LOOP
  else
    if x >= E1
      TextureAnalysis(I)
    else
      return Discard(I)
    end if
  end if
end LOOP
return I
```

3 Experimental Results

Experiments were conducted on both palmprint dataset and iris dataset. The most popular algorithm in palmprint recognition is CompCode while in iris recognition is IrisCode. Both the two algorithms use 2-D Gabor filter to extract texture features.

3.1 Blurring on Palmprint Database

The PolyU Palmprint Database [20] was employed to analyze the relationship between EAV mean and the recognition performance. This database contains 7605 grayscale palm images collected on two separate occasions at an interval of about two months. Session 1 of the database: 386 different palms were captured at the first time, about 10 images from each palm. Session 2 of the database: The same 386 were captured, about 10 images from each palm. The most popular palmprint recognition algorithm

CompCode [14] was adopted. After ROI extraction, a set of Gaussian filters with fixed window size 19*19 and different σ were applied to the ROI images. All the matching scores were computed as well as the Equal Error Rate (EER). In our experiments, EER is used to measure the recognition performance: the lower the EER is, the better the recognition performance is. The data of the "Original" column in Table 1 comes from original clear images. In Session 1, the best EER is 0.00031 % when the dataset is blurred to the EAV of 14.70, while the original EER is 0.0066 %; in Session 2, the best EER is 0.0364 % when the dataset is blurred to the EAV between 16.06 and 16.55, while the original EER is 0.0437 % (Table 2).

Table 1. Blurring experiment using Gaussian filter on PolyU Palmprint Database Session 1

NO	Original	1	2	3	4	5	6	7
σ	N/A	0.662	1.12	1.29	1.42	**1.635**	2.15	2.71
Mean (EAV)	42.36	23.54	17.09	16.06	15.45	**14.70**	13.53	12.74
EER (%)	0.0066	0.0014	0.000714	0.000844	0.00042	**0.00031**	0.000749	0.001
GAR, FAR = 0	0.9992	0.9996	0.9995	0.9996	0.9996	**0.9997**	0.9986	0.9977

Table 2. Blurring experiment using Gaussian filter on PolyU Palmprint Database Session 2

NO	Original	1	2	3	4	5	6	7
σ	N/A	0.662	0.935	1.12	**1.2**	**1.29**	1.4	1.635
Mean (EAV)	40.05	23.04	18.62	17.05	**16.55**	**16.06**	15.56	14.74
EER (%)	0.0437	0.0439	0.0438	0.038	**0.0364**	**0.0365**	0.0437	0.0437
GAR, FAR = 0	0.9978	0.9982	0.9982	0.9982	**0.9980**	**0.9983**	0.9981	0.9979

3.2 Blurring on Iris Database

The CASIA iris database [21] was also employed to verify the conclusion. The database contains 663 different irises and each iris has 6 images. The most popular iris recognition algorithm IrisCode [12] was adopted. After ROI extraction, a set of Gaussian filters with fixed window size 19*19 and different σ were applied to the ROI images. All the matching scores were computed as well as EER. In our experiments, the lower the EER is, the better the recognition performance is. Also, the data of the "Original" column in Table 1 comes from original clear images. When the dataset is blurred to the EAV of 16.83, the best EER of 0.1611 % can be obtained, while the original EER is 0.1809 % (Table 3).

Table 3. Blurring experiment using Gaussian filter on CASIA Iris Database

NO	Original	1	2	3	4	5	6	7
σ	N/A	0.31	0.345	0.364	**0.382**	0.415	0.45	0.495
Mean (EAV)	17.28	17.26	17.16	17.02	**16.83**	16.36	15.68	14.77
EER (%)	0.1809	0.1809	0.1823	0.1698	**0.1611**	0.1688	0.1887	0.1912
GAR, FAR = 0	0.9559	0.9559	0.9576	0.9565	**0.9614**	0.9626	0.9601	0.9567

4   Performance Analysis

To analyze why blurring images can improve Gabor texture feature based biometric recognition performance, further experiments have been conducted on the PolyU Palmprint Database. Gaussian filter window size, matching score and the distance between genuine class and impostor class are analyzed. The selection of optimal range is also discussed according to the analysis results.

4.1   Gaussian Filter Window Size

The Gaussian filter function has two main parameters: the window size and σ. In Sect. 3, all the experiments use a fixed window size of 19×19 because the window size barely affects the performance according to our previous experimental results in Table 4. The σ was fixed to 1.635 when the window size varied from 13×13 to 19×19. The results show the window size has very little influence on the performance.

Table 4. Gaussian filter window size experiment on PolyU Palmprint Database Session 1

Size	9×9	11×11	13×13	15×15	17×17
σ	1.655	1.638	1.635	1.635	1.635
EAV	14.7045	14.6973	14.6985	14.6978	14.6977
EER (%)	0.00025	0.00028	0.00033	0.00033	0.00031
GAR, FAR = 0	0.999703	0.999703	0.999778	0.999703	0.999703

4.2   Intra-class Distance and d_{prime}

Experimental results illustrate that the images after blurring become more blurred but the EER of the whole dataset become lower, which indicates that the inter-class distance or the intra-class distance or both the two distances become smaller. To analyze the reason why image blurring can improve the performance, the distribution of the genuine matching score and imposter matching score are compared before blurring and after blurring (Fig. 5). The genuine score distribution shifts towards 0 while the imposter score distribution remains almost the same. The matching score is actually the distance between two palmprint samples [14]. The distribution indicates that the decrease of the intra-class distance is larger than the decrease of the inter-class distance, which means the distance between the two classes becomes larger. To measure the distance, d_{prime} of the two classes after different blurring is calculated (Fig. 6).

$$d_{prime} = \left| \frac{mean(ImpostorScore) - mean(GenuineScore)}{\sqrt{var(ImpostorScore)/2 + var(GenuineScore)/2}} \right|. \tag{3}$$

From Fig. 6, d_{prime} increases as EAV mean decreases, and after the optimal EER is obtained, d_{prime} then decreases. As a conclusion, the main contribution to the

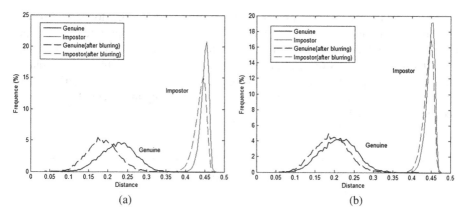

Fig. 5. Matching Scores. PolyU Palmprint Database (a): Session 1. (b): Session 2.

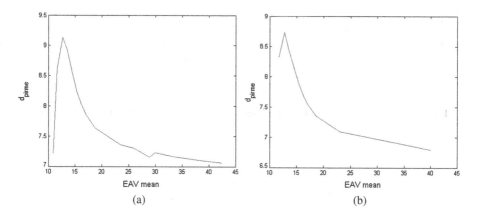

Fig. 6. d_{prime}-EAV curve. PolyU Palmprint Database (a): Session 1. (b): Session 2.

improvement of the performance is the decrease of the intra-class distance and the increase of d_{prime} caused by blurring.

4.3 Optimal Range

Given that all the original images are clear images, at the beginning, as σ increases, the EAV mean of the dataset decreases as well as the EER. When EAV mean reaches a specific value called EAV_{opt}, the EER will reaches the optimal point called EER_{opt}. After that, as σ goes on increasing, the EAV mean will continue decreasing while the EER will go up. Figure 7 describes the EER-EAV curve. We can define an $\alpha > 0$, and find two points on the curve where EER equals to $(EER_{opt} + \alpha)$. The corresponding EAVs of this two points are defined as E_1 and E_2. The optimal range is then defined as $[E_1, E_2]$ (Fig. 8). For a specific dataset using a specific blurring algorithm, the EAV_{opt}

and EER_{opt} are unique, but the optimal range can still vary according to the selection of different α. For different applications, α could be different. In palmprint recognition, the noise has little influence on ROI extraction and therefore ROI remains almost the same no matter it is extracted from a clear image or a blurred image. In this case, α can be small so that the optimal range is more accurate. After considering the experiments on both sessions of PolyUPalmprint Database, the suggested optimal range for palmprint recognition is as Table 5.

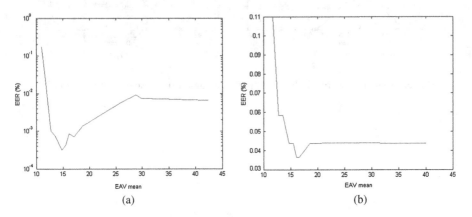

Fig. 7. EER-EAV curve. PolyU Palmprint Database (a): Session 1 and (b): Session 2.

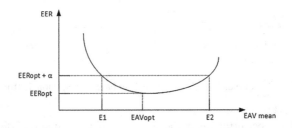

Fig. 8. Optimal range and the EER-EAV curve.

Table 5. Suggested α for palmprint recognition

EAV_{opt}	$EER_{opt}(\%)$	α	EERrange	Optimal range
15.0	0.036	0.007	[0.036,0.043]	[15.5,40.0]

5 Conclusion

This paper presents a finding that the performance of Gabor texture feature based biometric analysis can be improved by blurring images. Experiments were conducted on palmprint dataset and iris dataset using Gaussian filter as a blurring method.

Experimental results show that the performance of the two biometric recognitions is improved and there is an optimal range in each dataset. When all the sample images are blurred to this optimal range, the recognition performance reaches the best. A pre-processing scheme is also proposed to blur a clear image to the optimal range in order to improve the recognition accuracy. The decrease of intra-class distance and increase of d_{prime} between the genuine and impostor classes are the main factors of improving the performance. The selection of the optimal range varies according to different requirements and applications. By selecting an appropriate optimal range and following the blurring scheme, performance of Gabor texture feature based biometric analysis can be improved.

References

1. Tuceryan, M., Jain, A.K.: Texture analysis. In: Chen, C.H., Pau, L.F., Wang, P.S.P. (eds.) The Handbook of Pattern Recognition and Computer Vision, 2nd edn, pp. 207–248. World Scientific Publishing Co., Singapore (1998)
2. Daugman, J.G.: Complete discrete 2-D Gabor transforms by neural networks for image analysis and compression. IEEE Trans. ASSP **36**, 1169–1179 (1998)
3. Bovic, A.C., Clark, M., Geisler, W.S.: Multichannel texture analysis using localized spatial filters. IEEE Trans. Pattern Anal. Mach. Intell. **12**(1), 55–73 (1990)
4. Manjunath, B.S., Chellappa, R.: A unified approach to boundary perception: edges, textures, and illusory contours. IEEE Trans. Neural Netw. **4**(1), 96–108 (1993)
5. Daugman, J.G.: High confidence visual recognition of persons by a test of statistical independence. IEEE Trans. Pattern Anal. Mach. Intell. **15**(11), 1149–1161 (1993)
6. Lades, M., et al.: Distortion invariant object recognition in the dynamic link architecture. IEEE Trans. Comput. **42**(3), 300–311 (1993)
7. Manjunath, B.S., Chellappa, R.: A feature based approach to face recognition. In: Proceedings of IEEE Conference on CVPR 1992, pp. 373–378. Champaign, Ill., June 1992
8. Manjunath, B.S., Shekhar, C., Chellappa, R.: A new approach to image feature detection with applications. Pattern Recogn. (1996)
9. Jain, A., Hong, L., Bolle, R.: On-line fingerprint verification. IEEE Trans. Pattern Anal. Mach. Intell. **19**(4), 302–313 (1997)
10. Maio, D., Maltoni, D., Cappelli, R., Wayman, J.L., Jain, A.: FVC2000: fingerprint verification competition. IEEE Trans. Pattern Anal. Mach. Intell. **24**(3), 402–412 (2002)
11. Zhang, D.: Automated Biometrics: Technologies and Systems. Kluwer Academic Publishers, Dordrecht (2000)
12. Wildes, R.: Iris recognition: an emerging biometric technology. Proc. IEEE **85**(9), 1348–1363 (1997)
13. Flom, L., Safir, A.: Iris recognition system. US Patent 4,641,349 (1987)
14. Zhang, D., Kong, W.K., You, J.: Online palmprint identification. IEEE Trans. Pattern Anal. Mach. Intell. **25**(9), 1041–1050 (2003)
15. Zhang, D., Zuo, W., Yue, F.: Comparative study of palmprint recognition algorithms. ACM Comput. Surv. **44**(1), Article 2 (2012)
16. Daugman, J.: How iris recognition works. IEEE Trans. Comput. Syst. Video Technol. **14**(1), 21–30 (2004)
17. Daugman, J.: Probing the uniqueness and randomness of IrisCode: results from 200 billion iris pair comparisons. Proc. IEEE **94**, 1927–1935 (2006)

18. Daugman, J.: New methods in iris recognition. IEEE Trans. Syst. Man Cybern. B Cybern. **37**(5), 1167–1175 (2007)
19. Kong, A.W.K., Zhang, D., Kamel, M.S.: An analysis of IrisCode. IEEE Trans. Image Process. **19**(2), 522–532 (2010)
20. PolyUPalmprint Database. http://www.comp.polyu.edu.hk/~biometrics/
21. CASIA Iris Database. http://biometrics.idealtest.org/dbDetailForUser.do?id=4

Robust Face Recognition with Occlusion by Fusing Image Gradient Orientations with Markov Random Fields

Xiao-Xin Li[1]([✉]), Ronghua Liang[2], Yuanjing Feng[2], and Haixia Wang[2]

[1] College of Computer Science and Technology,
Zhejiang University of Technology, Hangzhou 310023, China
`mordekai@zjut.edu.cn`
[2] College of Information Engineering,
Zhejiang University of Technology, Hangzhou 310023, China
`{rhliang,fyjing,hxwang}@zjut.edu.cn`

Abstract. Partially occluded faces are very common in automatic face recognition (FR) in the real world. We explore the problem of FR with occlusion in the domain of Image Gradient Orientations (IGO) and center on the probabilistic generative model of occluded images. The existing works usually put stress on the error distribution in the non-occluded region but neglect the distribution in the occluded region for the unpredictability of occlusions. However, in the IGO domain, this distribution can be built simply and elegantly as a uniform distribution in the interval $[-\pi, \pi)$. We fully use this distribution to build the probabilistic error model conditioned on the occlusion support and construct a new error metric, which fully harnesses the spatial and statistical local information of two compared images and plays a very important role in initializing the occlusion support. In addition, we extend the definition of occlusions to other variations, such as highlight illumination changes, and suggest these occlusion-like variations should also be detected and excluded from further recognition. To detect the occlusion support accurately, the contiguous structure of occlusions is modeled using a Markov random field (MRF). By fusing IGO with MRF, we propose a new error coding model, called Double Weighted Error Coding (DWEC), for robust FR with occlusion. Experiments demonstrate the effectiveness and robustness of DWEC in dealing with occlusion and occlusion-like variations.

Keywords: Unconstrained face recognition · Face occlusion · Image gradient orientations · Markov random field

1 Introduction

In unconstrained settings [5], facial occlusions are very common. When there exists partial occlusions or disguises in the test face images, robust classifiers, such as SRC [12] and CESR [4], might fail to perform recognition accurately.

© Springer International Publishing Switzerland 2015
X. He et al. (Eds.): IScIDE 2015, Part I, LNCS 9242, pp. 431–440, 2015.
DOI: 10.1007/978-3-319-23989-7_44

Existing works [4, 6, 11, 12] show that it is not the missing discriminative information caused by occlusion but the common high order statistical structures (localization, orientation and bandpass) shared by occlusions and faces that mainly account for the FR performance drop. Therefore, the primary task for FR with occlusion is to effectively eliminate the influence of occlusions.

In uncontrolled settings, however, the spatial locations of the occluded pixels in a face image are usually priori unknown. Researchers suggest a variety of methods to estimate the occlusion locations. One main solving clue is to build an occlusion probability estimation map according to the error image between the occluded face image and its reconstruction, referred to as reconstruction error. Within the framework of sparse coding, the probabilistic generative model of the reconstruction error has been extensively explored [1, 8, 11, 12]. These models usually assume the reconstruction error follows a specific distribution. However, recent researches [13, 14] show that the reconstruction error may be far from any specific distribution, and pay attention to the weighted or separated error distribution. The weighted error distribution models, such as the robust sparse coding (RSC) [14] and the correntropy-based sparse representation (CESR) [4], try to weaken the effect of the error caused by outliers and focus on the error distribution in the non-occluded region. Nevertheless, the error distribution in the occluded region is also very important since it indicates how the occlusions are understood by the model. Tzimiropoulos et al. [9] observe that the gradient orientation differences of dissimilar images follow a *uniform distribution* on the interval $[-\pi, \pi)$ with a high significant level. With this observation, they further show that local orientation mismatches caused by outliers (such as occlusions) can be canceled out when the cosine kernel is applied, and propose an **IGO** (*Image Gradient Orientations*) subspace learning framework, such as IGO-PCA and IGO-LDA. However, no matter the models bias the error distribution in the non-occluded region or the ones bias the error distribution in the occluded region cannot deal with high level occlusions or complex occlusions very well. It might be more reasonable to build the error distributions in the two regions separately and explicitly. Li et al. [6] suggest a structured sparse error coding (SSEC) model, which describes the two regions of the reconstruction error using two independent exponential distributions. However, SSEC requires many training samples and any variation in the non-occluded region of the test image, such as illumination change, which is absent from the training samples, might lead to a false recognition of SSEC.

Another clue to detect occlusion is to explore the spatial structure of the occlusion support. With the prior knowledge that natural occlusions are usually local and contiguous in location, Zhou et al. [15] explore the spatial contiguous structure of occlusions using the Markov Random Fields (MRF) and propose a robust algorithm, called Sparse Error Correction with MRF (SEC_MRF). Constrained by the local property of MRF and the employed non-structured error metric, SEC_MRF, as with many existing methods, cannot deal with high-level occlusion effectively. Li et al. [6] also explore the shape structure of occlusion in their proposed SSEC method using a new graph model (i.e., the morphological graph).

The above two research clues are unified into the framework of the joint probabilistic generative model of the reconstruction error and the occlusion support by SEC_MRF [15] and SSEC [6]. In this work, we continue to study this framework in the uncontrolled settings, where there generally exists two main problems: *(i)* the number of training images is relatively small; *(ii)* other variations except for occlusions might also exist. To solve these problems and to enhance the robustness of FR with occlusion, we incorporate the outlier elimination solution implied in the IGO methods [9] into the framework of the joint probabilistic generative model. Experiments demonstrate the effectiveness and robustness of the proposed method in dealing with occlusions in the uncontrolled settings.

2   Probabilistic Generative Model for Occluded Images

Suppose we have a set of labeled training images $A = [A_1, A_2, \cdots, A_K] \in \mathbb{R}^{m \times n}$ of K subjects, where $A_k = \left[a_1^k, a_2^k, \cdots, a_{n_k}^k\right] \in \mathbb{R}^{m \times n_k}$ is a data matrix consisting of n_k training samples from subject k and $n = \sum_{k=1}^{K} n_k$. What is critical to recognize a new occluded face image $y \in \mathbb{R}^m$ from these training images is to detect and exclude its occluded region. We denote its occlusion support by $s \in \{-1, 1\}^m$, where $s_i = -1$ indicates pixel y_i is non-occluded and $s_i = 1$ indicates pixel y_i is occluded. For convenience, we further let $\dot{\mathcal{P}} = \{i \,|\, s_i = -1\}$ and $\ddot{\mathcal{P}} = \{i \,|\, s_i = 1\}$ denote the index set of the non-occluded pixels and that of the occluded ones, respectively, and then $\mathcal{P} = \dot{\mathcal{P}} \cup \ddot{\mathcal{P}}$ denotes the index set of the whole image pixels. We denote by $\dot{e} = e_{\dot{\mathcal{P}}}$ the error in the non-occluded region, and by $\ddot{e} = e_{\ddot{\mathcal{P}}}$ the error in the occluded region.

From the view of statistics, an effective way to detect the occlusion support s of the test image y is to model an effective probabilistic generative model about s and its related factors, such as the reconstruction error e between y and its reconstruction $\hat{y} \in \mathbb{R}^m$ with respect to (w.r.t.) the training set A. As the work in [6,15], we formulate the probabilistic generative model as the joint probability density function (PDF) $p(e, s)$. As discussed in the introduction, the error $\dot{e}$ and $\ddot{e}$ might follow some specific distribution, respectively, which lead to the following conditional PDF

$$\begin{aligned}
p(e|s) &= p(\dot{e}|s)\, p(\ddot{e}|s) \\
&= \prod_{i \in \dot{\mathcal{P}}} p(e_i|s_i = -1) \prod_{i \in \ddot{\mathcal{P}}} p(e_i|s_i = 1) \\
&= \prod_{i \in \mathcal{P}} \dot{p}(e_i)^{\dot{s}_i}\, \ddot{p}(e_i)^{\ddot{s}_i},
\end{aligned} \tag{1}$$

where we define $\dot{p}(e_i) \triangleq p(e_i|s_i = -1)$ and $\ddot{p}(e_i) \triangleq p(e_i|s_i = 1)$ as the conditional PDF in the non-occluded region and in the occluded region, respectively, and let $\dot{s} \triangleq \frac{1-s}{2}$ and $\ddot{s} \triangleq \frac{1+s}{2}$.

3   Markov Random Field for Occlusion Support

We first review the priori probabilistic mass function (PMF) $p(s)$ modeled by a Markov Random Field (MRF). In general, a pixel surrounded with occluded pixels might be also occluded with a high probability. This property is called the contiguity of occlusion in recognition literature [2,6,15], and can be modeled by an MRF $p(s) \propto \prod_{i \in \mathcal{P}} \mu^{\mathring{s}_i} (1-\mu)^{\mathring{s}_i} \prod_{j \in \mathcal{N}(i)} \exp\left(-\frac{d_{ij}^2}{\sigma_d^2}(1-s_i s_j)\right)$, where d_{ij} is the Euclidean distance between two pixels i and j, and $\mathcal{N}(i)$ is pixel i's neighborhood. As d_{ij} is almost constant for all $j \in \mathcal{N}(i)$, substitute $\frac{d_{ij}^2}{\sigma_d^2}$ with λ_s, and $p(s)$ can be reduced to

$$p(s) \propto \exp\left(\sum_{i \in \mathcal{P}} \lambda_\mu s_i + \sum_{i \in \mathcal{P}} \sum_{j \in \mathcal{N}(i)} \lambda_s s_i s_j\right), \tag{2}$$

where $\lambda_\mu = \frac{1}{2} \log \frac{\mu}{1-\mu}$. Note that (2) is in accordance with the form of the Ising model, where $\sum_{i \in \mathcal{P}} \lambda_\mu s_i$ is the data energy and $\sum_{i \in \mathcal{P}} \sum_{j \in \mathcal{N}(i)} \lambda_s s_i s_j$ is the smooth energy.

4   Conditional Probabilistic Error Model in IGO Domain

We now concentrate on the conditional PDF $p(\mathring{e}|s)$ in the IGO domain, where $\mathring{e} = \mathring{y} - \mathring{\hat{y}} = \mathring{y} - \mathring{A}x = \Phi(y) - \Phi(A) \cdot x$. Here, $\Phi(\cdot)$ denotes the IGO transformation function, $\mathring{y}$ denotes the IGO feature of y, and $x \in \mathbb{R}^n$ is the coding coefficient of the reconstruction $\mathring{\hat{y}}$ w.r.t. $\mathring{A}$.

4.1   Uniform Error Distribution in the Occluded Region

In [9], Tzimiropoulos *et al.* statistically verify that *the gradient orientation differences $\mathring{d} \in \mathbb{R}^m$ of any two pixel-wise dissimilar images $a \in \mathbb{R}^m$ and $b \in \mathbb{R}^m$ follow a uniform distribution in the interval $[-\pi, \pi)$ with a high significant level.* According to this experimental result, we naturally confirm that the error distribution in the occluded region follow the uniform distribution in $[-\pi, \pi)$ with a high significant level, that is,

$$\mathring{p}(\mathring{e}_i) = p(\mathring{e}_i|s_i = 1) = \frac{1}{2\pi}. \tag{3}$$

In spite of its simplicity and reliability in experiments, the conditional PDF (3) should satisfy that the two comparative images, y (or $\mathring{y}$) and its reconstruction $\hat{y}$ (or $\mathring{\hat{y}}$), should be point-wise dissimilar in the occluded region, which requires the occluded image y (or $\mathring{y}$) should be rightly coded w.r.t. the training set A (or $\mathring{A}$). We will discuss the coding scheme of y in Sect. 5.

4.2 Weight-Conditional Gaussian Error Distribution in the Non-Occluded Region

We now consider the PDF $\dot{p}\left(\mathring{e}_i\right) = p\left(\mathring{e}_i|s_i = -1\right)$ in the non-occluded region, where the reconstruction error $\mathring{e}_{\dot{p}}$ might be caused by Gaussian white noise, expression and illumination change, misalignment and etc. These variations existing between $y_{\dot{p}}$ and $\hat{y}_{\dot{p}}$, unlike occlusions, are irregular and not point-wise dissimilar. To explore the distribution of the errors incurred by irregular outliers, researchers usually resort to the regularized weighted error coding model [4,14]: $\min_{e,w} \sum_{i\in\mathcal{P}} \left((w_i e_i)^2 + \lambda_w \omega\left(w_i\right)\right)$, where $\lambda_w \geq 0$ is a regularized constant, $w_i \geq 0$ is a weight variable, and $\omega\left(w_i\right)$ is a cost function imposed on w_i[1]. In the IGO domain, according to this regularized weighted error coding model, we can formulate the PDF of the reconstruction error $\mathring{e}_i$ in the non-occluded region as follows

$$\dot{p}\left(\mathring{e}_i, w_i\right) = \dot{p}\left(\mathring{e}_i|\,w_i\right)p\left(w_i\right),$$

$$\propto \exp\left(-\frac{(w_i\mathring{e}_i)^2}{2\sigma^2} - \lambda_w\omega\left(w_i\right)\right). \tag{4}$$

5 The Proposed Model

Double Weighted Error Coding Model. By combining (3), (4) and (2), we have the following joint generative model

$$p\left(\mathring{e}, w, s\right) = p\left(\mathring{e}, w|\,s\right)p\left(s\right),$$

$$\propto \exp\left(-\sum_{i\in\mathcal{P}}\frac{1}{2}\left(1 - s_i\right)\left((w_i\mathring{e}_i)^2 + \tilde{\lambda}_w\omega\left(w_i\right)\right)\right.$$

$$\left.\sum_{i\in\mathcal{P}}\lambda_d s_i + \sum_{i\in\mathcal{P}}\sum_{j\in\mathcal{N}(i)}\lambda_s s_i s_j\right) \tag{5}$$

where $\lambda_d = \lambda_\mu - \tilde{\lambda}_\sigma$. In order to estimate the optimal $\mathring{e}$, w and s from the generative model, we adopt the criterion of maximizing the joint PDF $p\left(\mathring{e}, w, s\right)$. Note that the error $\mathring{e}$ is actually measured from $\mathring{y}$ and its linear reconstruction $\hat{y}$ w.r.t. the training set $\mathring{A}$. Hence, the linear coding scheme $\mathring{A}x$ is very important to determine the reconstruction quality of $\hat{y}$. As sparse coding has been shown to be robust to outliers [4,12], we confine the coding scheme within the frame of sparse representation. Then, the proposed model can be summarized as follows

$$\left(\mathring{e}^*, w^*, s^*\right) = \arg\max_{\mathring{e},s} p\left(\mathring{e}, w, s\right)$$

$$s.t.\mathring{e} = \mathring{y} - \mathring{A}x, x \geq 0. \tag{6}$$

---

[1] In [4], $\omega\left(w_i\right)$ is set to a convex conjugate function of the Gaussian function $g\left(x\right) = \exp\left(-\frac{x^2}{2\sigma^2}\right)$, which induces that $w_i = \sqrt{g\left(e_i\right)}$.

Since both the occlusion support s and the weight w can be interpreted as the weight imposed on the error $\mathring{e}$, we refer the optimization problem (6) as *Double Weighted Error Coding* (DWEC).

Optimization and Initialization. Since there are 3 variables $\mathring{e}$, w and s in the objective function (6) of the DWEC model, it is difficult to optimize them simultaneously. We therefore adopt the alternating maximization way to optimize (6). In general, the alternatively iterative algorithm can only recover a locally optimal solution of $(w, \mathring{e}, s)$. Nevertheless, a reasonable initialization might lead to a better solution. Among the 3 variables, what is critical is to initialize the reconstruction error $\mathring{e}$. To fully utilize the spatial and statistical local information of any two compared images, we suggest initializing $\mathring{e}$ as $\mathring{e}_i^{(0)} = \mathrm{CIM}\left(\mathrm{CEM}\left(\mathring{y}_i, \mathring{\bar{y}}_i\right)\right)$, where $\bar{y}$ is the mean face of all training images, $\mathrm{CIM}\left(x_i\right) \triangleq 1 - k_\sigma\left(x_i\right)$ [4] and $\mathrm{CEM}\left(\mathring{y}_i, \mathring{\bar{y}}_i\right) \triangleq \sum_{j \in \mathcal{N}(i)}\left(1 - \cos\left(\mathring{y}_j - \mathring{\bar{y}}_j\right)\right)$.

Classifier of DWEC. Once we obtain the final estimates of $(\mathring{e}, w, s)$ or (x, w, s), we still need to identify the test image from existing subjects based on some measure of goodness-of-fit within the non-occluded region. We adopt a subject specific reconstruction classifier similar to the sparse classifier proposed by [12] (see detail from step 6 to step 7 in Algorithm 1). Algorithm 1 summarizes the full procedure of the DWEC model.

Algorithm 1. Double Weighted Error Coding (DWEC)

Input: training data A, test sample y.
Output: estimated error $\mathring{e}$, estimated occlusion support s, identity (y)

1. Calculate the mean face $\bar{y}$ of the training image A;
2. Transform y, $\bar{y}$ and A into the IGO domain: $\mathring{y} = \Phi(y)$, $\mathring{\bar{y}} = \Phi(\bar{y})$, $\mathring{A} = \Phi(A)$;
3. Initialize the error $\mathring{e}^{(0)}$, where $\forall i \in \mathcal{P}$, $\mathring{e}_i^{(0)} = \mathrm{CIM}\left(\mathrm{CEM}\left(\mathring{y}_i, \mathring{\bar{y}}_i\right)\right)$;
4. Initialize the occlusion support $s^{(0)} = \mathcal{K}\left(\mathring{e}^{(0)}\right)$;
5. Solve (w, x, s) by alternatively iterative algorithm;
6. For $k = 1, \cdots, K$, compute the residuals $r_k = \left\| w \cdot \frac{1-s}{2} \cdot \left(\mathring{y} - \mathring{A}\delta_k(x)\right) \right\|_2^2$, where $\delta_k(x)$ is a new vector whose only nonzero entries are the ones in x that correspond to subject k;
7. identity $(y) = \arg\min_k r_k$.

---

6   Simulations and Experiments

To evaluate the performance of the proposed DWEC model, we compare it with three related state-of-the-art methods for robust FR with occlusion: CESR [4], SSEC [6] and IGO-PCA [9]. The parameters of our DWEC are selected as: $\tilde{\lambda}_w = 0$, $\lambda_d = 0$, $\lambda_s = 2$ and $T = 4$. The parameters of the other methods are set according to the strategy suggested in their papers.

We conduct a set of experiments on the Extended Yale B database [3] and the AR database [7]. The Extended Yale B database contains 2414 frontal face images of 38 persons under 64 different illumination conditions. In order to simulate different levels (from 0 % to 90 %) of contiguous occlusion, we replace a randomly located square patch from each test image with a mandrill image which has similar structure with the human face and has been widely used in robust FR testing [4,6,12,14,15]. The AR database, which is one of the very few databases that contain real disguise, consists of over 4,000 color images corresponding to 126 persons' frontal view faces with different facial expressions, illumination conditions, and occlusions (sunglasses and scarves). 26 pictures were taken for each person in two sessions (separated by two weeks).

6.1   Synthetic Occlusion with Various Illumination

In this experiment, we use the Extended Yale B database to test the robustness of our algorithm against various level synthetic occlusions with various illuminations. The illumination conditions of the Extended Yale B database are partitioned to 5 subsets: from normal illumination variations to extreme ones. For training, we use clean images from Subset I and II (717 images, with normal-to-moderate illumination conditions); for testing, we use images with synthetic occlusion from Subset III (453 images, with extreme illumination conditions), Subset IV (524 images, with more extreme illumination conditions) and Subset V (712 images, with the most extreme illumination conditions), respectively. Note that the training set and all the test sets are non-overlapping. The examples of the test images are shown as Fig. 1a-d. All images are cropped and resized to 96 × 84 pixels. Note that this experimental setting is akin to the one in [10].

Figure 2 compares the recognition rates of DWEC with other related approaches on the 3 different test subsets, respectively. Clearly, the nearest competitor of DWEC is IGO-PCA, which always performs as well as DWEC when the occlusion level is lower than some critical point. This critical point,

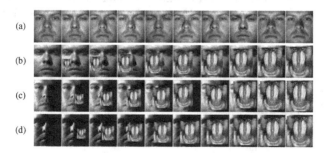

Fig. 1. Examples of the training and test images of Subject 1 from the Extended Yale B database: (a) the training samples from Subset I & II; (b)~(d) the test samples with 0 % to 90 % contiguous occlusion from different illumination subsets: (b) Subset III, (c) Subset IV, and (d) Subset V.

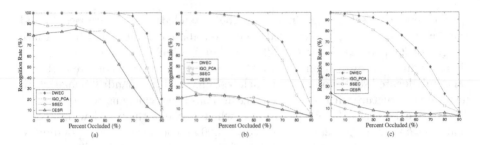

Fig. 2. Recognition against mandrill occlusion with various levels (0 %~90 %) on the Extended Yale B database. The synthetic occluded images are imposed on 3 different illuminations subsets, respectively: (a) Subset III, (b) Subset IV, and (c) Subset V.

however, rapidly declines when the illumination conditions of the test images become worse. This illustrates that the illumination conditions greatly enhance the influence of occlusions in FR. Specifically, for the most extreme illumination conditions, as seen from Fig. 2c, the critical point almost decreases to 0 and DWEC distinctly outperforms IGO-PCA. CESR and SSEC, which perform very well when the illumination conditions are not very worse, also have dramatically decreasing recognition performance when the illumination conditions worsen.

6.2   Real-World Occlusion with Non-uniform Illumination

We next test our algorithm on real disguises using the AR Face database. We select a subset of the database that consists of 65 male subjects and 54 female subjects. The grayscale images are resized to 112×92. For training, we use 952 non-occluded frontal view images (about 8 for each subject) with varying facial expressions but normal illuminations. For testing, we use images that simultaneously contain illumination variations and disguises, such as sunglasses and scarves. Figure 3 gives the training and test examples of Subject 2.

Figure 4 compares the FR performance of different methods using different downsampled images of dimensions 154, 644, 2576, and 1,0304, which correspond

Fig. 3. Examples of the training and test sets of Subject 2 from the AR database: (a) the 8 training images from two sessions with normal illumination; (b)/(c) the test images simultaneously containing illumination variations and disguises (sunglasses/scarves).

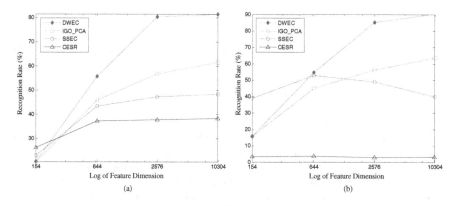

Fig. 4. Recognition rates of various algorithms under various feature spaces against real-world occlusions in the AR database: (a) sunglasses occlusion, (b) scarf occlusion.

to downsampling ratios of 1/8, 1/4, 1/2, and 1, respectively. Note that all of the 4 compared methods are within the framework of sparse coding. As indicated by [12], the choice of features is no longer critical for the sparse coding based classifiers. Nevertheless, as seen from Fig. 4, the recognition rates of the 4 compared methods do not converge to the same point as indicated in [12]. This shows that the choice of features becomes very critical, if bad features, such as occlusions and illumination changes, only exist in the test set (but not in the training set). With feature dimensions increasing, the FR performance of our DWEC significantly outperforms the other 3 methods. Specially, at the dimension 1,0304, DWEC outperforms IGO-PCA about 20 % for the sunglasses disguises and about 27 % for the scarf disguises.

7   Conclusions

Inspired by the work of [9] on the IGO features, we explore the probabilistic generative model of the occluded face images in the IGO domain. We stress that the error distribution in the occluded region is very critical to understand and detect occlusion. To detect occlusion accurately, we also use MRF to model the contiguous structure of occlusions. By effectively fusing the IGO feature with MRF, we propose the Double Weighted Error Coding (DWEC) model for robust FR with occlusion. The experiments show that both the robust IGO features and the occlusion detection are very critical for robust FR with occlusion, especially for extreme variations, such as highlight illumination changes.

Acknowlegment. This work is partially supported by National Science Foundation of China (61402411, 61379020), Zhejiang Provincial Natural Science Foundation (LY14F020015, LY14F020014), and Program for New Century Excellent Talents in University of China (NCET-12-1087).

References

1. Chen, S.S., Donoho, D.L., Saunders, M.A.: Atomic decomposition by basis pursuit. SIAM Review **43**, 129–159 (2001)
2. Dahua, L., Xiaoou, T.: Quality-driven face occlusion detection and recovery. In: Proceedings IEEE International Conference Computer Vision and Pattern Recognition, pp. 1–7 (2007)
3. Georghiades, A.S., Belhumeur, P.N., Kriegman, D.J.: From few to many: illumination cone models for face recognition under variable lighting and pose. IEEE Trans. Pattern Anal. Mach. Intell. **23**(6), 643–660 (2001)
4. He, R., Zheng, W., Hu, B.: Maximum correntropy criterion for robust face recognition. IEEE Trans. Pattern Anal. Mach. Intell. **33**(8), 1561–1576 (2011)
5. Hua, G., Yang, M.H., Learned-Miller, E., Ma, Y., Turk, M., Kriegman, D.J., Huang, T.S.: Introduction to the special section on real-world face recognition. IEEE Trans. Pattern Anal. Mach. Intell. **33**, 1921–1924 (2011)
6. Li, X.X., Dai, D.Q., Zhang, X.F., Ren, C.X.: Structured sparse error coding for face recognition with occlusion (2013)
7. Martínez, A.: The ar face database. Technical report Computer Vision Center (1998)
8. Tibshirani, R.: Regression shrinkage and selection via the lasso. J. Roy. Stat. Soc. B **58**(1), 267–288 (1996)
9. Tzimiropoulos, G., Zafeiriou, S., Pantic, M.: Subspace learning from image gradient orientations. IEEE Trans. Pattern Anal. Mach. Intell. **34**(12), 2454–2466 (2012)
10. Wei, X., Li, C.T., Hu, Y.: Robust face recognition under varying illumination and occlusion considering structured sparsity. In: 2012 International Conference on Digital Image Computing Techniques and Applications (DICTA), pp. 1–7 (2012)
11. Wright, J., Ma, Y.: Dense error correction via ℓ^1-minimization. IEEE Trans. Inf. Theory **56**(7), 3540–3560 (2010)
12. Wright, J., Yang, A., Ganesh, A., Sastry, S., Ma, Y.: Robust face recognition via sparse representation. IEEE Trans. Pattern Anal. Mach. Intell. **31**(2), 210–227 (2009)
13. Yang, M., Zhang, L., Yang, J., Zhang, D.: Regularized robust coding for face recognition. IEEE Trans. Image Process. **22**(5), 1753–1766 (2013)
14. Yang, M., Zhang, L., Yang, J., Zhang, D.: Robust sparse coding for face recognition. In: Proceedings IEEE International Conference Computer Vision and Pattern Recognition, pp. 625–632 (2011)
15. Zhou, Z.H., Wagner, A., Mobahi, H., Wright, J., Ma, Y.: Face recognition with contiguous occlusion using markov random fields. In: Proceedings IEEE International Conference Computer Vision, pp. 1050–1057 (2009)

3D Image Sharpening by Grid Warping

Andrey S. Krylov$^{(\boxtimes)}$ and Andrey V. Nasonov

Faculty of Computational Mathematics and Cybernetics,
Laboratory of Mathematical Methods of Image Processing,
Lomonosov Moscow State University, Moscow, Russia
{kryl,nasonov}@cs.msu.ru
http://imaging.cs.msu.ru/

Abstract. A method for sharpening of 3D volume images has been developed. The idea of the proposed algorithm is to transform the 3D neighborhood of the edge so that the neighboring pixels move closer to the edge, and then resample the image from the warped grid to the original pixel grid. The proposed technique preserves image textures while making the edges sharper. The effectiveness of the proposed method is demonstrated with synthetic volume images and real micro CT images.

Keywords: 3D image sharpening · Grid warping · Deblurring · micro CT · Volume images

1 Introduction

Image blur occurs in numerous types of 2D and 3D images, e.g. photographs, medical images of different modalities, telescopes, microscopes, satellite sensors, etc. As a consequence, the deblurring problem (also called deconvolution) has been widely investigated for the simpler 2D case and then extended to the 3D case. The deblurring methods require the complete knowledge or estimation of the blurring kernel, usually called as Point Spread Function (PSF). Image deconvolution is an ill-posed problem, and even the slightest error in the blur kernel may introduce strong artifacts in the resulting image.

Classical deblurring approaches are Wiener filtering, regularized linear least square algorithms, maximum likelihood and nonlinear iterative constrained methods like total variation regularization [1,3,13]. When the PSF is not known, blind deconvolution methods capable to identify the PSF and restore the image in the same time have been developed [5]. Inaccurate PSF estimation usually results in serious artifacts and distortions in the resulting image.

Blurred image enhancement methods that are not based on the PSF concept can be referred to as image sharpening methods. The commonly used method of image sharpening is the unsharp masking method [15,16]. The main problem of the existing sharpening methods is unpleasant overshoot artifact and noise amplification that may appear in the output image [7].

The work was supported by Russian Science Foundation grant 14-11-00308.

© Springer International Publishing Switzerland 2015
X. He et al. (Eds.): IScIDE 2015, Part I, LNCS 9242, pp. 441–450, 2015.
DOI: 10.1007/978-3-319-23989-7_45

To overcome the drawbacks of existing sharpening methods, an idea of grid warping was introduced. Instead of accurately estimating the blur kernel of the whole image and performing deconvolution, the pixel grid of the image is warped so that pixels in the neighborhood of the blurred edge move closer to the edge. This approach does not introduce artifacts like noise, overshooting, ringing effect because pixel values are not changed.

Warping approach for image enhancement was been introduced in [2]. The grid warping is performed according to the solution of a differential equation that is derived from the warping process constraints. Due to global nature of the method, the distortion of edges may appear.

In [8] the warping map is computed using simple local measures of the image. The measures are computed separately for rows and for columns of the image with restrictions that prevent two consecutive samples from interchanging their order in the 1D sequence. This approach does not introduce edge overshoot and does not amplify the noise, but it can introduce small local changes in the direction of edges.

In [18] a method is proposed that preserves the contours during enlargement, the method combines the warping of the coordinate point with the biasing of the signal amplitude.

The warping approach approach is related to the morphology-based sharpening [17] and shock filters [6,14,19]. These methods make the image appear piecewise constant which is effective mostly for cartoon-like images.

Special attention to edge model and preserving image textures was made in [9–11]. The 2D warping algorithm proposed in [10] has the following advantages:

1. It can be efficiently used as a post-processing method for global image deblurring methods.
2. No artifacts like ringing effect or noise amplification are introduced.
3. Unlike morphological methods and shock filters, the resulting images look natural and do not inevitably become piecewise constant.
4. It can be used as a standalone method of image sharpening. It is a good choice in the presence of strong noise and complex and non-uniform blur kernel.

In this work we propose an algorithm based on pixel grid warping [9–11] for edge sharpening of volume images.

2   Volume Image Blur Model

There are three main blur types: motion blur, defocus blur and camera sensor blur. The first blur type has irregular PSF which is hard to estimate while the latter two types have circular PSFs that can be approximated using Gauss filter kernel. Volume images are created for still objects so there is no motion blur and we can use Gauss filter as the approximation of real blur:

$$PSF(x,y,z) = \frac{1}{(2\pi)^{3/2}\sigma_x\sigma_y\sigma_z} \exp\left(-\frac{x^2}{2\sigma_x^2} - \frac{y^2}{2\sigma_y^2} - \frac{z^2}{2\sigma_z^2}\right), \tag{1}$$

where $\sigma_x, \sigma_y, \sigma_z$ are Gauss filter parameters that can be estimated using edge width estimation [12] by analyzing edge profiles corresponding to X, Y and Z axes.

Volume image dimensions usually have different resolution and different blur levels. Let h_x, h_y, h_z be the pixel grid steps. To enable using grid warping algorithm, we need to make Gaussian blur uniform $\sigma_x = \sigma_y = \sigma_z$. It can be done by simple scaling pixel step values which leads to proportional change of corresponding Gauss filter parameters. So we assume the following PSF

$$PSF(x, y, z) = \frac{1}{(2\pi)^{3/2}\sigma^3} \exp\left(-\frac{x^2 + y^2 + z^2}{2\sigma^2}\right).$$

(2)

3   One-Dimensional Edge Sharpening

The idea of the proposed 3D image sharpening algorithm is taken from [10]. Consider one-dimensional edge profile $g(x)$ centered at the point $x = 0$ (see Fig. 1), $d(x)$ — displacement function and $h(x) = g(x + d(x))$ — warped edge profile.

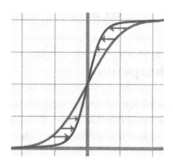

Fig. 1. The idea of edge sharpening using grid warping. Red line is the original edge profile, blue line is the edge profile sharpened using grid warping (Color figure online).

Simple scaling $f(x) = kx$ will give sharper edge but it will also shrink the entire image. To make the edge sharper without changing image size, only the area near the edge center should be shrunk while the area outside the edge should be stretched proportionally.

The warped grid should remain monotonic (i.e. for any $x_1 < x_2$ new coordinates should be $x_1 + d(x_1) \le x_2 + d(x_2)$), so the displacement function should match the following constraint:

$$d'(x) \ge -1.$$

(3)

Another constraint localizes the area of warping effect near the edge center:

$$d(x) \to 0, \qquad |x| \to \infty. \tag{4}$$

The displacement function $d(x)$ greatly influences the result of the edge warping. On the one hand, the edge slope should become steeper. On the other hand, the area near the edge should not be stretched over some predefined limit to avoid wide gaps between adjacent pixels in the discrete case.

The work [10] constructs the displacement function $d(x)$ using the proximity function

$$p(x) = 1 + d'(x)$$

$$d(x) = \int_{-\infty}^{x} (p(y) - 1)dy. \tag{5}$$

The proximity is the distance between adjacent pixels after image warping. This value is inverse to the density value. If the proximity function $p(x)$ is less than 1, then the image area is shrunk at the coordinate x. If the proximity is greater than 1, then the image is stretched. The identity transform has the value $p(x) \equiv 1$.

The constraint (3) leads to non-negativity of the proximity function. Also high values of the proximity function should be avoided to preserve image textures.

4   Volume Image Sharpening

For the problem of volume image sharpening, we approximate the proximity function (5) with the difference of two Gaussian functions in order to control the areas of shrinkage and stretching independently [10]:

$$p(x) = 1 + \kappa(G_{\sigma_1}(x) - G_{\sigma_2}(x)), \quad \sigma_2 > \sigma_1,$$
$$\kappa = 1/(G_{\sigma_1}(0) - G_{\sigma_2}(0)). \tag{6}$$

Parameter σ_1 controls the width of the densification area while parameter σ_2 controls the width of the rarefication area. We use $\sigma_1 = \sigma$ and $\sigma_2 = 2\sigma$. Fig. 2 illustrates using grid warping for edge sharpening using this proximity function.

To apply the warping algorithm to volume images, we adopt the two-dimensional extension [10] to the three-dimensional case. The displacement is a 3D vector field $\boldsymbol{d}(x, y, z)$ with the following constraints similar to the 2D case:

1) The shapes of the edges cannot be warped, so $\boldsymbol{d}(x_e, y_e, z_e) = 0$ for each edge point (x_e, y_e, z_e).
2) There cannot be any turbulence: rot $\boldsymbol{d} = 0$. Since rot$\nabla u \equiv 0$, the displacement field is assumed to be gradient of some scalar function $u(x, y, z)$: $\boldsymbol{d}(x, y, z) = \nabla u(x, y, z)$.

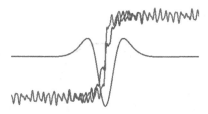

Fig. 2. Edge sharpening by grid warping using the proximity function (6). Blue function is the blurred edge with some noise. Red function is the result of warping. Green line is the proximity function (Color figure online).

3) The constraint (3) takes the form

$$\text{div}\boldsymbol{d} \geq -1 \tag{7}$$

and the proximity function looks as

$$p(x, y, z) = 1 + \text{div}\boldsymbol{d}(x, y, z). \tag{8}$$

Since $\text{div}\nabla \equiv \Delta$, where Δ is a Laplacian, the warping problem can be posed as a Dirichlet problem for the Poisson equation in the area of the image:

$$\begin{cases} \Delta u & = p(x, y, z) - 1, \\ u(x, y, z) & = 0 \text{ at image borders.} \end{cases} \tag{9}$$

The second constraint here is the boundary condition: the displacements at image borders should be equal to zero.

In order to get the same results as in the 1D case and to keep the edge pixels unwarped, the proximity value should be equal to the 1D proximity function depending on the distance to the edge. However, the distance to the closest edge ρ as an argument of the proximity function $p(\rho)$ is not efficient as it may produce gaps between close edges. Also it blurs edge ends.

We suggest the following method for calculating the proximity function:

$$p(x_0, y_0, z_0) = \frac{\displaystyle\sum_{(x,y,z)\in E(x_0,y_0,z_0)} p(x_n)G_\sigma(x_t)|\boldsymbol{g}(x, y, z)|}{\displaystyle\sum_{(x,y,z)\in E(x_0,y_0,z_0)} |\boldsymbol{g}(x, y, z)|} \tag{10}$$

where $E(x_0, y_0, z_0)$ is the set of edge points in the neighborhood of (x_0, y_0, z_0). The 3D edge point set is obtained using 3D Canny edge detector [4].

The value x_n is the projection and the value x_t is the length of the rejection of the vector $(x_0 - x, y_0 - y, z_0 - z)$ on the edge gradient vector $\boldsymbol{g}(x, y, z)$.

The function $p(x_n)$ is the 1D proximity function, weighting function $G_\sigma(x_t)$ is Gauss function with standard deviation equal to the edge's blur σ.

We solve the partial differential equation (9) using Fourier transform technique.

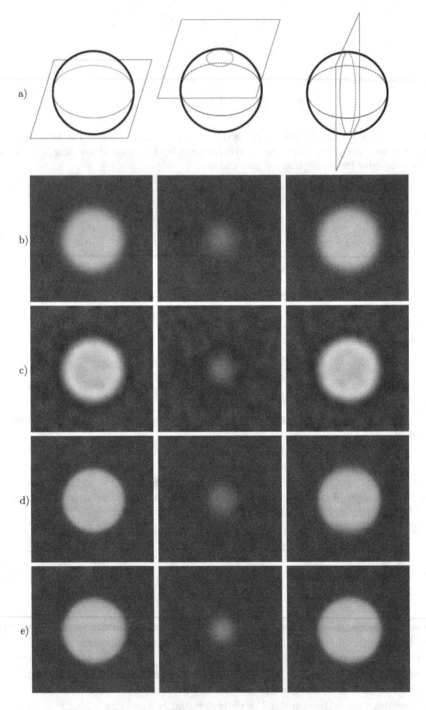

Fig. 3. Application of the proposed method to synthetic ball image: (a) Slices; (b) Blurred images; (c) 3D unsharp mask result; (d) Sharpening using 2D warping applied to each slice separately; (e) Proposed 3D image warping algorithm.

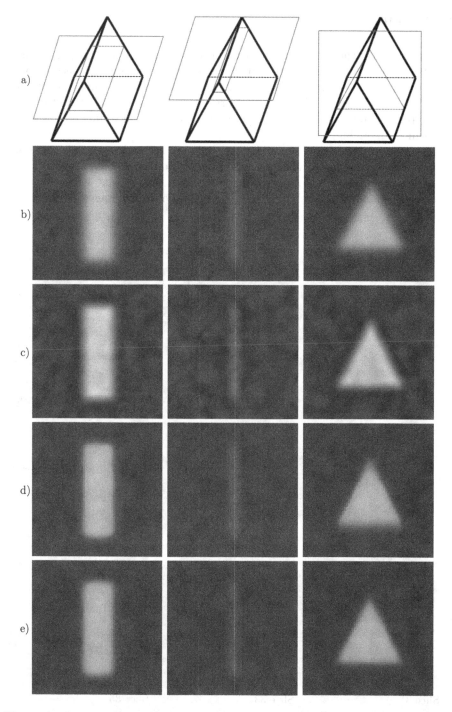

Fig. 4. Application of the proposed method to synthetic prism image: (a) Slices; (b) Blurred images; (c) 3D unsharp mask result; (d) Sharpening using 2D warping applied to each slice separately; (e) Proposed 3D image warping algorithm.

5   Results

We demonstrate the effectiveness of the proposed image sharpening method with two synthetic images of ball and prism (see Fig. 3 and Fig. 4). These images were corrupted by additive uniform noise and blurred using Gauss filter with $\sigma = 2$ and then deblurred by 3D unsharp mask with the same σ and $\alpha = 3$, 2D sharpening by image warping [10] applied to every slice and proposed 3D sharpening algorithm.

It can be seen that the proposed algorithm makes 3D edges sharper and does not produce any artifacts unlike unsharp mask. Two-dimensional algorithm was unable to improve the sharpness of edge surfaces that cannot be detected as edges using planar images. PSNR results for the synthetic images are shown in Table 1.

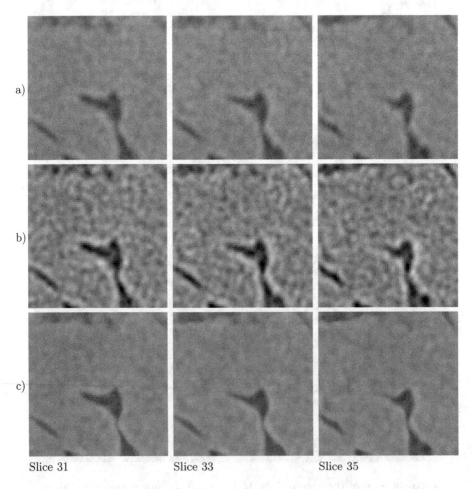

Slice 31   Slice 33   Slice 35

Fig. 5. Application of the proposed method to real micro CT image: (a) Source image slices; (b) 3D unsharp mask; (c)Proposed 3D image warping algorithm.

Figure 5 demonstrate the application of the proposed method to real micro CT image.

Table 1. PSNR results of the proposed 3D image sharpening method for the synthetic images

	Ball	Prism
Blurred image	32.02	29.87
Unsharp mask	33.14	31.89
2D warping	35.43	32.17
Proposed 3D warping	35.58	32.76

6   Conclusion

The two-dimensional warping approach for image sharpening has been effectively extended to the three-dimensional case. The proposed algorithm sharps even image edge corners that cannot be properly treated by independent application of two-dimensional sharpening methods to 2D slices. Like the 2D warping method, the developed 3D warping method enhances only image edges neighborhood and practically does not affect image textures.

References

1. Almeida, M., Figueiredo, M.: Parameter estimation for blind and non-blind deblurring using residual whiteness measures. IEEE Trans. Image Process. **22**, 2751–2763 (2013)
2. Arad, N., Gotsman, C.: Enhancement by image-dependent warping. IEEE Trans. Image Proc. **8**, 1063–1074 (1999)
3. Babacan, S.D., Molina, R., Katsaggelos, A.K.: Variational bayesian blind deconvolution using a total variation prior. IEEE Trans. Image Process. **18**, 12–26 (2009)
4. Canny, J.: A computational approach to edge detection. IEEE Trans. Pattern Anal. Mach. Intell. **8**, 679–698 (1986)
5. Conte, F., Germani, A., Iannello, G.: A Kalman filter approach for denoising and deblurring 3-D microscopy images. IEEE Trans. Image Process. **22**(12), 5306–5321 (2013)
6. Gilboa, G., Sochen, N.A., Zeevi, Y.Y.: Regularized shock filters and complex diffusion. In: Heyden, A., Sparr, G., Nielsen, M., Johansen, P. (eds.) ECCV 2002, Part I. LNCS, vol. 2350, pp. 399–413. Springer, Heidelberg (2002)
7. Gui, Z., Liu, Y.: An image sharpening algorithm based on fuzzy logic. Optik - Int. J. Light Electron Opt. **122**(8), 697–702 (2011)
8. Prades-Nebot, J., et al.: Image enhancement using warping technique. IEEE Electron. Lett. **39**, 32–33 (2003)

9. Krylov, A., Nasonova, A., Nasonov, A.: Grid warping for image sharpening using one-dimensional approach. In: Proceedings of International Conference on Signal Processing (ICSP2014), pp. 672–677. IEEE, Hangzhou (2014)

10. Nasonova, A., Krylov, A.: Deblurred images post-processing by Poisson warping. IEEE Signal Process. Lett. **22**(4), 417–420 (2015)

11. Nasonova, A., Nasonov, A., Krylov, A., Pechenko, I., Umnov, A., Makhneva, N.: Image warping in dermatological image hair removal. In: Campilho, A., Kamel, M. (eds.) ICIAR 2014, Part II. LNCS, vol. 8815, pp. 159–166. Springer, Heidelberg (2014)

12. Nasonova, A.A., Krylov, A.S.: Determination of image edge width by unsharp masking. Comput. Math. Model. **25**, 72–78 (2014)

13. Oliveira, J., Bioucas-Dia, J.M., Figueiredo, M.: Adaptive total variation image deblurring: a majorization-minimization approach. Signal Process. **89**, 1683–1693 (2009)

14. Osher, S., Rudin, L.I.: Feature-oriented image enhancement using shock filters. SIAM J. Numer. Anal. **27**(4), 919–940 (1999)

15. Ramponi, G.: A cubic unsharp masking technique for contrast enhancement. Signal Process. **67**, 211–222 (1998)

16. Ramponi, G., Strobel, N., Mitra, S., Yu, T.: Nonlinear unsharp masking methods for image contrast enhancement. J. Electron. Imaging **5**, 353–366 (1996)

17. Schavemaker, J., Reinders, M., Gerbrands, J., Backer, E.: Image sharpening by morphological filtering. Pattern Recognit. **33**(6), 997–1012 (2000)

18. Shimura, A., Taguchi, A.: Digital image interpolation with warping of coordinate point and biasing of signal amplitude. Electron. Commun. Jpn **88**, 10–21 (2005)

19. Weickert, J.: Coherence-enhancing shock filters. In: Michaelis, B., Krell, G. (eds.) DAGM 2003. LNCS, vol. 2781, pp. 1–8. Springer, Heidelberg (2003)

Interactive Visual Analysis for Comprehensive Dataset

Jihyoun Park, Yang Liu[✉], Carlos K.F. Tse, Minjing Mao,
Mengte Miao, Kangheng Wu, and Zhibin Lei

Hong Kong Applied Science and Technology Research Institute,
ASTRI, Shatin, Hong Kong
{jhpark,yangliu,carlostse,minjingmao,
mtmiao,khwu,lei}@astri.org

Abstract. In the big data era, handling the volume, velocity and variety of data is the prime requirement for analyzing an event. This paper presents our work for interactive visual analysis software with comprehensive format data input to solve such issues. There are three subsystems to process different types and formats of public and personal data at the same time. A detailed case study shows that the tool efficiently finds the target people and location from various data sources without any offline training or manual search.

Keywords: Interactive · Visual analysis · Investigation · Comprehensive data

1 Introduction

As the data collected and supplied for analysis of a social event are getting bigger and more diverse, it is increasing the demand of proper tools to assist human analyst to process those vast and various data comprehensively as well as in time. That is the challenge of the big data era that we are already living in. The big data can be defined by three main characteristics: volume, velocity and veracity [1]. The amount of data that we need to scan to reach to all related information of a social event becomes huge (volume). They are updated continuously in online. Moreover, the online issues are volatile and talking about immediate events at the time, so analysts are required to interpret the messages in near-real time (velocity). As they come from various channels, the types and formats of data are all different (variety). Therefore, it is crucial to provide data analysts a right tool that is capable to synthesize pieces of scattered data together and reconstruct the whole truth without much delay, in order to respond the data affluence phenomena of these days. Visually displaying the summary and relationship of collected data can be very useful for users to get comprehensive view of the situation effectively and efficiently rather than a simple classification. The application area of such visual analysis tools encompasses from marketing companies, which study their market and target customers, and to police officers who investigate the crime scenes.

In this paper, we propose an interactive visual analysis tool, which takes a comprehensive set of data as the input. The input data set includes public data such as

© Springer International Publishing Switzerland 2015
X. He et al. (Eds.): IScIDE 2015, Part I, LNCS 9242, pp. 451–461, 2015.
DOI: 10.1007/978-3-319-23989-7_46

online news and public announcement, and personal data such as emails, personal profiles, GPS information, credit card consumption records, and twitter-like text stream. The purpose of this tool as an analysis assistant is bringing up interesting targets for further investigation by combining all the available information. The tool is divided into three interacted subsystems to process different media (time and location-related records, static documents and stream texts) separately. Additionally, the information of one subsystem can interactively trigger the operations of other subsystems. To reduce the response time, we applied pattern recognition techniques to automatically find and recommend interesting targets without any prior knowledge or training. In addition, the related events to the targets are also identified and provided to users. In this way, this tool is expected to save the expensive human labor, which use to be used to manually check the relationship of raw data, for more sophisticated analysis.

The rest of this paper is composed of five parts. In Sect. 2, we review related works that studied heterogeneous input data set issues for comprehensive data analysis. In Sect. 3, the detail design of our visual analysis tool and its three subsystems to deal with a particular data type is presented. Section 4 shows a case study to investigate a crime incident using our tool. Finally, Sect. 6 summarizes our work.

2   Related Works

Visual analytics is defined as the science of analytical reasoning helped by software using interactive visual interfaces [2]. As a combination of visualization and automatic analysis, visual analytics can be more effective and efficient than each solution alone, if the data set provided to solve a problem contains many different types of data, thus requires a different level of human and machine involvement to properly deal with them [3].

Public data (online news and micro blogs) and personal data (GPS records, credit card records, and smart phone logs) are studied extensively for separate purposes. Generally, text from public data was analyzed to find topic evolution over time [4, 5], while information of personal data was seized for business such as advertisement and sales [5, 6]. Accordingly, as far as we can see, there is still a practice gap in the integration analysis of public data and personal data for investigation for specific events, such as clues to solve criminal case and civil case. The limited research results we found in investigation is simple, such as using suspect information as index, and focusing on one function (document management, forensic sample) [7, 8]. On the other hand, with the explosion of comprehensive data, more evidence should be available while it is harder to find the gold manually. Therefore, there is a need for an automatic analysis tool to lead the investigators to dig gold efficiently. In addition, it is expected the tool will help people in other areas, such as to understand the social network behaviors better by detecting the abnormal patterns automatically [6].

Regarding text analysis, Hogenboom et al. [9] classified methods to extract events from text roughly into two categories: data-driven and knowledge-driven. The latter has again two sub-methods: lexico-syntactic patterns and lexico-semantic patterns according to the characteristics of text. As our data sources have different formats and types, we employed the hybrid method. For example, we used the ontology modeling, which is a lexico-semantic pattern, to build relations of organizations and people, and lexico-syntactic-based keyword extraction methods and data-driven methods to summarize news with a set of important keywords.

3 Data

The data that this paper used for the case study is referred to the theoretic frame in [14]. The data sources consist of mixed types of both public and private data shown in Table 1. Public data means those data available for anybody, while private data is defined as those that an individual account holder can only access to. We categorized the data in this case study into three groups: time and location based data, static reference documents and short real-time stream texts from online social network services.

Table 1. Case study dataset.

Data type	Public data	Private data
Time and location		GPS, credit card records
Static documents	News articles	Emails, employee records
Stream text	Call center notice	Social network messages

4 Visual Analysis Tool Design

This section explains how we designed three subsystems of our visual analysis tool, which are specialized to handle each type of data identified in Table 1. The subsystems also work together interactively, so that they together can construct a total solution to reconstruct the target event wholly by combining piece of evidences found in different data sources.

Table 2. Subsystem design for different data types.

Data type	Analysis strategy	Tool
Time and location	– Location map – Pattern analysis	GPS analyzer (4.1)
Static document	– Natural language processing – Ontology analysis	Static document analyzer (4.2)
Stream text	– Keyword cloud – Timeline analysis	Stream text analyzer (4.3)

The target data type and analysis strategies of three subsystems are briefly summarized in Table 2. The first subsystem, GPS analyzer, is used to deal with the time and location-related GPS data and credit card transaction records using location map and pattern analysis methods. The second subsystem, static document analyzer, reads a variety type of different static documents such as Excel and Word and extracts important keywords, relations and events from them. Lastly, the third subsystem, stream text analyzer, identifies the time-sensitive events for further investigations from the streaming data.

4.1   GPS Analyzer

From the given data set, we obtained two types of time and location-related data. One is GPS data from related persons' cars. It contains the date and time, latitude and longitude but does not have the location name. The other is credit card records for shopping and meals. This type of data contains only date, that is, no time is given. We linked up the credit card transaction records with the GPS data to identify locations. If GPS data discontinues for a reasonable amount of time, for example 15 min, we assume that the car is parked at some place for shopping, meals, work or sleep. People normally stay at their houses at night and the place where most people regularly go and stay during the daytime on weekdays is the company. Otherwise, the parking places can be possibly linked to those places with credit card records.

After finding the locations of the restaurants, shops, offices and houses, we analyzed the daily patterns of a person in two ways. First, we simplified the locations appeared in a person's daily life to three categories: restaurant, company and home (Fig. 1). In this way, we could identify a certain person's abnormal traffic pattern compared with usual daily patterns of others and narrow down our targets for further investigation. For example, if a person's location is found somewhere else in the middle of night, it is abnormal.

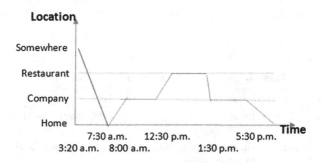

Fig. 1. Location classification and pattern analysis.

Second, we drew the latitude and longitude for all routes of every person in the target group with Python imaging tools. Every day all related person's movements are summarized in one page of map. This helped to find out who accompanied whom to which place. If two people have overlapped routes in the same time interval, they probably meet each other around the given time. Therefore, it is convenient to identify multiple patterns for further investigation by checking the figures generated. Also, the consumption records at the place have been visualized in the view of consumer and/or location.

4.2   Static Document Analyzer

This module is designed to read static data sources written in common document formats such as Excel, Word, plain text or web contents. The static document analyzer parses the contents and creates an ontology model to construct the link between them. Ontology is a semantic analysis method to derive indirect relations among objects or individuals from facts [11]. The advantage of the ontology model is its flexibility to define object structure, so users can use their expert knowledge more easily to address important terms for a particular data set.

Most of ontology modeling in our tool was done in automatic processes. Besides, we implemented a review and editing interface for users to manually add or correct information. For example, structured documents such as employment records and emails can be handled automatically to analyze the contents. For semi-structured documents such as countries' fact sheets and resumes, which have clearly divided sections but some contents were written in a free-style text format, we applied two steps to complete the ontology modeling. After roughly creating the initial ontology model using the automation process, users are required to review the structure manually. We also had more free-style reports, which are unstructured documents. In this case, we decided to input the ontology model of the data manually for more accurate analysis.

Lastly, we summarized the news contents. The semi-structured parts like published date and journal name were gathered separately, and then we applied natural language processing algorithms to automatically extract important keywords from the news text. To determine the importance of a keyword in a news report, we used Term-Frequency Inverse-Document-Frequency (TFIDF) measure. This measure highlights unique events of a certain news contents from all the other news contents by eliminating common terms. Furthermore, as a combined word (collocation keyword) can give us more meaningful and relevant information than each individual word, we used Part of Speech (POS) tags and custom dictionary to register particular people, organizations and places that we want to trace after [12]. Important persons, places and organizations identified from the previous ontology analysis are saved in our custom dictionary to extract keywords from news articles more accurately. In addition, keywords are aggregated according to the date of when the original news was published and calculated the co-occurrence count among keywords to find the linked set of keywords. Users can review the list of news in a certain day in the left window and quickly check the important events of the day in the D3 bubble chart [13] window on the right hand side.

4.3   Stream Text Analyzer

In order to unearth an event and its timeline in twitter like blogs, two major issues need to be considered in the design and implementation of our tool.

One issue is how to abstract the key components of an event from the messages: who, how, where and when. First, we apply the natural language processing method as in the static document analyzer to preprocess every message to eliminate the stop-words, stem and tag the remained words. Second, we calculate the co-occurrence of any two words in every minute to aggregate their relationship. We then measure the importance of each word by counting the number of messages that contain it. This is because the more messages contain the word, the more attraction it receives. After aggregating the data, we put the result into the database for future visual analytics. Finally, we calculate the Levenshtein distance between any two messages, and remove the same and similar messages, to clean up the result for display.

The other issue is how to properly display the key components summarized from the previous processing. To solve this, we mainly use three charts: SUBJECT, ACTION and LOCATION. Specifically, we put time on x-axis and word count on y-axis; the word itself appears on the chart. Notably, all three charts are synchronized. To facilitate the better analytics, we set a marker to indicate the time that the user is currently viewing. Through all the word charts, we can easily tell the high frequent subjects, actions and locations, respectively. By dragging the scrollbar below the charts, we can view all the data that are already processed. The charts are updated periodically when newly aggregated data are available. In this way, we can see the timeline and progress of events. For those words which are overlapped, we can either zoom in the chart, or check the details of messages in the right panel. On the right hand side of charts, there is a panel to list up the messages around the current time being viewed.

5 Interactive Analysis Case Study

The subject of the case study is to investigate a crime incident. The Tethys-based gas company GAStech has made huge profits in the island country of Kronos. However, Kronos people suffered from pollution related to the GAStech commercial activities. On January 20th 2014, several employees of GAStech were missing after the company's management group annual meeting. Protectors of Kronos (POK), one local environment protection organization seemed to be the suspect [14]. This paper focuses on the interaction among three subsystems to understand the incident more comprehensively in the investigator's point of view.

As shown in Fig. 2, the strategy for the analysis starts with the personal data such as GPS and credit card records. In the studied case, by visualizing the GPS records and

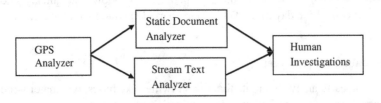

Fig. 2. Analysis flowchart.

consumption records in the GPS analyzer, people with abnormal patterns and the related locations are easily identified. After the abnormal pattern in personal data is determined, the abnormal list is then input in the static document analyzer and the stream text analyzer to find further suspects and related events. Additionally, the related public data offers more information as suspect candidate. Finally, the confirmed suspects and locations are supposed to be submitted to human investigators in this scenario as the clues for their further in-depth analysis.

5.1 Personal Data: Patten Identification

We start the analysis from those employees with abnormal traffic patterns since we quickly found some employees have went out at midnight using our GPS analyzer. For example, Isak Baza, Linnea Bergen, Nils Calixito, Axel Calzas, Lidelse Dedos Adra Nubarron, and Brand Tempestad had single late out at January 10 midnight. But, since logically several meetings are needed for planning the kidnapping event; we consider multiple late outs should be more important. In addition, although Lucas_Alcazar has multiple late outs, the GPS data shows he always stayed at the company, which indicates he is just working overtime. Therefore, the final suspect list has been narrowed down to Isia Vann, Loreto Bodrogi, Hennie Osvaldo and Minke Mies.

To confirm the employees with multiple going out in the midnight are really having meetings, a distance calculator program is also integrated in the GPS analyzer. The input is the employees and the time, and output is the corresponding distance between them. Limited by the discrete GPS data, the calculator can only find out the nearest location of the input time and calculate the distance between them. The calculator tells us that some employees did have meetings in the mid-night.

Accordingly, the places visited in the middle night frequently by the suspects are plotted in Fig. 3. The identified places are also matched to the tourist map given in the data set for further investigation.

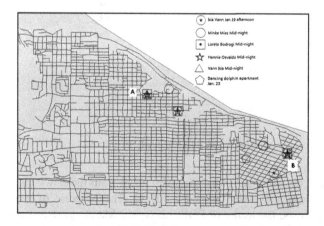

Fig. 3. GPS analyzer interface.

For the consumption pattern, we checked the difference of credit cards and loyalty cards per location. However, the difference is not evident. Therefore, we use the combined records only for simplicity. In addition, with the suspects found in traffic patterns, we studied the corresponding consumption patterns. For example, we see Isia Vann has an abnormal large bill at Frydos Autosupply n' More at January 17. But since he used a company car, the pattern does not clearly indicate if he is a suspect in the kidnapping. Finally, the consumption record charts by location may reveal people gathering information if we assume they are sharing the bill. But, again, it is hard to tell it is a normal office gathering or kidnapping plan meeting. Accordingly, we do not use consumption pattern analysis result for event identification input.

5.2   Personal Data to Public Data: Event Identification

The stream text analyzer is used to find the people's community and hot topics and events at this moment. As shown in Fig. 4, the chart of subject gives us some top frequent subjects: "POK Rally", "Sylvia Marek", "Abila City Park", "Prof. Stefano", "Lucio Jacab", "Victor-E" and "Dr. Newman". The verbs corresponding to the above subjects can be found in the chart of action. Together with using the word net in D3 bubble chart [13], the related actions are: "open", "begin", "introduce", "speak" and "play". From the chart of location, we can see the place of "Egeou St/Parla St". With studying the call centre record that contains the place (as listed on the right panel), we can confirm it is the event location.

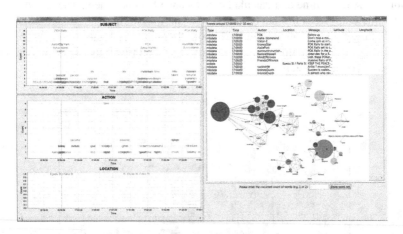

Fig. 4. Stream text analyzer interface.

In this way, we can abstract an important event from the raw data: POK Rally. It started at 17:00:00 and ended around 19:05:54. The progress of the event can be described as follows: the leader of POK, i.e. Sylvia Marek, gave an opening speak to start the rally in Abila City Park; a special guest Dr. Newman was invited to give a talk with Prof. Stefano and Lucio Jacab; the band Victor-E was invited to play music.

Similarly, we abstracted three other important events: the fire at the Dancing Dolphin Apartment, the vehicle accident between a black van and a bicycle, and the gunfire and subsequent events after the black van was stuck at the Gelato Galore.

With the location input from the GPS analyzer, we think the fire at the Dancing Dolphin Apartment is most important to the GAStech disappearances. In the stream text analyzer, we find a witness "dangermice" reporting the whole progress of fire. This people's micro blog data include the longitude and latitude. With such information, we can check the fire location and see that the location of vehicle accident and the fire are both closed to B in Fig. 3.

For the static document analyzer, Isia Vann is our start point since he is considered a strong suspect in 5.1. Figure 5 shows all personal profile, news articles, people and organizations, which are identified as having direct or indirect connections with Isia Vann during the ontology and keyword analysis processes. Accordingly, Isia Vann's email traffic is checked. The suspects list is then extended to the receipts of Email circle "RE: FW: ARISE - Inspiration for Defenders of Kronos": Isia Vann, Inga Ferro, Loreto Bodrogi, Hennie Osvaldo, Minke Mies. In addition, the abnormal multiple middle night traffics check in the GPS analyzer returns a list includes Isia Vann, Minke Mies, Loreto Bodrogi and Hennie Osvaldo.

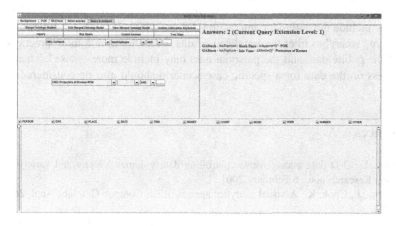

Fig. 5. Static document analyzer interface.

5.3   In-Depth Analysis Suggestions

The final step of the case study is to offer some suggestions to police officers, who will investigate this case further based on our findings.

First of all, we think the disappeared GAStech employees include executive group of GAStech and the kidnapper suspects. The kidnapper suspects include Isia Vann, Inga Ferro, Loreto Bodrogi, Hennie Osvaldo, Minke Mies, as indicated in 5.2. Therefore, we suggest that the police trace down whereabouts of those suspects.

Secondly, we roughly reconstructed the timeline of the kidnapping incidents as follows. The employee disappearance happened at January 20, around 10:00 am.

Before that, the executive group was having the annual meeting at GAStech building starting from 9:00 am. After 10:00 am, the suspects made a call to claim a false fire alarm, and took the victims away in the evacuation. Probably, the executive group and some kidnappers left Abila by private planes in the afternoon of January 20. The suspects may plan the actions several time at the locations listed in Fig. 3 during January 7–January 20. Point A and Point B in Fig. 3 are the most visited places by the suspects. And, at January 23, some of the suspects may try to set a real fire at dolphin apartment (point B in Fig. 3) to remove the evidence. After that, the suspects flee caused a traffic accident, as well as a subsequent gunfire with local police. Therefore, we recommend point A and point B in Fig. 3 for further investigation.

6 Summary

To provide a consolidated single analysis tool, which integrates various data sources to help reduce manual effort of human data analysts in this big data era, we proposed an interactive visual analysis tool, which is composed of three subsystems that takes part in analyzing a particular type of data. Our tool could successfully identify the abnormal patterns in the personal data automatically. Accordingly, the suspect people, location, and events found in the public data set are listed for later investigation efficiently.

Currently, the interaction between subsystems has to be done manually. The future work may include the automatic interaction design, which means the subsystem should work more "smart" to filter the unrelated results. In addition, in a more practical case, both of the public data and the personal data may include more "noise". Therefore, a pre-process on the data for a specific case/scenario should also be considered.

References

1. Laney, D.: 3-D data management: controlling data volume, velocity and variety. META Group Research note, 6 February 2001
2. Thomas, J., Cook, K.: A visual analytics agenda. IEEE Comput. Graphics Appl. **26**, 10–13 (2006)
3. Keim, D., et al.: Visual analytics: how much visualization and how much analytics? ACM SIGKDD Explor. Newsl. **11**(2), 5–8 (2009)
4. Dou, W., et al.: LeadLine: interactive visual analysis of text data through event, identification and exploration. In: IEEE Conference on Visual Analytics Science and Technology 2012, Seattle, WA, 14–19 October 2012
5. Wanner, F., et al.: State-of-the-art report of visual analysis for event detection in text data streams. http://bib.dbvis.de/uploadedFiles/3_submission.pdf
6. Slingsby, A., et al.: Visual analysis of social networks in space and time. Mobile Data Challenge Workshop 2012, Newcastle (2012)
7. Criminal analysis: new prospects for investigation using i2 software. https://visualanalysis.com/Downloads/CaseStudies/ANB-IBASEOCRVP_UK_Q2%202011_ICP_Low.pdf
8. Top law enforcement software tools. http://www.capterra.com/law-enforcement-software/

9. Hogenboom, F., et al.: An overview of event extraction from text. In: van Erp, M., et al. (eds.) Proceedings of Detection, Representation, and Exploitation of Events in the Semantic Web, pp. 48–57, Bonn (2011)

10. Kandel, S., Paepcke, A., Hellerstein, J.M., Heer, J.: Enterprise data analysis and visualization: an interview study. IEEE Trans. Visual Comput. Graphics **18**(12), 2917–2926 (2012)

11. Kietz, J.U., Maedche, A., Volz, R.: A method for semi-automatic ontology acquisition from a corporate intranet. In: EKAW-2000 Workshop "Ontologies and Text", Juan-Les-Pins (2000)

12. Arazy, O., Woo, C.: Enhancing information retrieval through statistical natural language processing: a study of collocation indexing. MIS Q. **31**(3), 525–546 (2007)

13. Data Driven Documents. http://d3js.org/

14. VAST Challenge (2014). http://www.vacommunity.org/VAST+Challenge+2014

Non-negative Locality-Constrained Linear Coding for Image Classification

GuoJun Liu[✉], Yang Liu, MaoZu Guo, PeiNa Liu, and ChunYu Wang

School of Computer Science and Technology,
Harbin Institute of Technology, Harbin, China
`hitliu@hit.edu.cn`

Abstract. The most important issue of image classification algorithm based on feature extraction is how to efficiently encode features. Locality-constrained linear coding (LLC) has achieved the state of the art performance on several benchmarks, due to its underlying properties of better construction and local smooth sparsity. However, the negative code may make LLC more unstable. In this paper, a novel coding scheme is proposed by adding an extra non-negative constraint based on LLC. Generally, the new model can be solved by iterative optimization methods. Moreover, to reduce the encoding time, an approximated method called NNLLC is proposed, more importantly, its computational complexity is similar to LLC. On several widely used image datasets, compared with LLC, the experimental results demonstrate that NNLLC not only can improve the classification accuracy by about 1–4 percent, but also can run as fast as LLC.

Keywords: Locality-constrained Linear Coding (LLC) · Non-negative constraint · Spatial Pyramid Matching (SPM) · Image classification

1 Introduction

As a fundamental problem in image processing and computer vision, image classification has attracted more and more attention in recent years. Generally, the image classification system consists of two major parts: bag of features (BoF) [1–3] and spatial pyramid matching(SPM) [4,5]. The BoF method represents an image as a histogram of local features. It is robust against spatial translations of features, but discards the spatial order of local features. In order to overcome this problem, Lazebnik et al. [4] have proposed a new method, called SPM. It partitions an image into increasingly finer spatial sub-regions and computes the histograms of local features in each sub-region. Typically, SPM partitions an image into three layers, and the number of sub-regions from each layer is set to $2^l \times 2^l$, $l = 1, 2, 3$.

Recently, it has been shown that the SPM method based on BoF with sparse representation is very effective and has achieved the state of the art performance on image classification. As shown in Fig. 1, the entire framework includes the

© Springer International Publishing Switzerland 2015
X. He et al. (Eds.): IScIDE 2015, Part I, LNCS 9242, pp. 462–471, 2015.
DOI: 10.1007/978-3-319-23989-7_47

Images   Descriptors   Coding   SPM

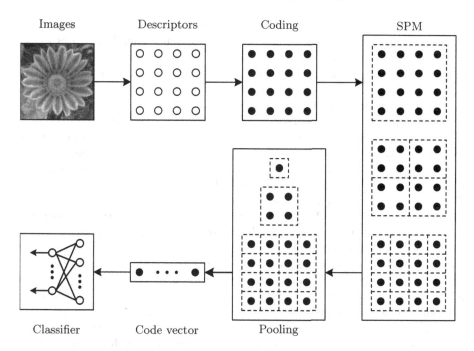

Classifier   Code vector   Pooling

Fig. 1. The entire framework of image classification.

following major parts. First, feature points are detected on the input image, and some descriptors such as SIFT, HOG are extracted from each feature point, each descriptor is a D-dimensional vector, e.g., for SIFT, $D = 128$. Then, via some dictionary training methods such as K-means, a $\mathbb{R}^{D \times M}$ dictionary is obtained, where M is the size of dictionary. Afterwards, a code $\mathbf{c} \in \mathbb{R}^M$ corresponding to each descriptor is computed by using the dictionary. Next, the SPM method is applied on each image to partition into three layers, L_0, L_1, L_2. Codes from each sub-region in each layer are pooled together by max pooling, and the $\mathbb{R}^M$ code of each sub-region is obtained. Finally, the codes of each sub-region in all layers are concatenated together to generate the final representation of the image, it is a $L \times M$ dimensional code vector, where L is the number of all sub-regions, $L = \sum_l 2^l \times 2^l$, $l = 0, 1, 2$. After that, the code vector can be inputted into the classifier to get its label information.

Effectiveness and robustness of the coding scheme play an important role in the classification performance. Therefore, many researchers mainly focus on the code step. The aim of different coding schemes is to convert a descriptor into a M dimensional code. Under the condition of different constraints, a variety of coding schemes have been proposed, e.g., vector quantization (VQ) [4], SPM based on sparse codes (ScSPM) [6,7] and locality-constrained linear coding (LLC) [8]. VQ is used in the traditional SPM method, thus each code only has one non-zero element. In practice, the single non-zero element is found by searching the nearest neighbor. Unlike VQ, to reduce reconstruction error, ScSPM uses the sparse coding method, but this coding scheme may be unstable, it means that the code for similar descriptors may be very different. Therefore, Yu et al. [9] proposed that the locality is more essential than the sparsity. In LLC, each descriptor is more accurately represented by multiple local bases, in addition, similar descriptors will have similar codes by sharing local bases. Thus, LLC can further improve the stability of codes in ScSPM.

While encoding the descriptor under the shift-invariant requirements as in LLC, the code may include some negative elements in order to minimize the reconstruction error. If the absolute difference of some negative and positive elements is extremely large, the coding scheme may be unstable. Therefore, different non-negative coding methods [10,11] have been proposed in recent years. Zhang et al. [12] proposed the non-negative sparse coding, low-rank and sparse matrix decomposition techniques (LR-Sc$^+$SPM) to establish the classification framework, and achieve the state of the art results on several datasets.

In this paper, a novel coding scheme called non-negative locality-constrained linear encoding is proposed to improve the stability of LLC. In comparison with traditional coding schemes, the main contribution is as the following. First, the proposed method adds an extra non-negative constraint based on the objective function of LLC, which make the coding scheme more stable than LLC. Second, the above constrained optimization problem has not the exact analytical solution, so we explore the solution of iterative optimization methods and analysis its computational complexity. Last, an approximate method called NNLLC is proposed as a trade-off. It can not only speed up the encoding process, but also achieve better performance than traditional coding methods on several widely used image datasets.

The rest of the paper is organized as follows: Sect. 2 introduces the basic idea of the proposed method and its stable theoretical analysis. Section 3 gives an approximate method called NNLLC to replace the general iterative optimization method. In Sect. 4, image classification results and coding time comparison on three widely used datasets are reported. Finally in Sect. 5 conclusions are made.

2   Non-negative Locality-Constrained Linear Coding

Let $\mathbf{H}$ denote a set of D dimensional descriptors extracted from an image, $\mathbf{H} = [\mathbf{h}_1, \mathbf{h}_2, \ldots, \mathbf{h}_N] \in \mathbb{R}^{D \times N}$, and $\mathbf{h}_i \in \mathbb{R}^N$ is the ith descriptor. Let $\mathbf{B}$ denote the dictionary generated on all descriptors, and $\mathbf{B}$ includes M bases, i.e., $\mathbf{B} = [\mathbf{b}_1, \mathbf{b}_2, \ldots, \mathbf{b}_M] \in \mathbb{R}^{D \times M}$. To reconstruct the descriptor $\mathbf{h}_i$ by utilizing the dictionary $\mathbf{B}$, i.e., $\mathbf{h}_i \approx \mathbf{B} \times \mathbf{c}_i$, different coding schemes can be used to convert $\mathbf{h}_i$ into its code $\mathbf{c}_i \in \mathbb{R}^M$.

2.1   Locality-Constrained Linear Coding

Theoretically, Yu et al. [9] pointed out that the locality is more essential than the sparsity under certain assumptions. Because the locality leads to the sparsity necessarily, not vice versa. LLC is based on the traditional sparse coding.

It achieves global sparsity by adding local restrictions. The LLC code uses the following criteria:

$$\text{minimize} \quad \sum_{i=1}^{N} ||\mathbf{h}_i - \mathbf{Bc}_i||^2 + \lambda ||\mathbf{d}_i \odot \mathbf{c}_i||^2 \tag{1}$$
$$\text{subject to} \quad \mathbf{1}^{\mathsf{T}} \mathbf{c}_i = 1$$

where $\mathbf{1}$ denotes the vector whose each element is 1, and the constraint $\mathbf{1}^{\mathsf{T}} \mathbf{c}_i = 1$ ensures the shift invariance requirements of the code. The locality regularization term $||\mathbf{d}_i \odot \mathbf{c}_i||^2$ can ensure that LLC generate similar codes for similar descriptors.

2.2 Approximated LLC

During the process of solving Eq. (1), the reconstructed descriptor $\mathbf{h}_i$ tends to select those local bases of the dictionary $\mathbf{B}$, the bases will form a local coordinate system, and $\mathbf{c}_i$ is the value of the coordinate. Therefore, Wang et al. [8] proposed a simple faster approximation of LLC to speed up the encoding process, more importantly, its performance is almost equivalent to LLC. Approximated LLC includes three parts:

1. Dictionary learning. A portion of descriptors are randomly extracted from all the descriptors for visual dictionary learning. In practice, K-means is often used to generate the dictionary $\mathbf{B} \in \mathbb{R}^{D \times M}$.
2. k nearest neighbors searching. The k nearest neighbors in the dictionary of $\mathbf{h}_i$ are selected as the local bases $\mathbf{B}_i = [\mathbf{b}_{[1]}, \ldots, \mathbf{b}_{[k]}]$.
3. Coding. $\mathbf{B}_i$ is utilized to reconstruct $\mathbf{h}_i$ by solving the following constrained least square fitting problem.

$$\text{minimize} \quad \sum_{i=1}^{N} ||\mathbf{h}_i - \mathbf{Bc}_i||^2 \tag{2}$$
$$\text{subject to} \quad \mathbf{1}^{\mathsf{T}} \mathbf{c}_i = 1$$

where the value of the $([1], \ldots, [k])$ elements are set to $(\mathbf{c}_{[1]}, \ldots, \mathbf{c}_{[k]})$, the remainders are 0.

More importantly, Eq. (2) has an analytical solution, and it reduces the computation complexity from $O(M^2)$ to $O(M + k^2)$, where $k \ll M$.

2.3 Non-negative Locality-Constrained Linear Coding

Compared with the approximated LLC, this paper presents the non-negative locality-constrained linear coding (NNLLC) by adding a non-negative constraint in the objective function:

$$\text{minimize} \quad \sum_{i=1}^{N} ||\mathbf{h}_i - \mathbf{Bc}_i||^2 \tag{3}$$
$$\text{subject to} \quad \mathbf{1}^{\mathsf{T}} \mathbf{c}_i = 1, \; \mathbf{c}_i \succeq 0$$

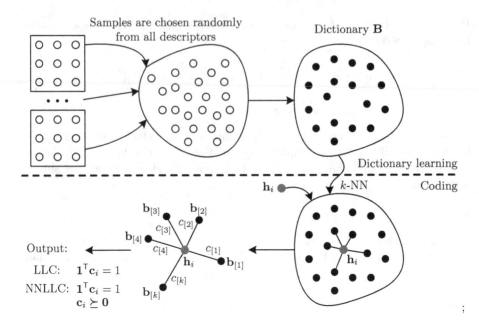

Fig. 2. The diagram of the encoding process.

where $\mathbf{b}_i$ denotes the non-negative code of $\mathbf{h}_i$, and the two constraints require $\mathbf{c}_i$ to lie within the convex hull of its neighbors in $\mathbf{B}_i$. As illustrated in Fig. 2, the encoding process between LLC and NNLLC is different, furthermore, the different code leads to the different result directly.

Generally, Eq. (3) can not be solved directly, and iterative optimization methods are used instead. In practice, to solve the constrained linear least squares problem, we can use the lsqlin function in the optimization toolbox of Matlab. For simplicity, this method to solve Eq. (3) is denoted by lsqlin.

3   Approximated NNLLC

The lsqlin method can solve Eq. (3) by using iterative optimization techniques, however, in contrast to LLC, it is too slower to be used in practical applications. In order to reduce the computational complexity of Eq. (3), an approximated coding method is proposed and denoted by NNLLC. It uses a two-step method to solve Eq. (3). The first step is to only consider the constraint $\mathbf{1}^{\mathsf{T}}\mathbf{c}_i = 1$, that is, LLC code can be obtained by solving Eq. (2), where LLC code is denoted as $\mathbf{c} = [c_1, c_2, \ldots, c_k]^{\mathsf{T}}$. The second step is to consider the constraint $\mathbf{c}_i \succeq \mathbf{0}$. If the LLC code contains negative elements, they are set to zero, then all elements of the code are normalized. Therefore, the final NNLLC code c_i^* is given as the following equation

$$c_i^* = \frac{\max\{0, c_i\}}{\sum_{c_i} \max\{0, c_i\}}, \quad i = 1, 2, \ldots, k \tag{4}$$

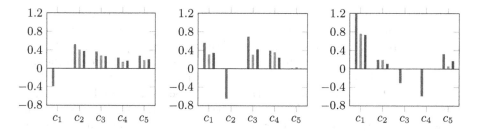

Fig. 3. The comparison of codes of LLC, lsqlin and NNLLC for three reconstructed descriptors with $k = 5$.

More importantly, the computation complexity is $O(M + k^2 + k)$, and equals to $O(M + k^2)$ of LLC, where $k \ll M$. Therefore, NNLLC can run as fast as LLC.

3.1 Stability of NNLLC

The comparative results of several descriptors on LLC, linlsq and NNLLC codes are shown in Fig. 3, respectively. Three reconstructed descriptors are chosen, each one corresponds to a code c_i, with $k = 5$, i.e., each code c_i includes five elements, represented as c_1, c_2, c_3, c_4, c_5, respectively. In Fig. 3, the horizontal axis indicates the five elements in the code, and the vertical represents the value of each element. Each bar with different color shows the value of the corresponding element in the different coding method, red, green and blue bars denote LLC, lsqlin and NNLLC, respectively. While the empty bar means that the value of the corresponding element is 0. In Fig. 3, the difference between positive and negative value of elements of LLC code is bigger, especially, in the last figure, the value of c_1 is larger than 1.2, at the same time, the difference between c_1 and c_4 is close to 2. Such big difference among all elements in one code will be possible to make LLC unstable.

By contrast, NNLLC can be more stable. From the point of view of mathematics, Eq. (3) has two constraints which make the code c_i lie in the convex hull. In other words, it means that the reconstructed descriptors h_i can be viewd as the weighed average of some bases of the dictionary B_i, these bases are k nearest neighbors in the dictionary, and the corresponding weighted coefficient is the code c_i. Most importantly, the biggest difference of the value of all elements is less than 1, it can ensure the stability of NNLLC.

4 Results

In this section, we evaluate the proposed non-negative locality-constrained linear coding method (lsqlin) and its approximated method (NNLLC) on three widely used datasets: The Scene-15 [4] dataset, the Caltech-101 dataset [13] and the Caltech-256 dataset [14].

In our experiments, we use the same setup as in [6], the SIFT features are extracted from patches densely located by every 6 pixels on the image, each patch has 16×16 pixels, where the dimension of the SIFT descriptor is $D = 128$. As to the dictionary size, it is set to 1024, as in [6,8,15]. For SPM, we use the first three layers as illustrated in Fig. 1, where L is the number of all sub-regions, $L = \sum_l 2^l \times 2^l$, $l = 0, 1, 2$. During the coding process, the number of nearest neighbors k is set to 5 as in [8]. After the code of each image is obtained, the liblinear toolbox [16] was applied for linear classification. In addition, to compare with the common benchmarking, we repeat five times each experiment, for each run, the training and testing images are chosen randomly. Finally, we record the mean and the standard deviation of the classification accuracy. All experiments were conducted on a server with 64G memory and 2.4Ghz Quad Core CPU.

4.1   Scene-15 Dataset

The Scene-15 dataset contains 4485 images falling into 15 categories, with the number of images each category varies from 200 to 400. The 15 categories vary from living room and kitchen to street and industrial. As suggested in [8], the whole dataset is partitioned into 10, 20, 40, 60, 80 and 100 training images per class, respectively.

Table 1 gives the performance comparison of the proposed methods and other traditional approaches [6,8,17] on the Scene-15 dataset. On all groups, the proposed methods outperform LLC and other methods by about 1 percent, it demonstrates the effectiveness of our method.

Table 1. The comparison of classification accuracy on the Scene-15 dataset.

	10	20	40	60	80	100
KC [17]	–	–	–	–	–	76.67±0.39
ScSPM [6]	–	–	–	–	–	80.28±0.93
LLC [8]	64.89±1.01	71.65±0.59	76.58±0.72	78.64±0.56	80.19±0.96	81.32±0.78
lsqlin	67.32±1.33	73.26±0.58	77.63±0.49	79.37±0.43	80.97±0.37	81.96±0.30
NNLLC	67.30±1.43	73.16±0.83	77.36±0.86	79.79±0.72	80.97±0.36	81.81±0.46

4.2   Caltech-101 Dataset

The Caltech-101 dataset contains 9144 images in 101 classes, where the number of images per class ranges from 31 to 800, including animals, flowers, vehicles, etc., with significant variance in shape. The whole dataset is partitioned into 5,10,15,20,25 and 30 training images per class, respectively.

As illustrated in Table 2, the performance of our method is superior to LLC by about 1–3 percent, and to other methods [4,6,12,14,17–21] more than 3 percent. Generally speaking, the performance of lsqlin and NNLLC is similar, however, lsqlin is more stable and effective than NNLLC while the size of the training images is small, as shown in the first column in Table 2.

Table 2. The comparison of classification accuracy on the Caltech-101 dataset.

	5	10	15	20	25	30
SVM-KNN [18]	46.60	55.80	59.05±0.56	62.00	–	66.23±0.48
KSPM [4]	–	–	56.40	–	–	64.60±0.80
SPM [14]	44.20	54.50	59.00	63.30	65.80	67.60
NBNN [19]	–	–	65.00±1.14	–	–	70.40
CORR [20]	–	–	61.00	–	–	69.60
KC [17]	–	–	–	–	–	64.16±1.18
LR-Sc⁺SPM [12]	–	–	69.58±0.97	–	–	75.68±0.89
ScSPM [6]	52.48±0.77	62.42±0.44	66.11±0.61	70.17±0.47	71.95±0.32	73.84±1.41
LLC [8]	55.65±1.45	65.03±0.67	70.65±0.37	74.25±0.63	75.82±0.35	78.56±0.23
lsqlin	58.28±1.03	66.85±0.36	72.64±0.33	75.33±0.36	77.39±0.23	79.26±0.35
NNLLC	56.00±1.62	67.42±0.75	72.23±0.66	75.78±0.22	78.23±0.26	79.91±0.26

Table 3. The comparison of classification accuracy on Caltech-256 dataset.

	15	30	45	60
SPM [14]	28.30	34.10	–	–
KC [17]	–	27.17±0.46	–	–
LScSPM [15]	30.00±0.14	35.74±0.10	38.54±0.36	40.43±0.38
ScSPM [6]	27.73±0.51	34.02±0.35	37.46±0.55	40.14±0.91
LLC [8]	30.17±0.19	37.24±0.16	41.68±0.38	45.03±0.36
lsqlin	34.37±0.21	41.32±0.20	45.02±0.29	48.31±0.22
NNLLC	34.51±0.27	41.29±0.48	45.45±0.16	48.19±0.23

4.3   Caltech-256 Dataset

The Caltech-256 dataset contains 30607 images in 256 classes. Each class includes at least 80 images. Compared with the Caltech-101 dataset, this dataset has much more class variability, it includes bats, flags, etc. The whole dataset is partitioned into 15, 30, 45 and 60 training images per class (Table 3).

Obviously, the performance of our method is better. More interestingly, the classification accuracy of lsqlin and NNLLC is very close. It means that the approximated method NNLLC can almost obtain the same result of lsqlin which must use some iterative optimization techniques.

4.4   Comparison of Encoding Time

For a practical image classification system, the performance of running time and the accuracy are equally important. As shown in Table 4, the encoding time of NNLLC is almost the same as LLC since their computational complexity are $O(M + k^2 + k)$ and $O(M + k^2)$ respectively, where $k \ll M$. However, due to the use of some iterative optimization techniques, lsqlin requires much more encoding time than LLC and NNLLC. Therefore, as a trade-off , NNLLC can not only obtain the same superior result as lsqlin, but also run as fast as LLC.

Table 4. The comparison of encoding time for different methods on Scene-15, Caltech-101 and Caltech-256 datasets, respectively (Unit: hour).

	Scene-15	Caltech-101	Caltech-256
LLC [8]	0.97	2.50	11.90
lsqlin	7.45	14.53	55.27
NNLLC	0.98	2.53	11.98

5   Conclusion

This paper presents a novel coding scheme by adding an extra non-negative constraint based on LLC. The proposed method is more stable and robust than LLC. At the same time, to avoid using the iterative optimization methods to solve the constrained least square fitting problem, an approximated non-negative coding method is proposed to reduce the encoding time. On several widely used image datasets, compared with LLC, the experimental results demonstrate that NNLLC can not only improve the classification accuracy by about 1–4 percent, but also run as fast as LLC.

Acknowledgments. This work was supported by the Fundamental Research Funds for the Central Universities (Grant No. HIT.NSRIF.2014069), Heilongjiang Province Science Foundation for Youths (Grant No. QC2014C071), National Natural Science Foundation of China (Grant No. 61173087, 61171185, 61271346), and Specialized Research Fund for the Doctoral Program of Higher Education of China (Grant No. 20112302110040).

References

1. Sivic, J., Zisserman, A.: Video google: a text retrieval approach to object matching in videos. In: Ninth IEEE International Conference on Computer Vision, Nice, France, October 2003
2. Csurka, G., Dance, C.R., Fan, L., Willamowski, J., Bray, C.: Visual categorization with bags of keypoints. In: Workshop on Statistical Learning in Computer Vision, ECCV, Prague, Czech Republic, pp. 1–22, May 2004
3. Zhao, Z.Q., Ji, H.F., Gao, J., Hu, D.H., Wu, X.D.: Image classification of multiscale space latent semantic analysis based on sparse coding. Chin. J. Comput. **37**(6), 1251–1260 (2014)
4. Lazebnik, S., Schmid, C., Ponce, J.: Beyond bags of features: spatial pyramid matching for recognizing natural scene categories. In: IEEE Conference on Computer Vision and Pattern Recognition. vol. 2, pp. 2169–2178, New York, USA, June 2006

5. Li, Q.: Image classification research of improved non-negative sparse coding. Master's thesis, Nanjing University of Science and Technology (2014)
6. Yang, J., Yu, K., Gong, Y., Huang, T.: Linear spatial pyramid matching using sparse coding for image classification. In: IEEE Conference on Computer Vision and Pattern Recognition, Miami, USA, 1794–1801, June 2009
7. Yang, J., Wang, J., Huang, T.: Learning the sparse representation for classification. In: IEEE International Conference on Multimedia and Expo, Barcelona, Spanish, pp. 1–6 (2011)
8. Wang, J., Yang, J., Yu, K., Lv, F., Huang, T., Gong, Y.: Locality-constrained linear coding for image classification. In: IEEE Conference on Computer Vision and Pattern Recognition, San Francisco, USA, pp. 3360–3367, June 2010
9. Yu, K., Zhang, T., Gong, Y.: Nonlinear learning using local coordinate coding. In: Advances in Neural Information Processing Systems, Vancouver, Canada, pp. 2223–2231, December 2009
10. Hoyer, P.O.: Non-negative sparse coding. In: IEEE Workshop on Neural Networks for Signal Processing, pp. 557–565 (2002)
11. Lin, T.H., Kung, H.T.: Stable and efficient representation learning with non-negativity constraints. In: Proceedings of the 31st International Conference on Machine Learning, Beijing, China, pp. 1323–1331 (2014)
12. Zhang, C., Liu, J., Tian, Q., Xu, C., Lu, H., Ma, S.: Image classification by non-negative sparse coding, low-rank and sparse decomposition. In: IEEE Conference on Computer Vision and Pattern Recognition, Colorado Springs, USA, pp. 1673–1680, June 2011
13. Li, F.F., Fergus, R., Perona, P.: Learning generative visual models from few training examples: an incremental Bayesian approach tested on 101 object categories. Comput. Vis. Image Underst. 106(1), 59–70 (2007)
14. Griffin, G., Holub, A., Perona, P.: Caltech-256 object category dataset. Technical report CNS-TR-2007-001, California Institute of Technology (2007)
15. Gao, S., Tsang, I., Chia, L.T., Zhao, P.: Local features are not lonely-laplacian sparse coding for image classification. In: IEEE Conference on Computer Vision and Pattern Recognition, San Francisco, USA, pp. 3555–3561, June 2010
16. Fan, R.E., Chang, K.W., Hsieh, C.J., Wang, X.R., Lin, C.J.: Liblinear: a library for large linear classification. J. Mach. Learn. Res. 9, 1871–1874 (2008)
17. van Gemert, J.C., Geusebroek, J.-M., Veenman, C.J., Smeulders, A.W.M.: Kernel codebooks for scene categorization. In: Forsyth, D., Torr, P., Zisserman, A. (eds.) ECCV 2008, Part III. LNCS, vol. 5304, pp. 696–709. Springer, Heidelberg (2008)
18. Zhang, H., Berg, A., Maire, M., Malik, J.: Svm-knn: discriminative nearest neighbor classification for visual category recognition. In: IEEE Conference on Computer Vision and Pattern Recognition, vol. 2, pp. 2126–2136, New York, USA, June 2006
19. Boiman, O., Shechtman, E., Irani, M.: In defense of nearest-neighbor based image classification. In: IEEE Conference on Computer Vision and Pattern Recognition, Anchorage, Alaska, USA, pp. 1–8, June 2008
20. Kulis, B., Jain, P., Grauman, K.: Fast similarity search for learned metrics. IEEE Trans. Pattern Anal. Mach. Intell. 31(12), 2143–2157 (2009)
21. Yuan, X.T., Yan, S.: Visual classification with multi-task joint sparse representation. In: IEEE Conference on Computer Vision and Pattern Recognition, San Francisco, USA, pp. 3493–3500, June 2010

Blind Image Quality Assessment in Shearlet Domain

Yuling Ren, Wen Lu[✉], Lihuo He, and Xinbo Gao

School of Electronic Engineering, Xidian University, Xi'an 710071, China
renyuling@stu.xidian.edu.cn,
{luwen,lhhe,xbgao}@mail.xidian.edu.cn

Abstract. The available blind image quality assessment (BIQA) criteria usually involve a large amount of human scored images to train a regression model used to judge image quality, which makes the results are closely related to the amount of training data. In this paper, a valid BIQA algorithm based on shearlet transform without using human scored images is presented. This is mainly based on that degradation of image induces considerable deviation in the distributed discontinuities in different directions. However, the distributed discontinuities can be localized by shearlet transform effectively. The nature scene statistics (NSS) of shearlet coefficients are capable for exhibiting the alteration of image quality. Experimental results on two benchmarking databases (LIVEII and TID2008) indicate the rationality and validity of the proposed method.

Keywords: Blind image quality assessment · Human scores free · Nature scene statistics · Shearlet transform

1 Introduction

Blind image quality assessment (BIQA) algorithms are designed with the aim of evaluating the quality of degraded images without any prior knowledge from the nature image [1], it has become a topic of high interest. The vast majority of BIQA algorithms have been appeared during the past several years. By characterizing the un-naturalness of distorted images using nature scene statistics (NSS), DIIVINE [2] employs the classical two-step structure to assess the image quality. Given NSS features, a simple Bayesian inference model is introduced in BLIINDS-II [3] to acquire image quality scores. A general metric of quality is produced in BRISQUE [4] by quantifying the damage of "naturalness" in the distorted image. In SHANIA [5], the mean of shearlet coefficient amplitudes (MSCA) in coarse scale of degraded image is exploited to predict the MSCA in fine scales by using a sparse auto-encoder. The difference between the predicted and real MSCA is employed as the criterion to perceive the distorted image quality. NIQE [6] adopts a set of nature images from which to estimate the parameters of NSS model statistics, and takes the difference between the parameters of NSS model and of degraded image as a measure of the image quality.

The above mentioned BIQA algorithms [2–5] all require plenty of human scored images when training the regression model, which makes the performance of proposed algorithm training database dependent and closely related to the amount of training data.

© Springer International Publishing Switzerland 2015
X. He et al. (Eds.): IScIDE 2015, Part I, LNCS 9242, pp. 472–481, 2015.
DOI: 10.1007/978-3-319-23989-7_48

Furthermore, these methods usually map the extracted image features to a quality score by learning a mapping function, which makes the relationship between features and quality score ambiguous and the BIQA process invisible [7]. However, the significant information of distributed discontinuities in different directions of distorted image is absence in NIQE [6]. Based on the fact that shearlet transform [8, 9] is capable for capturing the variation of discontinuities, an efficient blind image quality assessment algorithm is proposed based on shearlet transform without using human scored image. In the proposed framework, the NSS features of distorted images are first extracted in the shearlet domain; and then the mean and covariance of the extracted features are calculated to model their distribution; and finally a simple difference metric between the distribution of the features extracted from the training process and it of distorted image is employed to quantify the image quality.

The remainder of the paper is organized as follows. Section 2 illustrates the detailed implementation of the proposed method. Section 3 presents the experimental results and a though analysis. Finally, conclusion is made in Sect. 4.

2 Blind Image Quality Assessment in Shearlet Domain

Figure 1 illustrates the structure of the proposed algorithm. In the first procedure, the image is segmented into equally sized blocks and a local shearlet transform is applied on each block to capture the multi-scale and great change in the distribution discontinuities of image structure. The second procedure of the proposed algorithm is to extracted quality aware features from the shearlet coefficients of each block. Finally, we calculate the mean and covariance of the extracted features to model their distribution and introduce a simple difference metric between the natural mean and covariance trained from the corpus of natural images and distorted mean and covariance of distorted image to measure the distorted image quality.

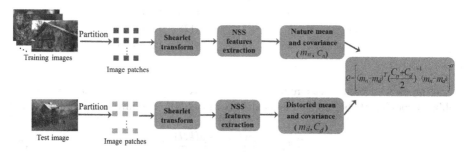

Fig. 1. Framework for the proposed BIQA

2.1 Shearlet Transform

The proposed BIQA algorithm is based on shearlet transform [8, 9]. Shearlet transform has a strong ability to acquire directional and anisotropic information of an image by applying a so-called shearing operator along with the anisotropic scaling operator.

The goal of shearlet transform used in the proposed method is to compute shearlet coefficients of each patch:

$$DST_{s,a,t}(f) = <\psi_{s,a,t}, f>,$$ (1)

where $\psi_{s,a,t}$ is a two-dimensional digital shearlet from the cone-adapted discrete shearlet system [9] indexed by a scale parameter s, a shearing parameter a, a translation parameter t and an image signal $f \in l_2(Z^2)$.

In summary, shearlet transform is capable for capturing the geometrical characteristic of image because of the basis elements of it with various shapes and high directional sensitivity. The degradation of image induces deviation in the distributed discontinuities in different orientations. However, shearlet transform has a great power to localize spread discontinuities. With those good properties, shearlet transform is very suitable to describe the variation of image generated by distortion.

2.2   NSS Features Extraction in Shearlet Domain

Assuming that the size of the input image is 2^m ($m \geq 7$ is a positive integer). Then, divide the image into patches of size 128×128, and apply shearlet transform on each patch to obtain its shearlet coefficients. Each patch is transformed into different directional subband over 4 different scales. The structure of shearlet coefficients of each patch is shown in Table 1. A set of NSS features is then extracted from the obtained subband coefficients.

Table 1.   Structure of shearlet coefficient of each patch.

Scale index	Orientation number	Matrix form
S1	10	$(128 \times 128) \times 10$
S2	10	$(128 \times 128) \times 10$
S3	18	$(128 \times 128) \times 18$
S4	18	$(128 \times 128) \times 18$

A process of local mean removal and divisive normalization is applied on the subband coefficients before the classical NSS model [10, 11] introduced in this paper.

$$\hat{S}(k,l) = \frac{S(k,l) - \mu(k,l)}{\sigma(k,l) + 1},$$ (2)

Where $k \in \{1, 2...K\}$, $l \in \{1, 2...L\}$ are indices, K and L are the subband dimensions, and $\mu(k,l)$, $\sigma(k,l)$ are the local mean and contrast of subband coefficients.

The coefficient $\hat{S}(k,l)$ of nature images follow a Gaussian distribution [10], however, this kind of distribution is disturbed by the distortion occurring in the nature images. Figure 2 presents the original image and the corresponding five types distorted images which are used to illustrate the effect of distortion on distribution of shearlet

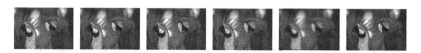

Fig. 2. Original image and five types distorted image of it from LIVE database II. From left to right: Original image, JP2K compressed image, JPEG compressed image, White noise image, Gaussian blur image, Fast fading image.

coefficients. The distributions of these coefficients for original image and the corresponding degraded images are shown in Fig. 3. The degree of deviation acts as a factor to indicate the image quality.

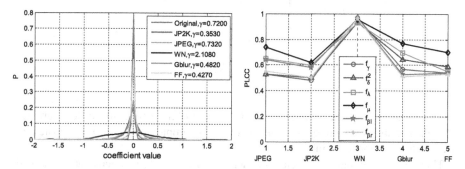

Fig. 3. Distribution of shearlet coefficients of images with different distortions and shape parameters.

Fig. 4. PLCC for each of the features proposed on the LIVE database II.

The generalize Gaussian distribution (GGD) is utilized to capture the distortion of subband coefficients, which is defined as:

$$f(x; \mu, \sigma^2, \gamma) = \frac{b\gamma}{2\Gamma(1/\gamma)} \exp(-(b(x - \mu))^\gamma), \tag{3}$$

where μ, σ^2, γ are the mean, variance, and shape-parameter of the distribution and

$$b = \frac{1}{\sigma}\sqrt{\Gamma(3/\gamma)/\Gamma(1/\gamma)}, \tag{4}$$

and $\Gamma(\bullet)$ is the gamma function:

$$\Gamma(x) = \int_0^\infty t^{x-1}e^{-t}dt \quad x > 0. \tag{5}$$

Since the mean of shearlet subands response are zero, we just compute two parameters (γ, σ^2) employing the moment-matching based technique appeared in [12] for each subband. The first group of features that is utilized to describe image distortion is formed.

$$f = [f_\gamma, f_{\sigma^2}]. \tag{6}$$

A regular structure is found in the distribution of transformed shealet coefficient with neighboring coefficients. However, this kind of regular structure becomes unregular due to the existing of distortion. This unregular structure is modeled by evaluating the distribution of adjacent coefficients on four directions: horizontal (H), vertical (V) and two diagonal directions (D1, D2).

The asymmetric generalized Gaussian distribution (AGGD) model [13] is applied to model the distribution of adjacent shearlet coefficients.

$$f(x; \lambda, \sigma_l^2, \sigma_r^2) = \begin{cases} \dfrac{\lambda}{(\beta_l + \beta_r)\Gamma(\frac{1}{\lambda})} \exp(-(\dfrac{-x}{\beta_l})^\lambda), & x < 0; \\ \dfrac{\lambda}{(\beta_l + \beta_r)\Gamma(\frac{1}{\lambda})} \exp(-(\dfrac{-x}{\beta_r})^\lambda), & x \geq 0. \end{cases} \tag{7}$$

$$u = (\beta_l - \beta_r)\Gamma(\frac{2}{\lambda}) \Big/ \Gamma(\frac{1}{\lambda}), \tag{8}$$

The parameters of the AGGD $(\lambda, \beta_l, \beta_r, \mu)$ controlling shape, mean, left variance and right variance respectively can be calculated by employing the moment-matching based technique appeared in [12].

Thus for each subband coefficient of each image patch, 16 parameters are generated which construct the next four sets of features:

$$f_\lambda = [f_\lambda^H, f_\lambda^V, f_\lambda^{D_1}, f_\lambda^{D_2}], \tag{9}$$

$$f_{\beta_l} = [f_{\beta_l}^H, f_{\beta_l}^V, f_{\beta_l}^{D_1}, f_{\beta_l}^{D_2}], \tag{10}$$

$$f_{\beta_r} = [f_{\beta_r}^H, f_{\beta_r}^V, f_{\beta_r}^{D_1}, f_{\beta_r}^{D_2}], \tag{11}$$

$$f_\mu = [f_\mu^H, f_\mu^V, f_\mu^{D_1}, f_\mu^{D_2}]. \tag{12}$$

Let f_{SNSS} denotes all the features extracted from an image.

$$f_{SNSS} = [f_\gamma, f_{\sigma^2}, f_\lambda, f_{\beta_l}, f_{\beta_r}, f_\mu] \times I \times J, \tag{13}$$

where $I \in R$ denotes the number of directional subband of each patch, and $J \in R$ represents the number of patch with the size of contained in an image.

Until now, all the features based on NSS have been defined and their relationships with various distortions have been introduced. Figure 4 shows the polynomial linear correlation coefficient (PLCC) of the proposed features of five types distortions extracted from LIVE database II verse the mean opinion score. The plot is in order to illustrate the different importance and validate the rationality of proposed features as indicators of image quality.

2.3 Quality Pooling

Given the NSS features extracted in the previous section, we calculate the mean and covariance of the extracted features of distorted images to model their distribution.

$$m = mean(f_{SNSS}), \tag{14}$$

$$c = covariance(f_{SNSS}). \tag{15}$$

The nature mean and covariance are trained from a varied set of 154 undistorted images with size ranging from 480 * 320 to 1280 * 720 as the ground truth. The difference between the distorted mean and covariance and nature mean and covariance act as the criterion to perceive image quality.

$$Q = \left[(m_n - m_d)^T (\frac{c_n + c_d}{2})^{-1} (m_n - m_d) \right]^{\alpha}, \tag{16}$$

Where m_n, m_d and c_n, c_d are the mean vectors and covariance matrices of nature images and distorted image, and $\alpha = 0.25$.

3 Experiments and Results

In this section, tests are done on the LIVE database II [14] and the TID2008 database [15] to verify the rationality and validity of proposed. 29 distortionless images and corresponding 779 distorted images are contained in the LIVE database II. The TID2008 database covers 1700 distorted images, which are originated from 25 reference images. Each reference image has 17 categories of distortion and there are four levels for each distortion. Both the spearmans rank ordered correlation coefficient (SROCC) and the linear correlation coefficient (LCC) are taken into account in the experiment. The values of SROCC and LCC closer to 1 imply superior consistency with human perception.

3.1 Consistency Experiment

In this part, we conduct the consistency experiment to compare the validity of the proposed with the existing BIQA metrics like DIIVINE [2], BLIINDS-II [3], BRIS-QUE [4], SHANIA [5], and NIQE [6]. The values for SROCC and LCC of all BIQA metrics mentioned above are given in Tables 2 and 3. Figure 5 presents the nonlinear fitting of the objective quality score achieved by proposed versus difference mean opinion score (DMOS) on LIVE database II. Consistency experiment shows that the proposed achieves a competed performance with the state-of -the-art techniques [2–5] which heavily employ the human scored images in training and performs better than NIQE [6] in which the IQA model trains without using human scored images.

Table 2. LCC and SROCC of different metrics on the LIVE database II

Metric	JPEG2K		JPEG		WN		Gblur		FF	
	LCC	SROCC	LCC	SROCC	LCC	SROCC	LCC	SROCC	LCC	SROCC
DIIVINE	0.9220	0.9319	0.9210	0.9483	0.9880	0.9821	0.9230	0.9210	0.8680	0.8714
BLIINDS-II	0.9386	0.9323	0.9426	0.9331	0.9635	0.9463	0.8994	0.8912	0.8790	0.8519
BRISQUE	0.9229	0.9139	0.9734	0.9647	0.9851	0.9786	0.9506	0.9511	0.9030	0.8768
SHANIA	0.9135	0.861	0.9380	0.8918	0.9731	0.9582	0.9790	0.9674	0.9413	0.9169
NIQE	0.9370	0.9172	0.9564	0.9382	0.9773	0.9662	0.9525	0.9341	0.9128	0.8594
PROPOSED	0.9043	0.8967	0.9054	0.8544	0.9809	0.9805	0.9393	0.9420	0.8879	0.8833

Table 3. LCC and SROCC of different metrics on the TID2008 database

Metric	JPEG2K		JPEG		WN		Gblur	
	LCC	SROCC	LCC	SROCC	LCC	SROCC	LCC	SROCC
DIIVINE	0.896	0.907	0.899	0.871	0.828	0.834	0.844	0.859
BLIINDS-II	0.919	0.911	0.889	0.838	0.714	0.715	0.825	0.826
BRISQUE	0.906	0.904	0.950	0.911	0.810	0.823	0.873	0.874
SHANIA	0.859	0.838	0.920	0.871	0.708	0.684	0.904	0.668
NIQE	0.893	0.894	0.861	0.749	0.779	0.753	0.909	0.825
PROPOSED	0.893	0.871	0.897	0.840	0.815	0.826	0.869	0.862

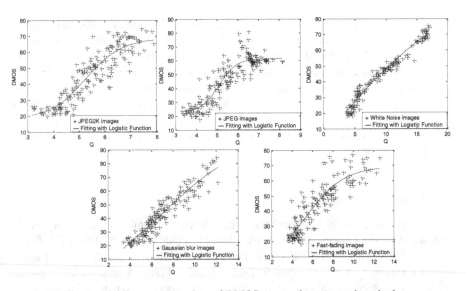

Fig. 5. Nonlinear scatter plots of DMOS versus the proposed method.

3.2   Rationality Experiment

The proposed is tested on the Einstein images with four types of distortion to verify its
rationality. The four types of distortion contained in Einstein images are blurring

(smoothing window = $W \times W$), additive Gaussian noise (mean = 0, variance = V), JPEG compression (compression data = R) and impulsive salt-pepper noise (density = D). Figure 6 shows the proposed prediction trend of the Einstein images with different quality. What we can see from Fig. 6 is that the prediction results of proposed trend to rise with the increasing of distortion degree. The higher values of prediction results and DMOS mean poorer visual perception quality. It is found that the proposed has a high consistency with the tendency of the decreasing image quality.

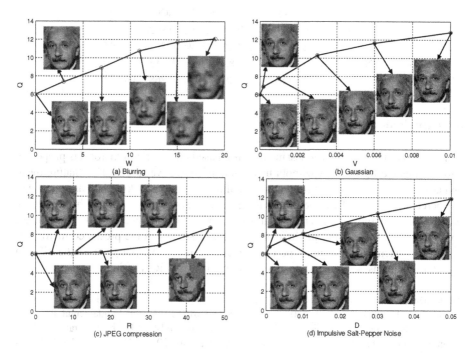

Fig. 6. Results of rationality experiment

3.3   Multiscale Experiment

It is common knowledge that images are naturally multiscale and that the mostly IQA systems involve decompositions over scales. As a consequence, a multiscale feature extraction approach based on shearlet transform is implemented in this paper. The

Table 4. LCC and SROCC on LIVE database II

Type	Two scales		Three scales		Four scales	
	LCC	ROCC	LCC	ROCC	LCC	ROCC
JPEG2K	0.8962	0.9014	0.8944	0.8974	0.9043	0.8967
JPEG	0.8647	0.8574	0.9044	0.8629	0.9054	0.8544
WN	0.9805	0.9787	0.9807	0.9788	0.9809	0.9805
Gblur	0.8992	0.9111	0.9230	0.9397	0.9393	0.9420
FF	0.8408	0.8376	0.8700	0.8648	0.8879	0.8833

comparisons of prediction results on two benchmarking databases for 2 scales, 3 scales, and 4 scales feature extraction are shown in Tables 4 and 5. The performance of the proposed is generally improved with the increasing of image scale utilized in feature extraction and the optimal results are achieved on the fourth scale.

Table 5. LCC and SROCC on TID2008 database

Type	Two scales		Three scales		Four scales	
	LCC	ROCC	LCC	ROCC	LCC	ROCC
JPEG2K	0.8690	0.8633	0.8880	0.8698	0.8926	0.8713
JPEG	0.8932	0.8608	0.8964	0.8394	0.8974	0.8396
WN	0.8833	0.8906	0.8382	0.8463	0.8152	0.8260
Gblur	0.8529	0.8529	0.8529	0.8454	0.8687	0.8623

4 Conclusion

A novel blind image quality assessment metric based on shearlet transform is proposed, which trains the IQA model without using human subjective scores. The "nature mean and covariance" are obtained from a corpus of nature images lacking human scores and the perception score of the degraded image is achieved by measuring the difference between the "nature mean and covariance" and "distorted mean and covariance". Experimental results on the public databases demonstrate that the presented metric has a high consistency with human perception. Furthermore, the potential of the NSS in shearlet domain is deserved to explore to implement the IQA task.

Acknowledgments. This research was supported partially by the National Natural Science Foundation of China (No. 61125204, 61372130, and 61432014), the Fundamental Research Funds for the Central Universities (No. BDY081426, and JB140214), the Program for New Scientific and Technological Star of Shaanxi Province (No. 2014KJXX-47), and the Project Funded by China Postdoctoral Science Foundation (No. 2014M562378).

References

1. Wang, Z., Bovik, A.C.: Modern Image Quality Assessment. Morgan and Claypool Publishing Company, New York (2006)
2. Moorthy, A.K., Bovik, A.C.: Blind image quality assessment: from natural scene statistics to perceptual quality. IEEE Trans. Image Process. **20**(12), 3350–3364 (2011)
3. Saad, M.A., Bovik, A.C., Charrier, C.: Blind image quality assessment: a natural scene statistics approach in the DCT domain. IEEE Trans. Image Process. **21**(8), 3339–3352 (2012)
4. Mittal, A., Moorthy, A.K., Bovik, A.C.: No-reference image quality assessment in the spatial domain. IEEE Trans. Image Process. **21**(12), 4695–4708 (2012)
5. Li, Y., Po, L.M., Xu, X., Feng, L.: No-reference image quality assessment using statistical characterization in the shearlet domain. Sig. Process. Image Commun. **29**(7), 748–759 (2014)

6. Mittal, A., Soundararajan, R., Bovik, A.C.: Making a completely blind image quality analyzer. IEEE Sig. Process. Lett. **20**(3), 209–212 (2013)
7. Xue, W., Zhang, L., Mou, X.: Learning without human scores for blind image quality assessment. In: The IEEE Conference on Computer Vision and Pattern Recognition, pp. 995–1002. IEEE Press, Portland (2013)
8. Easley, G., Labate, D., Lim, W.Q.: Sparse directional image representations using the discrete shearlet transform. Appl. Comput. Harmonic Anal. **25**(1), 25–46 (2008)
9. Lim, W.Q.: Nonseparable shearlet transform. IEEE Trans. Image Process. **22**(5), 2056–2065 (2013)
10. Ruderman, D.L.: The statistics of natural images. Netw. Comput. Neural Syst. **5**(4), 517–548 (1994)
11. Srivastava, A., Lee, A.B., Simoncelli, E.P., Zhu, S.C.: On advances in statistical modeling of natural images. J. Math. Imaging Vis. **18**(1), 17–33 (2003)
12. Sharifi, K., Leon-Garcia, A.: Estimation of shape parameter for generalized Gaussian distributions in subband decompositions of video. IEEE Trans. Circ. Syst. Video Technol. **5**(1), 52–56 (1995)
13. Lasmar, N.E., Stitou, Y., Berthoumieu, Y.: Multiscale skewed heavy tailed model for texture analysis. In: The IEEE International Conference on Image Processing, pp. 2281–2284. IEEE Pressing, Cairo (2009)
14. Sheikh, H.R., Wang, Z., Cormack, L., Bovik, A.C.: Live image quality assessment database, release 2. http://live.ece.utexas.edu/research/quality
15. Ponomarenko, N., Lukin, V., Zelensky, A., Egiazarian, K., Carli, M., Battisti, F.: Tid 2008 - a database for evaluation of full-reference visual quality assessment metrics. Adv. Mod. Radioelectron. **10**(4), 30–45 (2009)

A Novel Image Quality Assessment for Color Distortions

Shuyu Yang, Wen Lu[✉], Lihuo He, and Xinbo Gao

School of Electronic Engineering, Xidian University, Xi'an 710071, China
yangshuyu@stu.xidian.edu.cn,
{luwen,lhhe,xbgao}@mail.xidian.edu.cn

Abstract. Most of the existing assessment methods for color image quality consider little about the image content, which plays an essential role in indicating the distortion level with the color and structure information it contains. By incorporating the color image content with human color perception, we present a novel assessment metric for color distortions. The proposed method standardizes input images with a color perception transformation, and then the transformed images are divided into lightness part and chroma part. For each part, a region separation strategy based on image content is implemented. By calculating and pooling the similarity of each region using fuzzy integral, the final index is achieved. Experimental results on color-related distortions of TID2013 database show the superiority of this new approach and comparative experiments reveal the rationality and the validity of our method.

Keywords: Color image quality assessment · Color distortion · Content of color image

1 Introduction

Over the years, color images have been extensively applied to visualization techniques due to the advancement of multimedia technology and imaging devices. A significant advantage of color images over the gray-level counterparts is that color contains a higher information level, which could provide a more accurate and visually appealing reproduction of the natural scene. This superiority makes color a crucial role in practical applications. For instance, it brings great convenience to agricultural industry as the tools for implementing automatic classification or quality inspection of fruits can be achieved by detecting the peel color. However, color distortions take place frequently during the collection and transmission of color images. Since the distorted images fail to provide a precise description of the objective world, it's of great importance to evaluate the quality of the color images before utilizing them for practical application. As subjective quality assessment is costly and time-consuming, objective assessment methods that can predict image quality automatically and accurately are in much demand. Since most of the existing methods are based on gray-level images, where color information is totally ignored, it's inappropriate to apply them to assess color image quality. It can be seen from Fig. 1 that the results of PSNR [1] and SSIM [2] on color images with different distortion levels stay almost constant when the

© Springer International Publishing Switzerland 2015
X. He et al. (Eds.): IScIDE 2015, Part I, LNCS 9242, pp. 482–492, 2015.
DOI: 10.1007/978-3-319-23989-7_49

human eye can easily notice the changes of color. It's probably due to the fact that the pixel of a gray-level image is a scalar while that of a color image is a vector, which makes assessment algorithms based on scalar values inadequate for color images.

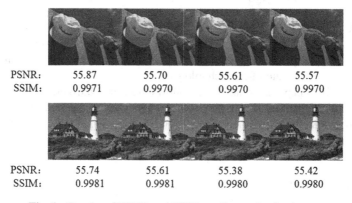

PSNR:	55.87	55.70	55.61	55.57
SSIM:	0.9971	0.9970	0.9970	0.9970

PSNR:	55.74	55.61	55.38	55.42
SSIM:	0.9981	0.9981	0.9980	0.9980

Fig. 1. Results of PSNR and SSIM on distorted color images

In recent years, researchers have taken different approaches to solve this problem and the previous methods for color images can be roughly divided into the following three categories: (1) Statistical characteristics-based paradigms, which take advantage of mathematical models to represent the color information. Reference [3] adopts MGGD to model the statistics of color components' coefficients and the KLD between two MGGDs is calculated to quantify the visual degradation. In [4], quaternion vector space is utilized to define a color pixel and measure the multiplication similarity between two color vectors. Reference [5] takes local variance distribution of RGB channels as a quaternion and then calculates the mid-point distance between quaternion matrices to obtain the final assessment result. These methods consider the dependencies between individual color components, but the results on distorted images with color contrast change have poor consistency with HVS. (2) Feature-based paradigms, where the features of color images are extracted to assess quality. In [6], features based on the color correlogram are extracted and trained to identify distortion type and yield an objective quality score. It's required only a few data but it can't deal with a new kind of distortion without extra training. In [7], phase congruency and gradient magnitude are employed as two features in YIQ color space to derive a quality assessment score. In [8], features based on fundamental quality parameters perceived by HVS of three color components are synthesized to assess quality. The features they use are actually mainly on gray-level rather than color, so its progress is limited. (3) Human color vision-based paradigms, which comprises properties of color perception and geometrical distortion measurements to predict the image quality. Reference [9] incorporates color perceptual model with existing metrics to assess quality. Reference [10] normalizes the input images with an image-appearance model before extracting image-difference features to predict image quality. Reference [11] improves [10] by adding two chromatic features. Although these methods are effective on certain types of distortion, there is still room for improvement.

In addition to the strengths and weaknesses of previous methods analyzed above, it's noteworthy that none of them has taken the effect of color distortion on image content into account. In fact, image content plays an important role in indicating the distortion level with the color, position and quantity information it contains. Therefore, we propose a novel assessment metric for color distortions in this paper. The proposed framework consists of color perception transformation, color content-based region separation and fuzzy fusion to predict the image quality. Experimental results show the effectiveness of the proposed method on different kinds of color distortions and the competitiveness with state-of-art technologies.

The rest of this paper is organized as follows. Section 2 introduces color distortion. In Sect. 3, the proposed metric for color image quality assessment is presented. Section 4 demonstrates experimental results and discussions. Finally, Sect. 5 concludes the paper and suggests directions for future research.

2   Color Distortion

This section presents a brief introduction of color distortion to address the key distinction between it and traditional distortion such as spatial blurring, which in reality exerts no influence on image's color attribute. Color distortion is generally regarded as the direct deviation of color information and it can be categorized into the following parts according to factors, known as luminance, chrominance and external effects in image processing, which lead to the change of color.

(1)   The luminace of a color image represents the brightness of the color. With the decreasing of luminance intensity, the overall color of an image tends to appear darker, as shown in Fig. 2(a). Moreover, change of luminance also affects image contrast, which describes the relation between the brightest and darkest area of an image. Only appropriate contrast can display a wide range of colors vividly. Large contrast makes color appear too colorful while small contrast makes it tend to be gray, which is shown in Fig. 2(b).

(2)   Chrominance is generally characterized as the combination of hue and saturation, which describes the type and purity of a color, respectively. Color varies with the change of hue. As for the same kind of color, the more white light it diluted with, the lower saturation it has and more faded it appears. Since chrominance contains both color's classification and shade, slight mutual shifting of R, G, and B components will lead to color aberrations, which may even generate the color that doesn't exist in original image. Distorted images caused by the chrominance attribute are represented in Fig. 2(c)–(e).

(3)   In addition to the two factors mentioned above, some external effects may also cause color distortions in image processing. For instance, when transmit images using JPEG compression, the color of abrupt changing in images would be removed. That is, the compression processing deletes the color around object's edges and fills it with the color closest to it. See Fig. 2(f) for details. Moreover, the perception sensitivity to color variance of human visual system is affected by noise in transmission, as shown in Fig. 2(g) and (h).

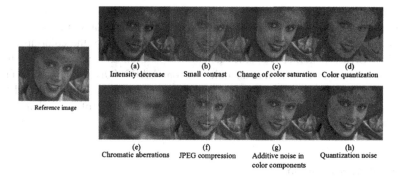

(a) Intensity decrease	(b) Small contrast	(c) Change of color saturation	(d) Color quantization
(e) Chromatic aberrations	(f) JPEG compression	(g) Additive noise in color components	(h) Quantization noise

Fig. 2. Reference image and distorted images with different distortion types

What's analyzed above indicates that color distortions have more complexity than traditional distortions and greatly degrade the image quality. Hence it's of great importance to design quality assessment models especially for color images.

3 Proposed Method

Figure 3 illustrates the framework of our method. The input images are transformed by a color perception model and then divided into a lightness part and a chroma part. For each part, structure information of pixels is extracted and a region separation based on color content is implemented. By calculating the similarity of each region using fuzzy integration, we propose a metric for color image quality assessment. The detailed steps of the metric are provided in following sections.

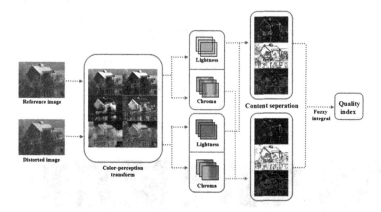

Fig. 3. General overview of proposed method

3.1 Color-Perception Transform

Color perception is affected by such factors as illuminate conditions, viewing distance, and background environment, meaning that human perception of the color with same CIE tristimulus varies with different viewing conditions. Thus a color-perception transformation is needed to simulate the perceptual processing of human eye. The method we adopt is to process input images with a color-appearance model and transform them into a working color space. A feasible choice of the color-appearance model is S-CIELAB model [12] since its color separation and spatial filtering imitate perception characteristics of human color vision. LAB2000HL color space [13] is chosen for the working space as it has better perceptual uniformity compared to the traditional CIELAB color space. Figure 4 shows the example of each channel after color-perception transformation.

3.2   Content Separation

The motivation behind this step is due to fact that different regions of the image affect the image quality in varying degrees. Since an image's lightness part and chroma part exert different influence on color distortion, we firstly divide the color-perception transformed images into lightness and chroma separately. As an image pixel in the working color space consists of a lightness and two chromatic values, the chroma part of the image is denoted as

$$C = \sqrt{a^2 + b^2}. \tag{1}$$

Therefore, we decompose the image into two parts, lightness (L) and chroma (C). After that, a content based separation is implemented, dividing each part into edge area (A_E), texture area (A_T) and flat area (A_F) respectively. Because the gradient magnitude of these three areas decreases regularly [14], we utilize the value of gradient magnitude to get the above three areas. By the maximum value (gm_{max}), two separation thresholds are determined denoted as $ST_1 = 0.12\ gm_{max}$ and $ST_2 = 0.06\ gm_{max}$. For a pixel $(x, y) \in A_E$, the separation complies with the following rules when the gradient magnitude of reference image is represented as $p_r(x, y) > ST_1$ and that of distorted image as $p_d(x, y)$. The separation results of example images are illustrated in Fig. 5.

$$R1: \quad p_r(x,y) > ST_1 \ or \ p_d(x,y) \le ST_1, \quad (x,y) \in A_E \tag{2}$$

$$R2: \quad p_r(x,y) > ST_2 \ and \ p_d(x,y) \le ST_1, \quad (x,y) \in A_T \tag{3}$$

$$R3: \quad ST_2 \le p_r(x,y) > ST_1 \ and \ p_d(x,y) \le ST_1, \quad (x,y) \in A_F \tag{4}$$

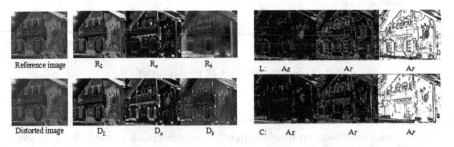

Fig. 4. Results of the transformation   **Fig. 5.** Results of content separation

3.3 Information Fusion

Since the level of distortion is mainly determined by the error of position, structure and quantity information between a reference image and a distorted image, we apply SSIM' to each area to get the structure information.

$$\text{SSIM}'(x,y) = [l(x,y)]^{\partial} \cdot [c(x,y)]^{\beta} \cdot [s'(x,y)]^{\gamma}, \tag{5}$$

$$s'(x,y) = \frac{|\sigma_{xy}| + C_3}{\sigma_x \sigma_y + C_3}. \tag{6}$$

The reason for using above equations instead of the original SSIM is that $s'(x,y)$ preserves all the structure information with a range of [0, 1], which is conducive to the following information fusion. Sugeno fuzzy integral is utilized to fuse three areas as SSIM' of each position can be regarded as a generator for evaluating the quality of an area. Details of fuzzy integral are provided in [15]. By assembling all values of SSIM' in a certain area, denoted as P, the overall similarity of this area is obtained.

$$P = \{p_i = \text{SSIM}'(x_i, y_i) | i = 1, 2, \ldots N\}, \tag{7}$$

$$\text{CSIM}(P) = (S) \int_P \text{SSIM}'(x_i, y_i) dg = \sup[\min(p_i, g(F_{P_i}(M)))]. \tag{8}$$

In this way, the similarity of A_E, A_T and A_F can be calculated respectively, denoted as CSIM(E), CSIM(T) and CSIM(F).

3.4 Quality Estimation

In this section, final quality index is defined by weighted strategy of three areas in both lightness and chroma part of the image. The formulas are as followed:

$$\text{PM} = \sum_{M \in [E,T,F]} a_M \cdot \text{CSIM}(M), \tag{9}$$

$$Q = \omega_1 \cdot \text{PM}_L + \omega_2 \cdot \text{PM}_C. \tag{10}$$

where a_M weight the contribution of each area to the global image quality prediction and ω are the weights attributed to the perceptual distortions in each of the two parts. Since image's lightness part and chroma part both affect color distortions, weighting parameters can be assumed to be similar, i.e., $\omega_1 = \omega_2 = 0.5$. Table 1 shows the evaluation results for example images given in Fig. 4.

Table 1. Results for example images

	CSIM(T)	CSIM(E)	CSIM(F)	PM	Q
L	0.8736	0.8532	0.8865	0.8693	0.8190
C	0.6773	0.7880	0.8608	0.7515	

4 Experiments and Discussion

In this section, we compare the performance of the proposed metric with several existing IQA methods for color images on public database and conduct sensitivity experiment and rationality experiment to verify the effectiveness of our method.

4.1 Experimental Data

The database used in the experiment is TID2013 [16], a modified version of the popular database TID2008. Compared to TID2008, seven new types and one more level of distortions are included in TID2013, meaning that the new database contains 25 reference color images and 3000 corresponding distorted images with 24 types of distortions(five levels for each distortion). Large number of images, various distortion types and a rather wide range of distortions levels make TID2013 a better choice for assessing quality metrics. More importantly, it contains several types of color distortions, which are not well represented in most of the existing databases such as LIVE database II, IVC database and TID2008. As the method proposed in this paper is especially for evaluating color images, it seems more reasonable to use this database. Table 2 shows the color-related distortions presented in TID2013.

Table 2. Types of color-related distortions in TID2013

No.	Type of distortion	No.	Type of distortion
#2	Additive noise in color components	#7	Quantization noise
#10	JPEG compression	#16	Mean shift (intensity shift)
#17	Contrast change	#18	Change of color saturation
#22	Image color quantization with dither	#23	Chromatic aberrations

Table 3. Performance of different metrics on color-related distortions

	Proposed	iCID	CID	QSSIM	S-SSIM	FSIM
Additive noise	0.8280	0.8388	0.7455	0.7755	0.8812	0.8139
Quantization noise	0.8816	0.8460	0.7760	0.8311	0.8949	0.8739
JEPG compression	0.9743	0.9295	0.9385	0.9497	0.9598	0.9701
Mean shift	0.8044	0.8311	0.8386	0.8029	0.7779	0.7054
Contrast change	0.7068	0.6450	0.6007	0.6010	0.6075	0.7486
Color saturation	0.8182	0.7358	0.6586	0.6894	0.6412	0.3538
Color quantization	0.8739	0.8955	0.8368	0.8310	0.9032	0.8863
Chromatic aberrations	0.9645	0.9340	0.9412	0.9643	0.9207	0.9769

4.2   Consistency Experiment

The performance of objective quality assessment is reflected by consistency between the mean opinion score (MOS) and the results predicted by proposed method. In this section, we adopt the Pearson linear correlation coefficient (CC), which indicates the accuracy of prediction, as the evaluation criteria. A CC value closer to 1 means a higher correlation with human perception. We compare the performance of our method with several existing technologies like iCID [11], CID [10], QSSIM [4], S-SSIM [9] and FSIM [7] on TID2013 database. As the assessment method proposed in this paper is for color image, we focus on analyzing the results of color-related distortions mentioned above. Table 3 shows the experimental results and nonlinear scatter plots of the subjective MOS versus the proposed method on some distortions are given in Fig. 6. The results of the CC criteria clearly show the strength and weakness of each method. In general, our metric outperforms CID, QSSIM, S-SSIM and FSIM in all color-related distortions. As for the comparison between iCID and the proposed method, our method achieves better performance in 5 types of distortions and greatly improves the results on contrast change and color saturation. It's due to the fact that our method considers the structure of chromatic part of the image. The consistency experiment suggests that our method is best correlated with subjective assessment.

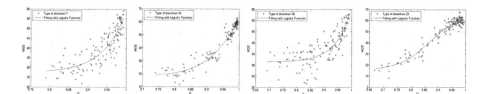

Fig. 6. Nonlinear scatter plots of the subjective scores versus the proposed method

4.3   Rationality Experiment

To verify the rationality of the proposed metric, we choose four sets of images with different distortions, which are JPEG compression, Change of color saturation, Image color quantization with dither and Chromatic aberrations. Figure 7 illustrates the relationship between images with a gradually decreasing level of distortion (MOS increasing) and metric prediction trend corresponding to them. It can be observed that the proposed method prediction trend rises with the increasing of MOS on different types of distortions. It is suggested that our method is consistent with the subjective assessment. So the rationality of the proposed framework is proved.

4.4   Effectiveness Test of the PM_c

In this section, we compare the performance of the chromatic feature (PM_c) used in our framework with IDF6 (image-difference feature, Chroma-contrast) and IDF7 (image-difference feature, Chroma-structure), two color information-based features

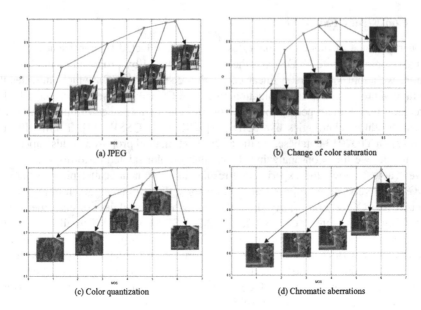

(a) JPEG

(b) Change of color saturation

(c) Color quantization

(d) Chromatic aberrations

Fig. 7. Results of rationality experiment

proposed in [11] lately which are derived from SSIM index. Figure 8 shows the results of each feature corresponding to a sequence of images with quality decreasing in turn. It is found that IDF6 and IDF7 contribute little in measuring the variation of image quality while PM_c is sensitive to the change of color. Therefore, the chromatic feature in our framework is impactful to color quality assessment.

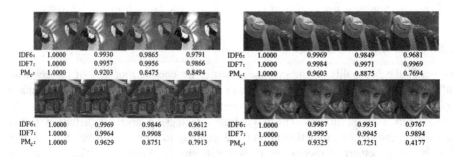

IDF6:	1.0000	0.9930	0.9865	0.9791	IDF6:	1.0000	0.9969	0.9849	0.9681
IDF7:	1.0000	0.9957	0.9956	0.9866	IDF7:	1.0000	0.9984	0.9971	0.9969
PM_c:	1.0000	0.9203	0.8475	0.8494	PM_c:	1.0000	0.9603	0.8875	0.7694

IDF6:	1.0000	0.9969	0.9846	0.9612	IDF6:	1.0000	0.9987	0.9931	0.9767
IDF7:	1.0000	0.9964	0.9908	0.9841	IDF7:	1.0000	0.9995	0.9945	0.9894
PM_c:	1.0000	0.9629	0.8751	0.7913	PM_c:	1.0000	0.9325	0.7251	0.4177

Fig. 8. Results of effectiveness test

5   Conclusion

In this paper, we propose a novel color image quality assessment framework which incorporates human color vision and the content of color image. Experimental results on the latest public database demonstrate that the presented method has a higher consistency with MOS than state-of-the-art metrics. In addition, the effectiveness test suggests that the chromatic content of image is instrumental to color image quality assessment. However, the proposed framework is still limited in assessing change of contrast, other features, like visual saliency could further be incorporated to improve this issue.

Acknowledgments. This research was supported partially by the National Natural Science Foundation of China (No. 61125204, No. 61372130, No. 61432014), the Fundamental Research Funds for the Central Universities (No. BDY081426, No. JB140214), the Program for New Scientific and Technological Star of Shaanxi Province (No. 2014KJXX-47), and the Project Funded by China Postdoctoral Science Foundation (No. 2014M562378).

References

1. Eskicioglu, A.M., Fisher, P.S.: Image quality measures and their performance. IEEE Trans. Image Commun. **43**(12), 2959–2965 (1995)
2. Wang, Z., Bovik, A.C., Sheikh, H.R., Simoncelli, E.P.: Image quality assessment: from error visibility to structural similarity. IEEE Trans. Image Process. **13**(4), 600–612 (2004)
3. Omari, M., Abdelouahad, A.A., Hassouni, M.E., Cherifi, H.: Color image quality assessment measure using multivariate generalized Gaussian distribution. In: International Conference on Signal-Image Technology and Internet-Based Systems, Japan, pp. 195–200 (2013)
4. Kolaman, A., Yadid-Pecht, O.: Quaternion structural similarity: a new quality index for color images. IEEE Trans. Image Process. **21**(4), 1526–1536 (2012)
5. Wang, Y., Zhu, M.: Color image quality assessment based on quaternion representation for the local variance distribution of RGB channels. In: 2nd International Congress on Image and Signal Processing, Tianjin, China, pp. 1–6 (2009)
6. Redi, J.A., Gastaldo, P., Heynderickx, I., Zunino, R.: Color distribution information for the reduced-reference assessment of perceived image quality. IEEE Trans. Circ. Syst. Video Technol. **20**(12), 1757–1769 (2010)
7. Zhang, L., Zhang, D., Mou, X.: FSIM: a feature similarity index for image quality assessment. IEEE Trans. Image Process. **20**(8), 2378–2386 (2011)
8. Xie, Z.X., Wang, Z.F.: Color image quality assessment based on image quality parameters perceived by human vision system. In: International Conference on Multimedia Technology, (ICMT), Ningbo, China, pp. 1–4 (2010)
9. He, L., Gao, X., Lu, W., Li, X., Tao, D.: Image quality assessment based on S-CIELAB model. Sig. Image Video Process. **5**(3), 283–290 (2011)
10. Lissner, I., Preiss, J., Urban, P., Lichtenauer, M.S.: Image-difference prediction: from grayscale to color. IEEE Trans. Image Process. **22**(2), 435–446 (2013)
11. Preiss, J., Fernandes, F., Urban, P.: Color-image quality assessment: from prediction to optimization. IEEE Trans. Image Process. **23**(3), 1366–1378 (2014)
12. Zhang, X., Silverstein, D.A, Farrell, J.E., Wandell, B.A.: Color image quality metric S-CIELAB and its application on halftone texture visibility. In: Proceedings of Compcon 1997, pp. 44–48. IEEE (1997)
13. Lissner, I., Urban, P.: Toward a unified color space for perception-based image processing. IEEE Trans. Image Process. **21**(3), 1153–1168 (2012)
14. Wang, T., Gao, X., Zhang, D.: An objective content-based image quality assessment metric. J. Image Graph. **12**(6), 1002–1007 (2007)

15. Tahani, H., Keller, J.M.: Information fusion in computer vision using the fuzzy integral. IEEE Trans. Syst. Man Cybern. **20**(3), 733–741 (1990)
16. Ponomarenko, N., Ieremeiev, O., Lukin, V., Egiazarian, K., Jin, L., Astola, J., Kuo, C.-C. J.: Color image database TID2013: peculiarities and preliminary results. In: Proceedings of 4th European Workshop on Visual Information Processing EUVIP 2013, Paris, France, pp. 106–111 (2013)

Efficient Video Quality Assessment via 3D-Gradient Similarity Deviation

Changcheng Jia, Lihuo He[(✉)], Wen Lu, Lei Hao, and Xinbo Gao

School of Electronic Engineering, Xidian University, Xi'an 710071, China
{ccjia,lhao}@stu.xidian.edu.cn,
{lhhe,luwen,xbgao}@mail.xidian.edu.cn

Abstract. The description of the spatio-temporal distortion is highly important for video quality assessment (VQA). However, traditional VQA metrics do not have enough ability to capture the spatio-temporal distortion in the video and suffer high computational complexity. Hence, this paper proposed a novel efficient VQA metric via 3D-gradient similarity deviation. Firstly, 3D-gradient is introduced to extract features of the spatio-temporal distortion in the video. In that, 3D-gradient kernels in three directions are constructed to convolute a group of frames in a video sequence to obtain 3D-gradient blocks. And then the 3D-gradient similarity indices between the reference and the distorted video are calculated to describe the local degradation of video quality. After that, the standard deviation of the local gradient similarity map is calculated to predict perceptual video quality of a group of frames. Finally, the worst-case pooling strategy is applied to pool all the quality indices of the groups of frames into a final quality score. Experimental results show that the proposed metric has a good consistency with the subjective perception and perform better than traditional metrics.

Keywords: Video quality assessment · 3D-gradient · Worst-case pooling

1 Introduction

Video acts as a significant information carrier and is more vivid than other media, such as text, image and audio. With the tendency of converging services delivered on wired and wireless networks, video has exploded over the past years. Meanwhile, the consumer expectancy is involving live video with high quality of experience (QoE). Furthermore, the quality of videos can optimize the video streaming of media service to obtain stunning visual experience. Hence, the video quality assessment (VQA) becomes an essential but challenging requirement.

Traditional VQA methods consider video as a volume of still images, thus a natural approach is to exert classical image quality assessment (IQA) metrics on a frame-by-frame basis and then pool the frame quality indices into a single score, such as structural similarity index (SSIM) [1], just noticeable differences metric (JND) [2], and visual signal-to-noise ratio (VSNR) [3]. However, this kind of approaches ignores the temporal relation between adjacent frames as well as the motion information contained in the video sequence. As a result, the performance of these VQA methods is limited. After

© Springer International Publishing Switzerland 2015
X. He et al. (Eds.): IScIDE 2015, Part I, LNCS 9242, pp. 493–502, 2015.
DOI: 10.1007/978-3-319-23989-7_50

all, video is different from volumes of still images. Advanced VQA methods take temporal information or motion information into consideration. Early VQA methods predict the temporal distortion by temporal filter response [4] and temporal decorrelation [5]. Due to lack of modeling of spatio-temporal distortion, they don't have a good consistency with human perception. Then the temporal trajectories [6] and the temporal evolution of spatial distortions [7] are considered to construct VQA model, which improved the performance. Recently, time-varying features [8], natural scene statistics [9], and human visual characteristics [10–12] are utilized to design VQA model to improve the consistency of objective prediction and human perception. Especially, motion-based video integrity evaluation index (MOVIE) [13] obtained a good performance by integrating both spatial and temporal (and spatio-temporal) aspects of distortion assessment. However, it suffers the high computational complexity due to the three-dimensional optical flow computation. Therefore, effective (high quality prediction accuracy) and efficient (low computational complexity) VQA is essential and has attracted increasing attentions. And this paper proposed a novel VQA metric based on 3D-gradient similarity deviation, named 3D-GSD.

In this paper, we consider video as a three-dimensional matrix, building a novel video quality assessment framework. 3D-gradient kernels are utilized to extract the spatio-temporal features of the video sequence to evaluate the degradation of video. Given a video sequence, the adjacent several frames, which are named a group of frames (GOF), are combined along the temporal dimension to obtain a three-dimensional video block. Secondly, 3D-gradient kernels in three directions are applied to a GOF in order to obtain the 3D-gradients in each pixel. For each GOF, local quality indices are employed by calculating the 3D-gradient similarity between the reference and the distorted video. Then, the standard deviation of the entire 3D-gradient similarities is calculated in a GOF as the quality index for the GOF. Finally, the worst-case pooling strategy is introduced to pool all the quality indices of all GOFs into a final quality score of the video sequence. Experimental result shows that the proposed metric can reflect the subjective perception accurately and are computationally efficient.

The rest of this paper is organized as follows. Section 2 gives an introduction of 3D-gradient. Section 3 details the proposed 3D-GSD metric. Section 4 presents the experimental results and analysis on the LIVE database. Finally, Sect. 5 concludes this paper and suggests directions for future research.

2   3D-Gradient

The gradient as a popular feature can effectively capture the local structures [14–18]. The most commonly encountered distortions, including noise corruption, blurring and compression artifacts, will lead to highly visible structural changes that "pop out" of the gradient domain. Therefore, the gradient is introduced as a significant structure feature to design image quality assessment models [15–18]. In this paper, the effective two-dimensional gradient [19] is extended to construct three-dimensional gradient (3D-gradient) kernels to extract local structure features of videos.

As shown in Fig. 1, x and y refer to the size of the frame image, and z represents the third dimension temporal axis. $f(x,y,z)$ refers to the luminance value of the pixel. The 3D-gradient $\nabla f(x, y, z)$ of the pixel is calculated by the 3D convolution.

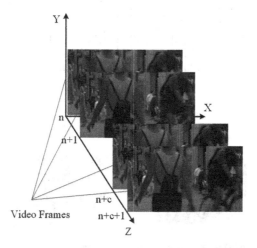

Fig. 1. Three-dimensional coordinate system for video sequences.

In this paper, for a pixel in the video, the three-dimensional gradient [21] is defined as

$$\nabla f(x, y, z) = \begin{bmatrix} F \otimes h_x \\ F \otimes h_y \\ F \otimes h_z \end{bmatrix}, \tag{1}$$

where

$$\begin{cases} F \otimes h_x = \sum_{i=-2}^{2} \sum_{j=-2}^{2} \sum_{k=-2}^{2} f(x + i, y + j, z + k) \cdot h_x(i, j, k) \\ F \otimes h_y = \sum_{i=-2}^{2} \sum_{j=-2}^{2} \sum_{k=-2}^{2} f(x + i, y + j, z + k) \cdot h_y(i, j, k) . \\ F \otimes h_z = \sum_{i=-2}^{2} \sum_{j=-2}^{2} \sum_{k=-2}^{2} f(x + i, y + j, z + k) \cdot h_z(i, j, k) \end{cases} \tag{2}$$

Figure 2 shows the 3D-gradient kernels designed for the proposed method. Through 3D-convolution with kernels at x, y and z direction, each video block is transformed to three gradient blocks. Each element in the gradient block refers to the gradient computed through convolution.

Figure 3 shows the x, y and z gradient blocks. They represent spatio-temporal features in different directions, respectively. It can be found that different gradient blocks in different directions represent different structural features. Hence, gradient similarity deviation in x, y and z directions can then be used to assess the quality degradation of the reference video.

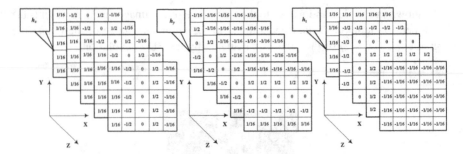

Fig. 2. 3D-gradient kernel.

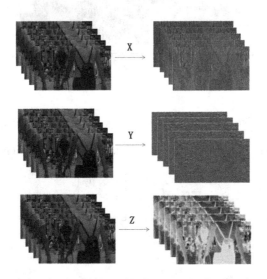

Fig. 3. The 3D-gradient blocks at x, y and z direction.

3 Efficient VQA via 3D Gradient Similarity Deviation

At present, many image quality assessment methods are based on 2D gradient trans-form. They take advantage of different kernels to obtain the gradient information at every pixel and pool them together to get the quality indices. Inspired by this way, many VQA methods regard video as volumes of images, and then use the image quality assessment method to evaluate the quality of video. However, such VQA metrics ignore the important temporal distortions in video, resulting in poor performance. Therefore, this paper regards video sequences as three-dimensional signal for later computational process. And the 3D-gradient features are extracted from the reference and distorted video to evaluate the local spatio-temporal degradation of video. Finally, standard deviation pooling and appropriate worst-case pooling strategy are utilized to obtain the quality score.

3.1 Spatio-temporal Features Extraction

Preprocessing is needed for efficiency. Every frames of video are filtered with 2×2 average kernel according to the method in Ref. [1].

Given a video sequence of $M \times N \times L$ pixels, the video is divided into equally spaced overlapped sub-videos of $M \times N \times s$ ($s \ll L$) pixels with overlapping step-size $t = 1$ along the temporal axis. Video consists of a group of frames. Note that the first frames of adjacent GOFs are adjacent, as the overlapping step-size $t = 1$. Secondly, the x, y and z three-dimensional gradient filters are applied to all the blocks in order to obtain the gradient blocks from the video in x, y and z direction. Through 3D-convolution with kernels at x, y and z direction, each video block is transformed to three gradient blocks. Each element in the three-dimensional gradient block represents the gradient computed through convolution. For each gradient block in x, y and z direction, the gradient similarity between the reference and the distorted video is calculated as local quality indices. Then, the x, y and z gradient indices are pooled together to one gradient similarity indices and its standard deviation is calculated as the quality index for the GOF. Lastly, according to the perceptual characteristic of the human visual system, worst-case pooling strategy is used to pool all the quality indices of all GOFs into a final quality score of the video sequence. The proposed framework is shown in Fig. 4.

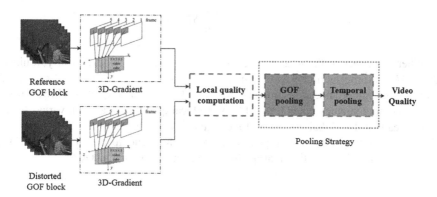

Fig. 4. The flowchart of the proposed FR-VQA metric.

3.2 Local Quality Computation

For each GOF, we employ the three-dimensional convolution with the x, y and z kernels (h_x h_y h_z) to obtain the gradient of each pixel (g_x g_y g_z) in the x, y and z directions. r refers to the GOF from the reference video and d refers to the distorted.

$$
\begin{aligned}
x \text{ direction} &: g_{rx} = r \otimes h_x, \ g_{dx} = d \otimes h_x \\
y \text{ direction} &: g_{ry} = r \otimes h_y, \ g_{dy} = d \otimes h_y \\
z \text{ direction} &: g_{rz} = r \otimes h_z, \ g_{dz} = d \otimes h_z
\end{aligned}
\tag{3}
$$

Here $\otimes$ means convolution. Then we calculate the gradient similarity $STGSD(i)$ between the reference and the distorted using the gradients in x, y and z direction.

$$STGSD(i) = (\frac{2g_{rx}g_{dx} + T_1}{g_{rx}^2 + g_{dx}^2 + T_1})^\alpha (\frac{2g_{ry}g_{dy} + T_2}{g_{ry}^2 + g_{dy}^2 + T_2})^\beta (\frac{2g_{rz}g_{dz} + T_3}{g_{rz}^2 + g_{dz}^2 + T_3})^\gamma \qquad (4)$$

Where, T_1, T_2 and T_3 are positive constants to keep the stability of the result. In order to make the algorithm not so cumbersome, we just set the parameter $T_1 = T_2 = 2000$ and $T_3 = 3000$ in our experiments. α, β and γ are weights of the x, y and z directional local quality indices. Here we set $\alpha = 1$, $\beta = 1$, and $\gamma = 1$. A more sophisticated model to obtain the best sets of parameters can be investigated as potential work. Till now, we have obtained the quality base $STGSD(i)$ for the i-th GOF.

3.3   Standard Deviation Pooling Strategy

The quality score of GOF can then be estimated from the $STGSD(i)$ via a novel pooling strategy. The most commonly used pooling strategy is average pooling, i.e., simply averaging the element values in the three-dimensional matrix named $STGSD(i)$ as the final quality score for the i-th GOF.

$$STGSDM = \frac{1}{N} \sum_{i=1}^{N} STGSD(i). \qquad (5)$$

Where, N refers to the total number of pixels in the GOF block. Clearly, a higher $STGSDM$ score means higher video quality. Average pooling assumes that each pixel has the same importance in estimating the overall video block quality. However, the average pooling strategy cannot reflect how the local quality degradation varies. Based on the idea that the global variation of local quality degradation of the GOF block can reflect its overall quality, we propose to compute the standard deviation [18] of the GOF block and take it as the final quality index for the particular GOF.

$$STGSD_{GOF} = \sqrt{\frac{1}{N} \sum_{i=1}^{N} (STGSD(i) - STGSDM)^2} \qquad (6)$$

Where, the value $STGSD_{GOF}$ reflects the degree of the degradation considering the reference and distorted video blocks. Therefore, the larger the value is, the worse quality of the distorted video has.

3.4   Temporal Pooling

It has been proved in [20] that the relatively bad frames in the video sequence seem much important to subjective perception. Thus, in the temporal pooling stage, we take the worst-case pooling strategy to pool just the worst p % GOFs quality score

$$Q = \sum_{i=1}^{K} STGSD_{GOF}(i)/K. \tag{7}$$

Where, K is the number of the worst p % of all GOFs and $K = p$ % × N. N refers to the number of the total GOFs, and GOFs are placing in descending order according to their quality score. We set $p = 42$ in our implementation.

4   Results and Analysis

4.1   Experimental Results

In order to evaluate the performance of the proposed 3D-GSD, experiments are conducted on the LIVE video quality assessment database from TEXAS. The database contains ten reference videos and 150 distorted videos whose spatial resolution is 768 × 432 spanning four categories of distortions, including compression artifacts due to MPEG-2 and H.264, errors introduced due to transmission over IP networks, and errors imported by transmission over wireless networks. Six videos contain 250 frames at 25 f/s, one video contains 217 frames at 25 f/s, and three videos contain 500 frames at 50 f/s. The subjective DMOS for each sequence is also included.

Table 1. Comparison of the performance by SROCC on LIVE database.

VQA	W	I	H	M	ALL
VSNR	0.7019	0.6894	0.6460	0.5915	0.6755
VQM	0.7214	0.6383	0.6520	0.7810	0.7026
PSNR	0.6574	0.4166	0.4585	0.3862	0.5397
SSIM	0.5233	0.4550	0.6514	0.5545	0.5257
MS-SSIM	0.7289	0.6534	0.7313	0.6684	0.7364
MOVIE	0.8019	0.7157	0.7664	0.7733	0.7890
3D-GSD	**0.7816**	**0.6672**	**0.7955**	**0.7755**	**0.8109**

The performance of the proposed 3D-GSD is compared to that of classical metrics, including PSNR, VSNR, SSIM, SW-SSIM, MS-SSIM, MOVIE and VQM. Table 1 shows the comparison of the VQA algorithms using SROCC on LIVE database. Table 2 shows the LCC comparison. Here W refers to wireless transmitted distortion; I refers to IP transmitted distortion; H refers to H.264 compressed distortion; and M refers to MPEG-2 compressed distortion.

From Tables 1 and 2, it can be concluded that the proposed 3D-GSD employs a good performance with 4 different distortions of the LIVE database, especially with the H.264 distortion type. Thus, it is well confirmed that 3D-gradient successfully describes the spatio-temporal structural information of video, and the 3D-gradient deviation describes the degradation of the quality of the distortion video well. Figure 5 shows the results of the proposed method testing on LIVE database. We can draw a conclusion that our estimator matches the subjective assessment in a high degree.

Table 2. Comparison of the performance by LCC on LIVE database.

VQA	W	I	H	M	ALL
VSNR	0.6992	0.7341	0.6216	0.5980	0.6896
VQM	0.7324	0.6480	0.6459	0.7860	0.7236
PSNR	0.6689	0.4645	0.5492	0.3891	0.5621
SSIM	0.5401	0.5119	0.6656	0.5491	0.5444
MS-SSIM	0.7157	0.7267	0.7020	0.6640	0.7379
SW-SSIM	0.5867	0.5587	0.7206	0.6270	0.596
MOVIE	0.8386	0.7622	0.7902	0.7595	0.8116
3D-GSD	**0.7955**	**0.7141**	**0.8237**	**0.7900**	**0.8152**

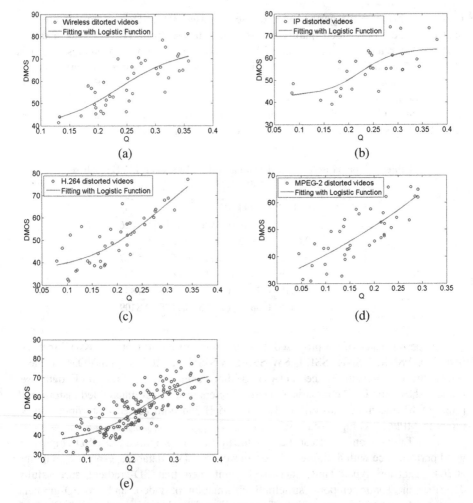

Fig. 5. Performance of the proposed method testing on LIVE database. (a) Wireless transmitted distortion; (b) IP transmitted distortion; (c) H.264 compressed distortion; (d) MPEG-2 compressed distortion; and (e) the complete LIVE database.

4.2   Parameter Sensitivity Analysis

The size and value of the 3D convolution kernel are important for the proposed metric. Hence, the parameter sensitivity analysis is presented in this section. The analysis is divided into two steps.

Step 1: The $3 \times 3 \times 3$ kernels surely take less time than the $5 \times 5 \times 5$ and $7 \times 7 \times 7$ kernels, but its accuracy is relatively limited. Compared with $3 \times 3 \times 3$ kernels, the $5 \times 5 \times 5$ kernels utilize more information in the image to calculate the 3D gradient of each pixel and obtain better performances. The $7 \times 7 \times 7$ kernels takes too much time. Hence, balancing the accuracy and running time, the size of $5 \times 5 \times 5$ is selected in the experimental parameter settings.

Step 2: The parameters settings of nonzero elements of $5 \times 5 \times 5$ kernels will have tremendous influence on the performance. In experiments, the kernels with the value of 1/16 obtain the best performance. Therefore, the value of the kernels' nonzero element is set to 1/16 in all experiments.

5   Conclusions

This paper proposed a novel efficient video quality assessment metric named 3D-GSD, which is based on 3D-gradient similarity deviation. By efficient 3D gradient kernels, the video sequence is convoluted to obtain the 3D-gradient feature. In that, the local 3D gradient index is exploited in three directions to design structural similarity index to measure the local degradation of video. According to the perceptual characteristic of HVS, standard deviation pooling and worst-case pooling strategy are utilized to obtain the final video quality. Experimental results show that the proposed metric has a good consistency with the subjective perception and performs much better than traditional metrics. And the proposed metric has a relatively low complexity, which is very important in real-time applications. For potential promotion of the whole framework, we believe that the performance will surely be improved by employing related machine learning pooling strategy to determine the weights of the x, y and z quality blocks. And efficient and effective 3D-gradient features can surely be utilized in reduced-reference and no-reference VQA framework.

Acknowledgments. This research was supported partially by the National Natural Science Foundation of China (No. 61125204, 61372130, 61432014, 61501349), the Fundamental Research Funds for the Central Universities (No. BDY081426, JB140214), the Program for New Scientific and Technological Star of Shaanxi Province (No. 2014KJXX-47), the Project Funded by China Postdoctoral Science Foundation (No. 2014M562378).

References

1. Wang, Z., Simoncelli, E.P., Bovik, A.C.: Multiscale structural similarity for image quality assessment. In: Proceedings of 37th Asilomar Conference on Signals, Systems and Computers, pp. 1398–1402 (2003)

2. Sarnoff Corporation. JND metrix Technology (2003). http://www.sarnoff.com/productsservices/videovision/jndmetrix/downloads.asp

3. Chandler, D.M., Hemami, S.S.: VSNR: a wavelet-based visual signal-to-noise ratio for natural images. IEEE Trans. Image Process. **16**(9), 2284–2298 (2007)

4. Watson, A.B., Hu, J., McGowan, J.F.: DVQ: digital video quality metric based on human vision. J. Electron. Imag. **10**(1), 20–29 (2001)

5. Hekstra, A.P., Beerends, J.G., Ledermann, D., de Caluwe, F.E., Kohler, S., Koenen, R.H., Rihs, S., Ehrsam, M., Schlauss, D.: PVQM: a perceptual video quality measure. Signal Process Image Commun. **17**(10), 781–798 (2002)

6. Barkowsky, M., Bialkowski, J., Eskoer, B., Bitto, R., Kaup, A.: Temporal trajectory aware video quality measure. IEEE J. Sel. Topics Signal Process **3**(2), 266–279 (2009)

7. Ninassi, A., Le Meur, O., Le Callet, P., Barba, D.: Considering temporal variations of spatial visual distortions in video quality assessment. IEEE J. Sel. Topics Signal Process **3**(2), 253–265 (2009)

8. Chen, C., Choi, L.K., de Veciana, G., Caramanis, C., Heath Jr., R.W., Bovik, A.C.: Modeling the time-varying subjective quality of HTTP video streams with rate adaptations. IEEE Trans. Image Process. **23**(5), 2206–2221 (2014)

9. Saad, M.A., Bovik, A.C., Charrier, C.: Blind prediction of natural video quality. IEEE Trans. Image Process. **23**(3), 1352–1365 (2014)

10. You, J., Ebrahimi, T., Perkis, A.: Attention driven foveated video quality assessment. IEEE Trans. Image Process. **23**(1), 200–213 (2014)

11. Park, J., Seshadrinathan, K., Lee, S., Bovik, A.C.: Video quality pooling adaptive to perceptual distortion severity. IEEE Trans. Image Process. **22**(2), 610–620 (2013)

12. Zhang, F., Lin, W., Chen, Z., Ngan, K.: Additive log-logistic model for networked video quality assessment. IEEE Trans. Image Process. **22**(4), 1536–1547 (2013)

13. Seshadrinathan, K., Bovik, A.C.: Motion-tuned spatio-temporal quality assessment of natural videos. IEEE Trans. Image Process. **19**(2), 335–350 (2010)

14. Kim, D.O., Han, H.S., Park, R.H.: Gradient information-based image quality metric. IEEE Trans. Consum. Electron. **56**(2), 930–936 (2010)

15. Cheng, G.Q., Huang, J.C., Zhu, C., Liu, Z., Cheng, L.Z.: Perceptual image quality assessment using a geometric structural distortion model. In: Proceedings of 17th IEEE ICIP, pp. 325–328 (2010)

16. Chen, G.H., Yang, C.L., Xie, S.L.: Gradient-based structural similarity for image quality assessment. In: Proceedings of 13th IEEE International Conference on Image Process, pp. 2929–2932 (2006)

17. Zhang, L., Zhang, L., Mou, X., Zhang, D.: FSIM: a feature similarity index for image quality assessment. IEEE Trans. Image Process. **20**(8), 2378–2386 (2011)

18. Liu, A., Lin, W., Narwaria, M.: Image quality assessment based on gradient similarity. IEEE Trans. Image Process. **21**(4), 1500–1512 (2012)

19. Xue, W., Zhang, L., Mou, X., Bovik, A.C.: Gradient magnitude similarity deviation: a highly efficient perceptual image quality index. IEEE Trans. Image Process. **23**(2), 684–695 (2014)

20. Narwaria, M., Lin, W., Liu, A.: Low-complexity video quality assessment using temporal quality variations. IEEE Trans. Multimedia **14**(3), 525–535 (2012)

21. Li, Y., Zhu, E., Zhao, J., Zhao, X.: A fast simple optical flow computation approach based on the 3-D gradient. IEEE Trans. Circ. Syst. Video Technol. **24**(5), 842–853 (2014)

Image Quality Assessment Based on Mutual Information in Pixel Domain

Hongqiang Xu, Wen Lu$^{(\boxtimes)}$, Yuling Ren, and Xinbo Gao

School of Electronic Engineering, Xidian University, Xi'an 710071, China
{xuhongqiang, renyuling}@stu.xidian.edu.cn,
{luwen, xbgao}@mail.xidian.edu.cn

Abstract. The natural scene statistics (NSS) model is widely used in image quality assessment algorithms, the NSS based features in frequency domain provide a good approximation to image structure, but not to the image content. To get a metric which is effectively to both structural distortion and content distortion, a new image quality assessment framework in image pixel domain based on mutual information is proposed. First, a non-overlapping segmentation set is acquired to establish the relation with image pixels. Second, the saliency and specific information are measured to catch the image content changes, and entanglement to the image structure change. Finally, the differences of image content and structural information are used to measure image quality. The experimental results show that the proposed framework has good consistency with subjective perception values.

Keywords: Image quality assessment · Information theory · Mutual information

1 Introduction

In modern applications of digital image processing, visual quality is of a prime importance [1, 2]. The essence of various distortion generated in process of acquisition, compression, processing, transmission and restoration is the loss of the visual perceptive information. the more the image quality descended, the lesser the visual perceptive information kept, that means that we can measure image quality by calculating the descended degree of visual perceptive information, and the descended degree can been estimated by the mutual information in information theory. That means the information theory, as a subject of studying the measurement, transfer and transformation of information [3], has a great advantage when it is used in image quality assessment (IQA).

Recent information theoretic approaches to visual quality assessment come under the class of natural scene statistics (NSS) model based methods and attempt to quantify the distances between reference and distorted signals using information theoretic quantities. Many algorithms that are based on this approach attempt to quantify the amount of loss of visual information that occurs in the distorted visual signal with respect to a distortionless reference [3]. These algorithms can be categorized into three categories. The first category is the entropy difference based methods [4–6], in [4] the

© Springer International Publishing Switzerland 2015
X. He et al. (Eds.): IScIDE 2015, Part I, LNCS 9242, pp. 503–512, 2015.
DOI: 10.1007/978-3-319-23989-7_51

authors use the Gaussian Scale Mixtures fitting the wavelet decomposition coefficients and approaches the image distortion as an additive stationary zero mean white Gauss noise, getting the image quality by quantifying the information that is present in the reference image and how much of this reference information can be extracted from the distorted image. The second category is the Kullback-Liebler (KL) divergence based methods [7, 8]. In [7], the generalized Gaussian distribution is used to model wavelet coefficients in a subband of the reference image. The quality index involves computing the KL divergence between the empirical distribution of the distorted image and the reference distribution parametrized using the generalized Gaussian distribution. The last category is the mutual information based methods. In [9], the empirical mutual information is calculated between the reference and distorted after each image has been band pass filtered and then threshold using the contrast sensitivity function. The quality index is expressed as weighted combinations of the ratio of mutual information between the reference and the distorted to the mutual information of the reference with itself in each subband.

The existing methods have the feature that the natural scene statistics model in frequency domain are widely used, for example, the wavelet decomposition is used in [4–9]. The reason why frequency domain is widely used is that frequency domain is related to the model of human visual processing of visual stimuli, it simulates the early human visual pathways [10]. The NSS based features can provide a good approximation to perceive image structure [11], but not the image content.

To overcome these disadvantages, the mutual information in image pixel domain between the image luminance histogram and the image segmentation regions is established to catch the image content information and structure information. Three components [12] are introduced to measure how much information is conveyed in each pixel value because the mutual information specifies the average over all the image pixels. The first component is defined as saliency information to express how significant the pixel is. The second component is defined as specific information to express how much image content information is conveyed in each pixel when the mutual information is deemed a mean of pixels' information. The last one is the entanglement information, which can evaluate how informative image structural information is included in a region, is extracted if the mutual information was interpreted as a product of a region specific information and its conditional probability. Based on the proposed information, a new IQA algorithm based on mutual information in image pixel domain is proposed. First, a non-overlapping segmentation set is acquired to build the relation with image pixels. Second, the saliency and specific information are measured to catch the image content changes, and entanglement to the image structural changes. Then, the differences of image content and structural information are used to measure image quality. After finding out the image quality factors, the Support Vector Regression (SVR) is used to learn the mapping from quality factor space to image quality.

The remainder of the paper is organized as follows. In Sect. 2 presents the details of the proposed IQA algorithm framework. Experimental results are presented in Sect. 3, and Sect. 4 concludes.

2 The Proposed Image Quality Assessment Algorithm

In this section, we propose a mutual information in image pixel domain based image quality assessment algorithm after building the transformation relation between the image pixels and the image information. First, a non-overlapping segmentation set is acquired to build the relation with image pixels. Second, the saliency and specific information are measured to catch the image context changes, and entanglement to the image structural change. Then, the differences of image context and structural information are used to measure image quality factors. After finding out the image quality factors, the Support Vector Regression (SVR) is used to learn the mapping from quality factor space to image quality. The algorithm framework shown in Fig. 1.

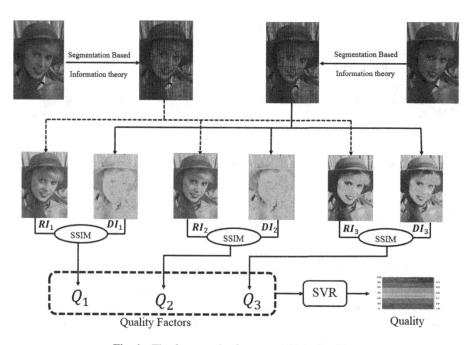

Fig. 1. The framework of proposed IQA algorithm

2.1 Image Partition

Two of the most basic elements of a photograph are composition and luminosity. In order to analyze their correlation, an information channel between the luminance histogram and the regions of the image is established by image partition, here, the image partition by the method introduced by Rigau et al. [13]. The portioning algorithm progressively splits the image by extracting the maximum information at each step. In the partition procedure, it takes image luminance histogram as the input and the set of regions R of the image as outputs and this process is defined as the information channel $B \rightarrow R$ and corresponding the information channel $R \rightarrow B$ is defined by taking

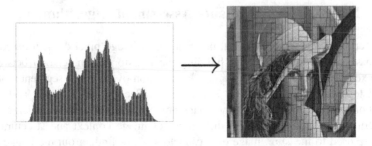

Fig. 2. The information channel $B{\rightarrow}R$, the left is input and the right is output

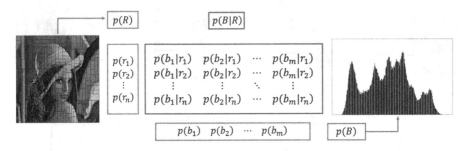

Fig. 3. The transformation relation between two information channels.

the set of regions R of the image as the input. Here use the Fig. 2 being a description and use the Fig. 3 to explain the relation between two information channels.

2.2   Measure the Image Information

Given an image I of N pixels, where $R = \{r_1, r_2, \ldots, r_t\}$ represent the set of region R with the number of the regions is t, and $B = \{b_1, \ldots, b_i\}$, $i = 0, 1, \ldots, 255$ represent the set of image pixel value, which b_i means a pixel value is i. N_b is the frequency of bin b ($N = \sum_{b \in B} N_b$) and N_r is the number of pixels of region r ($N = \sum_{r \in R} N_r$), according to the mutual information (MI) formula, the MI between B and R is given by

$$I(B; R) = \sum_{b \in B} p(b) \sum_{r \in R} p(r|b) \log \frac{p(r|b)}{p(r)}$$
$$= H(R) - H(R|B)$$
$$= H(B) - h(B|R) \tag{1}$$

Where $p(b) = N_b/N$, $p(r|b) = N_{b,r}/N_b$ and $p(r) = N_r/N$ in (1), and the MI represents the shared information or correlation between B and R.

After dividing the image, the set of regions R, the information channel $B{\rightarrow}R$ and $R{\rightarrow}B$ are got. Then the image information in image pixels is measured by measuring the saliency information, specific information and entanglement information which have been introduced in [12]. Specific process as following:

The Saliency Information I_1: From (1), the MI between luminance and regions can be expressed as

$$I(B; R) = \sum_{b \in B} p(b) \sum_{r \in R} p(r|b) \log \frac{p(r|b)}{p(r)} = \sum_{b \in B} p(b) I_1(b; R) \tag{2}$$

Where

$$I_1(b; R) = \sum_{r \in R} p(r|b) \log \frac{p(r|b)}{p(r)} \tag{3}$$

as the surprise associated with the luminance b and can be interpreted as a measure of its saliency. High values of express a $I_1(b;R)$ high surprise and identify the most salient luminance.

The Specific Information I_2: From (1), in the information channel $B \rightarrow R$, mutual information can be expressed as

$$I(B; R) = H(R) - H(R|B) = \sum_{b \in B} p(b)[H(R) - H(R|b)] = \sum_{b \in B} p(b) I_2(b; R) \tag{4}$$

Where

$$I_2(b; R) = H(R) - H(R|b) = - \sum_{r \in R} p(r) \log p(r) + \sum_{r \in R} p(r|b) \log p(r|b) \tag{5}$$

is the specific information of b and expresses the change in uncertainty about R when b is observed. A large value of $I_2(b;R)$ means that we can easily predict a region given luminance b.

Following a similar process for the reversed channel $R \rightarrow B$, in this paper, the specific information associated with region r is given by

$$I_2(r; B) = H(B) - H(B|r) = - \sum_{b \in B} p(b) \log p(b) + \sum_{b \in B} p(b|r) \log p(b|r) \tag{6}$$

and expresses the predictability of a luminance known the region.

The Entanglement Information I_3: From (1), in the information channel $B \rightarrow R$, the entanglement information is defined as

$$I_3(b; R) = \sum_{r \in R} p(r|b) I_2(r; B) \tag{7}$$

A large value of $I_3(b;R)$ means that the specific information $I_2(r;B)$ of the regions that contain the luminance b are very informative.

2.3   Measure the Quality Factors

After transforming image into the three information above, we can get three information matrixes which have the same size as the image, respectively are the saliency information matrix, the specific information matrix and the entanglement information matrix RI_i from the reference image, and DI_i from the distorted image where $i = 1,2,3$. As it has been said, the essence of image quality descended is the loss of visual perceptive information, the more the image quality descended, the lesser the visual perceptive information kept, that means that we can measure image quality by calculating image information matrix discrepancy between RI_i and DI_i. Here, we take RI_i and DI_i. as the input of the SSIM model and take the output of SSIM as quality factor q_i. where $i = 1,2,3$. We can represent it as

$$q_i = SSIM(RI_i, DI_i) \tag{8}$$

After finding out the image quality factors q_1, q_2 and q_3, the Support Vector Regression is used to learn the mapping from quality factor space to image quality. Since each image is represented by a single quality factor vector, the image quality assessment problem can be solved as a regression problem. A lot of regression techniques such as SVR and random forest can be used to learn the mapping. Here we use SVR with RBF kernel to do the regression.

3   Experimental Results

In this section, we compare the performance of the proposed framework with standard RR-IQA methods, i.e., RRVIF [14], FEDM [15], WNISM [16], RRED [6], RRED* [6], LBPs [17], RR-PCA [18] and RRIQV [19], based on the following experiments: the consistency experiment, the rationality experiment, and the sensitivity experiment. At the beginning of this section, we first brief the image database for evaluation.

The laboratory for image and video engineering (LIVE) database [20] has been recognized as the standard database for IQA measures performance evaluation. This database contains 29 high-resolution 24 bits/pixel RGB color images and 175 corresponding JPEG and 169 JPEG2000 compressed images, as well as 145 white noisy (WN), 145 Gaussian blurred (GB), and 145 fast-fading (FF) Rayleigh channel noisy images at a range of quality levels.

3.1   Consistency Experiment

In this subsection, we compare the performance of the proposed IQA framework with RRVIF, FEDM, WNISM, RRED, RRED*, LBPs, RR-PCA and RRIQV. The PLCC

Table 1. PLCC values the newest RR methods and the proposed method on whole LIVE Database and five data sets of different distortion categories the best one is noted with bold.

		JP2K	JPEG	WN	GB	FF	ALL
PLCC	RRVIF	0.932	0.895	0.957	0.955	0.944	0.725
	FEDM	0.921	0.875	0.925	0.902	0.875	–
	WNISM	0.924	0.876	0.890	0.888	0.925	0.710
	RRED	0.930	0.831	0.926	0.953	0.922	0.764
	RRED*	**0.963**	**0.979**	0.985	**0.969**	0.941	**0.939**
	LBPs	0.927	0.894	**0.990**	0.966	0.950	0.935
	RR-PCA	0.934	0.904	0.980	0.777	0.923	—
	RRIQV	0.944	0.909	0.886	0.882	0.919	—
	Proposed	0.954	0.927	0.961	0.967	**0.961**	0.924

Table 2. RMSE values the newest RR methods and the proposed method on whole LIVE Database and five data sets of different distortion categories the best one is noted with bold.

		JP2K	JPEG	WN	GB	FF	ALL
RMSE	RRVIF	**5.88**	7.14	4.65	4.66	5.42	17.1
	FEDM	6.31	7.73	6.06	6.78	7.96	—
	WNISM	6.17	7.71	7.28	7.22	6.24	18.4
	RRED	9.28	17.7	10.5	5.62	11.3	17.6
	RRED*	6.81	**6.44**	4.90	4.64	9.16	9.43
	LBPs	5.97	6.94	**2.25**	**3.98**	4.86	9.59
	RR-PCA	9.38	17.6	5.33	13.8	14.7	—
	RRIQV	9.26	15.8	14.0	8.22	11.0	—
	Proposed	6.02	6.65	5.49	4.11	**4.65**	**7.58**

and RMSE results for all IQA methods being compared are given as benchmark in Tables 1 and 2.

3.2 Rationality Experiment

To verify the rationality of the proposed framework, we choose the Einstein image with different distortions, which are blurring (with smoothing window of $W \times W$), additive Gaussian noise (mean = 0, variance = V), impulsive Salt-Pepper noise (density = D), and JPEG compression (compression rata = sR).

Figure 4 (all of the images are 8 bits/pixel; cropped from 512×512 to 128×128 for visibility) shows the Einstein image with different types of distortions and the metrics prediction trend to the corresponding image, respectively. It is found that the proposed framework prediction trend to image drop with the increasing intensity of different types of image distortions. It is consistent well with the tendency of the

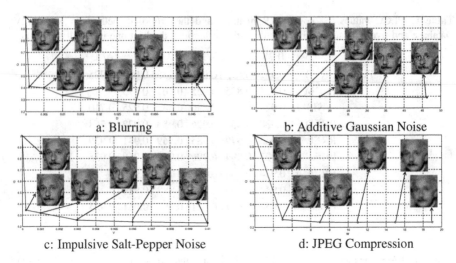

a: Blurring

b: Additive Gaussian Noise

c: Impulsive Salt-Pepper Noise

d: JPEG Compression

Fig. 4. Trend plots of Einstein with different types of distortion using the proposed framework.

decreasing image quality in fact. So the results demonstrate the rationality of the proposed framework.

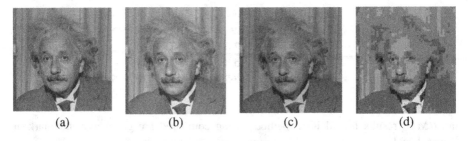

(a)   (b)   (c)   (d)

Fig. 5. Einstein image with the same PSNR but different perceived quality. (a) The reference image; (b) mean shift; (c) contrast stretching; (d) JPEG compression.

Table 3. Value of different metrics for images in Fig. 5

Metric	b	c	d
PSNR	24.8022	24.8013	24.8041
VIF	0.9895	0.9458	0.5709
MRMI	1	0.2154	0.1474
MIFG	1	0.6013	0.5304
Proposed	0.9782	0.8705	0.4262

3.3   Sensitivity Experiment

Currently, PSNR is one of the most popular objective quality metrics for images. However, it does not correlate well with the perceived quality measurement as shown in Table 3. Figure 5 shows three degraded "Einstein" images with different distortions but with almost identical PSNR. Another popular objective quality metric, VIF is also reported in Table 3, which shows VIF cannot perform well for the contrast stretching distortion, although the perceived quality of the contrast stretching distorted image is worse than that of mean shift transformed image. As shown in Table 3, the proposed framework can distinguish both the contrast stretching and JPEG compression distortions well. It is worth emphasizing that is sensitive to the changes of intensity, i.e., changes of grey values at image level can affect the image quality significantly. This is because the specific information is sensitive to the changes of intensity, so the proposed framework can catch this kind of distortion.

4   Conclusions

A novel general image quality assessment framework is proposed based on mutual information in image pixel domain. In order to analyze the correlation between the luminance histogram and the composition of the image, an information channel is established by a mutual information based image partition. In this channel, the saliency and specific information are measured to catch the image content changes, and entanglement to the image structural changes and the mean and variance to the changes between the pixels. Then the differences between these information are used to measure image quality. Experimental results illustrate that the proposed framework have good consistency with subjective perception values and the objective assessment results can well reflect the visual quality of images.

Acknowledgements. This research was supported partially by the National Natural Science Foundation of China (No. 61125204, No. 61372130, No. 61432014), the Fundamental Research Funds for the Central Universities (No. BDY081426, No. JB140214), the Program for New Scientific and Technological Star of Shaanxi Province (No. 2014KJXX-47), the Project Funded by China Postdoctoral Science Foundation (No. 2014M562378).

References

1. Wang, Z., Bovik, A.C.: Modern image quality assessment. Synth. Lect. Image Video Multimedia Process. **2**(1), 1–156 (2006)
2. Park, H., Har, D.H.: Subjective image quality assessment based on objective image quality measurement factors. IEEE Trans. Consum. Electron. **57**(3), 1176–1184 (2011)
3. http://en.wikipedia.org/wiki/Information_theory
4. Soundararajan, R., Bovik, A.C.: Survey of information theory in visual quality assessment. SIViP **7**(3), 391–401 (2013)
5. Sheikh, H.R., Bovik, A.C.: Image information and visual quality. IEEE Trans. Image Process. **15**(2), 430–444 (2006)

6. Soundararajan, R., Bovik, A.C.: RRED indices: reduced reference entropic differencing for image quality assessment. IEEE Trans. Image Process. **21**(2), 517–526 (2012)

7. Gabarda, S., Cristobal, G.: Blind image quality assessment through anisotropy. Opt. Soc. Am. **24**(12), 42–51 (2007)

8. Li, Q., Wang, Z.: Reduced-reference image quality assessment using divisive normalization-based image representation. IEEE J. Sel. Top. Signal Process. Spec. Issue Visual Media Qual. Assess. **3**, 202–211 (2009)

9. Zhu, H., Wu, H.: New paradigm for compressed image quality metric: exploring band similarity with CSF and mutual information. In: Geoscience and Remote Sensing Society, the International Geoscience and Remote Sensing Symposium 2005, pp. 2–4. IEEE (2005)

10. Sheikh, H.R., Bovik, A.C.: A visual information fidelity approach to video quality assessment. In: The First International Workshop on Video Processing and Quality Metrics for Consumer Electronics, pp. 23–25. IEEE (2005)

11. Wang, Z., Bovik, A.C., Sheikh, H.R., et al.: Image quality assessment: from error visibility to structural similarity. IEEE Trans. Image Process. **13**(4), 600–612 (2004)

12. Rigau, J., Feixas, M., Sbert, M.: Image information in digital photography. In: Koch, R., Huang, F. (eds.) ACCV 2010 Workshops, Part II. LNCS, vol. 6469, pp. 122–131. Springer, Heidelberg (2011)

13. Rigau, J., Feixas, M., Sbert, M.: An information theoretic framework for image segmentation. In: International Conference on Image Processing, pp. 1193–1196. IEEE (2004)

14. Wu, J., Lin, W., Shi, G., Liu, A.: Reduced-reference image quality assessment with visual information fidelity. IEEE Trans. Multimedia **15**(7), 1700–1705 (2013)

15. Zhai, G., Wu, X., Yang, X., et al.: A psychovisual quality metric in free-energy principle. IEEE Trans. Image Process. **21**(1), 41–52 (2012)

16. Wang, Z., Simoncelli, E.P.: Reduced-reference image quality assessment using a wavelet-domain natural image statistic model. In: Electronic Imaging 2005. International Society for Optics and Photonics, pp. 149–159. IEEE (2005)

17. Wu, J., Lin, W., Shi, G., et al.: Reduced-reference image quality assessment with local binary structural pattern. In: 2014 IEEE International Symposium on Circuits and Systems, pp. 898–901. IEEE (2014)

18. Uzair, M., Fayek, D.: Reduced reference image quality assessment using principal component analysis. In: IEEE International Symposium on Broadband Multimedia Systems and Broadcasting 2011, pp. 1–6. IEEE (2011)

19. Wang, Z., Simoncelli, E.P.: Reduced-reference image quality assessment using a wavelet domain natural image statistic model. In: Electronic Imaging 2005. International Society for Optics and Photonics, pp. 149–159 (2005)

20. Sheikh, H.R., Wang, Z., Cormack, L., et al.: LIVE image quality assessment database. http://live.ece.utexas.edu/research/quality/

Semisupervised Classification of SAR Images by Maximum Margin Neural Networks Method

Li Sun[1(✉)], Xiuxiu Li[2], Qun Zhang[1], and Miao Ma[3]

[1] Air Force Engineering University, Xi'an 710077, People's Republic of China
sl_lxa@mail.nwpu.edu.cn, zhangqunnus@gmail.com
[2] Xi'an University of Technology, Xi'an, People's Republic of China
lixiuxiu@xaut.edu.cn
[3] Shaanxi Normal University, Shaanxi, People's Republic of China
mmthp@snnu.edu.cn

Abstract. The proposed method is based on neural networks by modeling the data marginal distribution with the graph Laplacian built with both labeled and unlabeled samples, at the same time, optimizing neural networks layers in a single process, back-propagating the gradient of a Maximum Margin based objective function. Therefore, the proposed approach gives rise to an operational classifier, as opposed to previously presented semi-supervised scenarios. Results demonstrate the improved classification accuracy and scalability of this approach on SAR image classification problems.

Keywords: SAR image classification · Semi-supervised neural networks · Maximum margin

1 Introduction

SYNTHETIC aperture radar (SAR) is becoming more and more important in remote sensing applications. SAR classification is a challenging task not because the affection of speckle noise, but few number of labeled pixels are typically available, and thus classifiers tend to overfit the data [1, 2]. Recently, many classification methods, such as Hidden Markov model [3], Neural Networks [4] and SVM [5], have demonstrated very good performance in SAR image classification. However, when little labeled information is available, the underlying probability distribution function of the image is not properly captured and the problem of overfitting still persists. Semi-supervised learning naturally appears as a promising tool for combining labeled and unlabeled samples thus increasing the accuracy and robustness of class predictions [6–8].

The basic SSL methods self-training, generative models, transductive models and graph-based methods have been proposed in the literature. In recent years, more attention has been paid on graph-based approaches, in which each sample spreads its label information to its neighbors until a stable state is achieved on the whole data set. It can be described by using regularizing framework, which is adding regularization term on the loss function with labeled samples, utilizing the manifold hypothesis directly or indirectly. Multi-layer perceptron (MLP) of neural networks has been

X. He et al. (Eds.): IScIDE 2015, Part I, LNCS 9242, pp. 513–521, 2015.
DOI: 10.1007/978-3-319-23989-7_52

increasingly used for the classification in remote sensing images for many years, which achieved improved accuracy compared to (or at least similar as) other state-of-the-art approaches, such as SVM [9], Bayesian Neural Networks [10]. This success derives from MLP characteristics: An MLP can be seen as a hyperplane classifier in the RKHS implicitly defined by the MLP's input-to-hidden layer mapping [11]. In this paper, we advocate the use of semi-supervised neural networks based on the Maximum Margin principle to deal with large-scale SAR classification problems. The proposed method jointly optimizes both neural networks layers in a single process, back-propagating the gradient of a Maximum Margin based objective function. Through the output and hidden layers, in order to create a hidden-layer space that enables a higher margin for the output-layer hyperplane. The proposed Maximum Margin based objective function aims to stretch out the margin to its limit. In training methods the maximum value of Area Under ROC Curve (AUC) is applied as stop criterion.

The rest of the paper is outlined as follows. Section 2 fixes notation and briefly revises the main concepts and properties of the manifold-based regularization framework for SSL and presents the proposed method for semi-supervised image classification. Section 3 describes the data collection, experimental setup and the obtained results. Finally, Sect. 4 concludes with some remarks and further research directions.

2   Maximum Margin Training Algorithms for Semi-Supervised Neural Networks Using in the SAR Classification

One of the critical issues in the SAR classification is improving the accuracy of the classification process. Semi-supervised learning algorithm is based on assumption that the classification boundaries are naturally defined on the sub-manifold rather than the total ambient space, as well as exploiting both labeled and unlabeled samples. Regularization not only helps to form the semi-supervised framework, but also smooth decision functions and control the capacity of the classifier, especially when few labeled samples are available compared to the high dimensionality. In the case of Supervised Neural Networks, one of the problems that occur during network training is overfitting. The error on the training dataset is driven to a small value, however the error is large when new data are presented to the neural networks. It occurs because the neural networks memorizes the training examples [12]. To deal with this kind of problem, many studies propose different regularization methods for improving generalization are applied [13].

In this section, we first review the basics of this framework and fix notation. Then, a new algorithm: semi-supervised maximum margin neural networks is given. Experimental results are presented in the following section.

2.1   Manifold-Based Semi-Supervised Regularization Framework

Assuming the marginal distribution is known, we are given a set of l labels samples, $\{x_i, y_i\}_{j=1}^{l}$ with corresponding class labels y_i, and a set of u unlabeled samples

$\{x_i\}_{i=l+1}^{l+u}$, where $x_i \in \mathbb{R}^N$, $\mathrm{x}_i = [x_{i1}, x_{i2}, \ldots, x_{im}]^{\mathrm{T}}$ and $y_i \in \{-1, 1\}$. In our case, x_i represents one of the gray of the image. The regularized function to be minimized is defined as:

$$L = \frac{1}{l} \sum_{i=1}^{l} V(x_i, y_i, f) + \gamma_L \|f\|_{\mathcal{H}}^2 + \gamma_M \|f\|_{\mathcal{M}}^2 \tag{1}$$

Where V represents a generic cost function of the committed errors on the labeled samples, γ_L controls the complexity of f in the ambient space, and γ_M controls its complexity in intrinsic geometry of the marginal data distribution. $\|f\|_{\mathcal{M}}^2$ is a smoothness penalty corresponding to the probability distribution. We will develop many different algorithms from this framework by playing around with the loss function V and the regularization terms $\|f\|_{\mathcal{H}}^2$ and $\|f\|_{\mathcal{M}}^2$.

2.2 Maximum Margin Training Algorithms for Semi-Supervised Neural Networks

The idea of regularization has a rich mathematical history going back to Tikhonov [14], where it is used for solving ill-posed inverse problems. In the case of neural networks, the objective function is modified by adding a term that consists of the average of the sum of squares of the network weights and biases. Using this objective function causes the network response to be smoother and less overfitting. The new algorithm proposed in this paper also adopt a maximum margin objective function in order to improve the classification margin.

Maximum margin based classifiers are based on the idea that regularized large margin hypotheses had got good results [15]. MMS produce a classification that gives the lowest risk of making errors on future data. However, the manner in which the MMS objective is formulated in SVMs requires a solution to a complex quadratic optimization problem with constraints. It needs a batch training (all data samples available) that becomes very expensive computationally for large data sets, and it does not parallelize well. Therefore, a New MM-based MLP training algorithms is proposed in this section. On one hand, the cost function is defined as the maximal margin objective function based on MSE, which is the distance between each pattern to the output-layer hyperplane. On the other hand, we will add a regularization term to the loss function used for modeling semi-supervised neural networks. This methodology, named Semi-supervised Maximum-Margin Neural Networks (SMMNN), is different from other approaches.

A three layer MLP with sigmoid activation function model is:

$$f(w, \mathrm{x}) = W^{(2)} h\left(W^{(1)} \mathrm{x} + \mathrm{b}_1\right) + \mathrm{b}_2 \tag{2}$$

Where $f(w, \mathrm{x})$ is the output vector of the hidden layer, $W^{(k)}(k = 1, 2)$ is the synaptic weights matrix of the layer k, Assume the output layer has bias $\mathrm{b}_2 = 0$ and the hidden layer output vector is $y^{(1)} = h(W^{(1)}x + b_1)$, $h(\cdot)$ is the sigmoid function. Then the separating hyperplane is given by:

$$W^{(2)}y^{(1)} = 0 \tag{3}$$

The distance between a point in the hyperspace and the separating hyperplane d:

$$d = \frac{f(w, x)}{W^{(2)}} \tag{4}$$

As the sigmoid activation function bounds the hidden neuron output in the interval [0, 1], the distance d is bounded in the interval $[-\sqrt{n}, \sqrt{n}]$, where n is the number of hidden neurons.

The target output y_i assumes the values -1 or 1, for the case of multiple classes, the solution suggested by Belkin and Niyogi was to build a one-against-all classifier for each class, the margin between the target output and the MLP model output is:

$$e_i = \left(y_i\sqrt{n} - \frac{f(w, x)}{W^{(2)}} \right) \tag{5}$$

The cost function is defined as the maximal margin objective function based on MSE:

$$V(x_i, y_i, f(w, x)) = e_i^2 \tag{6}$$

$$f(w, x)_{\mathcal{M}} = \frac{1}{(l+u)^2} \sum_{i,j=1}^{l+u} S_{ij}\big(f(w, x_i) - f(w, x_j)\big)^2 = \mathbf{f}^T L \mathbf{f} \tag{7}$$

Where $L = D - S$ is the Laplacian, D is the diagonal degree matrix of S, given by $D_{ii} = \sum_{j=1}^{l+u} S_{ij}$, and $\mathbf{f} = [f(w, x_1), \cdots, f(w, x_{l+u})]^T$. Replace $V(x_i, y_i, f(w, x)) = e_i^2$ and $\|f(w, x)\|_{\mathcal{M}} = \mathbf{f}^T L \mathbf{f}$ into $\mathcal{L} = \frac{1}{l}\sum_{i=1}^{l} V(x_i, y_i, f(w, x)) + \gamma\|f(w, x)\|_{\mathcal{M}}$. Then, we can get the regularized objective function as:

$$\mathcal{L} = \frac{1}{l}\sum_{i=1}^{l}\left(y_i\sqrt{n} - \frac{f(w, x)}{W^{(2)}} \right)^2 + \frac{\gamma}{(l+u)^2}\mathbf{f}^T L \mathbf{f} \tag{8}$$

According to the above discussion, the MLP input dimension is m, the hidden neural number is n, and the output has one node.

The weight matrixes of the MLP network $W^{(k)}$ are defined as following:

$$W^{(1)} = \begin{bmatrix} w_{11}^{(1)} & w_{12}^{(1)} & \cdots & w_{1m}^{(1)} \\ w_{21}^{(1)} & w_{22}^{(1)} & \cdots & w_{2m}^{(1)} \\ \vdots & \vdots & \vdots & \vdots \\ w_{n1}^{(1)} & w_{n2}^{(1)} & \cdots & w_{nm}^{(1)} \end{bmatrix} \tag{9}$$

$$\mathbf{b}_1 = [b_1, \ldots, b_n] \tag{10}$$

$$W^{(2)} = \left[w_1^{(2)}, \cdots, w_n^{(2)} \right] \tag{11}$$

The weights update is based on the gradient descent method. We will calculate the derivatives of labeled and unlabeled cost function at first. Then, we get the derivatives of the objective function as follows:

$$\frac{\partial \mathcal{L}}{\partial w_{jk}^{(1)}} = \frac{-2w_j^{(2)}}{lW^{(2)}} \sum_{i=1}^{l} e_i h' \left(\sum_{p=1}^{m} w_{jp}^{(1)} x_{ip} + b_j \right) x_{ik}$$

$$+ \frac{2\gamma}{(l+u)^2} \mathbf{f}^{\mathrm{T}} \mathbf{L} \begin{bmatrix} w_j^{(2)} h' \left(\sum_{p=1}^{m} w_{jp}^{(1)} x_{1p} + b_j \right) x_{1k} \\ w_j^{(2)} h' \left(\sum_{p=1}^{m} w_{jp}^{(1)} x_{2p} + b_j \right) x_{2k} \\ \vdots \\ w_j^{(2)} h' \left(\sum_{p=1}^{m} w_{jp}^{(1)} x_{(l+u)p} + b_j \right) x_{(l+u)k} \end{bmatrix} \tag{12}$$

$$\frac{\partial \mathcal{L}}{\partial b_j} = \frac{-2w_j^{(2)}}{lW^{(2)}} \sum_{i=1}^{l} e_i h' \left(\sum_{p=1}^{m} w_{jp}^{(1)} x_{ip} + b_j \right)$$

$$+ \frac{2\gamma}{(l+u)^2} \mathbf{f}^{\mathrm{T}} \mathbf{L} \begin{bmatrix} w_j^{(2)} h' \left(\sum_{p=1}^{m} w_{jp}^{(1)} x_{1p} + b_j \right) \\ w_j^{(2)} h' \left(\sum_{p=1}^{m} w_{jp}^{(1)} x_{2p} + b_j \right) \\ \vdots \\ w_j^{(2)} h' \left(\sum_{p=1}^{m} w_{jp}^{(1)} x_{(l+u)p} + b_j \right) \end{bmatrix} \tag{13}$$

$$\frac{\partial \mathcal{L}}{\partial w_j^{(2)}} = -\frac{2}{l}\sum_{i=1}^{l} e_i \left(\frac{h\left(\sum_{p=1}^{m} w_{jp}^{(1)} x_{ip} + b_j\right)}{W^{(2)}} - \frac{w_j^{(2)} h\left(\sum_{p=1}^{m} w_{jp}^{(1)} x_{ip} + b_j\right) w_j^{(2)}}{W^{(2)3}} \right)$$

$$+ \frac{2\gamma}{(l+u)^2} f^{\mathrm{T}} \mathrm{L} \begin{bmatrix} h\left(\sum_{p=1}^{m} w_{jp}^{(1)} x_{1p} + b_j\right) \\ h\left(\sum_{p=1}^{m} w_{jp}^{(1)} x_{2p} + b_j\right) \\ \vdots \\ h\left(\sum_{p=1}^{m} w_{jp}^{(1)} x_{(l+u)p} + b_j\right) \end{bmatrix}$$

$$(14)$$

where $W_{l,k}^{(1)}$ is the k th row of matrix $W^{(1)}, W_j^{(2)}$ is the j th element of vector $W^{(2)}, b_j^{(1)}$ is the n th position of vector $b^{(1)}$, $h'(\cdot)$ is the derivative of the sigmoid function, The weights of layer ($l = 1; 2$) are updated as follows:

$$W_{l,k+1}^{(1)} = W_{l,k}^{(1)} + \alpha \frac{\partial \mathcal{L}}{\partial w_{l,k}^{(1)}} \tag{15}$$

$$W_{k+1}^{(2)} = W_k^{(2)} + \alpha \frac{\partial \mathcal{L}}{\partial w_k^{(2)}} \tag{16}$$

$$b_{k+1}^{(1)} = b_k^{(1)} + \alpha \frac{\partial \mathcal{L}}{\partial b_k^{(1)}} \tag{17}$$

Where k is the iteration, α is the learning rate. In short, each variable is adjusted according to the gradient descent. For each epoch, if the MM-based objective function $\mathcal{L}$ decreases to the goal, then the learning rate α is increased. If $\mathcal{L}$ increases by more than a given threshold, the learning rate is decreased by a given factor μ and the update of synaptic weights and biases that increased $\mathcal{L}$ in the current iteration are undone. During the training, the value of AUC is calculated at each τ epochs, over the validation dataset. If AUC increases, then a register of network parameters is updated. In the final, the registered network parameters, which correspond to the highest AUC, are adopted. The training section stops after ξ AUC checks without AUC improvement. Therefore, the MMGDX has time complexity $O(N)$. Similarly to other first-order optimization methods, the MMGDX does not need to calculate the Hessien matrix, therefore it has space complexity $O(N)$.

3   Experimental Results

This section reports on the experimental results obtained by the proposed LapMMNN in the semi-supervised classification of simulated data and SAR image. Figure 1 shows the classification maps obtained with the proposed Laplacian Maximum Margin Neural Networks (LapMMNN) with both TSVM and supervised SVM. The LapMMNN improves the results of the TSVM, while more homogenous areas and better classification results are observed for the LapMMNN.

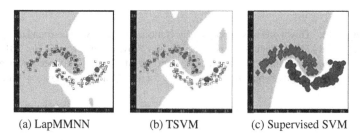

(a) LapMMNN   (b) TSVM   (c) Supervised SVM

Fig. 1. Segmentation with the SVM, TSVM and LapMMNN

The test site considered in our experiments is related to an area around Shaanxi, China. The image is 140 × 280 pixels, contains 6 surface types classes. This image constitutes a very challenging classification problem because of the strong mixture of the class signatures and small labeled samples. Figure 2 shows the classification maps obtained with the proposed LapMMNN. It is observed that numerical results are confirmed by the visual inspection of the image. LapMMNN produces accurate result, especially in gully and terrace area.

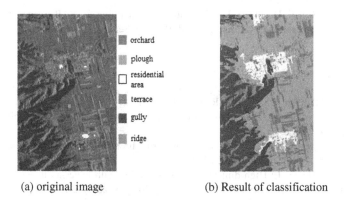

(a) original image   (b) Result of classification

Fig. 2. Classification with LapMMNN

4 Conclusion

This paper presented a semi-supervised neural networks method allows efficient SAR image classification that take advantage of unlabeled sample information for improving classification accuracy. We compared results to standard semi-supervised learning method (TSVM) as well as purely supervised algorithm (SVM). The proposed method demonstrates in general an improved classification accuracy and computational efficiency on simulated and SAR images. Further work is tied to analyze the improvement of semi-supervised classifiers as a function of the unlabeled samples and input dimensionality.

Acknowledgment. This work was supported by National Natural Science foundation of China under grant No: 61471386. Natural Science of Shaanxi Province under grant No: 2012JQ8036, 2013JQ8027. Coordinator innovative engineering project of Shaanxi Province under grant No: 2015KTTSGY0406. Fundamental Research Funds for the Central Universities Program No. GK200902018.

References

1. Hughes, G.F.: On the mean accuracy of statistical pattern recognizers. IEEE Trans. Inf. Theory **14**(1), 55–63 (1968)
2. Schölkopf, B., Smola, A.: Learning with Kernels – Support Vector Machines, Regularization, Optimization and Beyond. MIT Press Series, Cambridge (2002)
3. Nilubol, C., Merereau, R.M., Smith, M.J.T.: A SAR target classifier using random transform and hidden markov models. In: Proceedings of Defence of signal, vol. 12(2), pp. 274–283 (2004)
4. Vasuki, P., Roomi, S.M.M.: Man-made object classification in SAR image using Gabor wavelet and neural network classifier. In: 2012 International Conference on Devices, Circuits and Systems, pp.537–539 (2012)
5. Shiyong, C., Mihai, D., Pierre, B.: Cascadde active learning for SAR image annotation. In: 32nd IEEE international Geoscience and Remote Sensing Symposium, pp. 2000–2003 (2012)
6. Chapelle, O., Schölkopf, B., Zien, A.: Semi-Supervised Learning, 1st edn. MIT Press, Cambridge, Massachusetts and London, England (2006)
7. X. Zhu, "Semi-supervised learning literature survey". Computer Sciences,University of Wisconsin-Madison, USA, Tech. Rep. 1530(2005)
8. Belkin, M., Niyogi, P.: Semi-Supervised Learning on Riemannian Manifolds. Machine Learning **56**, 209–239 (2004)
9. Ruiz, A., Lopez-de-Teruel, P.E.: Nonlinear kernel-based statistical pattern analysis. IEEE Trans. Neural Netw. **12**(1), 16–32 (2001)
10. Neal, R.M., Zhang, J.: High dimensional classification with bayesian neural networks and dirichlet diffusion trees. Studies Fuzziness Soft Comput. **20**(7), 265–296 (2006)
11. Li, X., Bilmes, J., Malkin, J.: Maximum margin learning and adaptation of MLP classifiers. In Interspeech, Lisbon (2005)
12. Ludwig, O.: Novel maximum-margin training algorithms for supervised neural networks. IEEE Trans. Neural Netw. **21**(6), 972–984 (2010)

13. Prasad, B., Prasanna, S.R.M.: Images with regularized AdaBoosting of RBF neural networks. In: Speech, Audio, Image and Biomedical Signal Processing using Neural Networks, pp. 309–326 (2008)
14. Tikhonov, A.N.: Regularization of incorrectly posed problems. Sov. Math. Dokl **4**, 1624–1627 (1963)
15. Zhu, J., Chen, N., Perkins, H., Zhang, B.: Gibbs Max-Margin Supervised Topic Models with Fast Sampling Algorithms. In: To Appear ICML, Atlanta, USA (2013)

DT-CWT-Based Improved BAQ
for Anti-Saturation SAR
Raw Data Compression

Li Sun[1(✉)], Yuedong Zhang[2], Qun Zhang[1], and Miao Ma[3]

[1] Air Force Engineering University, Xi'an 710077, People's Republic of China
sl_lxa@mail.nwpu.edu.cn, zhangqunnus@gmail.com
[2] Division of PLA, Qingdao 95419, People's Republic of China
278797082@qq.com
[3] Shaanxi Normal University, Shaanxi, People's Republic of China
mmthp@snnu.edu.cn

Abstract. In this paper, an anti-saturation raw data compressing algorithm for Synthetic Aperture Radar (SAR) based on Dual-Tree Complex Wavelet Transform (DT-CWT) and improved Block Adaptive quantization (BAQ) is proposed. The DT-CWT has approximate shift-invariant and good directional selectivity. The improved BAQ rebuild data keeps the saturation signal. We apply DT-CWT to fractional saturation SAR raw data. The result put into improved BAQ quantizer. Experiment shows this method has a high compression rate and Signal to Noise Ratio (SNR).

Keywords: SAR · Raw data · BAQ · DT-CWT

1 Introduction

Synthetic Aperture Radar (SAR) is an imaging radar which can work in any weather and any time. Now SAR is widely used in military surveillance, geoscience and remote sensing. SAR could get high-resolution image, but SAR raw data is too large and difficult to be transmitted. In addition, target of city and coast area lead to signal saturation which means SAR raw data contain too much signal of saturation.

There have been many research to compress SAR raw data. For example, BAQ [1], BAVQ [2], VQ [3], LBG [4], Wavelet [5]. However these algorithm remain a problem that saturation SAR raw data cause a huge SNR loss. Navneet Agrawal and K. Venugopalan [6] presented to compensate power loss. Zhao [7] presented fractional saturation block adaptive quantization (FSBAQ) based on BAQ algorithm. Qi [8] improve Lloyd-max quantizer by add a correction term. Masanobu Shimada [9] also take about this question by radiometric correction. However, these promotions are change the quantization hardware, improved BAQ algorithm just change a quantization level in software. The speckle noise of SAR images in that situation remain to solve. Therefore, we will use the DT-CWT to reduce speckle noise. The advantage of preserve edge information and texture features of the original image [10] make DT-CWT better than the Lee filter [11], the Frost filter [12], the Gamma MAP filter [13], and their enhanced filters [14].

X. He et al. (Eds.): IScIDE 2015, Part I, LNCS 9242, pp. 522–530, 2015.
DOI: 10.1007/978-3-319-23989-7_53

In order to deal with the problem of signal saturation and data compression, this paper presents a new method that combine DT-CWT and improved BAQ. The improved BAQ use an equal and adaptive quantization level which is determined by the max data in each blocks to hold high SNR both for fractional saturation signal and normal signal. The approximate shift-invariant and good directional selectivity of DT-CWT is used to pre-denoising image in data domain. Two dimensional DT-CWT is used to remove the high frequency component, which could smooth signal and pre-denoising for final SAR image.

This paper is organized as follows. Section 1 is an brief introduction of fractional saturated data analysis, improved BAQ and DT-CWT. Section 2 explains the detail of the combined algorithm work. Section 3 shows the experimental results. Finally, a short conclusion is given in Sect. 4.

2 Anti-Saturation Algorithm Based on Improved BAQ and DT-CWT

The large area target which is distributed separately often lead to signal saturation because of the high echo power. Thus the data distribute no longer fits to the original BAQ algorithm. The improved BAQ algorithm solve the problem, and DT-CWT enhance the image quality by denoising.

2.1 SAR Raw Signal Analysis

SAR emission chirp and receive the addition echo of many scatters. Radar echo is down-converted to baseband and divided into I and Q channel, which represents the real part and imaginary part. I and Q signals separately digitized by A/D converters. Then signals are computed and encoded to 2 or more bits. The statistic are assumed to be Gaussian distribution, zero means, identical variances and unknown average power. A brief introduction is given here.

Echo in observation point can be written as:

$$A(x, y, z) = \sum_{i=1}^{k} A_k e^{j\psi k} \tag{1}$$

There $A(x, y, z)$ is represented the sum echo of scatters, A_k is the amplitude of echo, ψ_k is the phase delay due to the different paths and radar wavelength and is independent of A_k. Thus echo can be assumed to have such property: A_k and ψ_k are independent of each other, so does different echo. ψ_k are uniformly distributed on the interval $[-\pi, +\pi]$.

It shows that the real and imaginary parts of the signal have zero means, identical variances, and are uncorrelated. There is a large number of scatters because of the large illumination area by SAR. I and Q signal are the addition of a large number of independent random variables. According to the central limit theorem that $N_s \rightarrow \infty$, I and Q signal are submit to Gaussian distribution with unknown variances. According to the property, the real and the imaginary part of the signal are statistically independent.

Now many SAR systems are using 8-bit A/D converter. In some case, the data would be saturation because of a large dynamic range or an unsuitable receiver setting. The signal beyond the dynamic of A/D converter will reach the clipping point of A/D converter, it lead to an unknown SNR degradation. Apparently, this situation fit not the condition of optimization quantization.

According to the Fig. 1, fractional saturation data contain a lot of saturation signal, which lead to a breakage to the Gaussian distribution. We can see that the middle area of statistic submit to the Gaussian distribution. The maximum and minimum areas are contain more data than normal statistic.

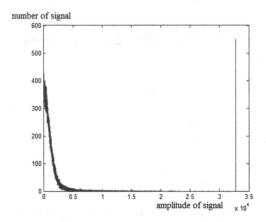

Fig. 1. Saturation SAR raw data

2.2 Improved BAQ Algorithm for Anti-Saturation SAR Raw Data Compression

The improved BAQ algorithm encode signal to 2 bits. Firstly divide raw data into suitable blocks, which could both keep similar-Gaussian distribution and slow-change variance. Then make each blocks Gaussian standardization by variance and average power. For the quantizer, the first bit encode bit 0 represent positive value, 1 represent negative value. Finding the biggest data as max rebuild value in each blocks, and divide this value into equal rebuild level.

The average between two rebuild values is a quantization level, all signals fall into two quantization level are quantized to a same quantization code. In this way, the quantization strategy could see like this.

According to the quantization areas and rebuild values in Figure x, the minimum and the maximum data would not influence the rebuild result.

The improved BAQ compress and rebuild SAR raw data in the following order:

(1) Divide data into blocks by range and azimuth, divide the data matrix to many small blocks in same scale. Calculate the average and variance of each block.

Convert the raw data into strict Gaussian distribution using the block average and variance.

(2) If the data is negative, the first code is 1, the first code of positive data is 0. Then find the max absolute value in the strict Gaussian distribution matrix, divide this max value equally into quantization level. Encode the rest of quantization bits by the absolute value of the data.

(3) Transmit code, quantization level, average and variance information to ground segment. Rebuild the strict Gaussian distribution by multiply the quantization level and code, and rebuild the data (Fig 2).

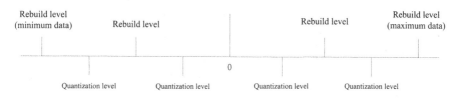

Fig. 2. Quantization level choosing

The distance between quantization levels and that between decode levels is equal. At the same time, this algorithm keep the biggest data in each blocks, decoded data could keep the original signal distribution, especially in the quantization bit higher than 3 bit. This method work well for fractional saturation data, enhance the SNR and improved image quality. For normal signal, this algorithm would not make image quality too bad. This algorithm would cost a bit more data, thus cost more downlink bandwidth. This algorithm has better adaptive quality than other promotion algorithm, notably increase the SNR for fractional saturation data.

The promotion algorithm could get higher SNR than original algorithm, and would not lose the original SNR for small data.

2.3 DT-CWT Algorithm

Kingsbury introduced a new kind of wavelet transform [15, 16], called the dual-tree complex wavelet transform, that exhibits shift invariant property and improved angular resolution. Compared with other complex transforms, it has the advantage of perfect reconstruction. Figure 3 shows the 1-D dual-tree complex wavelet transform. The success of the transform is because of the use of filters in two trees, A and B. It proposed a simple delay of one sample between the level 1 filters in each tree, and then the use of alternate odd-length and even-length linear-phase filters. Tree A can be interpreted as the real part of a complex wavelet transform, and tree B can be interpreted as the imaginary part.

As Fig. 3, extension of the DT-CWT to 2-D is achieved by separable filtering along columns and then rows. 2-D DT-CWT gives 4:1 redundancy and six distinct directions, which is strongly oriented at angles $\pm15°, \pm45°, \pm75°$. So DT-CWT is superior to DWT in shift-invariant and directional selectivity.

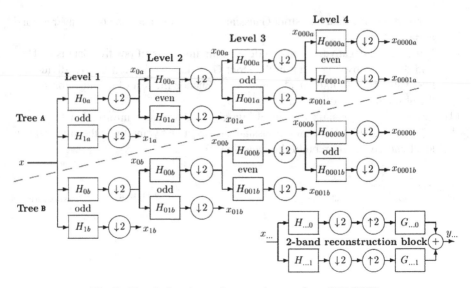

Fig. 3. The dual-tree complex wavelet transform (DT-CWT)

2.4   Anti-Saturation Algorithm Based on Improved BAQ and DT-CWT

SAR raw data is down-converted to I and Q channels, that is, real part and imaginary part. According to the signal analysis part, both real part and imaginary part submit to Gaussian distribution, zero means, identical variances and unknown average power. The imaginary part is uniformly distributed on the interval $[-\pi, +\pi]$. It represents the phase delay due to the different paths and radar wavelength and is independent of real part.

So the shift-invariant and good directional selectivity of DT-CWT is used to process the real part of SAR raw data. DT-CWT is used to remove the high frequency component, which could smooth signal and pre-denoising for final SAR image. Then improved BAQ algorithm is used to compress both real and image part of data.

Format (2) indicates the expression of signal in observation point, in the complex wavelet domain, the wavelet coefficients can be formulated as:

$$y = y_r + iy_i = x + n = (x_r + n_r) + i(x_i + n_i) \tag{2}$$

Where y is the complex wavelet coefficients of noisy signal, x is the original coefficients, x_r is the real part of x while x_i is the imaginary part of x, and n represents the noise, while n_r and n_i are both the independent Gaussian functions. From the theory of standard maximum a posteriori (MAP) estimator, the coefficient x in (2) can be calculated as follows

$$\hat{x} = \arg \max p_{x|y}(x|y) \tag{3}$$

According to the researches of details complex wavelet coefficients of nature data, PDF of the complex wavelet coefficients can be supposed as:

$$p_x(x) = 3/(2\pi\sigma^2)\exp(-\sqrt{2}\cdot\sqrt{x_r^2 + x_i^2}\Big/\sigma) = 3/(2\pi\sigma^2)\exp(-\sqrt{2}|x|/\sigma) \qquad (4)$$

Where x are complex wavelet coefficients, and σ is the standard deviation of x. And $p_n(y - x)$ is assumed to be a Gaussian density with variance σ_n^2.

Using (3) with (4), the estimator of x is derived to be:

$$\begin{cases} \hat{x}_r = \max(\sqrt{y_r^2 + y_i^2} - \sqrt{2}\sigma_n^2/\sigma, 0)\cdot y_r \Big/ \sqrt{y_r^2 + y_i^2} \\ \hat{x}_i = \max(\sqrt{y_r^2 + y_i^2} - \sqrt{2}\sigma_n^2/\sigma, 0)\cdot y_i \Big/ \sqrt{y_r^2 + y_i^2} \end{cases} \qquad (5)$$

It also can be written as follows

$$\hat{x} = \max(|y| - \sqrt{2}\sigma_n^2/\sigma, 0)\cdot\frac{y}{|y|} \qquad (6)$$

This estimator requires a prior knowledge of the noise variance σ_n^2, and marginal variance for each details wavelet coefficients. Generally speaking, (6) is a soft shrinkage function.

So the method is in follows:

Step 1: select a proper coefficient of DT-CWT,
Step 2: apply DT-CWT to the data matrix of real part,
Step 3: abandon the high frequency of the result,
Step 4: apply inverse DT-CWT,
Step 5: compress both processed real part of data and image part of data.

3 Experimental Result

Raw data of seven ships in Canada is used to test the method. The target is in a low saturation status. Mean Squared Error (MSE) and Peak Signal-to-Noise Ratio (PSNR) is used to evaluate the quality of the compressing algorithm.

In order to compare the practical effect, we use BAQ algorithm and Improved BAQ (IBAQ) algorithm to compare with this method. For PNSR and MSE, we get a compare table in following:

We can see from the Table 1, the promotion algorithm we propose have a PSNR of 28.9 dB, which is higher than BAQ and IBAQ algorithm. The MSE of the promotion algorithm has decrease of 0.0500e + 008 than BAQ algorithm. Because of this raw data has a low saturation degree, this algorithm would get better rebuild result of high degree saturation raw data.

Compare with the amplitude image of the 3 methods:

From Fig. 4, the rebuild amplitude image of the promotion algorithm and IBAQ is better than BAQ algorithm. We can see BAQ algorithm has an obvious decrease of peak signal, which has change the amplitude distribution of signal.

Compare the image quality of that tree algorithm, result is in following:

Table 1. Simulation result

Data	PSNR	MSE
BAQ	28.3733 dB	3.5376e + 008
IBAQ	28.8963 dB	3.4876e + 008
IBAQ + DTCWT	28.9321 dB	3.4876e + 008

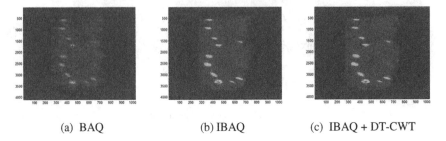

(a) BAQ   (b) IBAQ   (c) IBAQ + DT-CWT

Fig. 4. Comparison of data magnitude

From Fig. 5, we can see promotion algorithm keep more detail information than BAQ and IBAQ. BAQ algorithm has more noise dot. Thus, quality of image using promotion algorithm is better than which using the BAQ and IBAQ.

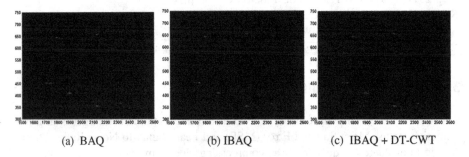

(a) BAQ   (b) IBAQ   (c) IBAQ + DT-CWT

Fig. 5. Image quality comparation

According to simulation result and image compare, we can see the promotion algorithm has better effect for saturation raw data processing. In addition, the raw data we use is from a low degree of saturation target. In the situation of high degree of saturation, this algorithm would have a greater promotion.

4   Conclusion

This paper presents a new algorithm to overcome the influence of raw data saturation. Two dimensional DT-CWT is used to remove the high frequency component, which could smooth signal and pre-denoising for final SAR image. Improved BAQ algorithm quantizer decrease data saturation. Simulation shows the promotion algorithm has a 0.03 dB of PSNR increase toward IBAQ algorithm for SAR raw data in low degree of saturation, and 0.55 dB for original BAQ algorithm. In addition, this algorithm could denoising for final SAR image. Simulation result indicate that the promotion algorithm suitable for fractional saturation data, would have a better rebuild result. Furthermore, promotion algorithm notably improve image quality for the target of city and coast areas. The image quality falls little for other observation object.

Acknowledgment. This work was supported by National Natural Science foundation of China under grant No: 61471386. Natural Science of Shaanxi Province under grant 2012JQ8036, 2013JQ8027. Coordinator innovative engineering project of Shaanxi Province under grant No: 2015KTTSGY0406. Fundamental Research Funds for the Central Universities Program No. GK200902018.

References

1. Kwok, R.: Adaptive quantizalion of Magellan SAR data [J]. IEEE Trans. Geosci. Remote Sens. **27**(7), 375–383 (1989)
2. Moreira, A., Blaeser, F.: Fusion of block adaptive and vector quantizer for efficient SAR. In: IGARSS, pp. 1583–1585. EIC, Tokyo (1993)
3. Giontsos, T.: Vector quantization use to reduce SAR data rates. SPIE Trans. Commun. **28** (1), 84–95 (1980)
4. Linde, Y., Buzo, A., Cray, R.M.: An algorithm for vector quanlizes design [J]. IEEE Trans. Commun **28**(1), 84–95 (1980)
5. Pascazio, V., Schitinzi, G.: Wavelet transform coding for SAR raw data compression. In: IGARSS, pp. 4225-12253, Hamburg (1999)
6. Agrawal, N., Venugopalan, K.: Restoration of saturated SAR data, evolution in networks and computer communications. In: A Special Issue from IJCA, pp: 20–23 (2011)
7. Yupeng, Z., Feng, W., Hong, L.: Compression on fractional saturation SAR RAW data. J. Electron. Inf. Technol. **26**(3), 438–494 (2004)
8. Haiming, Qi, Weidong, Yu.: Anti-saturation block adaptive quantization algorithm for SAR raw data compression over the whole set of saturation degrees. Prog. Nat. Sci. **19**, 1003–1009 (2009)
9. Shimada, M.: Radiometric correction of saturated SAR data. IEEE Trans. Geosci. Remote Sens. **37**(1), 467–478 (1999)
10. Gagnon, L., Jouan, A.: Speckle filtering of SAR images-a comparative study between complex-wavelet-based and standard filters. In: Proceedings of the SPIE (1997)
11. Lee, G.: Refined filtering of image noise using local statistics. Comput. Graph. Image Process. **15**(4), 380–389 (1981)
12. Frost, V.S., Stiles, J.A., Shanmugan, K.S., Holtzman, J.C.: A model for radar images and its application to adaptive digital filtering of multiplicative noise. IEEE Trans. Pattern Anal. Mach. Intell PAMI **4**, 373–383 (1980)
13. Lopes, A., Nezry, E., Touzi, R., Laur, H.: Maximum a posteriori filtering and first order texture models in SAR images. In: Proceedings of the IGARSS (1990)
14. Lopes, A., Touzi, R., Nezry, E.: Adaptive speckle filters and scene heterogeneity. IEEE Trans. Geosci. Remote Sens. **28**, 992–1000 (1990)

15. Kingsbury, G.: Complex wavelets for shift invariant analysis and filtering of signals. J. Appl. Comput. Harmonic Anal. **10**(3), 234–253 (2001)
16. Kingsbury. The dual-tree complex wavelet transform: a new efficient tool for image restoration and enhancement. In: Proceedings of the EUSIPCO 1998, Rhodes: EURASIP, pp: 319–322 (1998)

Non-Negative Kernel Sparse Coding
for Image Classification

Yungang Zhang[1]($\boxtimes$), Tianwei Xu[1], and Jieming Ma[2]

[1] Department of Computer Science, Yunnan Normal University,
No.1, Yuhua Higher Education Town, Chenggong District, 650000 Kunming, China
yungang.zhang01@gmail.com
[2] School of Electronics and Information Engineering, Suzhou University of Science
and Technology, No.1 Ke Rui Road, Suzhou 215009, People's Republic of China

Abstract. Sparse representation of signals have become an important
tool in computer vision. In many applications in computer vision, such as
image denoising, image super-resolution and object recognition, sparse
representations have produced remarkable performance. In this paper,
we propose a non-linear non-negative sparse coding model NNK-KSVD.
The proposed model extended the kernel KSVD by embedding the non-
negative sparse coding. Experimental results show that by exploiting
the non-linear structure in images and utilizing the 'additive' nature of
non-negative sparse coding, promising classification performance can be
obtained.

Keywords: Non-negative sparse coding · Kernel methods · Dictionary
learning · Image classification

1 Introduction

In recent years, sparse representation of signals have become an important tool in
computer vision. In many applications in computer vision, such as image denoising, image super-resolution and object recognition, sparse representations have
produced remarkable performance [1–3]. It has also been verified that sparse
representation or sparse coding can achieve good outcomes in many image classification tasks [4–6]. The reason of the success of sparse coding is that a signal
Y can be well represented by a linear combination of a sparse vector x and a
given dictionary D, namely, if x and D can be properly found, then Y can be
well approximated as: $Y \approx Dx$.

There are two methods to obtain a dictionary for sparse representation:
using an existing dictionary and dictionary learning. Using an existing dictionary means to choose a pre-computed basis such as wavelet basis and curvelet
basis. However, in different computer vision tasks, one often needs to learn a
dictionary from given sample images, then better image representation can be
expected from such a task-specific dictionary [7].

Y. Zhang – The project is funded by Natural Science Foundation China 61462097.

© Springer International Publishing Switzerland 2015
X. He et al. (Eds.): IScIDE 2015, Part I, LNCS 9242, pp. 531–540, 2015.
DOI: 10.1007/978-3-319-23989-7_54

Many algorithms have been proposed to tackle the problem of dictionary learning, the method of optimal directions (MOD) [8] and the KSVD algorithm [9] are two of the most well-known methods. The dictionaries learned by MOD and KSVD are linear representations of the data. However, in many computer vision applications, one has to face non-linear distributions of the data, using a linear dictionary learning model often leads to poor performance. Therefore several non-linear dictionary learning methods have been proposed. Among them, a recently proposed method kernel KSVD (KKSVD) [10] has shown its ability to obtain better image description than the linear models. The kernel KSVD includes two steps: sparse coding and dictionary update. In sparse coding stage, the Kernel Orthogonal Matching Pursuit (KOMP) is used for seeking the sparse codes. Once the sparse codes are obtained, the kernerlized KSVD is then used in the second stage for dictionary update. It is demonstrated that the kernel KSVD can provide better image classification performance than the traditional linear dictionary learning models.

Although the kernel KSVD can outperform its linear counterparts by introducing the non-linear learning procedure, the learned sparse vector and the dictionary yet comprise both additive and subtractive interactions. From the perspective of biological modeling, the existence of both the additive and subtractive elements in the sparse representations of signals is contrary to the non-negativity of neural firing rates [11,12]. Moreover, the negative and positive elements in the representations may induce the 'cancel each other out' phenomenon, which is contrary to the intuitive notion of combining non-negative representations [13]. Therefore, many researchers have claimed to use non-negative sparse representations in vision-related applications [14,15]. The non-negative sparse representations are based on the constraints that the input data Y, the dictionary D, and the sparse vector x are all non-negative.

The non-negative sparse representation has been successfully applied in many computer vision applications, such as face recognition [14], motion extraction [16], and image classification [17]. Nevertheless, these non-negative sparse representations are all based on linear learning models, therefore their ability to capture the non-linear distributions of data is limited.

Motivated by this drawback of the existing non-negative sparse representation models, in this paper, we propose a non-linear and non-negative sparse model: non-negative kernel KSVD (NNK-KSVD), which integrates the distinctive features of the non-linear models and the non-negative models. In NNK-KSVD, the non-negative constraints are embedded into the kernel KSVD for sparse coding and dictionary learning. With the non-negative constraints, the new update rules for sparse vector searching and dictionary learning of kernel KSVD are introduced. The proposed NNK-KSVD sparse model was tested on several benchmark image datasets for the task of classification, state-of-the-art results were obtained.

The rest of this paper is organized as follows. In Sect. 2, the proposed NNK-KSVD sparse model is introduced. The experiments and the results are given in Sect. 3. Finally, we draw the conclusions in Sect. 4.

2   The Non-negative Kernel KSVD

In this section, the proposed NNK-KSVD model is introduced. First, a brief description of the kernel KSVD is given, then the proposed NNK-KSVD model is discussed.

2.1   The Kernel KSVD Model

The goal of kernel dictionary learning is to obtain a non-linear dictionary $D = [d_1, \ldots, d_k]$ in a Hilbert space $\mathcal{F}$. Define $\phi : \mathbb{R}^n \to \mathcal{F} \in \mathbb{R}^m$ as a non-linear mapping from $\mathbb{R}^n$ to a Hilbert space $\mathcal{F}$, where $m \gg n$. To obtain the non-linear dictionary D, one has to solve the following optimization problem:

$$\underset{D,X}{\operatorname{argmin}} \parallel \phi(Y) - DX \parallel_F^2 \ s.t. \parallel x_i \parallel_0 \leq T, \parallel d_j \parallel_2 = 1, \forall i, j. \tag{1}$$

where $\parallel \cdot \parallel_F$ is the Frobenius norm of a matrix, for a matrix M has the dimension of $m \times n$, it can be obtained as: $\parallel M \parallel_F = (\sum_{i=1}^{m} \sum_{j=1}^{n} M(i,j)^2)^{\frac{1}{2}}$. $\phi(Y) = [\phi(Y_1), \phi(Y_2), \ldots, \phi(Y_n)]$. $X \in \mathbb{R}^{k \times n}$ is a sparse matrix, each column x_i in X is a sparse vector has maximum of T non-zero elements.

The dimension of the Hilbert space $\mathcal{F}$ can possibly be infinite, therefore the traditional dictionary learning methods such as KSVD and MOD are incapable for the optimization task. In order to tackle the optimization problem in Eq. (1), the kernel KSVD (KKSVD) dictionary learning [10] was proposed. KKSVD consists of two stages, first the kernel orthogonal matching pursuit (KOMP) is used for sparse coding, then the kernel KSVD is used for dictionary update.

In the sparse coding stage, by introducing kernel mapping, the problem in Eq. (1) can be rewritten as:

$$\parallel \phi(Y) - \phi(Y)AX \parallel_F^2 = \sum_{i=1}^{N} \parallel \phi(y_i) - \phi(Y)Ax_i \parallel_2^2 . \tag{2}$$

Note the dictionary D in Eq. (1) is changed to a coefficient matrix A, as it is proved that there exists an optimal solution D^*, which has the form $D^* = \phi(Y)A$ [10].

Therefore, now the problem in Eq. (2) can be solved by any pursuit algorithms. For example, a kernerlized version of the orthogonal matching pursuit algorithm (OMP) [18] is used in [10] for image classification task. In the sparse coding stage, the coefficient matrix A is fixed and the KOMP algorithm can be used to search for the sparse matrix X. The details of the KOMP algorithm can be found in [10].

In the dictionary update stage, the kernel KSVD is used. Let a_k and X_T^j represent the k-th column and the j-th row of A and X, respectively. The approximation error:

$$\parallel \phi(Y) - \phi(Y)AX \parallel_F^2 \tag{3}$$

can be written as:

$$\parallel \phi(Y)E_k - \phi(Y)M_k \parallel_F^2 \tag{4}$$

where $E_k = (I - \sum_{j \neq k} a_j x_T^j)$, $M_k = a_k x_T^k$. $\phi(Y)E_k$ demonstrates the distance between the true signals and the estimated signals when removing the k-th dictionary atom, while $\phi(Y)M_k$ specifies the contribution of the k-th dictionary atom to the estimated signals.

In dictionary update stage, as E_k is a constant for each k, then the minimization of (4) is indeed to find the best a_k and x_T^k for the rank-1 matrix $\phi(Y)M_k$ to produce the best approximation of $\phi(Y)E_k$. Using a singular value decomposition (SVD), one can obtain the solution. However, it is impossible to directly use SVD in this scenario. Using SVD here may greatly increase the number of non-zero elements in X. Moreover, the matrix may have infinitely large row dimension, which is computationally prohibitive.

Instead of working on all columns of M_k, one can only work on a subset of columns, as there is a fact that the columns of M_k associated with the zero-elements in x_T^k are all zero, these columns do not affect the objective function. Therefore, these zero columns can be discarded and only the non-zero elements in x_T^k are allowed to vary and therefore the sparsities are preserved [9].

Let S_k be the set of indices of the signals $\{\phi(y_i)\}$ which use the dictionary atom $(\phi(Y)A)_k$, then S_k can be defined as:

$$S_k = \{i | 1 \leq i \leq n, x_T^k \neq 0\}. \tag{5}$$

Denote Γ_k as a matrix of size $n \times |S_k|$, it has ones on the $(S_k(i), i)$-th entries and zeroes elsewhere. Using x_T^k multiplies Γ_k, by discarding zeroes, x_T^k will have the length of $|S_k|$. Then, the E_k and M_k in (4) are changed to:

$$E_k^R = E_k \Gamma_k; \quad M_k^R = M_k \Gamma_k. \tag{6}$$

By applying SVD decomposition, one can get:

$$(E_k^R)^T \mathcal{K}(Y, Y)(E_k^R) = V \Delta V^T, \tag{7}$$

where $\mathcal{K}(Y, Y)$ is a positive semidefinite matrix $[\mathcal{K}_{ij}] = [\kappa(y_i, y_j)]$, $\kappa(\cdot, \cdot)$ is a mercer kernel defined as $\kappa(x, y) = < \phi(x), \phi(y) >$. $\Delta = \mathbf{\Sigma}^T \mathbf{\Sigma}$, $\sigma_1 = \sqrt{\Delta(\mathbf{1}, \mathbf{1})}$. Use an iterative procedure, in each iteration, a_k can be updated as:

$$a_k = \sigma^{-1} E_k^R v_1, \tag{8}$$

where v_1 is the first vector of V corresponding to the largest singular value σ_1^2 in Δ.

The coefficient vector x_R^k is updated as:

$$x_R^k = \sigma_1 v_1^T. \tag{9}$$

For the overall procedure, in the first stage, the KOMP algorithm is used for sparse coding, then the KSVD is used in the second stage for dictionary update. The aforementioned two stages are repeated until a stopping criterion is met.

2.2   Non-negative Kernel KSVD Model

Based on the introduced kernel KSVD model, in this section, the proposed non-nagetive kernel KSVD model is discussed.

For some applications such as image recognition, using sparse representations and overcomplete dictionaries together with forcing non-negativity on both the dictionary and the coefficients, may lead to the 'ingredient' form which all training samples are built of [12]. With the non-negative constraint, the dictionary atoms become sparser and converge to the building blocks of the training samples [19]. However, the existing non-negative sparse models are all based on the linear learning algorithms, their inability to capture the non-linear data distribution inspires us to embed the non-negativity into the non-linear sparse models.

Although there are other non-linear dictionary learning models such as kernel MOD or kernel PCA can be selected for our task, we prefer to use the kernel KSVD, as in [10], the authors has demonstrated that the kernel KSVD outperforms other non-linear models in image classification tasks. As we try to make the kernel KSVD produce the non-negative dictionaries and coefficient matrices, it is necessary to vary the original kernel KSVD model. With the non-negative and non-linear constraints, now the goal of the sparse coding changed to:

$$\min_{X} \| \phi(Y) - \phi(Y)AX \| \ s.t. \quad X \geq 0. \tag{10}$$

In the sparse coding stage, a pursuit algorithm should be used in order to keep the coefficients non-negative. In [12], an iterative method for non-negative sparse coding is introduced:

$$x^{t+1} = x^t. * (D^T y)./(D^T D x^t + \lambda), \tag{11}$$

where t represents the iteration number. $.*$ and $./$ represent entry-wise multiplication and division. However, this iterative rule is designed for linear coding, in order to use it in the non-linear scenarios, using the kernel trick, (refeq10)) can be varied as:

$$x^{t+1} = x^t. * ((\phi(Y)A)^T \phi(Y))./ \left((\phi(Y)A)^T(\phi(Y)A)x^t + \lambda\right)$$
$$= x^t. * (A^T \mathcal{K}(Y,Y))./(A^T \mathcal{K}(Y,Y)Ax^t + \lambda). \tag{12}$$

It can also be proven that using the iterative update rule in (12), the objective (10) is nonincreasing. Moreover, it is guaranteed that x stay non-negative using this update rule, as the elements in x are updated by simply multiplying with some non-negative factors.

In the dictionary update stage, the dictionary atoms must be kept non-negative as well. Use the same techniques of the kernel KSVD, under the non-negative constraint, the minimization problem in (4) now has been changed to:

$$\min_{a_k, x_R^k} \| \phi(Y)E_k^R - \phi(Y)a_k x_R^k \|_F^2, \quad s.t. \quad a_k, x_R^k \geq 0. \tag{13}$$

In (13), one can see that it has the same nature with KSVD, we try to find the best rank-1 matrix which can approximate the error matrix E_k^R. However, in order to keep the non-negativity and to reach the local minima, the KSVD cannot be used here. An iterative algorithm is used here, as illustrated in Algorithm 1.

Algorithm 1. Iterative algorithm for non-negative approximation for E_k^R

Input: Set

$$a_k = \begin{cases} 0 & \text{if } u(i) < 0 \\ u(i) & \text{otherwise} \end{cases} , \quad x_R^k = \begin{cases} 0 & \text{if } v(i) < 0 \\ v(i) & \text{otherwise} \end{cases} ,$$

where $u = \sigma^{-1} E_k^R v_1$, $v = \sigma_1 v_1^T$, as has been introduced in (8) and (9).

Repeat step 1 to 2 for J times:

1: Update $a_k = \frac{E_k^R x_R^k}{(x_R^k)^T x_R^k}$. If $a_k(i) < 0$, set $a_k(i) = 0$, otherwise, keep $a_k(i)$ unchanged. i runs for the every entry of the vector.

2: Update $x_R^k = \frac{(a_k)^T E_k^R}{(a_k)^T a_k}$. If $x_R^k(i) < 0$, set $x_R^k(i) = 0$, otherwise, keep $x_R^k(i)$ unchanged. i runs for the every entry of the vector.

Output: a_k, x_R^k.

Given this non-negative approximation algorithm for the dictionary and the coefficient matrix, now the proposed non-negative kernel KSVD (NNK-KSVD)algorithm for learning the dictionary A and the sparse coefficient matrix X is given in Algorithm 2:

Algorithm 2. The NNK-KSVD algorithm

Input: Training sample set Y, kernel function κ.

Initialization: Find a random position in each column of $A^{(0)}$, set the corresponding element to 1. Normalize each column of $A^{(0)}$ to a unit norm. Set the iteration number $J = 1$.

1: *Sparse coding:* Use the iterative update rule in (12) to obtain the sparse coefficient matrix $X^{(J)}$ with the dictionary $A^{(J-1)}$ fixed.

2: *Dictionary update:* Use the kernel KSVD algorithm to obtain $a_k^{(J)}$ and $x_R^k(J)$.

3: Update $a_k^{(J)}$ and $x_R^k(J)$ with Algorithm 1.

4: Set $J = J + 1$ and repeat step 1 to 4 until a stopping criterion is met.

Output: A, X.

---

3   Experiments and Results

In this section, the experimental results using the proposed NNK-KSVD on image classification tasks are presented. The Caltech-101 object dataset [20] was used in our experiments. The dataset contains 101 object categories plus one random background image category. Each class has the size from 31 to 800, the size of images is about 300×300 pixels.

Table 1. Comparison of classification results on caltech-101 dataset

No. of training images	5	10	15	20	25	30
Yang [4]	-	-	67.0	-	-	73.2
LLC [21]	51.2	59.8	65.4	67.7	70.2	73.4
SRC [3]	48.8	60.1	64.9	67.7	69.2	70.7
KSVD [9]	49.8	59.8	65.2	68.7	71.0	73.2
NNSC [12]	42.4	46.5	58.9	62.5	63.1	64.7
LP-β [22]	54.2	65.0	70.4	73.6	75.7	78.1
K-KSVD [10]	56.5	67.2	72.5	75.8	77.6	80.1
NNK-KSVD (C)	53.2	54.2	64.8	67.0	70.8	76.9
NNK-KSVD (D)	**55.9**	**66.7**	**71.2**	**76.7**	**78.1**	**80.2**

In order to compare with other methods, we employed the same image features used in [10], totally 39 different features were used. The features includes spatial pyramid, the PHOG shape, the regions of covariance, the LBP, etc. We used the same kernel functions as in [10], which is: $\kappa = exp(-d(y_i, y_j)/\delta)$, d is the distance metric and δ is the mean of pairwise distances.

Different numbers of training images were used in the experiments, for each image category, we varied the training sample number from 5 to 30 with a step of 5 to train the dictionary. The proposed NNK-KSVD algorithm was used to learn a dictionary for each image category.

Denote the learned dictionary for the i-th image category as $D_i = \phi(Y)A_i$, $i = 1, 2, \ldots, 102$. For a new image y_t needs to be classified, using the NNK-KSVD algorithm, the corresponding sparse coefficients can be obtained as: $x_t = [x_{1,t}, \ldots, x_{102,t}]$, where $x_{i,t}$ is obtained by using the dictionary D_i from the i-th class. Then the reconstruction error for each dictionary is obtained as:

$$e_i = \| \phi(y_t) - \phi(Y_i)A_i x_{i,t} \|_2^2$$
$$= \mathcal{K}(y_t, y_t) - 2\mathcal{K}(y_t, Y_i)A_i x_{i,t} + x_i^T A_i^T \mathcal{K}(Y_i, Y_i)A_i^T x_i. \tag{14}$$

The classification result is given by the smallest reconstruction error. Note that we used a distribute form to separately obtain our dictionaries and sparse matrices. However, in [10], the dictionaries are concatenated to form one dictionary, then the dictionary is used in KOMP for sparse coding.

Table 1 gives the comparison of the classification results on caltech-101 dataset. The results from different numbers of training images are given.

In Table 1, several state-of-the-art methods are compared with our proposed NNK-KSVD. As aforementioned, in the sparse coding stage, the learned dictionary can be concatenated together to obtain the sparse coefficients, it also can be used distributively, we named these two different techniques as NNK-KSVD (C) and NNK-KSVD (D) in Table 1. In [10], the authors obtained better performance through the concatenated dictionary, however in our experiments the distributive method outperforms the concatenated technique. An explanation for this observation is that different from the kernel KSVD, the non-negative constraint is embedded in our NNK-KSVD, as the limited training samples are used, the distributive usage of the dictionaries can make the non-negative coding more closer to the given samples compared to the concatenated one.

Note that our proposed method outperforms other methods such as LLC, [4], NNSC and the traditional KSVD, it illustrates that using non-linear kernel to capture the non-linear structures in the data can be useful for classification, it also presents that the non-negative sparse coding can be improved by using non-linear data description.

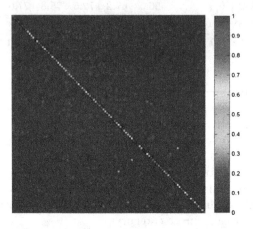

Fig. 1. Confusion matrix of the proposed NNK-KSVD on Caltech-101 dataset. The dictionary is learned from 30 images of each class.

Figure 1 shows the confusion matrix of the proposed NNK-KSVD algorithm on Caltech-101 dataset for our best performance (80.2 %). There are 14 categories achieve perfect classification, including the classes such as "Minaret", several classes obtained rather low performance, such as "Crocodile". The similar observation can be found in the results of [10] and [23]. Nevertheless, some categories like "Ant" obtain low performance in [23], while much better results are achieved by our method. This also illustrates that using non-linear and non-negative dictionary learning can help with the data description.

4   Conclusion

In this paper, we propose a non-linear and non-negative sparse coding model NNK-KSVD. The proposed model extended the kernel KSVD by embedding the non-negative sparse coding. The proposed model was tested on the Caltech-101 object dataset, experimental results show that by exploiting the non-linear structure in images and utilizing the 'additive' nature of non-negative sparse coding, better classification performance can be expected. Future work includes more theoretical analysis and evaluation of the model. The proposed model is expected to be applied on other tasks in image and video analysis such as image retrieval, video summarization, etc.

References

1. Elad, M., Aharon, M.: Image denoising via sparse and redundant representations over learned dictionaries. IEEE Trans. Image Process. **15**(12), 3736–3745 (2006)
2. Yang, J., Wright, J., Huang, T., Ma, Y.: Image super-resolution as sparse representation of raw image patches. In: Proceeding of IEEE Conference on Computer Vision and Pattern Recognition. CVPR 2008, pp. 1–8. IEEE (2008)
3. Wright, J., Yang, A.Y., Ganesh, A., Sastry, S., Ma, Y.: Robust face recognition via sparse representation. IEEE Trans. Pattern Anal. Mach. Intell. **31**(2), 210–227 (2009)
4. Yang, J., Yu, K., Gong, Y., Huang, T.: Linear spatial pyramid matching using sparse coding for image classification. In: Proceeding of IEEE Conference on Computer Vision and Pattern Recognition. CVPR 2009, pp. 1794–1801. IEEE (2009)
5. Huang, J., Nie, F., Huang, H., Ding, C.: Supervised and projected sparse coding for image classification. In: Proceedings of the Twenty-Seventh AAAI Conference on Artificial Intelligence. AAAI 2013, pp. 438–444. AAAI (2013)
6. Zhang, Y., Jiang, Z., Davis, L.: Learning structured low-rank representations for image classification. In: Computer Vision and Pattern Recognition (CVPR), 2013 IEEE Conference on. (June 2013) 676–683
7. Mairal, J., Bach, F., Ponce, J.: Task-driven dictionary learning. IEEE Trans. Pattern Anal. Mach. Intell. **34**(4), 791–804 (2012)
8. Engan, K., Aase, S.O., Husoy, J.H.: Method of optimal directions for frame design. In: Proceedings of the IEEE International Conference on Acoustics, Speech, and Signal Processing, vol. 5, pp. 2443–2446, April 1999
9. Aharon, M., Elad, M., Bruckstein, A.M.: The k-svd: an algorithm for designing of overcomplete dictionaries for sparse representation. IEEE Trans. Signal Process. **54**(11), 4311–4322 (2006)
10. Nguyen, H.V., Patel, V., Nasrabadi, N., Chellappa, R.: Design of non-linear kernel dictionaries for object recognition. IEEE Trans. Image Process. **22**(12), 5123–5135 (2013)
11. Lee, D.D., Seung, H.S.: Learning the parts of objects by non-negative matrix factorization. Nature **401**(6755), 788–791 (1999)
12. Hoyer, P.O., Hyvärinen, A.: A multi-layer sparse coding network learns contour coding from natural images. Vis. Res. **42**(12), 1593–1605 (2002)
13. Hoyer, P.O.: Non-negative sparse coding. In: Proceedings of IEEE Workshop on Neural Networks for Signal Processing, pp. 557–565 (2002)
14. Hoyer, P.O.: Non-negative matrix factorization with sparseness constraints. J. Mach. Learn. Res. **5**, 1457–1469 (2004)
15. Zass, R., Shashua, A.: Nonnegative sparse pca. In: Neural Information Processing Systems, pp. 1–7 (2007)
16. Guthier, T., Willert, V., Schnall, A., Kreuter, K., Eggert, J.: Non-negative sparse coding for motion extraction. In: The 2013 International Joint Conference on Neural Networks (IJCNN), pp. 1–8, August 2013
17. Zhang, C., Liu, J., Liang, C., Xue, Z., Pang, J., Huang, Q.: Image classification by non-negative sparse coding, correlation constrained low-rank and sparse decomposition. Comput. Vis. Image Underst. **123**, 14–22 (2014)

18. Chen, S., Donoho, D., Saunders, M.: Atomic decomposition by basis pursuit. SIAM J. Sci. Comput. **20**(1), 33–61 (1998)
19. Aharon, M., Elad, M., Bruckstein, A.M.: K-svd and its non-negative variant for dictionary design. Proc. SPIE Wavelets XID **5914**, 327–339 (2005)
20. Li, F.F., Fergus, R., Perona, P.: Learning generative visual models from few training examples: an incremental Bayesian approach tested on 101 object categories. In: 2004 Conference on Computer Vision and Pattern Recognition Workshop. CVPR 2004, p. 178, June 2004
21. Wang, J., Yang, J., Yu, K., Huang, T., Gong, Y.: Locality-constrained linear coding for image classification. In: Proceeding of IEEE Conference on Computer Vision and Pattern Recognition. CVPR 2010, pp. 3360–3367. IEEE (2010)
22. Gehler, P., Nowozin, S.: On feature combination for multiclass object classification. In: 2009 Conference on Computer Vision and Pattern Recognition Workshop. CVPR 2009, pp. 221–228, October 2009
23. Lazebnik, S., Schmid, C., Ponce, J.: Beyond bag of features: spatial pyramid matching for recognizing natural scene categories. In: 2006 Conference on Computer Vision and Pattern Recognition Workshop. CVPR 2006, pp. 2169–2178 (2006)

Practical Computation Acceleration for Large Size Single Image Filtering Based Haze Removal

Heng Liu[1,2(✉)], Hongshen Liu[1], and Jing Xun[2]

[1] School of Computer Science and Technology,
Anhui University of Technology, Maanshan, China
hengliusky@aliyun.com
[2] Institute of System Technology, Hisense R&D Center,
Qingdao, China
xunjing@hisense.com

Abstract. The main defect of traditional filtering based image dehazing approaches is huge time consuming, especially for large size image haze removal. In the work, a serial of practical computation acceleration strategies for bilateral filtering based haze removal are proposed. Our approach strength lies in its wide practicability: better dehazing with very fast computation speed even for large size images. Since our approach requires less time consuming corresponding to less CPU load, it will be promising for some real time application, such as smart TV image enhancement.

Keywords: Image dehazing · Bilateral filtering · Computation acceleration

1 Introduction

In recent ten years, many researchers have made big efforts on single image haze removal and get some significant progress. Work in [1] assumes that the transmittance and surface projection are not relevant in the local area, and adopts independent component analysis and Markov Random Field model to recover haze image. But this method could not handle dense fog and can only process gray image. He et al. [2] presents a physical statistical assumption - dark channel prior and take the rule to estimate atmospheric transmittance directly. This method can produce good effect in most cases. However since it needs to refine transmission map by guild filtering which requires intensive computation, the processing speed is rather slow. And due to lacking individual reference image, incomplete haze removing phenomenon especially in sky or over bright region will appear. Tarel [3] utilizes a median filter for haze removing. Without considering depth information, although this algorithm is very efficient but the dehazing effect is not satisfactory in terms of dense fog or in small edge regions. Thanks to Chunxia and Jiajia [4], they proposed a wonderful image dehazing method which utilizes joint bilateral filter to refine atmosphere veil and derives transmission map for final recovery. Although the fast dehazing problem has been be addressed rather well in the work, according to our practical experiments, for large size images

© Springer International Publishing Switzerland 2015
X. He et al. (Eds.): IScIDE 2015, Part I, LNCS 9242, pp. 541–550, 2015.
DOI: 10.1007/978-3-319-23989-7_55

(for example, larger than 1024*768), the algorithm still works a little slow due to huge processing cycles.

Stimulate by the idea of fast atmosphere veil reconstructing based recovery, in this work, we propose a set of bilateral filtering acceleration strategy for fast dehazing. Generally speaking, the motivation of our approach is to get an image containing the edge information of input foggy image as the cue of the scene depth information, and take this image as the reference image to filter initial atmosphere veil [3, 4]. In our practical processing, a fast sorting algorithm [5] is firstly used to compute the minimum component map of original three channels foggy image. Smoothed by Gauss filter following with bilateral filter, the minimum map becomes as the reference image. And after that, the guided joint bilateral filtering is taken to refine the atmosphere veil. As the filtering operation not only removes much redundant texture details in the atmosphere veil but adds useful edge information of original image into it, this will lead to better dehazing results at places where depth changes abruptly. And during the procedure, a computation acceleration strategy is adopted for fast filtering. Then in order to overcome incomplete haze removing phenomenon in sky and some bright regions, special treatment of intensity difference to atmosphere light band control is introduced. Finally, color bias correction is also taken after image recovery for vivid vision perception.

The rest of the paper is organized as follows. Section 2 introduces our technology diagram of this work. Section 3 describes our principles and the details of the proposed approach step by step. Section 4 presents the abundant experiments and comparative analysis. Section 5 finally makes a short conclusion and gives future research perspective.

2   Technology Diagram

The mode widely used to describe the formation of a hazy image is also called atmosphere imaging equation:

$$I(x) = J(x)t(x) + A(1 - t(x)) \tag{1}$$

where $I(x)$ is the observed haze image, $J(x)$ is the scene irradiance (the haze free image to be acquired), $t(x)$ is the medium transmission describing the portion of the light that reaches the camera, when the atmosphere is homogenous, $t(x) = e^{-\beta d(x)}$. Here, β is the scattering coefficient of the atmosphere, and $d(x)$ is the scene depth, A is the overall atmosphere light. The goal of haze removal is to recover $J(x)$, A and $t(x)$ from haze image $I(x)$. If we define the atmospheric veil $V(x) = A(1 - t(x))$, then last equation can be simplified as $I(x) = J(x)t(x) + V(x)$. Then the clean haze free image $J(x)$ can be acquired as

$$J(x) = \frac{I(x) - V(x)}{t(x)} \tag{2}$$

In our work, instead of estimating $t(x)$ directly as He et al. did [2], we manage to acquire the atmosphere veil $V(x)$ firstly by filtering, then to acquire $t(x)$ according to

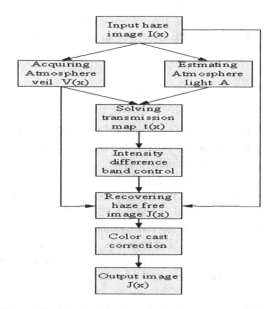

Fig. 1. The proposed image dehazing technology diagram

the definition of atmosphere veil. The overall technology diagram of our proposed approach is illustrated in the Fig. 1.

In order to compute atmosphere veil $V(x)$ fast, we can take a filtering strategy as [3] does. Here the desired $V(x)$ should meet the following two constraints: (1) the value $V(x)$ is positive ($V(x) \geq 0$) at each pixel; (2) the value of $V(x)$ is not higher than the minimum components of $I(x)$, $V(x) \leq M(x)$, where $M(x) = \min_{c \in \{r,g,b\}}(I^c(x))$. Under these two constraints, we then use Gauss filter for local image operation to yield desirable function $V(x)$.

Because Gauss filtering is a weighted smoothing operation in a local region, some small edge information in the initial acquired image $V(x)$ can be lost after two twice smoothing. As known, edges or the large jumping where rich depth information is contained in the atmosphere veil is important for fine image recovery. If the edge information is missing, the algorithm cannot detect the haze at these locations, which will result in incomplete removal. This defect naturally appeals to guided joint bilateral filtering which will be discussed in the following in detail.

3 Computation Acceleration for Bilateral Filtering Based Dehazing

3.1 Fast Atmosphere Light Estimation

How to estimate the overall atmosphere light A usually is a technical task. Many single image haze removal methods regard the highest intensity of haze image as the intensity of sky. In the work, we choose using the dark channel map [2] to estimate atmosphere light A. The method assumes that the intensity of the dark channel is a rough approximation of the thickness of the haze. For an arbitrary image I, its dark channel I_{dark} is given by

$$I_{dark}(x) = \min_{y \in \Omega(x)} \left(\min_{c \in \{r,g,b\}} (I_c(y)) \right) \tag{3}$$

where I_c is a color channel of I and $\Omega(x)$ is a local patch centered at x. A dark channel is the outcome of two minimum operators: $\min_{c \in \{r,g,b\}}$ is performed on three channels of each pixel, and $\min_{y \in \Omega(x)}$ is a minimum filter.

Actually in practical experiments, for larger size image, for example 1280*960, acquiring the dark channel I_{dark} will inevitably need huge time consuming (always more than 10 s configured with Intel i3 CPU and 2G memory). Here, we present two computation acceleration strategies for fast minimum filtering operation: (1) decomposing 2D window filter into two 1D filters, e.g. row and column minimum filters, respectively; (2) adopting a very fast stream minimum filtering algorithm [5] with O(1) time complexity which requiring no more than three comparisons per pixel in worst case.

For row or column 1D filtering, the stream minimum acquiring algorithm is described as follows.

```
Alogrithm 1: Fast stream minimum filter algorithm
1) Input: a[n] represents a row or a column of object im-
age to be filtered, indexed from 1 to n;
2) Input: B is the width size of filter; L is double end-
ed queue with initial value equaling to 1;
```
3) for i in $\{1, \cdots, n\}$ do
4)   if $i \geq B$ then
5)   Output: $a[front(L)]$ as minimum of range $[i-B,i)$
6)   if $a[i] < a[i-1]$ then
7) $pop(L)$ from back
8)   While $a[i] < a[back(L)]$ do
9) $pop(L)$ from back
10)   append i to L
11)   if $i = B + a[front(L)]$ then
12) $pop(L)$ from front

Taking above two acceleration strategies will improve about nine times filtering speed than that of original 2D window comparison based minimum acquiring.

3.2   Fast Filtering Based Atmosphere Veil Refinement

The goal of bilateral filter which was firstly proposed in [6] is to preserve edges and reduce noise for processed image. The Eq. (4) shows that utilizing bilateral filter to filter atmosphere veil $V(x)$ actually equals to do convolution of the function taking the weights of Gaussian on the difference between the values of neighboring pixels.

$$V^b(x) = \frac{\sum_{y \in \Omega(x)} f(x - y) \cdot g(V(x) - V(y)) \cdot V(x)}{\sum_{y \in \Omega(x)} f(x - y) \cdot g(V(x) - V(y))} \qquad (4)$$

where y is a pixel of local patch $\Omega(x)$ centered at x, and $f(x - y)$ is spatial factor (distance difference) and $g(V(x) - V(y))$ is range value factor (the difference of pixels' values) respectively, defined as $f(x - y) = e^{-\frac{\|x - y\|^2}{2\sigma_S^2}}$ and $g(V(x) - V(y)) = e^{-\frac{(V(x) - V(y))^2}{2\sigma_R^2}}$.

If we introduce the reference image $G(x)$, a difference measurement $h(V(y) - G(y))$ can be utilized to control deviation consideration in joint filtering, which could guide the result image corrected toward the reference image. Then this guided joint bilateral filtering can be defined as

$$V_G^b(x) = \frac{\sum_{y \in \Omega} f(x - y) \cdot g(G(x) - G(y)) \cdot h(V(y) - G(y)) \cdot V(x)}{\sum_{y \in \Omega} f(x - y) \cdot g(G(x) - G(y)) \cdot h(V(y) - G(y))} \qquad (5)$$

where $h(V(y) - G(y)) = e^{-\frac{(V(y) - G(y))^2}{2\sigma_T^2}}$.

And more worth considering, spatial and range domain computation of guided joint bilateral filters can be accelerated by piecewise linear approximation framework [7]. Denote the minimum and the maximum of $V(x)$ as $\min(V)$ and $\max(V)$ respectively, then $V(x) \in \{\min(V), \ldots, \max(V)\}$. And assume that the sampling number is N and the discretization intensity values set is $\{L_i\}, i \in \{0, \cdots, N\}$. In practice, we can take $N = (\max(V) - \min(v))/\sigma_R$. Then for each pixel y and each intensity value $k \in \{\min(V), \ldots, \max(V)\}$, the guided bilateral filtering can be approximated as piecewise linear filtering described as follows.

Algorithm 2: Piecewise linear bilateral filtering algorithm

1) **Input:** atmosphere veil $V(x)$, reference image $G(x)$, spatial kernel f_{σ_S}, range kernel g_{σ_R}, and difference measurement kernel h_{σ_T}

2) **Input:** array $D[N]$ represents the pre-filtered samples value, indexed from *1* to N with initialized value *0*;

3) for i in $\{0, \cdots, N\}$

4) $L_i = \min(V) + i \times (\max(V) - \min(V)) / N$

5) $W_i = g_{\sigma_R}(L_i - G(y)) \times h_{\sigma_T}(V(y) - G(y))$

6) $K_i = f_{\sigma_S} \otimes W_i$

7) $H_i = W_i \times V(y)$

8) $H_i^* = f_{\sigma_S} \otimes H_i$

9) $D[i] = H_i^* / K_i$ and record $D[i]$

10) For any $k = G(x)$, determine the order number j, such that $k \in [L_j, L_{j+1}]$

11) **Output:** D_k interpolated from $D[j]$ and $D[j+1]$ as the filtered result

In above bilateral filtering acceleration procedure, samples value array D can be calculated and stored in advance. The convolution with f_{σ_S} can be accelerated by Fast Fourier Transformation (FFT) in frequency domain with time complexity $O(n)$. And if using Deriche's recursive method [8] to approximate Gauss filter, we can acquire time complexity $O(1)$ but a little accuracy loss.

For further accelerate bilateral filtering, we note that all linear filtering operations except the final interpolation are low-pass filtering. Thus, we can safely use a down sampling version image in filtering procedure with little quality loss. However, the final interpolation must be performed using the full-scale image, otherwise edges would not be respected, resulting in visible artifacts. That means, we should do up sampling after the step of calculating samples filtered value $D[i]$. In practice, for big size image, we take down sampling and up sampling factor to be 2.

3.3   Image Recovery Based Haze Removal

Then the clean haze free image can be recovered with $J(x) = \frac{I(x) - V(x)}{t(x)}$. However, since the tiny transmission map $t(x)$ lies in dominator, its small varying will always lead the intensity of recovered sky region to be over too bright. Actually, the main reason is that the transmission map in sky region is acquired smaller than the real one. Thus, an intensity difference (to atmosphere light A) bind control factor B is introduced to make up inadequacy estimation of transmission map. Then, the haze free image recovery equation can be redefined as following. B can be adjusted from 80 to 100 according to real needs.

$$
J(x) = \begin{cases} \dfrac{I(x)-V(x)}{\min(\max(\frac{B}{|I(x)-A|},1)\cdot t(x),1)} & |I(X)-A|<B \\[3mm] \dfrac{I(x)-V(x)}{\tau(x)} & other \end{cases} \tag{6}
$$

In addition, in order to overcome color bias problem in some dense fog image recovering, while balance correction is also introduced after recovery. White balance approach is adopted to improve haze removal effect and get more natural visual perception. The basic processing is: (1) count the maximum 10 % pixels in three channels respectively; (2) calculate the average values of these pixels all in three channels and record them as avr_R, avr_B, avr_G; (3) for each pixel of recovery image, take the formula $J_{R,G,B}{}'(x) = J_{R,G,B}(x) \cdot \frac{p \cdot 255}{avg_{R,G,B}}$ to correct three channel intensity values, respectively. p is a coefficient belonging to $[0.9, 1.1]$.

4   Experimental Results and Analysis

In this section, we will show and discuss the results of proposed method in a variety of fog scene images. We will also compare with several state of art image dehazing methods both in recovery quality and the time consuming. Our operations are implemented in C++ on a Intel i3-Core CPU M370@2.40 GHz with 2 GB RAM.

In Fig. 2, it shows the intermediate process result of proposed approach compared with Tarel et al. [3]. By using guided joint bilateral filter to refine the atmosphere veil, we can find that the edge information is recovered effectively and some dirty texture details are removed either.

In Fig. 3, we compare with several state of the art dehazing methods, such as He et al. [2], Tarel [3], Xiao et al. [4]. We can find that in most case, our method can get adequate satisfactory dehazing results in most cases, even in distant scenes or places where the depth changes abruptly, as shown in the last column.

For the building image in Fig. 3 (second row first column), Table 1 shows the dehazing quality object evaluation comparison with He et al. [2], Tarel [3], Xiao et al. [4], and ours under fixed image size 640*480. We take objective image evaluation approaches - stand deviation, mean gradient and information entropy as practical dehazing quality measures. These three criteria reflect the real characteristics of image dehazing in three aspects - image contrast, texture clarity, and details richness. The larger every evaluation value, the better the dehazing image quality is. From the table, we can easily see that our approach gets the best quality measure.

Table 2 demonstrates the time consumption comparisons with He et al. [2], Tarel [3], and Xiao et al. [4] in different size images. We can find that the speed of our

Fig. 2. From top to down: (a) original haze images; (b) the derived transmission map; (c) the dehazed images

Fig. 3. From left to right: original foggy images, recovered images using He et al. [2], Tarel [3], Xiao et al. [4] and the proposed approach, respectively

method is far faster than that of [2] and [4]. Especially the bigger the size, the greater the speed advantage takes. Even the running performance of our approach is surprise much faster than that of Tarel [3] for larger size image. The reason is that though more

Table 1. Dehazing quality evaluation comparison

Objective evaluation	Original image	He et al. [6]	Tarel [3]	Xiao et al. [4]	The proposed
Stand deviation	35.1	54.6	48.7	56.3	61.8
Mean gradient	7.5	10.1	8.6	10.7	11.4
Entropy	11.3	16.7	13.9	17.2	18.9

Table 2. Time consumption comparison with He et al. [2], Tarel [3], Xiao et al. [4] (s: seconds)

Image size	He et al. [2]	Tarel [3]	Xiao et al. [4]	The proposed
640*480	8.343 s	1.112 s	1.866 s	0.783 s
1024*768	26.71 s	2.148 s	4.654 s	1.287 s
1280*960	147.8 s	4.462 s	7.479 s	2.030 s
1920*1080	313.4 s	5.189 s	12.13 s	2.845 s

Fig. 4. Different types foggy images removal. (a) dense fog removal; (b) rain fog removal; (c) uneven fog removal; (d) aerial fog removal

filtering operations are taken, a serial of acceleration strategies are also exerted effectively while Tarel's method do not take any one.

Foggy images of more types are recovered and illustrated in Fig. 4. The first row is dealing with dense fog. We can see that after recovery with color correction, the haze free image is close to natural one. The Second row is uneven fog recovery. We can find that most fuzzy parts in original images are restored clear but the parts in far corner are missed. All experimental results show the proposed method is effective and can be handled with practical different foggy scene.

5 Conclusion

Towards practical application motivation, in this work, we proposed a set of computation acceleration approaches for bilateral filtering based haze removal. We firstly present fast minimum filtering method to acquire dark channel map for atmosphere light estimation. Then we accelerate guided joint bilateral filtering to refine atmosphere veil for fast and accurate image recovery. And to restrain over recovery phenomenon in sky region and correct color bias, intensity difference band control and white balance processing are exerted in final haze free image. Diverse kinds of foggy image recovery experiments and comparisons have show the proposed approach can not only achieve better haze removal effects, but also have the potential to satisfy large size scene fast dehazing requirements.

Future work focuses on transplanting the proposed approach to smart TV embedded platform real time video applications.

Acknowledgements. This work is partly supported by National Natural Science Foundation of China (Grant No.61105020), by Foundation for Key Program of Education Bureau of Anhui Province (2015), and The Natural Science Foundation of Anhui Province (Grant No. 1308085Qf113).

References

1. Namer, E., Schechner, Y.Y.: Advanced visibility improvement based on polarization filtered images. In: Proceedings of the Polarization Science and Remote Sensing II, pp. 36–45. SPIE, San Diego (2005)
2. He, K., Sun, J., Tang, X.: Single image haze removal using dark channel prior. IEEE Trans. PAMI **33**(12), 1956–1963 (2011)
3. Tarel, J., Hautiere, N.: Fast visibility restoration from a single color or gray level image. In: ICCV 2009, pp. 2201–2208. IEEE Press, New York (2009)
4. Xiao, C., Gan, J.: Fast image dehazing using guided joint bilateral filter. Visual Comput. **28**, 713–721 (2012)
5. Daniel, L.: Streaming maximum-minimum filter using no more than three per element. arxXiv:cs/0610046v5 (22 March 2007)
6. Tomasi, C., Manduchi, R.: Bilateral filtering for gray and color images. In: ICCV 1998, pp. 839–846. IEEE Press, New York (1998)
7. Yang, Q., Tan, K., Ahuja, N.: Real-time O (1) bilateral filtering. In: CVPR 2009, pp. 557–564. IEEE Press, New York (2009)
8. Deriche, R.: Recursively implementing the Gaussian and its derivatives. Research report 1893, INRIA (1993)

Fast Image Filter Based on Adaptive-Weight and Joint-Histogram Algorithm

Zhenhua Wang[(✉)], Fuyuan Hu, Shaohui Si, Yajun Gu, Ze Li,
and Zhengtian Wu

School of Electronic and Information Engineering,
Suzhou University of Science and Technology, Suzhou, China
875889022@qq.com

Abstract. Adaptive-weight operators are ubiquitous in numerous computer vision applications. The structure of general adaptive-weight models, however, are hard to accelerate with high speed to large or complex images. In this paper, the proposed adaptive-weight image filter algorithm is mainly on a new joint-histogram representation, median value searching, and a new data structure that contributes to fast data access. The effectiveness of these schemes is demonstrated on estimation of median position, which not only better preserves edges, but also reduces computation complexity from $O(mnr^2)$ to $O(mnr)$ using histogram, where $m * n$ and r denote image size and radius of the mask window respectively. The results of our experiments demonstrate that our approach is effective to image filtering and image enhancement.

Keywords: Adaptive-weight · Stereo matching · Joint-histogram · Median filtering

1 Introduction

Image filter is one of the most important tools in image processing and is widely used in computer vision. Adaptive-weight operator distributes the different weight for current element to analyze the different environment and its effect on image filter has attracted much attention.

Efficient edge-preserving stereo matching [1] exploits a novel paradigm, namely separable successive weighted summation (SWS) among horizontal and vertical directions with constant operational complexity, providing effective connected 2D support regions based on local color similarities. The intensity adaptive aggregation enables crisp disparity maps which preserve object boundaries and depth discontinuities. In [2], an adaptive-weighted bilateral filtering is proposed to offer the desirable property of preserving texture within the OCT images and this method performs well in removing speckles. Zhang et al. [3] estimates a single latent sharp image given multiple blurry or noisy observations using an adaptive Bayesian-inspired penalty function which couples the unknown latent image, blur kernels, and noise levels together in a unique way. Despite of many advantages of the adaptive-weight in image filter, it has some inherent drawbacks. Algorithms proposed above spend much time in calculating adaptive-weight due to complex data structures. To address these problems, we propose

© Springer International Publishing Switzerland 2015
X. He et al. (Eds.): IScIDE 2015, Part I, LNCS 9242, pp. 551–563, 2015.
DOI: 10.1007/978-3-319-23989-7_56

a fast adaptive-weight image filter using histogram. We replace the average value used to calculate the adaptive irregular region in Cross-based local multipoint filtering (CBLMF) [4] with the median value to preserve the image edges. The median value can be calculated via histogram which reduces the computation complexity from $O(mnr^2)$ to $O(mnr)$. In order to improve the performance of image filtering, we combine gray levels based on both geometric closeness and color similarity.

In following sections, we discuss related studies and our proposed method. We also demonstrate that our method works more effectively.

2   Technical Background

WMF [5] the core idea of weighted median filter (WMF) is to replace the current pixel with the weighted median value of neighboring pixels, and the median value is given by:

$$p* = \min k \tag{1}$$

$$\text{s.t.} \quad \sum_{q=1}^{k} w_{pq} \geq \frac{1}{2} \sum_{q=1}^{n} w_{pq}$$

where $q \in R(p)$, $R(p)$ denotes the local window radius r of center pixel p, w_{pq} is a typical influence function between neighboring pixels, which can be in Gaussian $\exp\{-\|f(p) - f(q)\|\}$ or other forms. Our algorithm sets the median value of neighboring pixels to be the threshold of our adaptive-weight structure according to the histogram. Therefore, we can compute the median value by:

$$p* = \min k \tag{2}$$

$$\text{s.t.} \quad \sum_{q=1}^{k} H[q] \geq \frac{1}{2} \sum_{q=1}^{255} H[q]$$

where $H[q]$ counts the number of gray value in histogram.

CBLMF [4] this method estimates an upright cross region with average value which helps to capture the local image structures adaptively, the filtering process of this method can be simply divided into two steps as shown in Fig. 1:

(1) Multi-point estimation. Calculating the estimation Y_s^p for a series of points $s \in U(p)$ within a locally adaptive region of center pixel p, where $U(p)$ describes an arbitrarily-shaped, connected local support region for anchor pixel p as shown in Fig. 1(a).

(2) Aggregation. Fusing all the estimates $\{Y_p^k\}$ derived from each k-centered local region ($k \in U(p)$) to reckon the final estimation of Y_p for each pixel p, which was described in Fig. 1(b). For example, $U(s)$ in Fig. 1(b) denotes the s-centered local region $s \in U(p)$.

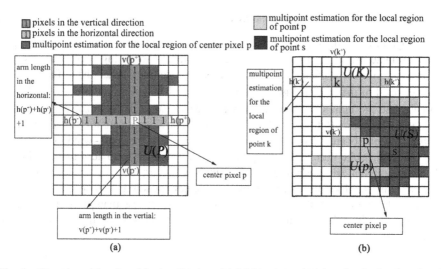

Fig. 1. Cross-based local multipoint filtering. (a) Multipoint estimation for the local region U (p) of center pixel p. (b) Aggregation of multiple estimation by each $k \in U(p)$

3 Our Framework

Our method is different from previous CBLMF and other filters. Here, we present three important steps to illustrate our algorithm (Fig. 2).

(1) Searching median value using histogram. We use histogram for not only storing pixel count but also calculating median value quickly.
(2) Estimating the adaptive irregular region with improved adaptive-weight of CBLMF. We improve the cross model in CBLMF replacing the average value with median value, and the improved model is used to identify local non-regular self-similar support regions in this paper.
(3) Filtering the pixels in terms of geometric closeness and color similarity. We combine the closeness function and similarity function to improve the quality of input images.

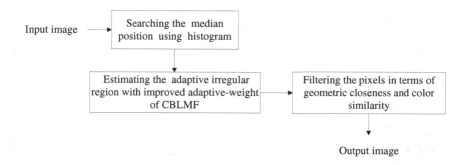

Fig. 2. Block diagram of the proposed method

3.1   Searching Median Using Histogram

The histogram is a function of grayscale, its abscissa means the grayscale of an image, and its ordinate means the frequency of each grey level. According to the features, we estimate the position of median pixel quickly using histogram. When the ordinate value of center pixel in the input image is $\frac{r+1}{2}$, $p*$ can be calculated according to Eq. (2), we also calculate the number of the pixels which are equal to or less than the $p*$ in the whole mask window. Before we shift the mask window we delete the first column so that the new pixels are all in the new column if we shift the mask window. Accordingly, we reduce the number of the pixels which are equal to or less than $p*$ in the deleted column and add the number of the pixels which are equal to or less than $p*$ in the new column. The displacement of the median position can be estimated via comparing the number of the pixels which are equal to or less than $p*$ in the updated mask window with half of the mask window size. The details of our method are shown in Algorithm 1.

Algorithm 1 Proposed method using histogram search median position

Input: filtering input image p, image size $m*n$, mask window size $r*r$, the number of the gray value k in the histogram $H[k]$, the number of the pixels which are equal to or less than the median value p^{*} in the whole mask window s_1, the number of the pixels which are equal to or less than the median value at the first column s_2, the number of the pixels which are equal to or less than the median value at the new column s_3

Output: median value p^{*}

1:Initialization

$$\begin{cases} s_1 = s_1 - s_2 & P_x = m - \dfrac{r+1}{2} \\[2mm] s_1 = s_1 + s_3 & P_x > m - \dfrac{r+1}{2} \end{cases}$$

2: **while** $s_1 > \dfrac{r^2}{2} + 1$ **do**

   $s_1 = s_1 - H[p^{*}]$

   $p^{*} = p^{*} - 1$

  end while

  $p^{*} = p^{*} + 1$

3: **while** $s_1 \leq \dfrac{r^2}{2} + 1$ **do**

   $p^{*} = p^{*} + 1$

   $s_1 = s_1 + H[p^{*}]$

  end while

As shown in Fig. 3, we calculate $p*$ at the beginning and calculate the number of pixels which are equal to or less than $p*$ in the updated mask window according to step 1 and step 2 in Algorithm 1. Then we compare s_1 with half of the mask window size. In fact, Fig. 3 represents the case that the number of the pixels which are equal to or less than $p*$ hasn't up to half the size of updated mask window. Therefore, we calculate the new median value according to step 4. The proposed method has an explicitly advantage that the scheme we consider the first column and the new column after we shifted the mask window can be finished in complexity $O(r)$, and $p*$ can be calculated in complexity $O(r) * O(1)$. Although there are some searching calculating, its average time is limited to a small range because of high degree similarity between neighboring two pixels.

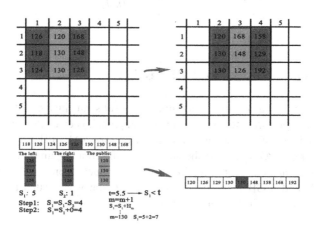

Fig. 3. Searching median value using histogram

3.2   Improving Adaptive-Weight of CBLMF and Filtering

To better preserve edges during filtering and reduce the time spent in calculating adaptive-weight, we improve the adaptive-weight of CBLMF replacing the average value in CBLMF with median value calculated in Sect. 3.1. Our motivation originates from bilateral filtering [6] which enforces the policy that applied to geometric closeness and color similarity. In terms of the geometric closeness:

$$c_{ij}(k,l) = exp(-\frac{(i-k)^2 + (j-l)^2}{2\partial_c^2})\qquad(3)$$

where (i, j) denotes the location of center pixel, and (k, l) denotes the current pixel in the irregular region.

In terms of the color similarity:

$$s_{ij}(k,l) = exp(-\frac{\|p^* - f(k,l)\|^2}{2\partial_s^2}) \qquad (4)$$

where $f(k, l)$ denotes the gray value of the pixel in adaptive irregular region at center pixel p, we construct weight function according to (3) and (4):

$$w_{ij}(k,l) = c_{ij}(k,l) * s_{ij}(k,l) \qquad (5)$$

Then the output can be expressed by

$$g(i,j) = \frac{\sum\limits_{(k,l) \in U(p)} f(k,l)w_{ij}(k,l)}{\sum\limits_{(k,l) \in U(p)} w_{ij}(k,l)}. \qquad (6)$$

4 Experimental Results and Discussion

In this part, we discuss some experimental results in image filter and image enhancement. The performance of the proposed approach is evaluated by the measures of average gradient (AG), average information entropy (AE) [7], and average peak signal to noise ratio (APSNR). At the same time, the running time of different algorithms is also compared. Our experiments are conducted on a PC with an Inter i5-4200M CPU and 8 GB memory, our implementation was written in C#. The following evaluation is on 500 + natural images.

4.1 Image Filter

The proposed algorithm is compared with other popular enhancement algorithms like bilateral filter, CBLMF, and 100 + WMF [8] in Fig. 4 to validate its effectiveness in image filter. We add a Gaussian noise with variance 30 to input images, and the results are shown in Fig. 4.

As can be observed, some hair-like gradient reversals appear in edge areas based on bilateral filter due to the instability of Gaussian weighted. In addition, some image artifacts are produced by isolated pixels, such as the black region at the bottom of windowsill in Fig. 4(b5) and the white line on the sweater in Fig. 4(b3). For the CBLMF, it is observed that local smooth areas are not clear such as the spots on the leopard in Fig. 4(c1) and cells under the microscope in Fig. 4(c4), because the output images of CBLMF are mean values of a linear function, which may destroy image texture and details. While 100 + WMF clearly highlights the edge information of high-frequency areas, it ignores the textures which in flat areas such as the white lines of triangle in Fig. 4(d2), and the noise pixels are not significantly reduced based on this method in Fig. 4(d4). Comparing with the methods above, our approach can take full

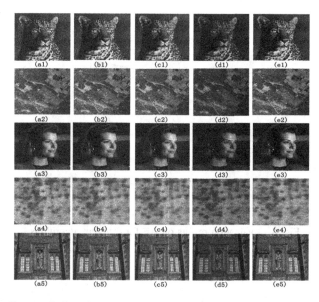

Fig. 4. Image filter results by other methods and ours. (a) Input gray-scale images. (b) Results of Bilateral filter. (c) Results of CBLMF. (d) Results of 100 + WMF. (e) Our results

consideration of local texture varieties while filtering the noise pixels, and has generally improved the quality of input images.

Figure 5 shows the enlarged local texture regions highlighted in red in Fig. 4. It can be observed that the noise pixels in smooth areas such as the leopard's ear in Fig. 5(1), woman's clothes in Fig. 5(3) and building's window in Fig. 5(4) are less than others.

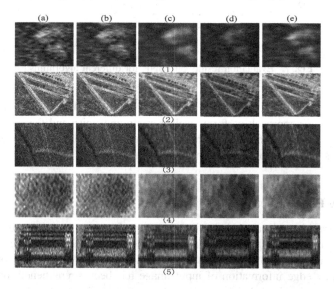

Fig. 5. Details of Image filter results by other methods and ours. (a) Input gray-scale images. (b) Results of Bilateral filter. (c) Results of CBLMF. (d) Results of 100 + WMF. (e) Our results

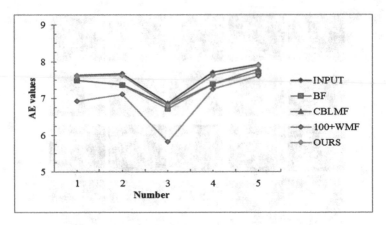

Fig. 6. Evaluation results on AE for different algorithms

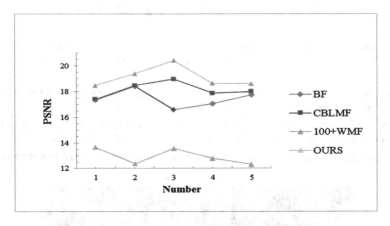

Fig. 7. Evaluation results on PSNR for different algorithms

The same conclusion can also be obtained from experimental data. As shown in Figs. 6 and 7, our approach performs well in removing noise and the AE values and PSNR values of our algorithm are both the maximum which also proves that our algorithm has better performance than others in image filter.

4.2   Image Enhancement

In order to provide a visual evidence in enhance performance based on our algorithm, we give the enhancement results in Fig. 8. The vertical projections of the image gray level values are also shown in Fig. 8(3) and (4). It is obviously that not only the high-frequency edge information of input image has been strengthened, but also the details in flat region have been stretched. As shown in Fig. 8(3) and (4), additional peaks are added to the projection envelope of the enhanced image which proves that

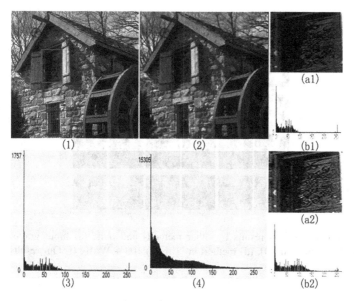

Fig. 8. Comparison between input image and enhanced image based on our operator: (1) Original. (2) Our result. (3) Vertical projection of input image. (4) Vertical projection of our result. (a1) Texture of image (1). (a2) Texture of image (2). (b1) Vertical projection of (a1). (b2) Vertical projection of (a2).

our method has achieved nonlinear enhancement. Moreover, the color and luster of waterwheel in Fig. 8(2) are more brighter than which in Fig. 8(1), which illustrates that our output image has a better contrast, and it is possible to notice the pattern of the window is more clearly based on our method in Fig. 8(a2).

We present a filter for image enhancement in Fig. 9. Although image enhancement methods [9, 10] perform well in the high frequency regions, their performances are instability when they are applied to smooth regions. Rich texture in smooth regions is basically removed, such as the blurry wrinkles of the hat in Fig. 9(c2), and the gray values of the corresponding pixels appear dark, such as the black buildings in aerial photograph in Fig. 9(d1). What's more, it's obviously that algorithm proposed in [9] excessively bright enhanced images in Fig. 9(c), and algorithm proposed in [10] almost has lost the background information in Fig. 9(d). Even worse, many noise pixels were added to original one. Both the 100 + WMF and CBLMF have some problems in overexposure, many wide white edges appear in Fig. 9(e3) and the words on the label are significant distortion as can be seen in Fig. 9(b3). After we evaluate the adaptive irregular region, we calculate the output image in terms of geometric closeness and color similarity, which helps to preserve image edges. Therefore, our results have better effects in enhancement of the contrast and texture.

It can be observed from the enlarged local textures of experiment's images in Fig. 10 that the details around the flowers in Fig. 10(f2) or the letters on the wing in Fig. 10(f4) are more conspicuous than those of other methods. In addition, the overall

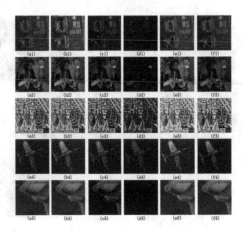

Fig. 9. Image enhancement results by other methods and ours. (a) Input gray-scale image, (b) CBLMF, (c) method in [9], (d) method in [10], (e) 100 + WMF. (f) Our results.

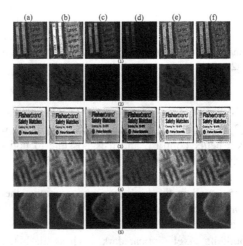

Fig. 10. Details of Image enhancement results by other methods and ours. (a) Input gray-scale images. (b) Results of CBLMF. (c) Results of method in [9]. (d) Results of method in [10]. (e) Results of 100 + WMF. (f) Our results.

intensity of the enhanced images are neither too high nor too low based on our method. Table 1 summarizes the evaluation results by the way of AG and AE for the different methods. It is obvious that our approach has higher values of AG and AE than other methods, which demonstrates the advantages of our method in enhancing texture details and preserving texture information.

Table 1. AE values and AG values with respect to different algorithm.

	1		2		3		4		5	
	AG	AE	AG	AE	AG	AE	AG	AE	AG	AE
INPUT	6.0131	0.0118	6.7575	0.0092	7.0973	0.0617	4.7578	0.0046	5.6139	0.0073
CBLMF	6.9341	0.0221	6.8099	0.0137	7.8434	0.0845	5.4980	0.0088	5.7461	0.0097
Method in [9]	4.5844	0.0062	5.5811	0.0061	7.6483	0.0707	4.9947	0.0063	6.0720	0.0015
Method in [10]	1.5795	0.0127	2.0380	0.0094	6.9221	0.0569	4.7095	0.0069	4.6905	0.0062
100 + WMF	6.6372	0.0092	7.0973	0.0158	7.6241	0.0926	6.6372	0.0092	5.9801	0.0149
OURS	7.1024	0.0304	7.1034	0.1209	7.9816	0.0985	6.9623	0.0106	6.2179	0.0125

4.3   Calculation Time

The typical solution of searching the median position is to sort all data in complexity $O(r^2)$, and then find the median value in complexity $O(1)$. Therefore, the time complexity of whole work for median filter is $O(mnr^2)$. However, our method can finish this work in complexity $O(mnr)$. As illustrated in Sect. 4.1, we seek the histogram and calculate the number of the pixels in the last column whose intensity is equal to or less than the median value to judge the direction we need to move the former median pixel with $O(r)$ complexity, and then find the new median position in complexity $O(1)$. Table 2 presents the computational time cost by median filter and our algorithm in calculating median values. The median filter spends 134 s when the window size is 31 * 31 pixels and image size is 256 * 256 pixels, and our method only spends 28.90 s which achieves about 78.57 % faster than the median filter at the same condition.

Table 2. Execution time for median filter and our method

	Mask window size								
	11 * 11			21 * 21			31 * 31		
Image size	Median filter	Ours	Save time	Median filter	WMF	Save time	Median filter	Ours	Save time
128 * 128	1.50 s	0.97 s	35.33 %	9.42 s	3.20 s	66.02 %	29.91 s	6.35 s	78.76 %
256 * 256	5.97 s	3.70 s	38.02 %	34.47 s	12.94 s	62.46 %	134..91 s	28.90 s	78.57 %

We also compare the calculation time of our algorithm with others. As shown in Table 3, our algorithm is faster than others for the experiments in Sect. 4.1. Bilateral filter and CBLMF are finished in complexity $O(mnr^2)$. 100 + WMF spends much time in seeking histogram during the process of calculating median value despite it uses joint-histogram. Our method spends less time than Guided filter (GF) [11] which because our algorithm can achieve the median value with complexity $O(r)$ during the process of image filter.

Table 3. Execution time for different algorithms

Algorithms	Image 1	Image 2	Image 3	Image 4	Image 5
BF	6.1380 s	7.5723 s	7.6138 s	4.0398 s	8.1523 s
CBLMF	7.6195 s	7.3804 s	7.8695 s	4.6679 s	8.7524 s
100 + WMF	6.0168 s	6.3596 s	7.0413 s	3.9852 s	7.6012 s
GF	6.0021 s	5.9764 s	5.9687 s	3.6145 s	6.9430 s
OURS	5.9324 s	5.6836 s	5.4827 s	3.4406 s	6.9132 s

5 Conclusion

In this work, a new algorithm has been presented for image filter. The method uses an improved cross-based Local multipoint filtering to better preserve edges during filtering. We also present a joint-histogram to search the median position that improves the speed of searching median values. The future work would consider how to introduce adaptive threshold.

Acknowledgement. This research is supported by the National Natural Science Foundation of China under Grant No. 61472267 and No. 61203048, Nature Foundation of Jiangsu Province under Grant No. BK2012166, the Open Foundation of Modern Enterprise Information Application Supporting Software Engineering Technology R&D Center of Jiangsu Province under Grant No. SK201206, and the Innovation Project of Graduate Student Training under Grant No. CXZZ13_0854.

References

1. Cigla, C., Alatan, A.A.: Efficient edge-preserving stereo matching. In: International Conference on Computer Vision Workshops, pp. 696–699 (2011)
2. Anantrasirichai, N., Nicholson, L., Morgan, J.E., et al.: Adaptive-weighted bilateral filtering for optical coherence tomography. In: International Conference on Image Processing, pp. 1110–1114 (2013)
3. Zhang, H., Wipf, D., Zhang, Y.: Multi-image blind deblurring using a coupled adaptive sparse prior. In: IEEE Conference on Computer Vision and Pattern Recognition, pp. 1051–1058 (2013)
4. Lu, J.B., Shi, K.Y., Min, D.B., et al.: Cross-based local multipoint filtering. In: IEEE Conference on Computer Vision and Pattern Recognition, pp. 430–437 (2012)
5. Ma, Z., He, K., Wei, Y., et al.: Constant time weighted median filtering for stereo matching and beyond. In: IEEE International Conference on Computer Vision, pp. 49–56 (2013)
6. Tomas, C., Madnuchi, R.: Bilateral filtering for gray and color images. In: International Conference on Computer Vision, pp. 839–846 (1998)
7. Hu, F.Y., Si, S.H., San Wong, H., et al.: An adaptive approach for texture enhancement based on a fractional differential operator with non-integer step and order. Neurocomputing **158**, 295–306 (2015)
8. Zhang, Q., Xu, L., Jia, J.Y.: 100 + times faster weighted median filter. In: IEEE Conference on Computer Vision and Pattern Recognition (2014)

9. Xiang, Z.J., Ramadge, P.J.: Edge-preserving image regularization based on morphological wavelets and dyadic trees. IEEE Trans. Image Process. **21**(4), 1548–1560 (2012)
10. Tanaka, G., Suetake, N., Uchino, E.: Image enhancement based on nonlinear smoothing and sharpening for noisy images. J. Adv. Comput. Intell. Intell. Inform. **14**(2), 200–207 (2010)
11. He, K.M., Sun, J., Tang, X.O.: Guided image filtering. IEEE Trans. Pattern Anal. Mach. Intell. **35**(6), 1397–1409 (2013)

2.5D Facial Attractiveness Computation Based on Data-Driven Geometric Ratios

Shu Liu[1,2(✉)], Yangyu Fan[1], Zhe Guo[1], and Ashok Samal[2]

[1] School of Electronics and Information,
Northwestern Polytechnical University,
Xi'an, Shannxi, China
liushu0922@mail.nwpu.edu.cn,
{fan_yangyu,guozhe}@nwpu.edu.cn
[2] Department of Computer Science and Engineering,
University of Nebraska-Lincoln,
Lincoln, NE, USA
samal@cse.unl.edu

Abstract. Computational approaches to investigating face attractiveness have become an emerging topic in facial analysis research. Integrating techniques from image analysis, pattern recognition and machine learning, this subarea aims to explore the nature, components and impacts of facial attractiveness and to develop computational algorithms to analyze the attractiveness of a face. In this paper we develop an attractiveness computation model for both frontal and profile images (2.5D). We focus on the role of geometric ratios in the determination of facial attractivenss. Stepwise regression is used as the feature selection method to select the discriminatory variables from a huge set of data-driven ratios. Decision tree is then used to generate an automated classifier for both frontal and profile computation models. The BJUT-3D Face Database is pre-processed and tested as our experimental dataset. The low statistic errors and high correlation indicate the accuracy of our computation models.

Keywords: Facial attractiveness computation · 2.5D · BJUT-3D · Face ratios · Data-driven

1 Introduction

Beauty is a complex but universal aspect of human experience. Beautiful faces can delight the eyes, provoke pleasure, and gain more attention than ordinary ones [1]. The nature and components of facial attractiveness have been extensively studied by researchers since the time of Aristotle. It is important to note that while humans decode the attractiveness of a face effortlessly and instantaneously, it has been difficult to develop methods to determine it. With the widespread use of computer science and technology, there is a great interest in analyzing facial attractiveness objectively and accurately from a computational view. Research in this area aims to develop automated methods to reliably quantify the attractiveness of a face.

© Springer International Publishing Switzerland 2015
X. He et al. (Eds.): IScIDE 2015, Part I, LNCS 9242, pp. 564–573, 2015.
DOI: 10.1007/978-3-319-23989-7_57

Even though the research in face attractiveness is relatively recent, there has been considerable progress. Almost all the studies have focused on 2D face images, where almost of them are designed for 2D frontal faces. A number of geometric [2–8], appearance [3, 9], holistic features [10–12], and a combination of them [5, 9, 13] are extracted to determine their effect on frontal facial attractiveness. Geometric features, usually defined by putative beauty rules, are the most frequently used features in face attractiveness research. Most of the beauty rules are described as some kinds of ideal ratios, such as golden ratios, neoclassical canons, and symmetry indicators. While a set of face landmarks induces a large number of ratios, researchers generally choose some specific ratios based on the putative rules to find if they can fit their ideal value [2, 3, 7, 8]. Since the performance of the existing ratios used in the current literature is reported subpar in the general case [2, 8], this kind of rule-driven ratios may not be able to perform a systematic selection of best choice of ratios.

However, frontal images cannot accurately describe some prominent facial regions (e.g., forehead, nose, chin) as well as their anthropometric landmarks (e.g., glabella, nasion, subnasale, pogonion, etc.). Research shows that in addition to the frontal view, facial attractiveness is determined by the profile view as well as their combination [14]. Some studies [15–17] have been reported to choose face profiles as the experimental data in attractiveness determination.

The aim of this work is to get a more comprehensive understanding of face attractiveness, therefore, we develop an attractiveness computation model for both frontal and profile images (2.5D). We only focus on one of the geometric features, i.e., geometric ratio, and analyze its role in attractiveness determination. Unlike the previous rule-driven work, we conduct a data-driven method on facial ratios, i.e., a two-way ratio vector of all the possible ratios derived from the distances between all the landmarks. To our knowledge, this is the first time to deal with such a huge dimension of facial features in face attractiveness study. Stepwise regression is used as the feature selection method to select the most discriminatory ratios, and then decision tree is used to train a 2.5D prediction model.

2 Methods

2.1 Framework

Facial attractiveness computation is aimed at developing automated approaches to measure facial attractiveness from face images and their features, and revealing the quantitative relationship between attractiveness and facial features. The approaches in facial attractiveness computation literature share a similar framework (shown in Fig. 1). The computation of facial attractiveness typically involves six steps: face database acquisition, pre-processing, attractiveness score rating, feature extraction and selection, computation model development, and model evaluation. Here we describe the methods of feature extraction, selection, and model development used in this paper. The database pre-processing, attractiveness score rating, and model evaluation will be described in Sect. 3.

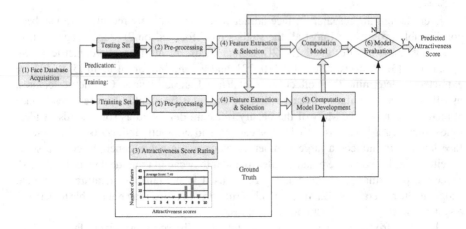

Fig. 1. The research framework of facial attractiveness computation.

2.2   Feature Extraction

Attractiveness features used to develop a predictive model can be geometric, shape, and texture based that may be either at local or holistic scale. Geometric features, derived from locations of the extracted landmarks, distances between these points, and ratios of these distances, are the most frequently used features in face attractiveness research. Here we choose 82 landmarks from frontal images and 40 from left-side images. We only focus on one of the geometric features, i.e., ratios derived from the distances between the landmarks, and analyze the relationship between facial ratios and face attractiveness.

In this work, we use Asmlibrary developed by Wei [18] based on active shape model (ASM) [19] to automatically extracted 82 landmarks from each frontal image and 40 landmarks from each left-side image (shown in Fig. 2). These landmarks are selected based on the existing literature [3, 20] and Chinese beauty canons (e.g., The Facial Fifths and Thirds). Manual adjustment may be applied when the locations of landmarks are far from accurate. After that, we use Procrustes analysis [21] to normalize each face into the pre-shaped space.

Unlike the previous work which only used specific face ratios defined by putative beauty rules [2, 3, 7, 8], we use a set of data-driven ratios, i.e., a two-way ratio vector of all the possible ratios derived from the distances between all the landmarks.

For 82 frontal landmarks, the number of distances that can be derived from them is 3321 (C_{82}^2, where C_n^i is the combination function), and the number of ratios that can be derived from these distances is 5512860 (C_{3321}^2). While the distance between landmark A and B is exactly the same with the distance between landmark B and A, the order of two distances in each ratio does matter. The ratio of distance C to D may have different effect from ratio of distance D to C on face attractiveness. Therefore, a two-way ratio vector of all 11025720 ratios is created. Similarly, a two-way ratio vector of all 607620 ratios is derived from 40 profile landmarks.

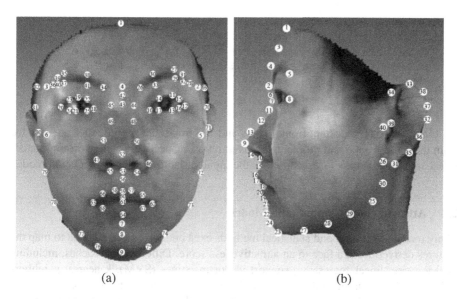

(a)	(b)

Fig. 2. Facial landmarks on a (a) frontal image; (b) left-side image.

2.3 Feature Selection

To our knowledge, this is the first time to deal with such a high dimension of facial features in face attractiveness study. While the dimension of these data-driven face ratios is really huge, the curse of dimensionality may arise. It is not surprising that most of the ratios are unlikely to be useful for attractiveness computation and many of them are likely to be correlated. Therefore, feature selection methods have to be used to determine the most discriminatory ratio features.

In this paper, we choose the stepwise regression [22] as the feature selection method to select the most discriminatory ones from the original set of data-driven ratios. Briefly, stepwise regression is an automatic procedure for statistical model selection where no initial variables in the model. At each step, with a new variable added, a selection criterion is used to check whether there are some variables that are already in the model can be deleted. The procedure will be stopped when adding a new variable into the model and removing the existing variables from the model cannot meet the selection criterion, or the number of existing variables in the model has reached the specified maximum number.

Here we use the 'hpreg' procedure in SAS [23] to run stepwise regression method. We specify the significance level as the selection criterion in the 'selection' statement, where the largest significance for a variable entering into the model is 0.25 and the smallest significance for a variable staying in the model is 0.25. The maximum number of variables in the model is set to 180 (the number of dataset samples, see Sect. 3.1). The core code of 'hpreg' procedure in SAS is as follows:

```
Proc hpreg data=Mydata;

    model attractivenessscore = all ratio variables;
    selection method = stepwise (select=sl sle=0.25
    sls=0.25 maxeffects=180);

run;
```

Finally, 11651 important frontal ratios and 797 important profile ratios are selected from the set of frontal and profile data-driven ratios, respectively. These discriminatory ratios are attractiveness features used as the input variables for the computation model.

2.4   Attractiveness Computation Model

Many machine learning and statistical methods exist and have been adapted to map the features derived from a face to an attractiveness score. Different approaches, including decision tree, linear regression, support vector machines (SVM), k nearest neighbors (KNN), and artificial neural networks (ANN), can be used for this purpose.

Here we use the decision tree to generate both frontal and profile computation model. Decision tree is a supervised symbolic classifier, requiring a set of pre-classified samples. Here the face samples are assigned the attractiveness scores (see Sect. 3.2), which are pre-classified labels. The important face ratios selected in Sect. 2.3 are a set of attributes characterizing faces and used as the input variables for the decision tree.

3   Experiments

3.1   Experiment Dataset

In this paper, we use the BJUT-3D Face Database [24] as our experimental database. It contains 250 Chinese Male and 250 Chinese Female 3D face models. The faces have a range of age from 16 to 49, with almost frontal view and neutral expression.

All the face models are stored in a binary file format. It can only be viewed by a specifically designed software named FDViewer and be transfered into a txt file (including vertex coordinates, texture value of vertices, and vertex indices in triangles of 3D faces), making it difficult to further process the face data. Therefore, we adopt the method in [25] to unwrap 3D face models into 2D texture images, and then convert the binary file format into one of the commonly used 3D representing formats, i.e., wrl file format.

To build our dataset for facial attractiveness computation, we have selected the good quality face models from the BJUT-3D database. Here we only choose female models for experiments. We manually remove the female faces with the expressions, bad details of eye and ear area, and terrible acne. The result is 180 clear neutral expression female faces. For 2.5D facial attractiveness computation, the frontal and left-side views of the selected faces are obtained. The final dataset is two sets of 180 images, shown in Fig. 3.

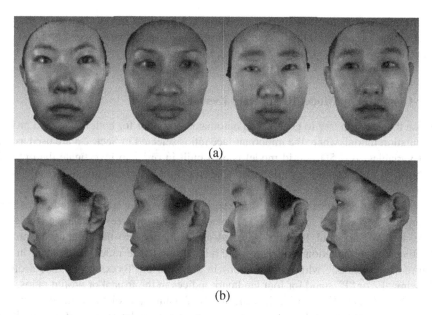

(a)

(b)

Fig. 3. Sample images from our dataset. (a) Frontal view samples; (b) left-side view samples.

3.2 Attractiveness Score Rating

One of the challenges in face attractiveness computation is the lack of universally accepted ground truth. The typical approach is to employ human raters to derive a proxy for the ground truth.

Using a partially balanced incomplete block design described in [2], we divide the 180 female images (including frontal and left-side view) into 4 blocks of 45 images. Each block has 23 duplicate images in order to check the consistency within each rater. Each rater is asked to assign attractiveness scores to two of these blocks, i.e., 136 images in total.

We recruit 48 raters (24 males and 24 females) from students and employees at the University of Nebraska-Lincoln, and people from Lincoln Nebraska. The age range of raters is 19 to 76. Each rater rate 136 face images in 10-point scale (1 is the least attractive, 10 is the most attractive) based on a specially designed rating interface.

After we get all the rating data, we analyze the rating distribution in order to verify that the statistical results can adequately represent the "collective attractiveness ratings". For each rater, a one-way analysis of variance (ANOVA) model is built to check the score variability between different images (measured by mean square of treatment (MST)) and the score variability within duplicate images (measured by mean square error (MSE)). Normally, when MST is more than two times larger than MSE, we can assume that the scores given by this rater are fairly reasonable. For those "bad" raters, we remove the scores they give and then recruit new raters to replace them. When the analysis of ratings is done, the average score for each image is treated as the ground truth for the development of the attractiveness computation model. In our results, the attractiveness scores of all images range from 3.083 to 7.333, with a mean of 5.060 and a standard deviation of 0.803.

3.3   Computation Model Results and Evaluation

We use the decision tree to generate an automated classifier for both frontal and profile computation model, where the important face ratios selected in Sect. 2.3 are used as the input variables. A 2.5D hybrid model is also built in order to investigate the optimal combination of frontal and profile ratios, and to obtain better computation results than either one of these models.

We randomly selected 90 % of the experiment dataset for training the tree and the remaining 10 % is used for testing. The generated decision trees are shown in Fig. 4. There are 15 ratios and 19 ratios selected as predictors into the frontal and profile attractiveness computation model, respectively. Frontal face ratios are notated in the form of "Fxxxxxxxx", while profile ratios in the form of "Pxxxxxxxx". The 8 digits following "F" or "P" stand for four landmark indices in the frontal or profile ratios. For example, "F33293723" in Fig. 4(a) means the frontal ratio of the distance between landmark 33 and landmark 29 to the distance between landmark 37 and landmark 23 (landmark indices are shown in Fig. 2(a)). Similarly, "P17102005" in Fig. 4(b) means the profile ratio of the distance between landmark 17 and landmark 10 to the distance between landmark 20 and landmark 5 (landmark indices are shown in Fig. 2(b)). The hybrid model shown in Fig. 4(c) contains 14 frontal ratios and only 2 profile ratios, indicating that frontal face features are the primary contributors in 2.5D facial attractiveness computation.

After the tree construction, we calculate the predicted attractiveness scores of testing images. The MSE and the correlation between the predicted scores and rater scores can be used to evaluate the performance of our built models. Table 1 gives details on the evaluation results. The small value of MSE and large value of the correlation can prove the accuracy of the computation models. The accuracy of the profile computation model is even better than that of the frontal model, which is in accordance with the findings in [14] that the profile is another significant factor in facial attractiveness determination. As we expected, the hybrid model achieved better results than the either of the two models.

Table 1. Model evaluation results.

Computation model	MSE	Correlation
Frontal	0.1386	0.8852
Profile	0.1351	0.8870
Hybrid	0.1171	0.9034

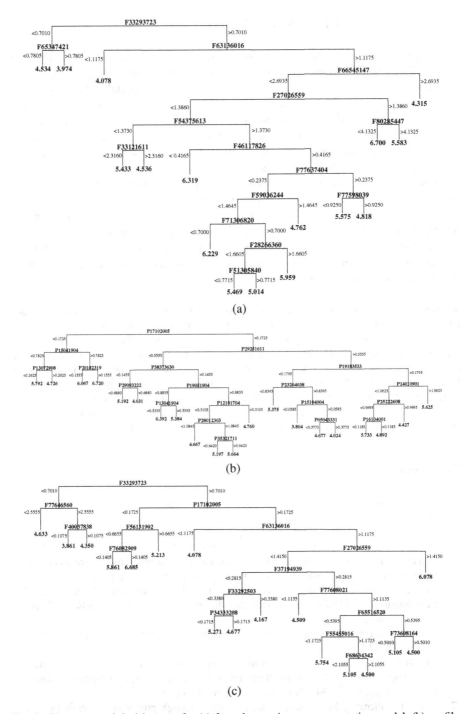

Fig. 4. The generated decision tree for (a) frontal attractiveness computation model; (b) profile attractiveness computation model; (c) hybrid attractiveness computation model.

4 Conclusion and Future Work

This paper develops an attractiveness computation model for both frontal and profile images (2.5D) based on data-driven geometric ratios. By pre-processing the BJUT-3D Face Database, 180 female face models are selected, and the frontal and left-side view of each face are used as our dataset. Facial landmarks are automatically extracted and a two-way ratio vector of all the possible ratios is created. Stepwise regression is used as a feature selection method to select the discriminatory ratios, and then decision tree is used to train a computation model. The small MSE value and high correlation indicate the accuracy of our prediction models.

Since facial attractiveness computation is still relatively new, comparison with other works probably out of reach without currently public database. Thus constructing benchmark database is inevitable for the future facial attractiveness studies. In the future, we are interested in exploring the facial attractiveness using other combinations of face features (i.e., geometric, texture, etc.), different prediction techniques, and in 3D domain for the further study. We would also like to analyze this issue in terms of different gender, ethnicity, and age groups.

Acknowledgments. This work is supported in part by China Scholarship Council (CSC) under Grant No. 201306290099, the National Natural Science Foundation of China under Grant 61402371, Science and Technology Innovation Engineering Plan in Shaanxi Province of China under Grant 2013SZS15-K02, Natural Science Basic Research Plan in Shaanxi Province of China under Grant 2013JQ8039, the Fundamental Research Funds for the Central Universities under Grant 3102014JCQ01060, Graduate Starting Seed Fund of Northwestern Polytechnical University under Grant Z2013064. Portions of the research in this paper use the BJUT-3D Face Database collected under the joint sponsor of National Natural Science Foundation of China, Beijing Natural Science Foundation Program, Beijing Science and Educational Committee Program.

References

1. Etcoff, N.: Survival of the Prettiest: The Science of Beauty. Anckpr Books, New York (1999)
2. Schmid, K., Marx, D., Samal, A.: Computation of face attractiveness index based on neoclassic canons, symmetry and golden ratio. Pattern Recogn. **41**(8), 2710–2717 (2008)
3. Kagian, A., Dror, G., Leyvand, T., et al.: A humanlike predictor of facial attractiveness. In: Advances in Neural Information Processing Systems, pp. 649–656 (2007)
4. Aarabi, P., Hughes, D., Mohajer K, et al.: The automatic measurement of facial beauty. In: IEEE International Conference on Systems, Man and Cybernetics, pp. 2644–2647. IEEE Press, Tucson (2001)
5. Eisenthal, Y., Dror, G., Ruppin, E.: Facial attractiveness: beauty and the machine. Neural Comput. **18**, 119–142 (2006)
6. Mao, H.Y., Jin, L.W., Du, M.H.: Automatic classification of Chinese female facial beauty using support vector machine. In: IEEE International Conference on Systems, Man and Cybernetics, pp. 4842–4846. IEEE Press, San Antonio (2009)
7. Gunes, H., Piccardi, M.: Assessing facial beauty through proportion analysis by image processing and supervised learning. Int. J. Hum Comput Stud. **64**(12), 1184–1199 (2006)

8. Fan, J., Chau, K.P., Wan, X., et al.: Prediction of facial attractiveness from facial proportions. Pattern Recogn. **45**(6), 232–2334 (2012)
9. Whitehill, J., Movellan, J.R.: Personalized facial attractiveness prediction. In: IEEE International Conference on Automatic Face and Gesture Recognition, pp. 17–23. IEEE Press, Amsterdam (2008)
10. Duan, H.S., Zhu, Z.F., Zhao, Y.: Dual subspace algorithm for facial attractiveness analysis. J. Data Acquisition Process. **27**(1), 105–110 (2012)
11. Davis, B.C., Lazebnik, S.: Analysis of human attractiveness using manifold kernel regression. In: IEEE International Conference on Image Processing, pp. 109–112. IEEE Press, San Diego (2008)
12. Gray, D., Yu, K., Xu, W., Gong, Y.: Predicting facial beauty without landmarks. In: Daniilidis, K., Maragos, P., Paragios, N. (eds.) ECCV 2010, Part VI. LNCS, vol. 6316, pp. 434–447. Springer, Heidelberg (2010)
13. Mu, Y.: Computational facial attractiveness prediction by aesthetics-aware features. Neurocomputing **99**, 59–64 (2013)
14. Liao, Q., Jin, X., Zeng, W.: Enhancing the symmetry and proportion of 3D face geometry. IEEE Trans. Visual Comput. Graphics **18**(10), 1704–1716 (2012)
15. Bottino, A., Laurentini, A.: The intrinsic dimensionality of attractiveness: a study in face profiles. In: Alvarez, L., Mejail, M., Gomez, L., Jacobo, J. (eds.) CIARP 2012. LNCS, vol. 7441, pp. 59–66. Springer, Heidelberg (2012)
16. Valenzano, D.R., Mennucci, A., Tartarelli, G., et al.: Shape analysis of female facial attractiveness. Vision Res. **46**(8), 1282–1291 (2006)
17. Karimi, K., Devcic, Z., Avila, D., et al.: A new approach in determining lateral facial attractiveness. Laryngoscope **120**(S4), 157 (2010)
18. Wei, Y.: Research on Facial Expression recognition and synthesis. Master Thesis, Nanjing University (2009)
19. Cootes, T.F., Taylor, C.J., Cooper, D.H., Graham, J.: Active shape models-their training and application. Comput. Vis. Image Underst. **61**(1), 38–59 (1995)
20. Farkas, L.G.: Anthropometry of the Head and Face, 2nd edn. Raven Press, New York (1994)
21. Dryden, I.L., Mardia, K.V.: Statistical Shape Analysis. Wiley, New York (1998)
22. Efroymson, M.A.: Multiple Regression Analysis. Wiley, New York (1960)
23. SAS Institute Inc.: SAS/STAT(R) 12.3 user's guide: high-performance procedures. http://support.sas.com/documentation/cdl/en/stathpug/66410/HTML/default/viewer.htm%23stathpug_hpreg_overview01.htm
24. The BJUT-3D Large-Scale Chinese Face Database. http://www.bjut.edu.cn/sci/multimedia/mul-lab/3dface/face_database.htm
25. Gong, X., Wang, G.: Automatic 3D face segmentation based on facial feature extraction. In: IEEE International Conference on Industrial Technology, pp. 1154–1159. IEEE Press, Mumbai (2006)

Spatial Resolution Enhancement of MWRI of FY-3 Satellite Using BGI

Yu Sun[1,2], Weidong Hu[1(✉)], Xiangxin Meng[1], and Xin Lv[1]

[1] Beijing Institute of Technology, Zhongguancun South Street 5,
Beijing 100081, China
48248524@qq.com, 3120100280@bit.edu.cn
[2] Academy of Military Transportation, Tianjin 300161, China

Abstract. Space-borne microwave radiometer has great potential in the field of space remote sensing and the observation of the earth, while it has bad spatial resolution because of the limitation of the antenna size. We choose Backus-Gilbert Inversion Method that used to enhance the resolution of FY-3 MWRI. The paper introduces the basic theory of BGI method and simulates the potential to improve the resolution of the 18.7 GHz and 23.8 GHz, then decide the optimum tuning parameter γ. Researching the actual brightness temperature of MWRI of the sea coast of South China and Kara Sea coastal areas in northern Russia. There is more rich information appearing in the resolved images, such as extension of the land, outline of the bay and the small island, they become clearer. The experiments verify BGI to improve the spatial resolution of MWRI.

Keywords: BGI · MWRI · Spatial resolution enhancement · Noise tuning parameter γ · Actual brightness temperature

1 Introduction

FY-3 (FENGYUN-3) is the second generation polar orbit meteorological satellite of China, mainly used in numerical weather prediction, weather and climate applications and environmental and natural disaster monitoring. Microwave Radiometer Imager (MWRI) can be monitored around the typhoons and other severe convective weather, access to important information of the total atmospheric precipitation, cloud liquid water content and the ground precipitation, providing important data for the severe weather monitoring, water cycle research, global climate change and environmental studies [1]. The main system parameters is shown in Table 1 [2]. As the limitation of antenna size and weak microwave radiation of the ground surface, MWRI has lower spatial resolution. For example, 10.65 GHz channel is mainly generating for global sea surface temperature, wind speed, soil moisture content and other geophysical parameters while its spatial resolution is 51 km × 85 km. The low spatial resolution restricts its broader applications; therefore some technical means are required to improve spatial resolution of MWRI.

Spatial resolution enhancement algorithm can be used to the ground processing system, the method is simple and easy to implement. Therefore this calls for the interests of researches.

© Springer International Publishing Switzerland 2015
X. He et al. (Eds.): IScIDE 2015, Part I, LNCS 9242, pp. 574–582, 2015.
DOI: 10.1007/978-3-319-23989-7_58

Table 1. Main system parameters of MWRI [2]

Frequency (GHz)	10.65	18.7	23.8	36.5	89
Polarization	V/H	V/H	V/H	V/H	V/H
Bandwidth/MHz	180	200	400	900	4600
Sensitivity/K	0.5	0.5	0.5	0.5	1.0
Calibration error/K	1.0	2.0	2.0	2.0	2.0
Dynamic Range/K	3 ~ 350				
Spatial Resolution/km × km	51 × 85	30 × 50	27 × 45	18 × 30	9 × 15
Scan style	Conical scan				
Aerial perspective/°	45°				
Width/km	1400				
Scan cycle/s	1.7 s				

Currently spatial resolution enhancement algorithms of microwave radiometers mainly has Backus-Gilbert inversion algorithm (BGI), image deconvolution technique and scatterometer image restoration (SIR). BG inversion algorithm was originally developed by Backus and Gilbert [3, 4] proposed to obtain accurate geophysical data certainly. Stogryn [5] introduced BGI into the field of microwave remote sensing firstly in 1978 achieving spatial resolution enhancement of SSM/I, and used a noise tuning parameter to balance resolution enhancement and noise error. Then BGI algorithm is applied to microwave radiometer resolution enhancement, spatial resolution matching of different channels, interpolation in the antenna pattern by some researchers. Image deconvolution technique originally is proposed by Sethmann et al. [6] to correct diffraction blur effects caused by antenna ground footprints overlapping and doesn't require the antenna pattern but the point spread function (PSF) at different positions. BGI algorithm can better deal for the case of deformation due to sampling window curvature of the earth and other factors than image convolution technique when applying to real brightness temperature. SIR is a maximizing entropy reconstruction algorithm originally developed by Long and Daum [7] resolving image restoration of scatterometer, then introduced into image enhancement of microwave radiometer to take an iterative process from initial brightness temperature in order to get close to the maximum entropy estimation of the real brightness temperature. But SIR algorithm does not introduce noise tuning parameters compared to BGI algorithm, so noise cannot be suppressed effectively, this paper selected BGI algorithm for spatial resolution enhancement of MWRI.

This paper analysis resolution enhancement potential of 18.7 GHz, 23.8 GHz channel of MWRI, the resolution variation with the tuning parameter Y and how to select the optimum Y then apply BGI algorithm to real brightness temperature data to provide more accurate data.

2 BG Inversion Algorithm

Microwave radiometer resolution enhancement algorithms depend on antenna gain function and geometric measurement model and use redundant information generated by the ground overlapping footprints of the antenna to achieve the purpose of resolution enhancement. BGI algorithm can be expressed as the linear combination of N adjacent measurements.

$$T_B(\rho_0) = \sum_{i=1}^{N} c_i T_{Ai} \tag{1}$$

$T_B(\rho_0)$ is inversion brightness temperature and T_{Ai} is the adjacent measurement, The purpose of BGI algorithm is requiring a set of coefficients to resolve $T_B(\rho_0)$. Consider a set of integral equation,

$$Q_R = \int \left[\sum_{i=1}^{N} c_i \overline{G}_i - F(\rho, \rho_0) \right]^2 J(\rho, \rho_0) dA \tag{2}$$

And the normalized constraint condition,

$$\int \sum_{i=1}^{N} c_i \overline{G}_i dA = 1 \tag{3}$$

Choosing F and J to minimize Q_R, setting J = 1 and F = 1/A$_0$ in the region A$_0$ or J = 0 out of the region A$_0$. This causes the instrument noise transfers to the resolved brightness temperature, the random noise is $(\Delta T_{rms})^2$, error of the resolved brightness temperature can be expressed as

$$Q_N = c^T \overline{E} c \tag{4}$$

The element of c is c_i. Noise is random and not related, $\overline{E}$ is matrix of errors and diagonal which element are $(\Delta T_{rms})^2$. Sethmann et al. [6] introduced noise tuning parameter γ to balance noise error and resolution enhancement.

$$Q = Q_0 \cos \gamma + \omega e^2 \sin \gamma \tag{5}$$

Q_0 is the resolution item, ω is scale factor, e^2 is the noise error item. Assuming the dimension of $\overline{\overline{G}}$ is $N \times N$,

$$G_{ij} = \int \overline{G}_i(\rho) \overline{G}_j(\rho) dA \tag{6}$$

Considering the constraint conditions, use Lagrange multiplier method to resolve the equation,

$$H(c_i) = \cos\gamma \int \left[\sum_{i=1}^{N} c_i \overline{G}_i - F(\rho, \rho_0)\right]^2 J(\rho, \rho_0) + \omega \sin\gamma \sum_{i=1}^{N} c_i^2 \varepsilon^2$$
$$- \lambda\left[\int \sum_{i=1}^{N} c_i \overline{G}_i dA - 1\right] \tag{7}$$

Setting $\frac{\partial H}{\partial c_i} = 0$ to get,

$$c = W^{-1}\left[\cos\gamma \cdot v + \frac{1 - \cos\gamma \cdot u^T W^{-1} v}{u^T W^{-1} u} u\right] \tag{8}$$

Parameters are expressed as,

$$u_i = \int_E \overline{G}_i dA \tag{9}$$

$$v_i = \int_E \frac{1}{A_0} \overline{G}_i dA \tag{10}$$

$$W = \cos\gamma \overline{G} + \omega \sin\gamma \overline{G} \tag{11}$$

Where u, v is N-dimensional column vector, the details are shown in Sethmann et al. [6].

We need to set a series of parameters when applying BGI algorithm. Firstly, we set the antenna gain function, some researchers chose Gaussian gain function, for example, Migliaccio and Gambardella [8] used Sinc function and Gaussian function dividedly, they achieved the similar results, and Chakraborty et al. [9] chose Gaussian fit gain function reaching the same purpose of resolution enhancement when applying to the real brightness temperature. Secondly, is the set of scale factor and the noise tuning parameter. This paper set $\omega = 0.001$ as same as Robinson et al. [10]. Farrar and Smith [11] propose the criterion for choosing optimum γ that maximizing the correlation coefficient between enhanced images and the original images. While Wang et al. [12] minimize root-mean-square error (RMSE) for choosing optimum γ. This paper balances the two criterions, then decides the optimum γ. Finally, the choice of adjacent measurements, some researchers set it to be a fixed value. For example, Poe [13], Chakraborty et al. [9] set N as 4 and 3 respectively. Robinson et al. [10] discussed the impact on the finally results when using different N, the result shown that as the N value increases the inversion brightness temperature is closer to the initial brightness temperature. When N reaches a certain value, the resolution enhancement is no longer obvious, but a significant increase appears in computation and this paper select N = 3.

3 Simulations

From the perspective of simulation, we analysis the potential of BGI algorithm used for resolution enhancement and the effect of different noise tuning parameters on the resolution enhancement firstly. The research steps are as follows, (1) simulating the initial brightness temperature images which have 126 × 126 pixels, represent 504 km × 504 km sense. (2) Simulating antenna gain function and sampling the initial images. (3) Applying BGI algorithm for the sampled images.

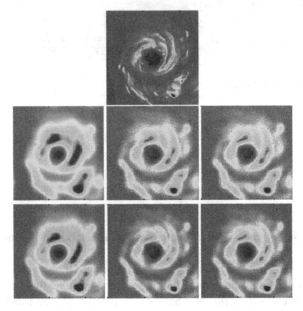

Fig. 1. Simulations (a) the initial image (b) 18.7 GHz non-enhanced image (c) 18.7 GHz enhanced image (γ = 0.018) (d) 18.7 GHz enhanced image (γ = 0.021) (e) 23.8 GHz non-enhanced image (f) 23.8 GHz enhanced image (γ = 0.018) (g) 23.8 GHz enhanced image (γ = 0.03)

Figure 2 shows the relationship of the correlation coefficient and RMSE among 18.7 GHz, 23.8 GHz respectively. Figure 2 shows the correlation coefficient and RMSE before and after BGI. Summaries are as follows: for 18.7 GHz, the correlation coefficient difference 0.0007 and RMSE difference 0.0055 when γ are equal to 0.018 when the correlation coefficient maximized and 0.021 when RMSE minimized, and the resolution is 21 km and 23 km respectively. Then select γ = 0.018 as the optimum γ value and the enhanced images are shown in Fig. 1(c), (d). For 23.8 GHz, the correlation coefficient difference 0.0024 and RMSE difference 0.1095 when γ are equal to 0.018 that the correlation coefficient is 0.9135 maximized and 0.021 whose RMSE is 2.3442 minimized, and the resolution is 18 km and 21 km respectively. Then select γ = 0.03 as the optimum γ value for reducing noise error and the enhanced images are shown in Fig. 1(f) and (g) (Table 2).

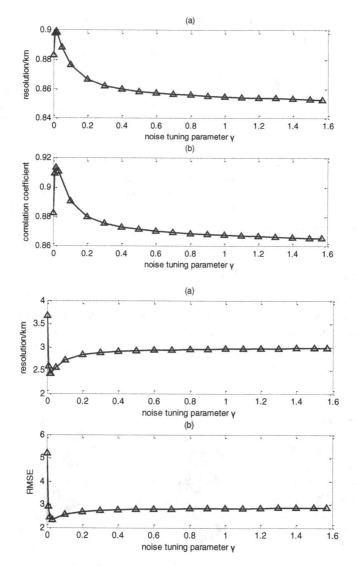

Fig. 2. The correlation coefficient and RMSE between the initial image and enhanced image (a) 18.7 GHz (b) 23.8 GHz

4 Applying BG Algorithm in MWRI

BG algorithm eventually be applied to the actual brightness temperature data, verify its effectiveness to improve the spatial resolution of the microwave radiometer. This paper selects L1-level brightness temperature data of MWRI, where L1-level data which include scaling, positioning information and are the standard product of HDF5 format used to calculate is the result of processing L0-level data [1]. To verify the validity of BG algorithm for improving the spatial resolution of microwave radiometers, the paper

Table 2. Parameters comparisons before and after BGI

Frequency/GHz	Before BGI			After BGI		
	Correlation coefficient	RMSE	Resolution/km	Correlation coefficient	RMSE	Resolution/km
18.7	0.8554	2.9261	40	γ = 0.018, 0.8992 γ = 0.021, 0.8985	γ = 0.018, 2.4310 γ = 0.021, 2.4255	γ = 0.018, 21 γ = 0.021, 23
23.8	0.8683	2.7977	36	γ = 0.018, 0.9135 γ = 0.03, 0.9111	γ = 0.018, 2.4537 γ = 0.03, 2.3442	γ = 0.018, 18 γ = 0.03, 21

selects the coastline, islands or lakes that brightness temperature change suddenly. Because for these places, the original brightness temperature images are blurred out of a lot of information and can't reflect the brightness temperature changing gradient, BG algorithm can better distinguish the brightness temperature changing [12]. This paper chooses three regions to verify BG algorithm effectively.

4.1   Southern China Coast Area

Southern China coastline is shown in Fig. 3, Leizhou Peninsula and Pearl River Estuary disappear blurs in (a) the original 10.65 GHz vertical polarization brightness temperature image. After BG algorithm processing, these regions generate these parts information in (b) BG enhanced image through comparing with (c) 18.7 GHz vertical polarization brightness temperature image.

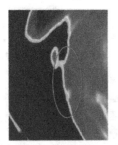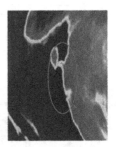

Fig. 3. Southern China coastline (a) 10.65 GHz-VV (b) 10.65 GHz-VV BG enhanced image (c) 18.7 GHz

4.2   Kara Sea Coastal Areas in Northern Russia

Kara Sea coastal areas in northern Russia are shown in Fig. 4. In (b) 18.7 GHz-VV BGI enhanced image, changes in brightness temperature of land and sea borders are more gradient and sharper. At the same time, (b) adds a wealth of detailed information and has a clearer outline of the bay compared with (a) such as Yenisei Bay circled with green color in the mouth of Yenisei River estuary.

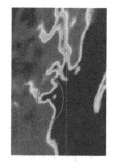

Fig. 4. 18.7 GHz-VV (b) 18.7 GHz-VV BGI enhanced image (c) 36.5 GHz-VV

5   Conclusions

This paper analysis 18.7 GHz, 23.8 GHz simulation brightness temperature images, resolution of the three channels has been enhanced 19 km and 15 km respectively. The optimum noise tuning parameter has been selected by optimum criterion through balancing maximizing the correlation coefficient and minimizing the root mean square error criterion, its values are 0.018 and 0.03 corresponding to two channels. When the noise tuning parameter γ varies from 0 to $\pi/2$, resolution changes from high to low. Choosing southern China coastline and Kara Sea coastal areas in northern Russia as research regions, the results show richer detail information, the outline of the coastline and bay closer to the high-frequency image, blurred islands in the low-frequency image appeared. However, for the analysis of the real measurements is only from the visual effects but not development of quantitative methods. In fact, the results after BG algorithm processing can be used to the inversion of specific physical parameters for the real brightness temperature, so that analysis the effectiveness of the algorithm qualitatively compared with non-enhanced brightness temperature inversion results. This is one of the applications of BG algorithm; such a method would be discussed in the after work.

Acknowledgment. The authors wish to express appreciation to the National Satellite Meteo-rological Center for providing L1-level brightness temperature and the antenna pattern data of MWRI.

References

1. Jun, Y., Chaohua, D.: A New Generation of Polar-Orbiting Meteorological Satellite FY-Business Products and Applications. Science Press, Beijing (2010). (in Chinese)
2. Xu, B.M., Li, Q., Chen, G.L., Sun, Y.Z.: Meteorological Satellite Payload Technology. China Aerospace Press, Beijing (2009). (in Chinese)
3. Backus, G., Gilbert, F.: Uniqueness in the inversion of inaccurate gross earth data. Philos. Trans. R. Soc. London, Ser. A. **266**, 123–192 (1970)

4. Backus, G., Gilbert, F.: The resolving power of gross earth data. Geophys. J. Int. **16**, 169–205 (1968)
5. Stogryn, A.: Estimates of brightness temperatures from scanning radiometer data. IEEE Trans. Antennas Propag. **26**, 720–726 (1978)
6. Sethmann, R., Burns, B.A., Heygster, G.C.: Spatial resolution improvement of SSM/I data with image restoration techniques. IEEE Trans. Geosci. Remote **32**, 1144–1151 (1994)
7. Long, D.G., Daum, D.L.: Spatial resolution enhancement of SSM/I data. IEEE Trans. Geosci. Remote. **36**, 407–417 (1998)
8. Migliaccio, M., Gambardella, A.: Microwave radiometer spatial resolution enhancement. IEEE Trans. Geosci. Remote Sens. **43**, 1159–1169 (2005)
9. Chakraborty, P., Misra, A., Misra, T., Rana, S.S.: Brightness temperature reconstruction using BGI. IEEE. Trans. Geosci. Remote **46**(6), 1768–1773 (2008)
10. Robinson, W.D., Kummerow, C., Olson, W.S.: A technique for enhancing and matching the resolution of microwave measurements from the SSM/I instrument. IEEE Trans. Geosci. Remote **30**(3), 419–429 (1992)
11. Farrar, M.R., Smith, E.A.: Spatial resolution enhancement of terrestrial features using deconvolved SSM/I microwave brightness temperatures. IEEE Trans. Geosci. Remote **20**(2), 349–355 (1992)
12. Wang, Y.Q., Shi, J.C., Jiang, L.M., Du, J.Y., Tian, B.S.: The development of an algorithm to enhance and match the resolution of satellite measurements from AMSR-E. Sci. China Ser. D. **54**, 410–419 (2011)
13. Poe, G.A.: Optimum interpolation of imaging microwave radiometer data. IEEE Trans. Geosci. Remote **28**, 800–810 (1990)

Multispectral Image Fusion Based on Laplacian Pyramid Decomposition with Immune Co-evolutionary Optimization

Xiuxiu Li[1(✉)], Haiyan Jin[1], Li Sun[2], and Bin Wang[1]

[1] Xi'an University of Technology, Xi'an, China
{lixiuxiu,jianhaiyan,wangbin}@xaut.edu.cn
[2] Air Force Engineering University, Xi'an, China
sl_lxa@mail.nwpu.edu.cn

Abstract. In this paper, a new multispectral image fusion algorithm is proposed by considering the abundant directional information and complex high frequency of source images. In this algorithm, Laplacian pyramid (LP) is constructed to filter the source images to obtain initial fusion coefficients, then an evolution computation idea, immune co-evolutionary optimization algorithm, is introduced into the image fusion to optimize the fusion coefficients. The optimized fusion coefficients can make better fusion result. Simulation experiments of clearly demonstrate the superiority of this proposed algorithm by comparing with conventional wavelets systems: the information entropy (IE) values keep at a similar level, the average grads (AG) values are higher, the standard deviation (STD) values increases 1.2 averagely, and the time efficiency increases averagely about 51 %.

Keywords: Image fusion · Immune co-evolutionary optimization · Laplacian decomposition · Wavelet transform

1 Introduction

Image fusion techniques have become a vital problem in image processing field. For 2-D images, the aim of fusion is to generate a single image which contains all the useful information from source images. The information in the fused image is integrated, which is more useful, exact, comprehensive and reliable for future human or machine perception. Recently, image fusion techniques have been widely used in multispectral image comprehension, video surveillance, and weather forecast and so on.

Generally, there are three levels of image fusion: pixel-based fusion, feature-based fusion and decision-based fusion. Currently, the image fusion of pixel-based level is an important topic in the field of digital image processing. The conventional methods include simple weighted image fusion methods, pyramid-based methods, wavelet-based methods [1–3] and so on. In simple weighted methods, because the fusion rules is to select average or weighted average values of pixels from multispectral images, the contrast of fusion image is lower and the results are not satisfying usually. In the fusion idea of conventional wavelet-based, due to the directionality and non-redundancy of the wavelet decomposition, it is popular in image fusion. However, this idea only considers

© Springer International Publishing Switzerland 2015
X. He et al. (Eds.): IScIDE 2015, Part I, LNCS 9242, pp. 583–591, 2015.
DOI: 10.1007/978-3-319-23989-7_59

the maximal absolute value of wavelet coefficients or local features of source images, so that the wavelet-based idea cannot approximate the edge structure and detail information well, that is, wavelet are not effective in representing objects with line singularities along. Especially for images with abundant texture and intricate background, the fused images with wavelet-based idea would generate serious "ringing" effects.

Multi-scale data representation is a powerful idea. It represents data in a hierarchical model where each level corresponds to an original reduced resolution approximation. Laplacian pyramid (LP) is one way of multi-scale decomposition, which is proposed by Burt and Adelson [4]. In image fusion, LP decomposes the source images into different level with different space band respectively, then the fusion is implemented in each band, which is beneficial to extrude the details and features of different resolution images. In contrast to the critically sampled wavelet scheme, the LP has the distinguishing feature that each pyramid level only generates one band-pass image which does not have "scrambled" frequencies.

In this paper, LP decomposition is applied to source images, and an optimized co-evolutionary idea—immune clonal selection (ICS) algorithm is introduced to find the optimized fusion coefficients. Experiment results show that the proposed algorithm makes the information of the fused image richer.

2 Fusion Algorithm Based on Laplacian Pyramid Decomposition with Immune Co-evolutionary (ICS_LP)

2.1 Constructing Laplacian Decomposition

The LP decomposition result is a band-pass image, which is carried out by generating a sampled low-pass version of the source and the difference between the source and the prediction at each step.

The basic analysis procedure of the LP is as the following (Fig. 1(a)): firstly, derive a coarse approximation of the source signal by low-pass filtering and down-sampling; secondly, predict the original (by up-sampling and filtering) based on the coarse approximation; finally, calculate the difference between the source and the prediction as the prediction error. Contrarily, for the reconstruction, the reconstructed signal can be obtained by adding the difference d to the prediction c from the coarse signal simply, which is known as the synthesis procedure to of the LP (Fig. 1(b)).

In this paper, a $d \times d$ nonsingular integer matrix $\mathbf{M}$ is used to represent the sampling operation. For an M-fold down-sampling, the input $x[n]$ and the output $x_d[n]$ are represented as in [5]:

$$x_d[n] = x[\mathbf{M}n] \tag{1}$$

For an M-fold up-sampling, the input $x[n]$ and the output $x_u[n]$ are related by

$$x_u[n] = \begin{cases} x[k] & \text{if } n = \mathbf{M}k, k \in Z^d \\ 0 & \text{otherwise} \end{cases} \tag{2}$$

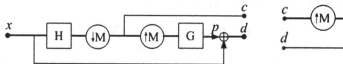 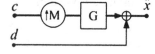

(a) The basic analysis procedure of the LP (c is the coarse approximation of the original signal, and d is the difference between the original signal and the prediction)

(b) The synthesis procedure to the LP

Fig. 1. Laplacian pyramid scheme (H is a low-pass filter, and G is a high-pass filter)

The coarse approximation signal is

$$c[n] = \sum_{k \in Z^d} x[k]h[\mathbf{M}n - k] = \langle x, \tilde{h}[\cdot - \mathbf{M}n] \rangle \tag{3}$$

where $\tilde{h}[n] = h[-n]$. The up-sampling and filtering operation results in:

$$p[n] = \sum_{k \in Z^d} c[k]g[n - \mathbf{M}k] = \sum_{k \in Z^d} \langle x, \tilde{h}[\cdot - \mathbf{M}k] \rangle g[n - \mathbf{M}k] \tag{4}$$

Writing signals as column vectors, for example, $x = (x[n] : n \in Z^d)^T$, the related analysis operations of Laplacian pyramid scheme is expressed as left matrix multiplications

$$c = \mathbf{H}x \quad \text{and} \quad p = \mathbf{G}c \tag{5}$$

where $\mathbf{H}$ and $\mathbf{G}$ correspond to ($\downarrow$M) H and G ($\uparrow$M), respectively. Furthermore, using this matrix notation and the Eq. (5), the difference signal of the LP can be written as:

$$d = x - p = x - \mathbf{G}\mathbf{H}x = (\mathbf{I} - \mathbf{G}\mathbf{H})x \tag{6}$$

where $\mathbf{I}$ is the identity matrices with appropriate size depending on the context.

For the synthesis procedure of the LP, the matrix notation can be wrote as follows by combining the Eqs. (5) and (6):

$$\underbrace{\begin{pmatrix} c \\ d \end{pmatrix}}_{y} = \underbrace{\begin{pmatrix} \mathbf{H} \\ \mathbf{I} - \mathbf{G}\mathbf{H} \end{pmatrix}}_{A} x \tag{7}$$

In the LP structure, each level generates only one band-pass signal, so the resulting band-pass signals of the LP do not suffer from the "scrambled" frequencies.

2.2 Immune Co-evolutionary Optimizing Coefficients

Artificial immune system [6–8] is an intelligent method with learning and memory functions, which simulates the functions of natural immune system and provides a new approach for information processing. In the immune co-evolutionary optimizing, based on conventional evolutionary computation, affinity maturation, clone and memory mechanisms are introduced to accelerate the convergent speed of the algorithm to the global optimization.

The clonal selection (CS) is a dynamic process of the immune system that is a self-adapting antigen stimulation. The cells are selected if they can recognize the antigens, and then proliferated. During this process, only simple replication occurs between the parent generation and the cloned child generation without communication of different information, that is, CS cannot promote the evolution of antibody population. Hence, further processing is necessary for the cloned child generation. In the artificial immune system, CS is an antibody random match derived according to the affinity factually.

The state transition of antibody population can be explained as Fig. 2.

$$\text{CS: } A(k) \xrightarrow{\text{clone}} A'(k) \xrightarrow{\text{clone mutation}} A''(k) \xrightarrow{\text{clone selection}} A''(k+1)$$

Fig. 2. Random process of antibody population state transition

The process in Fig. 2 includes three steps: clone, mutation and selection. During a certain generation evolution, a new antibody population can be obtained by mutation and selection: according to the affinity function $Aff(\cdot)$, any point in solution space as $a_i(k) \in A(k)$ splits into q_i same points $a_i'(k) \in A'(k)$, and the antibody population $A''(k)$ is obtained after the mutation and selection.

The essential of immune clone is to produce a population of mutated solution near the candidate solutions according to the affinity value. It enlarges the search scope and is helpful to avoid prematurity and trapping into local optimization, which has been described in a Lemma [9] as following:

Lemma. An antibody population sequence of clonal selection algorithm is convergent to the optimization at ratio 1.

2.3 The Fusion Algorithm

In this section, the fusion based on LP decomposition with the immune co-evolution is used to optimize the fusion coefficients.

Suppose the coefficient matrix LP_c, whose elements are gained from LP decomposition. Each element in LP_c is optimized with the immune co-evolution to maximize the affinity function $Aff(\cdot)$. The coefficient matrix, which makes the value of affinity function $Aff(\cdot)$ the largest, is the final fusion coefficients.

The detailed steps are as follows:

Step 1: Initialization. Take the coefficient matrix LP_c obtained from two source images according to the modulus maxima fusion rules [11] as the original population A(0), in which each element can be regarded as chromosome. The original number of generation is k = 1, and the maximal number of iterated generation is GS = 50.

Step 2: Judgment of terminal condition. Judge whether the terminal condition is satisfied. If the predefine number of iteration is finished, stop and confirm the current population composed of current individuals is the optimized solution population and turn to step 8; else turn to step 3.

Step 3: Clone operation. Clone operation is performed to the k-th generation parent population $A(k)$ to obtain $A'(k)$.

Step 4: Mutation operation. Gaussian mutation with variance 0.01 is performed to $A'(k)$ to obtain $A''(k)$.

Step 5: Affinity function computation. Compute affinity function value $Aff(A''(k))$ for each individual.

Step 6: Clonal selection operation. In the child generation, if there exists muted antibody $b = \max\{Aff(a_{ij})|j = 2,3,\ldots,q_i -1\}$, which make $Aff(a_i) < Aff(b)$, $a_i \in A(k)$, then choose b to enter the new parent population.

Step 7: $k = k+1$, turn to step 2.

Step 8: Acquirement of a group of optimized fusion coefficients.

Step 9: LP reconstruction. According to the fusion coefficients from Step 8, LP reconstruction is implemented as Fig. 1 (b), and the fusion result is output.

3 Experiments

3.1 The Evaluation Criterion of Fusion Effect

Currently, there is no a set of general and uniform criterion to evaluate the fusion effect. In this paper, the information entropy (IE), average grads (AG) and standard deviation (STD) are used to score the fusion results.

(1) Information entropy (IE). According to information theory, the IE of an 8-bit image can be denoted as E:

$$E = -\sum_{t=0}^{255} p_t \log_2 p_t \qquad (8)$$

where p_t is the occurrence probability of the pixel value t in an image. IE ban be used to indicate the change capability of information, and the greater the entropy is, the more information is included in images.

(2) Average grads (AG). AG reflect the contrast of an image, and it can be used to evaluate the resolution of the images. The AG is donated as $\bar{g}$ and the formula is

$$\overline{g} = \frac{1}{(M-1)(N-1)} \times \sum_{i=1}^{(M-1)} \sum_{j=1}^{(N-1)} \sqrt{\left(\left(\frac{\partial f(x_i, y_j)}{\partial x_i} \right)^2 + \left(\frac{\partial f(x_i, y_j)}{\partial y_i} \right)^2 \right) \Big/ 2} \quad (9)$$

where $f(x, y)$ represents pixel value of the fused image at the position (x,y). M and N are the number of row and column, respectively. In general, the greater it is, the resulting image is of higher quality.

(3) Standard deviation (STD). The STD is an important indicator to measure the information contained in images and the contrast value of images. The formula is

$$Std = \left(\frac{1}{n-1} \sum_{i=1}^{n} (x_i - \bar{x})^2 \right)^{\frac{1}{2}} \quad (10)$$

where n is the number of the pixel in the image, x_i is the pixel value, and $\bar{x}$ is the mean of pixel values, In general, the greater it is, the more additional information is included.

3.2   Numerical Experiments

In this section, the superiority of proposed fusion algorithm is measured by comparing with the wavelet-based method. In the comparison experiments, wavelet transform (WT) adopted Daubechies filters [10] and the number of decomposition level of WT is 3; and the proposed algorithm adopted the affinity function $Aff(\cdot)$, which is set to the sum of the 3 indicators in Sect. 3.1.

Some experiment results are shown in Figs. 3 and 4.

In Fig. 3, two source images are blurred images indoors at different position, respectively, and the fusion result are shown in Fig. 3(a) and (b). Values of measurable indicators are shown in Table 1.

In Fig. 4, two source images are an infrared image and a visual image outdoors, and the fusion result are shown in Fig. 4(a) and (b). Values of the measurable indicators are shown in Table 2.

Visually, in result images fused by the wavelet-based method (Figs. 3(c) and 4(c)), there are obvious "ringing" effect at the edges, the contrast of images is low, and the details and edges are not obvious. As a contrast, in the result of ICS_LP (Figs. 3(d) and 4(d)), edges and backgrounds of the fused images are prominent with high sharpness and clearness.

Numerically, as shown in Tables 1 and 2, IE, AG and STD of fused images are higher than source images. Comparing the ICS_LP with WT, in fused images, IE of ICS_LP is slightly higher than WT, and AG and STD of ICS_LP is higher, that is, fused images of ICS_LP can reflect the details better, which is beneficial and significant for further applications.

(a) The first blurred source image

(b) The second blurred source image

(c) 3-level WT fusion result

(d) ISC-LP fusion result

Fig. 3. Images fusion results indoors

Table 1. Numerical fusion results of images indoors

Images	IE	AG	STD	Time (S)
Source image 1	7.1702	5.1801	59.5479	/
Source image 2	7.2853	6.1394	63.4367	/
Fusion result by WT	7.5051	8.2764	66.8783	21.1
Fusion result by ICS-LP	7.5431	8.2518	67.1765	17.5

Table 2. Numerical fusion results of images outdoors

Images	IE	AG	STD	Time (S)
Source image 1	4.1612	2.1158	14.4112	/
Source image 2	5.9263	4.0691	18.7487	/
Fusion result by WT	6.0918	4.8453	19.2396	206.5
Fusion result by ICS-LP	6.0973	5.3061	21.3186	20.5

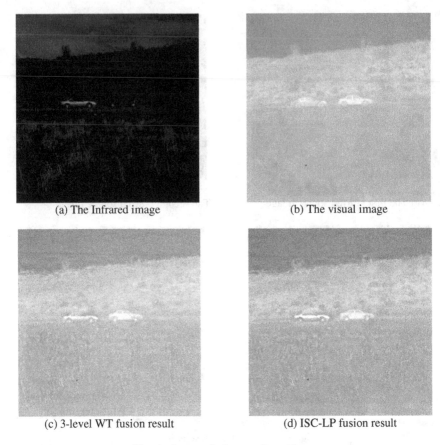

(a) The Infrared image   (b) The visual image

(c) 3-level WT fusion result   (d) ISC-LP fusion result

Fig. 4. Images fusion results outdoors

In time consumption, the time consumption of ICS-LP is lower than that of wavelets. Compared to wavelets, computation efficiency increases averagely 51 %. It is of great crucial to many important and practical applications.

In sum, the algorithm put forward in this paper is preponderant in the visual effect, the numerical indicators, and time consumption.

4   Conclusion and Discussion

In this paper, to fuse images well, based on LP decomposition, the immune clone selection is introduced to optimize the fusion coefficients. The proposed algorithm is efficient at edges with abundant directional information and complex high frequency information. The validity and practicability have been verified by comparing with the wavelet-based method.

However, the proposed algorithm is feasible after source images have been aligned by pixels. In future work, the image fusion technique with auto-aligned function will be investigated.

Acknowledgment. This work has been partially supported by the National Natural Science Foundation of China under grant Nos. 61201416, 61472204, China Postdoctoral Science Foundation under grant No. 2012M521897, the Natural Science Basic Research Project of Shaanxi Province under grant No. 2014JQ8340, Science Research Program Project of Educational Committee of Shaanxi Province under grant No. 2013JK1135, Beilin District Science and technology Project under grant No. GX1309, and Coordinator innovative engineering project of Shaanxi province under grant No. 2015KTTSGY0406.

References

1. Jin, H.Y., Wang, Y.Y.: A fusion method for visible and infrared images based on contrast pyramid with teaching learning based optimization. Infrared Phys. Technol. **64**, 134–142 (2014)
2. Lee, C.H., Chen, L.H., Wang, W.K.: Image contrast enhancement using classified virtual exposure image fusion. IEEE Trans. Consum. Electron. **58**(4), 1253–1261 (2012)
3. Jang, J.H., Bae, Y., Ra, J.B.: Contrast-enhanced fusion of multisensor images using subband-decomposed multiscale retinex. IEEE Trans. Image Process. **21**(8), 3479–3490 (2013)
4. Burt, P.J., Adelson, E.J.: The Laplacian pyramid as a compact image code. IEEE Trans. Commun. **31**(4), 532–540 (1983)
5. Viscito, E., Allebach, J.P.: The analysis and design of multidimensional FIR perfect reconstruction filter banks for arbitrary sampling lattices. IEEE Trans. Circ. Syst. **38**(1), 29–42 (1991)
6. Leandro, N., Fernado, J.: Learning and optimisation using the clonal selection principle. IEEE Trans. Evol. Comput. **6**(3), 239–251 (2002)
7. Felipe, C., Frederico, G., Hajime, I., et al.: A clonal selection algorithm for optimization in electromagnetics. IEEE Trans. Magn. **41**(5), 1736–1739 (2005)
8. Jiao, L.C., Du, H.F.: Development and prospect of the artificial immune system. Acta Electronica Sinica **31**(10), 1540–1548 (2003)
9. Jiao, L.C., Du, H.F.: Immune Optimization Computing, Learning and Identifying. Science Press, Singapore (2006)
10. Cohen, A., Daubechies, I., Feauveau, J.C.: Biothogonal base of compactly supported wavelets. Commun. Pure Appl. Math. **45**(5), 485–560 (1992)
11. Nikolov, S.G., Bull, D.R., Canagarajah, C.N., et al: Fusion of 2-D images using their multiscale edges. In: The 15th International Conference on Pattern Recognition, vol. 3, pp. 41–44 (2000)

Local Variational Bayesian Inference Using Niche Differential Evolution for Brain Magnetic Resonance Image Segmentation

Zhe Li[1], Zexuan Ji[2], and Yong Xia[1(✉)]

[1] Shaanxi Key Lab of Speech and Image Information Processing (SAIIP),
School of Computer Science,
Northwestern Polytechnical University, Xi'an 710072, China
yxia@nwpu.edu.cn
[2] School of Computer Science and Engineering,
Nanjing University of Science and Technology, Nanjing 210094, China

Abstract. Brain magnetic resonance (MR) image segmentation is pivotal for quantitative brain analyses, in which statistical models are most commonly used. However, in spite of its computational effectiveness, these models are less capable of handling the intensity non-uniformity (INU) and partial volume effect (PVE), and hence may produce less accurate results. In this paper, a novel brain MR image segmentation algorithm is proposed. To address the INU and PVE, voxel values in each small volume are characterized by a local variational Bayes (LVB) model, which is inferred by the niche differential evolution (NDE) technique to avoid local optima. A probabilistic brain atlas is constructed for each image to incorporate the anatomical prior into the segmentation process. The proposed NDE-LVB algorithm has been compared to the variational expectation-maximization based and genetic algorithm based segmentation algorithms and the segmentation routine in the widely used statistical parametric mapping package on both synthetic and clinical brain MR images. Our results suggest that the NDE-LVB algorithm can differentiate major brain tissue types more effectively and produce more accurate segmentation results.

Keywords: Image segmentation · Magnetic resonance imaging (MRI) · Gaussian mixture model · Variational Bayes inference · Niche differential evolution

1 Introduction

Segmentation of major brain tissues into grey matter, white matter, and cerebrospinal fluid (CSF) using magnetic resonance (MR) images plays an essential role in a wide spectrum of clinical and research applications. Since manual segmentation is time-consuming, expensive and subject to operator variability, automated algorithms have been developed in the literature [1–4]. Among them, statistical model based methods have attracted numerous research attentions.

Gaussian mixture model (GMM) [5] has been extensively used in brain MR image segmentation, where the distribution of voxel values from one tissue type is assumed to

© Springer International Publishing Switzerland 2015
X. He et al. (Eds.): IScIDE 2015, Part I, LNCS 9242, pp. 592–602, 2015.
DOI: 10.1007/978-3-319-23989-7_60

be Gaussian, and the prior probability of that tissue gives the weight of the Gaussian component. The parameters of GMM are traditionally estimated by maximizing the likelihood of the observed image using the expectation-maximization (EM) algorithm [4], and thereafter applied to a naive Bayes classifier to classify each voxel. Despite of its abundance applications, the EM-based GMM estimation is prone to over-fit the training data, and thus results in limited segmentation accuracy [5]. To overcome this drawback, evolutionary techniques, such as the genetic algorithm (GA) [6], have been used as a substitute for the EM algorithm [7]. Tohka et al. [8] proposed a real coded GA with new permutation operator specifically designed for GMM estimation. Although having the potential of global optimization, evolutionary estimation of GMM does not address the intrinsic difficulties of applying classical statistical models to brain MR image characterization.

Besides noise and limited contrast, MR images also suffer from intensity non-uniformity (INU) and partial volume effect (PVE), which render a challenging task for brain MR image segmentation. INU, also referred to as the bias field, arises from the imperfections in the radio-frequency coils, and may results in a shading effect across the whole MR image. Current strategies for bias field correction can be divided into three categories [9], including reference-points-based straightforward methods [10], simultaneous segmentation and bias field correction methods [11] and localized statistical methods [12, 13]. PVE is a combination of two factors, i.e. the limited resolution and image sampling. Image sampling refers to the fact that MR voxel has a definite volume, which may consist only partially of the desired tissue, reflecting underlying tissue heterogeneity. For each partial volume voxel, since multiple tissue types contribute to its measured hydrogen atoms concentration, its voxel value distribution is dependent on the parameters of pure voxel classes. Let pure voxel values follow a Gaussian distribution $N(\mu_k, \sigma_k^2)$, partial volume voxel values also follow a Gaussian distribution $N(\mu_{PV}, \sigma_{PV}^2)$, where $\mu_{PV} = \sum_k a_k \mu_k$, $\sigma_{PV}^2 = \sum_k a_k^2 \sigma_k^2$ and a_k is the abundance of the k-th tissue types that satisfies the constraint $\sum_k a_k = 1$ [8]. As a result, voxel values, though still follow Gaussian distributions, may have different means and variances that spatially vary on a voxel-by-voxel basis. To address PVE, variational Bayesian inference is hereby adopted as a substitute for the classical GMM estimation [14].

In this paper, we proposed a novel brain MR image segmentation approach based on evolutionary estimation of local variational Bayes (LVB) models. We take partial volume voxels into consideration, and hence assume the voxel values from each tissue type satisfy a Gaussian distribution, whose mean and variance are also random variables. To address the INU effect, we replacing the global GMM model with a cohort of LVB models learned on small data volumes in the original MR image based on the assumption that the bias field is the lowest frequency component in an image and can be ignored within a small volume. Instead of using the step-wised iterative variational EM algorithm that is prone to falling into local optima, we employ the niche differential evolution (NDE) [15] algorithm, a population based stochastic global minimizer, to infer each LVB model, where the prior probability of each voxel belonging to every tissue type is given by a constructed probabilistic brain atlas. After training, we linearly combine all LVB models to classify each voxel. The proposed NDE-LVB algorithm has been evaluated against three state-of-the-art brain image segmentation approaches on both synthetic and clinical brain MR images.

2 Variational Mixture of Gaussians

Assume each data x_s in an observed dataset $X = \{x_s \in R^D : s = 1, 2, \cdots, N\}$ to be drawn independently from one of K Gaussian distributions $\mathcal{N}(\mu_k, \Sigma_k)$ with a prior probability ω_k, where all statistical parameters, denoted by $\Theta = \{\omega_k, \mu_k, \Sigma_k : 1 \leq k \leq K\}$ are further assumed to be stochastic variables. If the latent class labels of X are denoted by $Z = \{z_s \in R : s = 1, 2, \cdots, N\}$, the complete-data likelihood $p(X, Z, \Theta)$ is

$$p(X, Z, \Theta) = p(X|Z, \mu, \Lambda)p(Z|\omega)p(\omega)p(\mu, \Lambda), \tag{1}$$

where $\Lambda_k = \Sigma_k^{-1}$ is the precision matrix.

Variational Bayesian inference aims to infer the posterior distribution $p(Z, \Theta|X)$ by introducing a variational distribution $q(Z, \Theta)$ to approximate it [14]. For any choice of $q(Z, \Theta)$, the following decomposition of the model evidence holds

$$\ln p(X) = KL(qp) + L(q), \tag{2}$$

where

$$KL(q\|p) = -\iint q(Z, \Theta) \frac{p(Z, \Theta|X)}{q(Z, \Theta)} dZ d\Theta \tag{3}$$

is the Kullback-Leibler (KL) divergence between $q(Z, \Theta)$ and $p(Z, \Theta|X)$, and

$$L(q) = \iint q(Z, \Theta) \frac{p(X, Z, \Theta)}{q(Z, \Theta)} dZ d\Theta \tag{4}$$

is a functional of $q(Z, \Theta)$ that gives a lower bound on $\ln p(X)$. Since $\ln p(X)$ is a constant, increasing the lower bound $L(q)$ will lead to decreasing the KL divergence $KL(qp)$ towards vanishing. Therefore, the desired variational distribution $q^*(Z, \Theta)$ that approximate $p(Z, \Theta|X)$ can be inferred as follows

$$q^*(Z, \Theta) = \arg \max_{q(Z, \Theta)} L[q(Z, \Theta)]. \tag{5}$$

Assuming $q(Z, \Theta)$ can be factorized between $q(Z)$ and $q(\Theta)$, Eq. (5) can be solved by using the two-stage iterative variational EM (VEM) algorithm [14, 16]

$$\begin{cases} \ln q^*(Z) = \mathbb{E}_\Theta[\ln p(X, Z, \Theta)] + \text{const} \\ \ln q^*(\Theta) = \mathbb{E}_Z[\ln p(X, Z, \Theta)] + \text{const} \end{cases} \tag{6}$$

3 NDE-LVB Algorithm

The proposed NDE-LVB algorithm consists of four major steps: (1) construction of probabilistic brain atlas, (2) extraction of small data volumes, (3) training a LVB model using the NDE algorithm on each small data volume, and (4) combination of trained LVB models for image segmentation.

3.1 Probabilistic Brain Atlas

For each brain MR image, we build its probabilistic atlas using the MR template, prior brain tissue maps and spatial normalization routine provided by the statistical parametric mapping (SPM) package [17]. First, we estimate the non-linear transform that register the MR template on to our MR image using the spatial normalization routine. Next, we apply the obtained transform to the prior maps of gray matter, white matter and CSF, respectively, which are co-aligned with the MR template. We then define the assembly of co-registered brain tissue maps as the probabilistic brain atlas $A = \{a_{sk} : s \in S, k \in \{1, 2, 3\}\}$, where $a_{sk} \in [0, 1]$ gives the prior probability of voxel s belonging to brain tissue type k, and S denotes the 3D lattice on which the MR image is defined.

3.2 Small MRI Data Volumes

Let the voxel size of MR images be $nr \times nc \times ns\,mm^3$. We extract partly overlapped small data volumes with a size of $\lceil \beta/nr \rceil \times \lceil \beta/nc \rceil \times \lceil \beta/ns \rceil$ voxels every $\lceil \gamma/nr \rceil$ rows, $\lceil \gamma/nc \rceil$ columns and $\lceil \gamma/ns \rceil$ slices. If, according to the brain atlas, more than 1/2 of the voxels in a sampled volume are located inside brain, we use this volume to train a LVB model; otherwise, we discard it.

3.3 Variational Bayesian Inference

Let each brain MR image be denoted by $X = \{x_s \in R : s \in S\}$. We assume that the voxel values satisfy a variational mixture of Gaussians, due to the existence of partial volume voxels. Specifically, each voxel value x_s is assumed to be drawn independently from one of K Gaussian distributions $\mathcal{N}(\mu_k, \Sigma_k)$ with the prior probability a_{sk} that is given by the probabilistic brain atlas A, and all statistical parameters, denoted by $\Theta = \{\mu_k, \Sigma_k : 1 \leq k \leq K\}$, are also stochastic variables whose conjugate prior distributions are the independent Gaussian-Wishart distribution over the mean and precision matrix of each Gaussian component. Based on these assumptions, the complete-data likelihood in our model can be calculated as follows

$$p(X, Z, \mu, \Lambda; A) = p(X|Z, \mu, \Lambda)p(Z; A)p(\mu|\Lambda)p(\Lambda). \tag{7}$$

Similarly, we also introduce a variational distribution $q(Z, \Theta)$ to approximate the posterior distribution $p(Z, \Theta|X)$. However, instead of using the variational EM algorithm, which is a local optimization technique, we employ the NDE algorithm to directly solve the maximization problem given in Eq. (5). To this end, we must evaluate the lower bound L(q). Based on the complete-data likelihood given in Eq. (7) and the assumption that $q(Z, \Theta)$ can be factorized between $q(Z)$ and $q(\Theta)$, the lower bound L(q) can be calculated as the sum of following six terms [14]

$$
\left\{
\begin{aligned}
&\mathbb{E}[\ln p(X|Z,\mu,\Lambda)] = \frac{1}{2}\sum\nolimits_{K=1}^{K} N_k \left\{ \begin{aligned} &\ln\tilde{\Lambda} - \beta_K^{-1} - v_k T_r(S_k W_k) - \\ &v_k(\bar{x}_k - m_k)^T W_k(\bar{x}_k - m_k) - \ln(2\pi) \end{aligned} \right\} \\
&\mathbb{E}[\ln p(Z;A)] = \sum\nolimits_s \sum\nolimits_k r_{sk}\ln a_{sk} \\
&\mathbb{E}[\ln p(\mu|\Lambda)] = \frac{1}{2}\sum\nolimits_k \left\{ \ln\left(\frac{\beta_0}{2\pi}\right) + \ln\tilde{\Lambda}_k\frac{\beta_0}{\beta_k} + \ln\tilde{\Lambda}_k - \frac{\beta_0}{\beta_k} - \beta_0 v_k(m_k - m_0)^T W_k(m_k - m_0) \right\}, \\
&\mathbb{E}[\ln p(\Lambda)] = K\ln B(W_0, v_0) + \frac{(v_0 - 2)}{2}\sum\nolimits_k \ln\tilde{\Lambda}_k - \frac{1}{2}\sum\nolimits_k v_k Tr(W_0^{-1}W_k) \\
&\mathbb{E}[\ln q(Z)] = \sum\nolimits_s \sum\nolimits_k r_{sk}\ln r_{sk} \\
&\mathbb{E}[\ln q(\mu,\Lambda)] = \frac{1}{2}\sum\nolimits_k \left\{ \ln\tilde{\Lambda}_k + \ln\left(\frac{\beta_k}{2\pi}\right) - 1 - 2H[q(\Lambda_k)] \right\}
\end{aligned}
\right.
\tag{8}
$$

where, according to the VEM algorithm, the responsibilities r_{sk} is

$$
\ln\rho_{sk} = \frac{1}{C_{sk}}\left\{ \ln a_{sk} + \frac{1}{2}\mathbb{E}(\ln|\Lambda_k|) - \frac{D}{2}\ln(2\pi) - \frac{1}{2}\mathbb{E}_{\mu_k,\Lambda_k}\left[(x_s - \mu_k)^T \Lambda_k(x_s - \mu_k) \right] \right\},
\tag{9}
$$

the statistics of the observed data evaluated with respect to the responsibilities are

$$
\left\{
\begin{aligned}
N_k &= \sum\nolimits_s r_{sk} \\
\bar{x}_k &= \frac{1}{N_k}\sum\nolimits_s r_{sk}x_s \\
S_k &= \frac{1}{N_k}\sum\nolimits_s r_{sk}(x_s - \bar{x}_k)(x_s - \bar{x}_k)^T
\end{aligned}
\right.,
\tag{10}
$$

the posterior model parameters are

$$
\left\{
\begin{aligned}
\alpha_k &= \alpha_0 + N_k \\
\beta_k &= \beta_0 + N_k \\
m_k &= \frac{1}{\beta_k}(\beta_0 m_0 + N_k\bar{x}_k) \\
W_k^{-1} &= W_0^{-1} + N_k S_k + \frac{\beta_0 N_k}{\beta_0 + N_k}(\bar{x}_k - m_0)(\bar{x}_k - m_0)^T \\
v_k &= v_0 + N_k
\end{aligned}
\right.
\tag{11}
$$

$H[q(\Lambda_k)]$ is the entropy of the Wishart distribution, and $B(W_0, v_0)$ is a constant.

3.4   NDE for Variational Bayesian Inference

NED maintains a population of N_p individuals, each encoding an assembly of statistics $\{N_k, \bar{x}_k, S_k : 1 \leq k \leq K\}$. The fitness of each individual is defined as the corresponding lower bound $L(q)$, which can be computed in three steps: (1) apply the statistics to Eq. (11) to generate posterior model parameters; (2) evaluate Eq. (9) to get responsibilities r_{sk}; and (3) evaluate Eq. (8) to obtain $L(q)$.

NDE evolves the population of individuals iteratively [15]. In each generation, a new population is generated by the mutation, crossover and combination operations. The mutation operator extracts the distance and direction information from the current individuals and adds random deviation for diversity. Let the G th generation of population be denoted by $\{\psi_i^G : i = 1, 2, \cdots, N_p\}$. The i th mutant vector can be generated by the following "DE/best/1/bin" mutation operator

$$V_i^G = \psi_{best}^G + F.\left(\psi_{r_1}^G - \psi_{r_2}^G\right), \tag{12}$$

where r_1 and r_2 are mutually exclusive integers taking random values from the range $[1, N_p]$, ψ_{best}^G is the best individual in the current population, and F controls the scaling of the difference vector. After mutation, the following crossover operation is applied to each mutant vector V_i^G and its parent ψ_i^G to generate a trial vector U_i^G

$$U_{i,j}^G = \begin{cases} V_{i,j}^G, & if\,(rnd < CR)\,or\,(j = j_{rnd}) \\ \psi_{i,j}^G, & otherwise \end{cases}, j = 1, \cdots, D, \tag{13}$$

where rnd is a uniform distributed random value within the range $[0, 1]$, j_{rnd} is a random integer within the range $[1, D]$, and CR is the user specified crossover rate. In the combination operation, if a trial vector U_i^G has an improved fitness than its parent ψ_i^G, it will replace ψ_i^G in the new generation. This evolution process continues until meeting the pre-specified stop criterion, which can be either a threshold of the change of highest fitness or a maximum number of generations.

Different from the conventional differential evolution, NDE simultaneously evolves a cohort of populations in a niche environment. Since many small data volumes share similar voxel value distributions, a good individual achieved on one volume may also perform well on other volumes. Therefore, after generating a new generation, the best individual in each population is ranked and best $\lceil N_p/10 \rceil$ among the best are broadcast to each population and used to replace randomly selected individuals. It is expected to accelerate the convergence of all populations in this way.

3.5   Voxel Classification

For each trained model $LVB^{(n)}$, we have the optimal statistics $\{N_k, \bar{x}_k, S_k : 1 \leq k \leq K\}$ that may lead to the maximum lower bound $L(q)$. We first apply these statistics to Eq. (11) to calculate the posterior model parameters, and then use Eq. (9) to the responsibilities $r_{sk}^{(n)}$ of each voxel value x_s. By combining all trained LVB models, we can classify voxel s as

$$z_s^* = arg \max_k \sum_n w_{sn} \cdot r_{sk}^{(n)}, \qquad (14)$$

where

$$w_{sn} = \frac{2}{1 + exp(\delta \cdot d_{sn})} \qquad (15)$$

is a non-linear transform of the Euclidian distance between voxel s and center of the n-th sampled volume.

4   Results

The proposed NDE-LVB algorithm was evaluated against the GMM-based unified registration-segmentation routine in SPM [2], GA-GMM algorithm [8], and variational EM-based Bayesian inference (VEM-BI) [14] on both simulated and clinical T1-weighted brain MR images. The segmentation accuracy was assessed by the percentage of correctly classified brain voxels. The parameters used in the proposed algorithm were empirically set as follows: volume size parameter $\beta = 6$, volume interval parameter $\gamma = 8$, NDE population size $N_p = 100$, NDE scaling factor $F = 0.5$, NDE crossover rate $CR = 0.5$, and non-linear transform parameter $\delta = 1.5$. Default parameter settings were adopted in the other three algorithms.

The first experiment was performed on four simulated T1-weighted brain MR images from the BrainWeb database [18]. Each image has a dimension of $181 \times 217 \times 181$ and a voxel size of $1 \times 1 \times 1 mm^3$, and is corrupted with 40 % intensity inhomogeneity and noise ranging from 1 % to 7 %. The 87th transverse slice of these four images and corresponding segmentation results obtained by using the proposed algorithm were shown in Fig. 1. It reveals that the proposed algorithm can appropriately handle both noise and INU effect. The accuracy of four segmentation algorithms on each 3D MR image was compared in Table 1. It is clear that the proposed algorithm achieved the highest accuracy on all four simulated brain MR images.

The second experiment was carried out on 20 clinical T1-weighted brain MR images from the internet brain segmentation repository (IBSR) [19], which were acquired on a 1.5T Siemens Magnetom MR System (10 FLASH scans) and a 1.5T GE Signa MR System (10 3D-CAPRY scans). All image were reduced to 8-bit from 16-bit by thresholding intensities above or below 99.99 % of the total number of different intensities and then by scaling the remaining range to $[0, 255]$. Each image was first spatially normalized into a standard 3D brain coordinate system by using rotation and non-deformable transformation, and then resliced into a dimension of $256 \times 63 \times 256$ and a voxel size of $1 \times 3 \times 1 mm^3$. The segmentation ground truth was generated by trained investigators using both image histograms and a semi-automated intensity contour mapping algorithm. Figure 2 shows the sagittal, coronal, and transverse slices

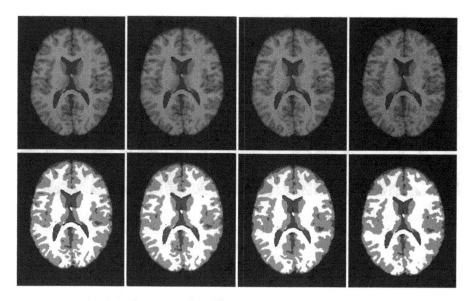

Fig. 1. 88th transverse slice of BrainWeb MR studies with 40 % INU and 1 % (1st column), 3 % (2nd column), 5 % (3rd column) or 7 % noise (4th column) and corresponding segmentation results obtained by using the proposed algorithm (bottom row)

Table 1. Accuracy of four segmentation algorithms on four brain MR images from BrainWeb

Algorithm	40 % INU 1 % Noise	40 % INU 3 % Noise	40 % INU 5 % Noise	40 % INU 7 % Noise	Average
SPM	95.79 %	95.43 %	93.70 %	91.69 %	94.15 %
GA-GMM	94.24 %	93.94 %	92.16 %	89.91 %	92.56 %
VEM-BI	93.27 %	92.13 %	90.53 %	89.54 %	91.37 %
NDE-LVB	97.47 %	96.62 %	94.32 %	92.85 %	95.32 %

of the MR image with index "12-3" and the corresponding segmentation results obtained by using four algorithms. It is clear that the result of the proposed algorithm is more similar to the ground truth. The accuracy of those four algorithms on 20 MR images was depicted in Fig. 3. It shows that the proposed algorithm steadily outperforms the other three algorithms on these clinical studies.

5 Discussions

In the NDE-LVB algorithm, each individual in a evolutionary population encodes the assembly of statistics. Its fitness, i.e. the lower bound $L(q)$, cannot be directly evaluated unless running the variational EM algorithm for one iteration. The differential evolution algorithm will gradually find the solution that could lead to a maximum of $L(q)$, and the one-iteration of variational EM will further increase the $L(q)$. Therefore, strictly speaking, the proposed NDE-LVB algorithm combines the differential evolution that

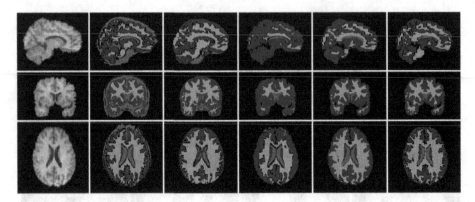

Fig. 2. Segmentation results of brain MR study "12-3" (1st column) obtained by applying the SPM (2nd column), GA-GMM (3rd column), VEM-BI (4th column) and proposed NDE-LVB algorithm (5th column), together with the ground truth (6th column)

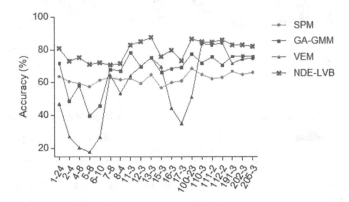

Fig. 3. Accuracy of four segmentation algorithms on 20 clinical brain MR images from IBSR

has global search ability and variational EM algorithm that is a greedy local maximizer to solve the maximization problem in Eq. (13). However, since evolutionary algorithms are usually time-consuming, the proposed algorithm has a relatively high computational complexity. Our experiments show that the average time cost for processing each 3D IBSR data (Intel Core i5-4430 CPU, 8 GB memory and Matlab R2012a) is 24.36 munities, which is significantly higher than the cost of non-evolutionary algorithm based approaches. To improve the efficiency of the proposed algorithm, we must accelerate the convergence of the NDE algorithm and reduce the number of small data volumes used for training LVB models by identifying the most representative volumes.

6   Conclusions

In this paper, we propose the NDE-LVB algorithm for automated brain MR image segmentation. We use the LVB model to characterize the statistical distribution of local voxel values and apply the NDE technique to LVB inference. Our experimental results suggest that the proposed algorithm outperforms not only the variational EM or GA based methods but also the brain image segmentation routine in the commonly used SPM package on both synthetic and clinical MR images. Our future work will focus on reducing the computational complexity of the proposed algorithm.

Acknowledgments. This work was supported in part by the National Natural Science Foundation of China under Grants 61401209 & 61471297, in part by the Natural Science Foundation Youth Project of Jiangsu Province, China, under Grant BK20140790, in part by China Postdoctoral Science Foundation under Grants 2014T70525 & 2013M531364, in part by the Fundamental Research Funds for the Central Universities under Grants 3102014JSJ0006, and in part by the Returned Overseas Scholar Project of Shaanxi Province, China.

References

1. Feng, D., Tierney, L., Magnotta, V.: MRI tissue classification using high-resolution Bayesian hidden Markov normal mixture models. J. Am. Stat. Assoc. **107**, 102–119 (2012)
2. Ashburner, J., Friston, K.J.: Unified segmentation. NeuroImage **26**, 839–851 (2005)
3. Zexuan, J., Yong, X., Quansen, S., Qiang, C., Deshen, X., Feng, D.D.: Fuzzy local Gaussian mixture model for brain MR image segmentation. IEEE Trans. Inf. Technol. Biomed. **16**, 339–347 (2012)
4. Van Leemput, K., Maes, F., Vandermeulen, D., Suetens, P.: A unifying framework for partial volume segmentation of brain MR images. IEEE Trans. Med. Imaging **22**, 105–119 (2003)
5. Dimitris, K., Evdokia, X.: Choosing initial values for the EM algorithm for finite mixtures. Comput. Stat. Data Anal. **41**, 577–590 (2003)
6. Goldberg, D.E.: Genetic Algorithms in Search, Optimization and Machine Learning. Addison-Wesley Professional, Reading (1989)
7. Pernkopf, F., Bouchaffra, D.: Genetic-based EM algorithm for learning Gaussian mixture models. IEEE Trans. Pattern Anal. Mach. Intell. **27**, 1344–1348 (2005)
8. Tohka, J., Krestyannikov, E., Dinov, I.D., Graham, A.M., Shattuck, D.W., Ruotsalainen, U., Toga, A.W.: Genetic algorithms for finite mixture model based voxel classification in neuroimaging. IEEE Trans. Med. Imaging **26**, 696–711 (2007)
9. Li, X., Li, L.H., Lu, H.B., Liang, Z.R.: Partial volume segmentation of brain magnetic resonance images based on maximum a posteriori probability. Med. Phys. **32**, 2337–2345 (2005)
10. Dawant, B.M., Zijdenbos, A.P., Margolin, R.A.: Correction of intensity variations in MR images for computer-aided tissue classification. IEEE Trans. Med. Imaging **12**, 770–781 (1993)
11. Wells III, W.M., Crimson, W.E.L., Kikinis, R., Jolesz, F.A.: Adaptive segmentation of MRI data. IEEE Trans. Med. Imaging **15**, 429–442 (1996)
12. Chunming, L., Rui, H., Zhaohua, D., Gatenby, J.C., Metaxas, D.N., Gore, J.C.: A level set method for image segmentation in the presence of intensity inhomogeneities with application to MRI. IEEE Trans. Image Process. **20**, 2007–2016 (2011)
13. Wang, L., He, L., Mishra, A., Li, C.: Active contours driven by local Gaussian distribution fitting energy. Sig. Process. **89**, 2435–2447 (2009)
14. Bishop, C.M.: Pattern Recognition and Machine Learning. Springer, New York (2007)

15. Sun, J.Y., Zhang, Q.F., Tsang, E.P.K.: DE/EDA: a new evolutionary algorithm for global optimization. Inf. Sci. **169**, 249–262 (2005)
16. Bruneau, P., Gelgon, M., Picarougne, F.: Parsimonious reduction of Gaussian mixture models with a variational-Bayes approach. Pattern Recogn. **43**, 850–858 (2010)
17. Members and collaborators of the Wellcome Trust Centre for Neuroimaging. "Statistical Parametric Mapping (SPM)". http://www.fil.ion.ucl.ac.uk/spm/
18. Kwan, R.K.S., Evans, A.C., Pike, G.B.: MRI simulation-based evaluation of image-processing and classification methods. IEEE Trans. Med. Imaging **18**, 1085–1097 (1999)
19. Center for Morphometric Analysis at Massachusetts General Hospital, "The Internet Brain Segmentation Repository (IBSR)". http://www.cma.mgh.harvard.edu/ibsr/index.html

Natural Image Dehazing Based on L_0 Gradient Minimization

Jingjing Qi, Wen Lu$^{(\boxtimes)}$, Shuyu Yang, and Xinbo Gao

School of Electronic Engineering, Xidian University, Xi'an 710071, China
{jjqi,YangShuyu}@stu.xidian.edu.cn,
{luwen,xbgao}@mail.xidian.edu.cn

Abstract. Haze removal is a challenging problem because of the unknown depth map. Most existing methods focus on deriving an estimate of scene transmission and then restore the true scene appearance from a single observation, which always suffer from halo artifacts as the scene transmission is not accurate. In this paper, a natural image dehazing method based on L_0 gradient minimization is introduced. Firstly, a rough transmission is obtained based on the inherent boundary constraint on the scene, then this constraint is modeled into an optimization problem incorporating L_0 gradient minimization to get a refined transmission. Finally, haze-free images are obtained from the atmospheric scatting model making use of the optimized transmission map and the global atmospheric light. Experimental results demonstrate that the proposed has a better ability to obtain haze-free scenes with abundant edges, smooth detail and vivid color.

Keywords: Haze removal · L_0 gradient minimization · Scene transmission · Haze-free image

1 Introduction

Outdoor images are often contaminated by suspended atmospheric aerosols such as haze, fog, smoke and others. Therefore, the reflected light from objects, before it reaches the camera, is attenuated and absorbed by these particles in the atmosphere. As a result, contrast, color and visibility of images are drastically degraded, which makes it unable to meet the requirements of most image perception systems and algorithms for surveillance, intelligent vehicles, object recognition, and target location. In recent years, haze removal has been a hot spot of research given its wider application range.

The goal of image dehazing is to remove the weather effect caused by suspended atmospheric particles, alleviate loss of contrast and color distortion, and eventually make images more visible. Some techniques have been introduced to tackle the problem. Briefly, based on whether hazy image formation model is used, they can be categorized into two classes in general: one [1, 2] is devoted to image enhancement algorithm, which can significantly enhance contrast and details of the scene, and improve visual quality of the image, but probably lead to a loss of salient details as it is not physically valid. The other [3] is dedicated to recovering the true scene based on the atmospheric scatting model. The restoration is an inverse problem, which is actually a

© Springer International Publishing Switzerland 2015
X. He et al. (Eds.): IScIDE 2015, Part I, LNCS 9242, pp. 603–610, 2015.
DOI: 10.1007/978-3-319-23989-7_61

compensation for distortions arising in degradation processes and whose effect looks more natural and vivid.

Single image dehazing based on the atmospheric scatting model could restore the real haze-free scene, which is being real-time and easiness to achieve, and is known to be severely ill-posed thus needs priors to tackle the problem. Considerable statistical priors have been explored for single image dehazing in previous work to recover the real haze-free scene. In [17], a high-contrast prior on haze-free image, which assumes haze-free images have higher contrast than hazy ones to maximize the number of edges in the dehazed image. Fattal [4] estimated the albedo of the scene and the medium transmission on the assumption that the transmission and the surface shading are locally uncorrelated. In [9], Fattal proposed the color-lines offset regularity in natural images, where pixels of small image patches typically exhibit a one-dimensional distribution in RGB color space, to restore the scene. He et al. [5] proposed the dark channel prior to estimate the rough image transmission map. This prior comes from a statistical observation that most local patches in haze-free images often contain some low intensity pixels.

Several optimized techniques have been investigated in recent work to refine the rough transmission map. Tan [17] regularized the smoothness constraint of atmospheric light to solve the ambiguity of airlight color using Markov random field. Although the method significantly enhances the visibility of hazy images, it usually leads to unrealistic color. Nishino et al. [11] regularized smoothness constraint in luminance field with natural image statistics of heavy-tail distribution on the gradient field. But, it needs more constraints to improve the accuracy. Carr and Hartley [10] encoded geometry constraint into potentials and a Laplacian smooth term is applied to a homogeneous random field. Local regularization is applied for both smoothing and edge-preservation constraint. Reference [12] modeled a fused depth map into an energy minimization problem incorporating a spatial Markov dependence. Meng et al. [14] presented an efficient regularization dehazing method to refine the rough scene transmission achieved from an inherent boundary constraint of the scene. In general, as most approaches are based on a local patch operation, they inevitably suffers from halo artifacts and edges remains insufficient.

To overcome these problems, we propose a natural image dehazing method based on L_0 gradient minimization. The L_0 gradient minimization [13], which targets globally preserving salient edges and structures, can dramatically alleviate the halo artifacts caused by the local patch. Moreover, small-resolution objects and thin edges can be faithfully maintained, which can efficiently alleviate loss of edges that most methods suffered. In our algorithm, we put the L_0 gradient minimization into refining the transmission map, which is curial to image dehazing. The proposed algorithm has a better ability to obtain an accurate scene transmission and recover a haze-free scene with abundant edges, smooth detail and vivid color.

This paper is organized as follows. Section 1 discusses several image dehazing works and problems they suffered. Section 2 gives an instruction of the atmospheric scatting model and details the main work of the algorithm. In Sect. 3, we show some results on real images and make a comparison with several methods. A conclusion is provided in Sect. 4.

2 Image Dehazing Based on L_0 Gradient Minimization

Most endeavors are devoted to find an estimate of the scene transmission, and recover the haze-free scene. The proposed is also following this idea. Given an hazy image, a rough transmission is obtained based on the inherent boundary constraint on the scene, then this constraint is modeled into an optimization problem incorporating L_0 gradient minimization to get a refined transmission. Subsequently, the haze-free images are obtained from the atmospheric scatting model making use of the optimized transmission map and the global atmospheric light. Figure 1 shows the framework of this algorithm.

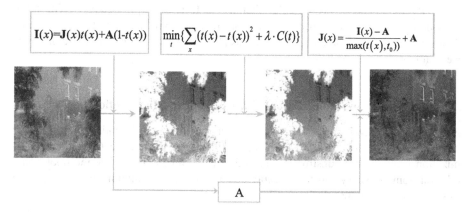

Fig. 1. The framework of the natural image dehazing method based on L_0 gradient minimization

2.1 The Scene Transmission Based on the Boundary Constraint of the Scene

In computer vision and computer graphics, the atmospheric scatting model [6–8] proposed by McCartney [16] and derived by Narasimhan [15], is described as:

$$\mathbf{I}(x) = \mathbf{J}(x)t(x) + \mathbf{A}(1 - t(x)) \tag{1}$$

Where x indexes the pixel of the observed hazy image $\mathbf{I}(x)$, $\mathbf{J}(x)$ is the haze-free image we desired, $\mathbf{A}$ is the atmospheric light, and the transmission $t(x)$ describes the portion of the light that is not scattered and reaches the camera. The goal of haze removal is to obtain $\mathbf{A}$, and $t(x)$, and eventually recover $\mathbf{J}(x)$ from Eq. (1).

The pixel of $\mathbf{I}(x)$ contaminated by fog will approach the global atmospheric light $\mathbf{A}$. The scene radiance $\mathbf{J}(x)$ meets an inherent boundary constraint of the scene [14], that is $\mathbf{C}_0 \leq \mathbf{J}(x) \leq \mathbf{C}_1, \forall x \in \Omega$, Ω is the local neighborhood of pixel x in the radiance cube. Then the transmission can be described as the ratio of two line segments as illustrated in (2), according to the atmospheric scatting model, which leads to the following boundary constraint on $t(x) : 0 \leq t_b(x) \leq t(x) \leq 1$. The $t_b(x)$ strands for the lower bounder of $t(x)$, given by (3). Under this condition, a patch-wise scene transmission $\tilde{t}(x)$

can be calculated by (4), allowing the transmissions in a local patch to be slightly different.

$$\frac{1}{t(x)} = \frac{\|\mathbf{J}(x) - \mathbf{A}\|}{\|\mathbf{I}(x) - \mathbf{A}\|} \tag{2}$$

$$t_b(x) = \min\{ \max_{c\in\{r,g,b\}} (\frac{A^c - I^c(x)}{A^c - C_0^c}, \frac{A^c - I^c(x)}{A^c - C_1^c}), 1\} \tag{3}$$

$$\tilde{t}(x) = \min_{y\in w_x} \max_{z\in w_y} t_b(z) \tag{4}$$

In (3), I^c, A^c, C_0^c and C_1^c are the color channels of $\mathbf{I}$, $\mathbf{A}$, $\mathbf{C_0}$, $\mathbf{C_1}$ respectively. The scene transmission $\tilde{t}(x)$ obtained from [14] is more reliable compared with the scene transmission derived by the dark channel prior, as [14] can handle the bright sky region very well and produce fewer halo artifacts.

2.2 Refining Scene Transmission with L_0 Gradient Minimization

For the local patch-based scene transmission $\tilde{t}(x)$ from the boundary constraint, it inevitably suffers from halo artifacts. In the literature, we put the L_0 gradient minimization measure into constructing an optimization function to refine the poor transmission from boundary constraint, expressed as (6). Through minimizing the objective function (6) to optimize $\tilde{t}(x)$.

$$C(t) = \#\{p\big|\big|\partial_x t(x)\big| + \big|\partial_y t(x)\big| \neq 0\} \tag{5}$$

$$\min_t\{\sum_x (t(x) - \tilde{t}(x))^2 + \lambda \cdot C(t)\} \tag{6}$$

Where x indexes pixel, $\tilde{t}(x)$ is the original transmission under the boundary constrain $t(x)$ is the refined transmission map, $C(t)$ represents the gradient measure along the x' and y' directions given as (5), which counts x whose magnitude $\big|\partial_x t(x)\big| + \big|\partial_y t(x)\big|$ is not zero and λ is a parameter controlling the significance of $C(t)$.

2.3 Estimating the Atmospheric Light

We filter each color channel of the hazy image by a local minimum filter with a moving window w_x, to obtain the dark channel, given as (7). And then pick the top 0.1 % brightest pixels in the dark channel. Among these pixels, the pixel with highest intensity in the input hazy image $\mathbf{I}$ is selected as the atmospheric light, as can be seen from (8) and (9). Where x and y index pixels in hazy image $\mathbf{I}$, and c presents R, G, B color channel.

$$\mathbf{I}_{dark} = \min_{y \in w_x} \{ \min_{c \in \{r,g,b\}} I(y) \} \tag{7}$$

$$\mathbf{A} = \max_{x \in \Phi} (\mathbf{I}(x)) \tag{8}$$

$$\Phi = \{x | \mathbf{I}(x) \text{ is the } 0.1\% \text{ of brightest pixels in dark channel } \mathbf{I}_{dark}\} \tag{9}$$

2.4   Recovering the Scene Radiance

With the atmospheric light and the refined scene transmission, we can recover the scene radiance according to (10):

$$\mathbf{J}(x) = \frac{\mathbf{I}(x) - \mathbf{A}}{[\max(t(x), \varepsilon)]^{\delta}} + \mathbf{A} \tag{10}$$

Where ε is a small constant for avoiding division by zero, and the exponent δ, serving as the role of the medium extinction coefficient β, is used for fine-tuning the dehazing effects.

3   Experimental Results

To demonstrate the effectiveness of our method, we select different types of outdoor scenes in our experiments, which have rich edge characters and depth information. In our algorithm, parameters are set respectively as follows, $\mathbf{C}_0 = (20, 20, 20)^T$, $\mathbf{C}_1 = (300, 300, 300)^T$, $\delta = 0.85$ and λ is allowed to be adjusted to obtain better effect.

As can be seen from the above, images contaminated by haze present several common characters: gray image color, low contrast, blurred scene information, serious distortion. Handling input hazy images by [14] and our method, haze-free images with salient edge information, considerable improved contrast and smooth detail can be got, seen from Fig. 2. Experimental results show that both can obtain a good visual effect. Comparing Meng' scene transmission map with the proposed in column (d), edges of scene transmission in discontinued depth region can be effectively kept, while fields with same depth change smoothly. As Meng' method is based on a local patch measure, they are inevitably affected by halo artifacts. Our method bases on global L_0 gradient minimization smoothing, counting on the number of the image of discrete gradient values, thus alleviate halo problems generated by local patch or average operation. What's more, edges in our scene transmission are sharper and clearer than edges in Meng'. In general, the proposed outperforms Meng' method in most cases.

In order to prove robustness of our algorithm, we further compare our method with more previous methods in Fig. 3. Tan' method is not based on the atmospheric scatting model, which leads recovered images to presenting edge halo effect on the edge of the trunk, seeing from the partial enlarged image. Also, Tan' method will produce color distortion (over supersaturated color). Our technique can effectively overcome halo

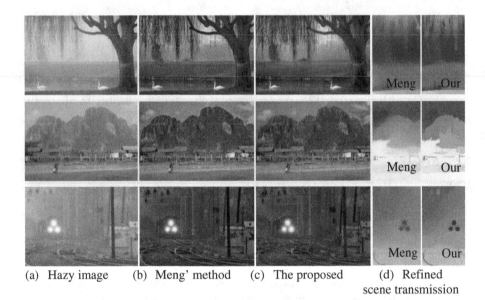

(a) Hazy image   (b) Meng' method   (c) The proposed   (d) Refined
                                                                scene transmission

Fig. 2. Results in [14] and the proposed results. (a) Hazy image (b) Meng' method (c) The proposed (d) Refined scene transmission map

(a) Hazy image   (b) Tan' method   (c) The proposed   (d) Partial
                                                            enlarged image

(a) Hazy image   (b) Tarel' method   (c) Fattal' method   (d) The proposed

Fig. 3. Comparisons with other methods

effects and maintain sufficient edges. Comparing with Tarel' and Fattal' methods, we can find that Tarel' result in edge regions will produce halo effect and remain residual mist, and Fattal' presents oversaturation color. However, our method makes improvement on suppressing halo effect and recovering color reduction.

4 Conclusion

We introduce an edge-preserving smoothing – L_0 Gradient Minimization operator to optimize the transmission under boundary constraint condition. Specifically, we globally maintain and enhance the most prominent set of edges by globally control how many non-zero gradients are counted. Experiments show that salient edges information of images are effectively preserved while smoothing the fields with similar depth, what's more important is that halo artifacts are significantly alleviated. As can be seen from the results, our method can recover haze-free images with vivid color and prominent structural information and the visual effects of our results are quite consistent with our intuitions.

Acknowledgments. This research was supported partially by the National Natural Science Foundation of China (No. 61125204, No. 61372130, No. 61432014), the Fundamental Research Funds for the Central Universities (No. BDY081426, No. JB140214), the Program for New Scientific and Technological Star of Shaanxi Province (No. 2014KJXX-47).

References

1. Zimmerman, J.B., Pizer, S.M., Staab, E.V., Perry, J.R., McCartney, W., Brenton, B.C.: An evaluation of the effectiveness of adaptive histogram equalization for contrast enhancement. IEEE Trans. Med. Imaging **7**(4), 304–312 (1988)
2. Hines, G.D, Rahman, Z., Jobson, D.J., et al.: Single-scale retinex using digital signal processors. In: Global Signal Processing Conference, vol. 27(30) (2004)
3. Narasimhan, S.G., Nayar, S.K.: Chromatic framework for vision in bad weather. In: IEEE Proceeding of Computer Vision and Pattern Recognition, New York, pp. 598–605 (2000)
4. Fattal, R.: Single image dehazing. ACM Trans. Graph. **27**(3), 1–9 (2008)
5. He, K., Sun, J., Tang, X.: Single image haze removal using dark channel prior. IEEE Trans. Pattern Anal. Mach. Intell. **33**(12), 2341–2353 (2011)
6. Nayar, S.K., Narashiman, S.G.: Vision in bad weather. In: International Conference on Computer Vision, vol. 2, pp. 820–827 (1999)
7. Narasimhan, S.G., Nayar, S.K.: Vision and the atmosphere. Int. J. Comput. Vis. **48**(3), 233–254 (2002)
8. Narasimhan, S.G., Nayar, S.K.: Contrast restoration of weather degraded images. IEEE Trans. Pattern Anal. Mach. Intell. **25**(6), 713–724 (2003)
9. Fattal, R.: Dehazing using color-lines. ACM Trans. Graph. **34**(1), 13 (2014)
10. Carr, P., Hartley R.: Improved single image dehazing using geometry. In: Proceedings of Digital Image Computing: Techniques and Applications, pp. 103–110 (2009)
11. Nishino, K., Kratz, L., Lombardi, S.: Bayesian defogging. Int. J. Comput. Vis. **98**(3), 263–278 (2012)

12. Wang, Y.-K., Fan, C.-T.: Single image defogging by multiscale depth fusion. IEEE Trans. Image Process. **23**, 4826–4837 (2014)
13. Xu, L., Lu, C., Xu, Y., et al.: Image smoothing via L_0 gradient minimization. ACM Trans. Graph. **30**(6), 174 (2011)
14. Meng, G., Wang, Y., Duan, J., et al.: Efficient image dehazing with boundary constraint and contextual regularization. In: IEEE International Conference on Computer Vision, pp. 617–624. IEEE, Sydney, NSW (2013)
15. Narasimhan, S.G., Nayar, S.K.: Removing weather effects from monochrome images. In: IEEE Conference on Computer Vision and Pattern Recognition, vol. 2, pp. II-186–II-193 (2001)
16. McCartney, E.J.: Optics of the atmosphere: scattering by molecules and particles. IEEE J. Quantum Electron. **14**(9), 698–699 (1976)
17. Tan, R.T.: Visibility in bad weather from a single image. In: IEEE Conference on Computer Vision and Pattern Recognition, pp. 1–8 (2008)

An Adaptive Approach for Keypoints Description Using Fractional Derivative

Shaohui Si[(⊠)], Fuyuan Hu, Zhenhua Wang, Ziqiang Bi,
Cheng Cheng, and Ze Li

School of Electronic and Information Engineering,
Suzhou University of Science and Technology, Suzhou 215011, China
305368722@qq.com

Abstract. Traditional image descriptors tend to utilize integral derivative characterizing local features, like orientation histogram. However, integral-based derivative has a disadvantage in describing image texture details in smooth area. In this paper, we propose a novel framework for reestablishing orientation histogram based on adaptive fractional derivative, which is better at representing local feature. Then a general weighting scheme for orientation histogram is developed, which improves the accuracy of keypoints description. Finally, we demonstrate the utility of our formulation in implementing solutions for various keypoints descriptions tasks. To exercise our framework we have created a new SIFT and SURF application over images.

Keywords: Local descriptors · Support cross skeleton · Adaptive fractional derivative · Orientation histogram · Bilateral-based weight

1 Introduction

Point detection plays a basic and important role in computer vision and pattern recognition. It has been widely used in many areas such as image registration, object recognition and tracking, motion target detection, etc. The classical and well-known image descriptor, SIFT, is the milestone for image feature researching [1]. The SIFT features are invariant to image scale and rotation, and shown to provide robust matching across a substantial range of affine distortion. And the descriptor has its own dominant orientation and magnitude, which is computed by a local image patch. The following a wide variety of detectors and descriptors have already been proposed in the literature (e.g. [2–6]). Bay et al. [2] have proposed the fast robust feature, SURF, considering integral images and Haar wavelet response. It reduced the computation time drastically while keeping it sufficiently distinctive. Fan et al. [3] combined intensity orders and gradient distributions in multiple regions to further improve the descriptor's discriminative ability. Note that most local descriptors are assigned a consistent magnitude and orientation to have better invariant performance. We know these vector descriptors are obtained depending on orientation histogram estimation with integral-based derivation among patch pixels. However, the integral differential in images is proved that it has some disadvantages in charactering texture detail in low-frequency areas [7].

© Springer International Publishing Switzerland 2015
X. He et al. (Eds.): IScIDE 2015, Part I, LNCS 9242, pp. 611–625, 2015.
DOI: 10.1007/978-3-319-23989-7_62

On the other side, since a fractional differential operator is capable of characterizing fractal-like structures [8] which are often found in the texture regions, this class of operator is considered as an effective tool for texture characterization or even enhancement in images. Pu et al. [7] have developed an $n \times n$ fractional differential operator mask, and it was noticed that the adoption of the mask results is in better enhancement of texture details compared with traditional integral-based differential operators. Nevertheless, the spatial step in the numerical implementation of the fractional differential operator usually advances by one. As a result, the high degree of self-similarity in images is not well characterized. In addition, it is not convenient to manually search for the optimal fixed fractional derivative order which matches the local texture details. In our recent work [9], a novel fractional differential operator with adaptive non-integral step and order is proposed to enhance texture according to local underlying texture patterns.

As a matter of fact, most gradient histogram-based descriptors, which are accustomed to computing the magnitude or orientation by first-order difference, may bring some errors in points matching. Due to the superiority of fractional derivative application in 2-D images, we reestablish the histogram orientation using fractional derivative to improve accuracy for local texture characterization. Additionally, the established histogram orientation will lead to descriptors accurate application. The contributions in this paper lie in three aspects as follows:

(1) A general framework for exploring orientation histogram based on adaptive fractional derivative is identified, so that the local gradient can be well characterized.
(2) A novel measure for orientation histogram establishment is presented, so that not only the distance but intensity similarity is both considered.
(3) We show how several local descriptors can be efficiently expressed in our proposed framework.

The paper is organized as follows. The related research is introduced in Sect. 2. The proposed algorithm is organized in Sect. 3. Then the experimental results are demonstrated with discussion in Sect. 4. Finally, our conclusions are summarized in Sect. 5.

2   Related Research

2.1   Orientation Histogram with Integral-Based Derivative Estimation

Most of image local descriptors tend to establish the orientation histogram, which is formed from gradient orientations of sample points within a region around the keypoint [1]. For the Gaussian smoothed image L, the gradient magnitude $m(x, y)$, and orientation $\theta(x,y)$ is precomputed using pixel integral-based differences:

$$m(x,y) = \sqrt{(L(x+1,y) - L(x-1,y))^2 + (L(x,y+1) - L(x,y-1))^2} \qquad (1)$$

$$\theta(x,y) = \tan^{-1}((L(x,y+1) - L(x,y-1))/(L(x+1,y) - L(x-1,y))) \qquad (2)$$

From Eq. (1), no matter horizontal or vertical direction, the magnitudes are represented by pixel integral-based differences. And each sample added to the histogram is weighted by its gradient magnitude and by a traditional Gaussian-weighted circular window.

2.2 Grumwald-Letnikov Defined Fractional Differential

The fractional differential in recent years is considered as an effective tool for texture enhancement in images, and the Grumwald-Letnikov (G-L) defined fractional differential is adopted frequently:

$$
{}^G_\alpha D^v_t = \lim h^{-v} \sum_{m=0}^{[\frac{t-a}{k}]} (-1)^m \frac{\Gamma(v+1)}{m!\Gamma(v-m+1)} f(t-mh) \tag{3}
$$

where Γ denotes the gamma function, $\Gamma(n) = \int_0^\infty e^{-t} t^{n-1} dt = (n-1)!$. If the duration of signal $f(t)$ is within the range $[a, t]$, dividing this range according to $h = 1$, as a result, $n = [\frac{t-a}{h}]^{h=1} = [t-a]$. Therefore, the approximate form of the fractional derivative of a signal $f(t)$ can be expressed as

$$
\frac{d^v f(t)}{dt^v}\bigg|_{G-L} \approx f(t) + (-v)f(t-1) + \frac{(-v)(-v+1)}{2} f(t-2) + \cdots + \frac{\Gamma(n-v+1)}{(n-1)!\Gamma(v)!} f(t-n+1) \tag{4}
$$

where coefficients here are used to establish filter operator.

3 Orientation Histogram with Fractional Derivative and Joint Weight

Generally, the gradient orientation that forms the orientation histogram is usually estimated based on integral-based derivative in Sect. 2.1. We reestablish the orientation histogram using the adaptive fractional derivative with robust joint weight to, which have the superiority in describing complex textures always characterized by irregular and disorderly patterns. Figure 1 provides an overview of the algorithm. As can be seen, with a suitable texture similarity measure, a local support cross skeleton domain on anchor point p, i.e. $\{h_p^i\}$, where $i \in \{0, 1, 2, 3\}$, can be defined. Based on the proposed skeleton domain, the local adaptive fractional order can be dynamically determined. With obtained adaptive order set, the gradient orientation can be redesigned. Finally, the orientation histogram is formed through gradient orientations added by joint weight.

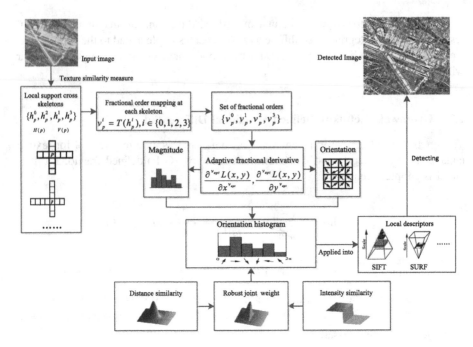

Fig. 1. Block diagram of the proposed method

3.1 Adaptive Selection Mechanism for Fractional Order Set

Before adaptive selecting fractional order, the local support cross skeleton [9] is adopted. Without establishing local non-regular region complexity, only the cross skeleton is applied to estimate underlying texture patterns. As seen in Fig. 2(c), W_p is a square window of radius r, and the cross skeleton, i.e., $\{h_p^0, h_p^1, h_p^2, h_p^3\}$ is obtained through texture similar measure.

Notice that the arm lengths reflect the dominant direction of the underlying texture patterns, we aim to emphasis the texture along these major directions, and weak the pattern along the minor directions which can be achieved by increasing the fractional order for the longer arms, and decreasing the order for the shorted arms.

As observed in [7], a suitable range of the fractional order v is from 0.4 to 0.7. To design a smooth mapping, we consider the following exponential model:

$$v_p^i = T(h_p^i) = \eta_1 \exp(-\frac{h_p^i}{r}) + \eta_2, \ s.t. \begin{cases} h_p^i \in [1, r] \\ v_{ideal-down} \leq v_p^i \leq v_{ideal-up} \\ v_p^i = v_{ideal-down}\big|_{h_p^i = r} \ or \ v_p^i = v_{ideal-up}\big|_{h_p^i = 1} \end{cases} \quad (5)$$

Based on Eq. (5), the set of fractional orders $\{v_p^0, v_p^1, v_p^2, v_p^3\}$ can be computed.

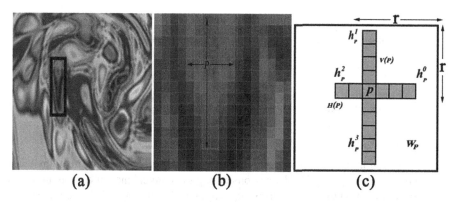

Fig. 2. Local support cross skeleton: (a) Original image, (b) Local support cross skeleton corresponding to the black box in (a), (c) Modal of the support cross skeleton

3.2 Magnitude and Orientation Recomputation with Adaptive Fractional Derivative

The magnitude and orientation for sample points in the image patch can be recomputed based on the local adaptive fractional derivative. For 2-D image signal $L(x, y)$, the fractional differential seen in Eq. (4) can rewrite in the following two expressions

$$
\frac{\partial^{v_p^i} L(x, y)}{\partial x^{v_p^i}} \bigg|_{\{v_p^0, v_p^2\}} \approx L(x, y) + (-v_p^i)L(x - 1, y)
$$
$$
+ \frac{(-v_p^i)(-v_p^i + 1)}{2} L(x - 2, y) + \cdots \tag{6}
$$
$$
+ \frac{\Gamma(n - v_p^i + 1)}{(n - 1)!\Gamma(-v_p^i)!} L(x - n + 1), y)
$$

$$
\frac{\partial^{v_p^i} L(x, y)}{\partial y^{v_p^i}} \bigg|_{\{v_p^1, v_p^3\}} \approx L(x, y) + (-v_p^i)L(x, y - 1)
$$
$$
+ \frac{(-v_p^i)(-v_p^i + 1)}{2} L(x, y - 2) + \cdots \tag{7}
$$
$$
+ \frac{\Gamma(n - v_p^i + 1)}{(n - 1)!\Gamma(-v_p^i)!} L(x, y - n + 1))
$$

where v_p^i denotes adaptive fractional differential order set, which depends on the directional arm length, i.e. $\{h_p^0, h_p^1, h_p^2, h_p^3\}$, in the cross skeleton.

Note that the arm length reflects the dominant direction of underlying texture. Therefore, we consider using pixels among the longer arm to well character the texture property, redefining the magnitude and orientation in Eqs. (1) and (2) as following:

$$m(x,y) = \sqrt{ \left(\frac{\partial^{v_p^i} L(x,y)}{\partial x^{v_p^i}} \bigg|_{v_p^i Max\{h_p^0, h_p^2\}} \right)^2 + \left(\frac{\partial^{v_p^i} L(x,y)}{v_p^i} \bigg|_{v_p^i Max\{h_p^1, h_p^3\}} \right)^2 } \tag{8}$$

$$\theta(x,y) = \tan^{-1} \left(\left(\frac{\partial^{v_p^i} L(x,y)}{\partial y^{v_p^i}} \bigg|_{v_p^i Max\{h_p^1, h_p^3\}} \right) \Big/ \left(\frac{\partial^{v_p^i} L(x,y)}{\partial x^{v_p^i}} \bigg|_{v_p^i Max\{h_p^0, h_p^2\}} \right) \right) \tag{9}$$

where the selection of v_{apt} depends on the longer arm at horizontal or vertical direction. Limited by the h_p, i.e., $h_p \in [1, r]$, and considering the convenience of computation and accuracy, we take first three items from Eqs. (6) and (7) when $h_p > 1$. If the longer arm h_p in any direction is equal to 1, we make a little adjustment for Eqs. (8) and (9) as:

$$\frac{\partial^{v_p^i} L(x,y)}{\partial x^{v_p^i}} \approx L(x-1,y) + (-v_p^i) L(x,y) + \frac{(-v_p^i)(-v_p^i + 1)}{2} L(x+1,y) \tag{10}$$

$$\frac{\partial^{v_p^i} L(x,y)}{\partial y^{v_p^i}} \approx L(x,y-1) + (-v_p^i) L(x,y) + \frac{(-v_p^i)(-v_p^i + 1)}{2} L(x,y+1) \tag{11}$$

which is capable of reflecting the gradient change more accurately.

3.3 A Robust Weighting Scheme

As many local descriptors described, a Gaussian-weight circular window is always added to the gradient orientation to establish orientation histogram. Gaussian-weight is well known to distinguish distance similarity. However, its ignorance in intensity similarity may introduce noise or low correlated pixels, bringing unappealing detecting results.

Inspired by the bilateral filter which enforces the policy considering not only geometric closeness but color similarity, we add color similarity to the weight measurement. By applying the appropriate joint weight w_d and w_c to our framework, the weakness mentioned earlier can be considerably mitigated. The joint weighting functions are defined as follows.

$$w_d = \frac{(x-i)^2 + (y-j)^2}{2\sigma_d^2} \tag{12}$$

$$w_c = \frac{\|I(x,y) - I(i,j)\|}{2\sigma_c^2} \tag{13}$$

where w_d denotes the distance similarity, and w_c reflects the color similarity. I is the intensity component from the HIS space. $I(i, j)$ is a neighbor pixel at the same scale with $I(x, y)$, and σ_d, σ_c are parameters to regulate similarity factors respectively. Pseudo-code for the proposed scheme is summarized in Algorithm 1.

Algorithm 1:	Orientation histogram with adaptive fractional derivative and robust joint weight

Input: input   image L, radius r,

distance similar factor σ_d, intensity similar factor σ_c

Output: output image G

For each pixel in local patch image around the key point

1: Compute the local support cross skeleton $\{h_p^0, h_p^1, h_p^2, h_p^3\}$;

2: Use Eq. (5) to estimate the set of fractional orders $\{v_p^0, v_p^1, v_p^2, v_p^3\}$;

3: Use Eq. (8) to compute the magnitude with fractional derivative

Use Eq. (9) to compute the orientation with fractional derivative;

4: Use Eq. (12) (13) to add the robust weight for orientation histogram.

End For

4   Discussion and Experiments

Our formulation can be added to any gradient-based estimation in local feature description. In this section, we apply the framework focusing on adaptive fractional-based derivative in SIFT and SURF as examples. It is well known that the standard SIFT or SURF description is the milestone for keypoints representation. Notice that both of them counts on the gradient estimation in patch images. We aims to demonstrate the performance for the new SIFT and SURF descriptions exerted by our scheme, comparing to the original algorithms on controlled experiments, the Graffiti[1]. To objectively quantify the capability for descriptors, the evaluation metrics is discussed. Our experiments are conducted on a PC with an Intel i5-4850HQ CPU and 4 GB RAM, and the system is implemented using C#.

4.1   An Evaluation Metric

Recall-Precision is the popular metric in the literature. It captures the fact that we want to increase the number of correct positives while minimizing the number of false positives. We choose to measure the performance of the new SIFT and SURF local descriptors representation on a keypoint matching problem. The keypoints localization for all of the images are identified (using the initial stages of the SIFT or SURF algorithm). All pairs of keypoints from different images are examined and presented

---

[1] http://lear.inrialpes.fr/people/mikolajczyk/Database/index.html.

based on developed scheme. If the Euclidean distance between the feature vectors for a particular pair of keypoints falls below the chosen threshold, this pair is termed a match. A correct-positive is a match where the two keypoints correspond to the same physical location. A false-positive is a match where the two keypoints come from different physical locations. The total number of positives for the given dataset is known a priori. From these numbers, we can determine recall and 1-precision:

$$recall = \frac{number\ of\ correct\ -\ positives}{total\ number\ of\ positives} \tag{14}$$

and

$$1 - precision = \frac{number\ of\ false\ -\ positives}{total\ number\ of\ matches(correct\ or\ false)} \tag{15}$$

4.2   Implementation in Standard SIFT Description

The classical local SIFT descriptor, as described in [1], consists of four major stages: (a) scale-space peak selection; (b) keypoints localization; (c) orientation assignment; (d) keypoints description. The third stage identifies the dominant orientations for each keypoint based on its local image patch. And the final stage builds a local image descriptor for each keypoint, based upon the image gradients in its local neighborhood.

As discussed formulation in Sect. 3, our scheme can be applied naturally into the SIFT. The orientation histogram can be redefined with fractional derivative, i.e., Eqs. (8) and (9), seen in Fig. 3. Comparing with integral differential, fractional derivative is capable of characterizing local texture detail more accurately. Additionally, our robust joint weight is also taken into consideration, not only from distance but color similarity, to judicially assign the contribution that sample points in image patch make. In this way, the orientation histogram is more superior to the previous one with respect to local texture characterization and feature distinctiveness. It is certain that this scheme is beneficial for the descriptors invariance to kinds of deformations. And obviously, SIFT-like descriptors, i.e., PCA-SIFT, ASIFT, are also suitable for our method.

4.3   Implementation in Standard SURF Description

The popular local descriptor SURF, as described in [2], whose keypoints localization is defined based on Hessian matrix and integral images. Differ from the SIFT, Haar wavelet response adopted in sub-regions to assign orientation and construct a descriptor. The dominant orientation in this method is estimated by calculating the sum of all horizontal and vertical responses within a sliding window.

The proposed scheme in this paper can also be applied in SURF. With our formulation, the Haar wavelet response in horizontal and vertical direction can be redefined respectively as

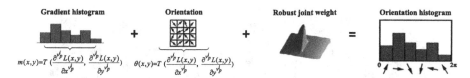

Fig. 3. Improved SIFT description: an orientation histogram is formed from the fractional-based gradient orientations of sample points within a region around the keypoint. Each sample added to the histogram is weighted by its gradient magnitude and by the robust joint weight.

$$V_x = \left(\left[-\frac{\partial^{v_p^i} L(x, y)}{\partial y^{v_p^i}} \right] - \left[\frac{\partial^{v_p^i} L(x+1, y)}{\partial y^{v_p^i}} \right] \right) * (w_d + w_c) \qquad (16)$$

$$V_y = \left(\left[-\frac{\partial^{v_p^i} L(x, y)}{\partial x^{v_p^i}} \right] - \left[\frac{\partial^{v_p^i} L(x, y+1)}{\partial x^{v_p^i}} \right] \right) * (w_d + w_c) \qquad (17)$$

Equations (16) and (17) are calculated through adaptive fractional derivative combined with joint weight, which helps to describe the dominant orientation more accurately. The detail experiments are shown in Sect. 4.4.

4.4   Controlled Transformations

We ran the various types of experiments to explore the difference between the standard SIFT representation (or SURF) and the fractional-based derivative SIFT (or SURF). Each descriptor's robustness to effects caused by the addition of noise, changes in illumination and the application of image transformation is examined. We collected a dataset of images and applied the following transformations to each image.

(a) Performance on images with noise addition

In order to demonstrate the performance of local texture characterization based on our proposed framework, we add the Gaussian noise with different σ to the input image as matching images. Figure 4 gives the matching results intuitively for algorithms with $\sigma = 0.02$ as an instance, and Table 1 shows the detail data for it. As seen in Fig. 4 and Table 1, fractional-based derivative descriptors find more corresponding points in terms of no matter total number or final correct matching points. It is because that our scheme has an advantage in characterizing local feature even influenced by Gaussian noise. Figure 5 presents the result for *recall* and 1-*precision* with σ increasing. Figure 5 also shows that fractional-based derivative descriptors are dramatically better at handling noisy images for almost all values of 1-*precision*, and that fractional-based derivative SIFT (on the orientation histogram) dominates the SIFT representation of the local image patch.

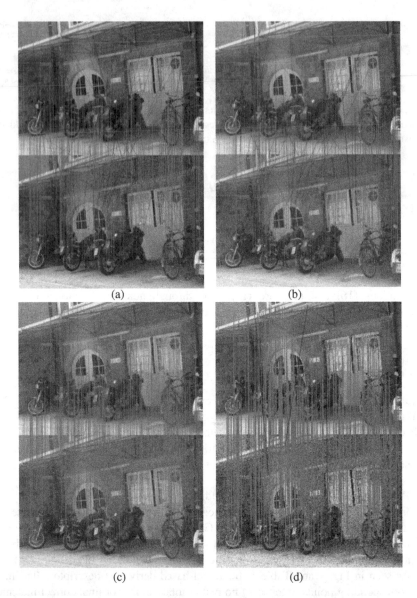

Fig. 4. Standard local descriptors vs. ours on matching task in addition of noise: (a) SIFT, (b) fractional derivative-based SIFT, (c) SURF, (d) fractional derivative-based SURF. The up image 1 is standard image, and the down image 2 is noisy image with σ = 0.02. The red lines represent correct matching key points, and the blue lines denotes false corresponding points (Color figureonline).

Table 1. The detail result data in Fig. 5

Descriptors	Total matches	Correct matching rate
SIFT	454	92.1 %
Improved SIFT	527	93.4 %
SURF	627	96.0 %
Improved SURF	694	96.7 %

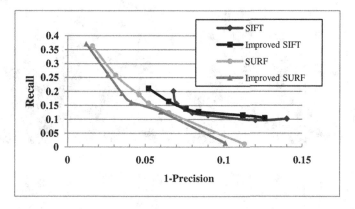

Fig. 5. Recall & 1-Precision for images matching with increasing Gaussian noise

(b) Performance on images with illumination changes

Most local descriptors consider the illumination influence, and are invariant to affine changes in illumination. However, non-linear illumination changes can also occur due to camera saturation. Noticed that the texture in smooth areas, especially in dark areas, is difficult to localize keypoints. Thus, inspired by the recent study [9], preprocessing has been adopted for image texture detail enhancement in low-frequency regions. The experimental results are shown in Fig. 6 and Table 2. As seen in Fig. 6, the number of wrong matches (blue matching lines) in Fig. 6(b) is significantly reduced compared to standard SIFT in Fig. 6(a) because well characterization in orientation histogram with our algorithm. As seen in Table 2, the number of keypoints detection in image 2 in (b) and (d) has been increased remarkably due to the stage of preprocessing. Furthermore, the correct matching rates for improved local descriptors with our formulation are both higher than standard ones. The fractional derivative-based SIFT is 3.1 % higher than SIFT, and the fractional derivative-based SURF is 1.4 % than SURF.

(c) Performance on images with viewpoint changes

The image features of the same object may present significantly difference in different viewpoints, which may bring difficulties for corresponding points matching. Therefore, we examine the robustness of the proposed local descriptors in terms of viewpoint changes seen in Fig. 7. And Table 3 gives the detail results. As seen in Fig. 7, the number of corresponding points decreased significantly with

(a)   (b)

(c)   (d)

Fig. 6. Standard local descriptors vs. ours on matching task in illumination changes: (a) SIFT, (b) fractional derivative-based SIFT, (c) SURF, (d) fractional derivative-based SURF. The up image 1 is standard image, and the down image 2 has been adjusted in brightness. The red lines represent correct matching keypoints, and the blue lines denotes false corresponding points.

Table 2. Matched results in Fig. 7

Descriptors	Key points in image 1	Key points in image 2	Total matches	Correct matching rate
SIFT	2220	1220	315	91.4 %
Improved SIFT	2220	1318	473	94.5 %
SURF	1952	1152	357	97.3 %
Improved SURF	1952	1239	396	98.7 %

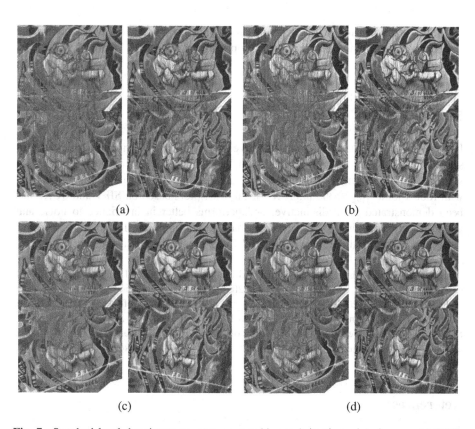

(a)   (b)

(c)   (d)

Fig. 7. Standard local descriptors vs. ours on matching task in viewpoint changes: (a) SIFT, (b) fractional derivative-based SIFT, (c) SURF, (d) fractional derivative-based SURF. The up image 1 is standard image, and the down image 2 has been rotated. The red lines represent correct matching keypoints, and the blue lines denotes false corresponding points.

viewpoints changing evidently. Table 3 shows that the correct matching rate for improved descriptors with our formulation is higher than that for original standard descriptors. It is because that the dominant orientation is well determined and orientation histograms are reestablished with our adaptive fractional-based derivation and robust joint weight. As accurate calculation in Table 3, our improved

Table 3. The detail result data in Fig. 7

Descriptors	Total matches in viewpoint 1 (left)	Correct matching rate in viewpoint 1 (left)	Total matches in viewpoint 2 (right)	Correct matching rate in viewpoint 2 (right)
SIFT	881	93.81 %	356	79.50 %
Improved SIFT	909	95.96 %	387	82.74 %
SURF	595	97.25 %	140	94.23 %
Improved SURF	625	98.45 %	227	95.17 %

SIFT is 2.15 % and 3.24 % higher than the standard SIFT in correct matching rate in two viewpoints. Similarity, the improved SURF is 1.2 % and 0.94 % than standard SURF.

5 Conclusion

In this work, a formulation for the adaptive fractional-based derivative has been presented for keypoints description. Through local fractional order selection based on support cross skeleton, we can do better in characterizing local texture patterns. And with robust joint weight, the improved local descriptors based on SIFT and SURF has been demonstrated more distinctive, and performs better in invariance to noise and affine distortion. In the future work, we shall further apply the formulation as a fundamental scheme to different areas that all focus on gradient measurement.

Acknowledgement. This research is supported by the National Natural Science Foundation of China under Grant No. 61472267 and No. 61203048, Nature Foundation of Jiangsu Province under Grant No. BK2012166, a grant from the City University of Hong Kong (Project No. 7004220), the Open Foundation of Modern Enterprise Information Application Supporting Software Engineering Technology R&D Center of Jiangsu Province under Grant No. SK201206, and the Innovation Project of Graduate Student Training under Grant No. CXZZ13_0854.

References

1. Lowe, D.G.: Distinctive image feature from scale-invariant keypoints. Int. J. Comput. Vis. **60** (2), 91–110 (2004)
2. Bay, H., Tuytelaars, T., Gool, L.V.: SURF: speed up robust features. In: Proceedings of the Europeans Conference on Computer Vision, pp. 404–417 (2006)
3. Fan, B., Wu, F.C., Hu, Z.Y.: Rotationally invariant descriptors using intensity order pooling. IEEE Trans. Pattern Anal. Mach. Intell. **34**(10), 2031–2045 (2012)
4. Cao, X., Zhang, H., Liu, S., et al.: SYM-FISH: a symmetry-aware flip invariant sketch histogram shape descriptor. In: Proceedings of the IEEE International Conference on Computer Vision, pp. 313–320 (2013)

5. Yi, K.M., Jeong, H., Heo, B., et al.: Initialization-insensitive visual tracking through voting with salient local features. In: Proceedings of IEEE International Conference on Computer Vision, pp. 2912–2919 (2013)
6. Zhang, H., Zhou, W., Reardon, C., et al.: Simplex-based 3D spatio-temporal feature description for action recognition. In: Proceedings of IEEE Conference on Computer Vision and Pattern Recognition, pp. 2067–2074 (2014)
7. Pu, Y.F., Zhou, J.L., Yuan, X.: Fractional differential mask: a fractional differential-based approach for multiscale texture enhancement. IEEE Trans. Image Process. **19**(2), 491–511 (2010)
8. Rocco, A., West, B.: Fractional calculus and the evolution of fractal phenomena. Phys. A Stat. Mech. Appl. **265**(3), 535–546 (1999)
9. Hu, F.Y., Si, S.H., Wong, H.S., et al.: An adaptive approach for texture enhancement based on a fractional differential operator with non-integer step and order. Neurocomputing, **158**, 295-306 (2015)

Second Order Variational Model for Image Decomposition Using Split Bregman Algorithm

Jinming Duan[1][(✉)], Wenqi Lu[2], Guodong Wang[2], Zhenkuan Pan[2], and Li Bai[1][(✉)]

[1] School of Computer Science, University of Nottingham, Nottingham, UK
jxd@cs.nott.ac.uk, bai@cs.nott.ac.uk
[2] College of Information Engineering, Qingdao University, Qingdao, China

Abstract. The classical first order Vese-Osher model is capable of decomposing an image into its structure and texture components. However, an undesirable feature of this model for the task is the 'staircase' side effect that appears in the structure component. In this paper, we propose a second order Vese-Osher model for image decomposition, which incorporates second order derivative information and is able to eliminate the side effect of the first order model. In order to avoid directly calculating the high order nonlinear partial differential equation (PDE) of the proposed model, the split Bregman algorithm is applied, which allows the use of fast Fourier transform and analytical generalized soft thresholding equation. Experiments are conducted to demonstrate the effectiveness and efficiency of the proposed model.

Keywords: Image decomposition · Vese-Osher model · Second order derivative · Split Bregman algorithm · Fast fourier transform

1 Introduction

Image decomposition into structure and texture is an important procedure for image understanding and analysis. Structure contains edges and hues of objects and texture is characterized as oscillations in the image. These individual components can be used, for example, for similarity analysis, texture synthesis, texture image segmentation [1], and structure or texture image inpainting [2,3].

A common way to decompose an image into structure and texture is to use the variational approach. Vese and Osher proposed the first order Vese-Osher (FOVO) variational model [4] that combines Meyer's oscillating function [5] for texture images with the total variation (TV) model [6]. However, the FOVO model suffers from the undesirable staircase effect - the decomposed structure image has a uneven appearance. This is because energy minimization in the bounded variation (BV) space [6] results in a piecewise constant objective function, leading to the staircase effect. High order variational models can be used to remedy this side effect. However, the high order models usually contain second order derivatives which lead to nonlinear fourth-order partial differential equations (PDEs) that are very difficult to discretize to solve computationally.

© Springer International Publishing Switzerland 2015
X. He et al. (Eds.): IScIDE 2015, Part I, LNCS 9242, pp. 626–636, 2015.
DOI: 10.1007/978-3-319-23989-7_63

In this paper, we propose a second order Vese-Osher (SOVO) model for image decomposition to overcome these problems with existing models mentioned above. The proposed model replaces the first order regularizer in original FOVO model with the second order regularizer used in Lysaker-Lundervold-Tai (LLT) model [7], which enables it to decompose an image into an oscillating texture component and a structure part without the staircase effect. Instead of solving high order nonlinear PDEs, the split Bregman algorithm [8], which has been successfully applied to optimise L1-based variational models, is adapted to transform the energy minimization problem of the proposed SOVO decomposition model into four subproblems. These subproblems are then efficiently solved by fast Fourier transform (FFT) and analytical soft thresholding equations without any iteration. We validate the new model through extensive experiments.

2   The Proposed SOVO Model

In order to extract the structure and texture components, Meyer defined the Banach space G as follows

$$G = \{v\,|v = div\,(g)\,,\ g = (g_1, g_2) \in L^\infty\}$$

where $div\,(\cdot)$ is divergence operator and $div\,(g) = \partial_x g_1 + \partial_y g_2$. The space G is equipped with the following norm

$$\|v\|_G = \inf_{g=(g_1,g_2)} \left\{ \left\| \sqrt{g_1^2 + g_2^2} \right\|_{L^\infty} \middle| v = div\,(g)\,,\ g \in L^\infty,\ |g| = \sqrt{g_1^2 + g_2^2} \right\}$$

A function belonging to space G may have large oscillation and a small norm. Thus the G norm can be adapted to capture the oscillations of a function in energy minimization. As such, Meyer proposed the following image decomposition model (2.1)

$$\min_{u \in BV(\Omega)} \left\{ E(u) = \int_\Omega |\nabla u| + \lambda \|v\|_G,\ f = u + v \right\} \tag{2.1}$$

where Ω is an open and bounded domain and an input image function f is defined on Ω. Function u is defined in BV space and represents the non-oscillating structure part. The first term in this model is total variation of u, which helps to preserve sharp edges or contours of objects. Function v in the second term belonging to space G denotes the oscillating part (texture or noise) of an image. The model only replaces the L^2 norm in the data fitting term of the TV/ROF model by the G norm. However, it is not possible to derive the Euler-Lagrange equation for the G norm to use a straightforward PDE method to solve it.

To implement the model numerically, Vese and Osher [4] first propose the FOVO model based on $div\,(L^p)$ norm

$$\min_{u,g} \left\{ E(u,g) = \frac{1}{2} \int_\Omega (f - u - div\,(g))^2 + \lambda \int_\Omega |\nabla u| + \gamma \left[\int_\Omega |g|^p \right]^{1/p} \right\} \tag{2.2}$$

where $p \geq 1$. λ and γ are two positive tuning parameters balancing the three energy terms. Vese and Osher also confirmed that there are no obvious numerical differences using different values of p, with $1 \leq p \leq 10$. However, the case $p = 1$ yields faster calculation. However, as the term $\int_\Omega |\nabla u|$ in (2.2) is defined in BV space, leading to piecewise constant result (i.e. staircase effect) in the decomposed structure.

To avoid this side effect without blurring edges of objects, we directly incorporate the second order derivative information into the original FOVO model, that is, we replace the first order TV term with the second order one. Our proposed SOVO model then becomes

$$\min_{u,g} \left\{ E\left(u, g\right) = \frac{1}{2} \int_\Omega \left(f - u - div\left(g\right)\right)^2 + \lambda \int_\Omega e\left(|\nabla u|\right) |\nabla^2 u| + \gamma \int_\Omega |g| \right\}$$
(2.3)

where $\nabla^2 u$ denotes the Hessian matrix of u over image domain Ω. And the diffusivity e is defined as follows

$$e\left(s\right) = 1 - \exp\left(\frac{-C_h}{\left(s/\lambda\right)^h}\right)$$

The constant C_h is automatically calculated in such way that $\left(\partial\Phi\left(s\right)/\partial s\right)\big|_{s=\lambda} = 0$, where $\Phi\left(s\right) = se\left(s\right)$. Parameter h determines how fast the diffusivity e changes, and λ controls smoothness of function u.

The new regularizer $\int_\Omega |\nabla^2 u|$ with diffusivity e in (2.3) has shown good performance in image decomposition producing smooth structures while preserving the edge features. Traditionally, the optimisation problem is solved directly by gradient decent flow leading to a fourth order nonlinear PDE which is very complicated to discretize. Even though the semi-implicit finite difference can be imposed, it is still computationally expensive. In addition, if directly deriving the Euler-Lagrange equations of the vector function g used to represent oscillating part of an image in the proposed model, we obtain another two PDEs with respect to g_1 and g_2, which are not analytical. To solve the equations, we need to use semi-implicit fixed-point iteration method to iterate g_1 and g_2 until convergence. Inevitably, this iterative process needs additional computing time. We will address the challenge of computational efficiency in next section.

3   The Split Bregman Algorithm

In Goldstein-Osher [8], fast iterative schemes were proposed and tested for the TV model. It is one of most efficient numerical schemes for solving the L1-based variational models [11,12]. There exist some other fast algorithms. For example, the fast dual projection algorithm was adopted [9,10,13] to efficiently solve the second order texture-extraction models. Moreover, the more recent augment Lagrangian method [14,15] also attracts attention due to its extensive applications to the variational models such as LLT model [7], Euler-elastic model [16,17], mean-curvature model [18], and TV-Stokes model [19]. In this section, we apply split Bregman algorithm to solve the proposed SOVO model. The idea is to first split the original minimization problem into several subproblems by introducing some auxiliary variables, and then solve each subproblem. We first transform the unconstrained minimization problem (2.3) into minimising the following constrained functional

$$E\left(u, g, w, m\right) = \frac{1}{2} \int_{\Omega} \left(f - u - div\left(g\right)\right)^2 + \lambda \int_{\Omega} e\left|w\right| + \gamma \int_{\Omega} \left|m\right| \qquad (3.1)$$

with constraints $w = \nabla^2 u$ and $m = g$. In (3.1), vector function $m = (m_1, m_2) \in \left(R^{M \times N}\right)^2$ is related to function g, and $w = \begin{pmatrix} w_1, w_2 \\ w_3, w_4 \end{pmatrix} \in \left(R^{M \times N}\right)^4$ is a matrix valued function related to the Hessian of the function u. $|w| = \sqrt{\sum_{1 \le n \le 4} \left(w_n\right)^2}$ stands for the Frobenius norm of the matrix w. The two constraints above can be enforced effectively via the Bregman distance technique. Two Bregman iteration parameters $b_1 = \begin{pmatrix} b_{11}, b_{12} \\ b_{13}, b_{14} \end{pmatrix} \in \left(R^{M \times N}\right)^4$ and $b_2 = (b_{21}, b_{22}) \in \left(R^{M \times N}\right)^2$ are introduced to transform the constrained functional (3.1) into following

$$\begin{aligned} E\left(u, g, w, m; b_1, b_2\right) = &\frac{1}{2} \int_{\Omega} \left(f - u - div(g)\right)^2 \\ &+ \lambda \int_{\Omega} e\left|w\right| + \frac{\theta_1}{2} \int_{\Omega} \left|w - \nabla^2 u - b_1\right|^2 \qquad (3.2) \\ &+ \gamma \int_{\Omega} \left|m\right| + \frac{\theta_2}{2} \int_{\Omega} \left|m - g - b_2\right|^2 \end{aligned}$$

where θ_1 and θ_2 are two positive penalty parameters. It is known that one of saddle points for functional (3.2) will give a minimizer for the constrained minimization problem (3.1). In practice, it is very difficult to directly solve the minimization problem (3.2), so we have developed an alternative optimization method. Specifically, we split (3.2) into four subproblems for u, g, w and m each of which can be solved quickly. The following section will introduce how to solve the four subproblems.

4   Solving the Subproblems

4.1   Minimization of Subproblem with Respect to u

$$u = \operatorname*{argmin}_{u \in R^{M \times N}} \left\{ E\left(u\right) = \frac{1}{2} \int_{\Omega} \left(f - u - div\left(g\right)\right)^2 + \frac{\theta_1}{2} \int_{\Omega} \left|w - \nabla^2 u - b_1\right|^2 \right\}$$

Deriving its Euler-Lagrange equation leads to following fourth order linear PDE

$$u + \theta_1 div^2 \left(\nabla^2 u\right) = f - div\left(g\right) - \theta_1 div^2 \left(b_1 - w\right) \qquad (4.1)$$

div^2 denotes second order divergence operator. By applying discrete Fourier transform to the both sides of equation (4.1), one can obtain

$$\underbrace{\left[1 + 4\theta_1\left(\cos\frac{2\pi s}{N} + \cos\frac{2\pi r}{M} - 2\right)^2\right]}_{\xi} \mathcal{F}(u) = \mathcal{F}(G) \qquad (4.2)$$

where $G = f - div\,(g) - \theta_1 div^2\,(b_1 - w)$. $i \in [1, M]$ and $j \in [1, N]$ are the indexes in discrete time domain. $r \in (1, M)$ and $s \in (1, N)$ are the frequencies in the discrete frequency domain. (4.2) provides us with a closed-form solution of u as

$$u = \Re\left(\mathcal{F}^{-1}\left(\frac{\mathcal{F}(G)}{\xi}\right)\right)$$

$\mathcal{F}$ and $\mathcal{F}^{-1}$ denote the discrete Fourier transform and inverse Fourier transform respectively. $\Re$ is the real part of a complex number. "—" stands for pointwise division of matrices.

4.2   Minimization of Subproblem with Respect to g

$$g = \operatorname*{argmin}_{g \in (R^{M \times N})^2}\left\{E\,(g) = \frac{1}{2}\int_\Omega (f - u - div\,(g))^2 + \frac{\theta_2}{2}\int_\Omega |m - g - b_2|^2\right\}$$

Its Euler-Lagrange equation is given by

$$-div\,(g) + \theta_2 g = \nabla\,(u - f) - \theta_2\,(b - m)$$

By applying discrete Fourier transform to the both sides of equation, we have

$$\begin{pmatrix} a_{11} & a_{12} \\ a_{21} & a_{22} \end{pmatrix}\begin{pmatrix} \mathcal{F}(g_1) \\ \mathcal{F}(g_2) \end{pmatrix} = \begin{pmatrix} \mathcal{F}(h_1) \\ \mathcal{F}(h_2) \end{pmatrix}$$

where the coefficients are

$$a_{11} = \theta_2 - 2\left(\cos\frac{2\pi s}{N} - 1\right)$$
$$a_{12} = -\left(-1 + \cos\frac{2\pi s}{N} + \sqrt{-1}\sin\frac{2\pi s}{N}\right)\left(1 - \cos\frac{2\pi r}{M} + \sqrt{-1}\sin\frac{2\pi r}{M}\right)$$
$$a_{21} = -\left(-1 + \cos\frac{2\pi r}{M} + \sqrt{-1}\sin\frac{2\pi r}{M}\right)\left(1 - \cos\frac{2\pi s}{N} + \sqrt{-1}\sin\frac{2\pi s}{N}\right)$$
$$a_{22} = \theta_2 - 2\left(\cos\frac{2\pi r}{M} - 1\right)$$

with

$$h_1 = \nabla_x\,(u - f) - \theta_2\,(b_{21} - m_1)$$
$$h_2 = \nabla_y\,(u - f) - \theta_2\,(b_{22} - m_2)$$

We have $M \times N$ numbers of 2×2 systems. The determinant of the coefficient matrix $\begin{pmatrix} a_{11}\ a_{12} \\ a_{21}\ a_{22} \end{pmatrix}$ is

$$D = \theta_2^2 - 2\theta_2\left(\cos\frac{2\pi s}{N} + \cos\frac{2\pi r}{M} - 2\right)$$

which is always positive for all discrete frequencies if $\theta_2 > 0$. After the systems of linear equations are solved for each frequency r and s over the discrete frequency domain, we use the discrete inverse Fourier transform to obtain

$$g_1 = \Re \left(\mathcal{F}^{-1} \left(\frac{a_{22}\mathcal{F}(h_1) - a_{12}\mathcal{F}(h_2)}{D} \right) \right)$$

$$g_2 = \Re \left(\mathcal{F}^{-1} \left(\frac{a_{11}\mathcal{F}(h_2) - a_{21}\mathcal{F}(h_1)}{D} \right) \right)$$

4.3   Minimization of Subproblem with Respect to w and m

The minimisation problems of w and m works in a same manner. Both of their solutions are form of analytical generalised soft thresholding equation. Specifically, we solve the following two problems

$$w = \underset{w \in (R^{M \times N})^4}{\operatorname{argmin}} \left\{ E(w) = \lambda \int_\Omega e |w| + \frac{\theta_1}{2} \int_\Omega |w - \nabla^2 u - b_1|^2 \right\}$$

$$m = \underset{m \in (R^{M \times N})^2}{\operatorname{argmin}} \left\{ E(m) = \gamma \int_\Omega |m| \, dx + \frac{\theta_2}{2} \int_\Omega |m - g - b_2|^2 \right\}$$

Their solutions respectively read

$$w = \max \left(\left| \nabla^2 u + b_1 \right| - \frac{\lambda e}{\theta_1}, 0 \right) \frac{\nabla^2 u + b_1}{|\nabla^2 u + b_1|}$$

$$m = \max \left(|g + b_2| - \frac{\gamma}{\theta_2}, 0 \right) \frac{g + b_2}{|g + b_2|}$$

with the convention that $0/0 = 0$. The above equation is known as the analytical soft thresholding equation or shrinkage.

After the minimizers of the four subproblems are found, we can update Bregman iterative parameters in (3.2) as follows

$$b_1 = b_1 + \nabla^2 u - w$$

$$b_2 = b_2 + g - m$$

5   Experimental Results

In this section, we use an example to illustrate the effectiveness of our proposed SOVO model. More experimental results will be performed in the forthcoming paper. Decomposition results on Fig. 1 (b) by the proposed SOVO model, TV (ROF) model [6], FOVO model [4], Chambolle's Inf-convolution model [21] and TGV model proposed by Bredies et al. [20] are shown from Figs. 2 to 5.

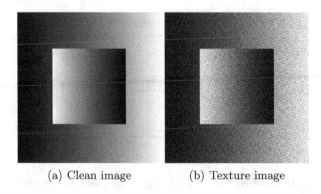

(a) Clean image   (b) Texture image

Fig. 1. Main test images (resolution 510 × 520).

The quality of the decomposed images is validated quantitatively by the SSIM index [22].

In Fig. 2, the decomposed structures and their corresponding SSIM calculated from the reference image Fig. 1 (a) are given. It is clear to see that some texture still remains in the result by TV model, leading to the lowest SSIM. FOVO model outperforms TV model as there is no texture left in its decomposed result. However, the staircase effect appears, which makes the result less appealing. From Fig. 2 (c) and (d). The high order Inf-convolution and TGV models give good and smooth output whilst preserving the edge features. As Inf-convolution slightly blur edges of the decomposed structure, and it does not outperform TGV model. The proposed SOVO model achieves the best visual and quantitative evaluation result. It decomposes all texture while removes the staircase effect and preserves the sharp edges.

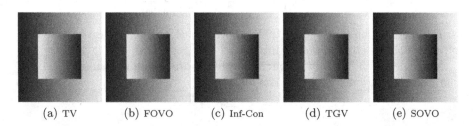

(a) TV   (b) FOVO   (c) Inf-Con   (d) TGV   (e) SOVO

Fig. 2. Decomposed structural components of Fig. 1 (b) by different methods. The resulting SSIMs from left to right are 0.3958, 0.6934, 0.7725, 0.8225, and 0.8551, respectively.

In Fig. 4, the rescaled decomposed textural components by different models are presented. Figure 4 (a) is the pure texture image computed by rescaling the result of Fig. 3 (a) minus Fig. 3 (b). We do not list the SSIM index in the Figure as it is not accurate after rescaling. However, one can observe Fig. 4 (b), (d) and (e) contain more structure of original image Fig. 1 (a) than these by FOVO and SOVO models. Thus, Fig. 4 (e) and (f) are more similar to the Fig. 4 (a). Note

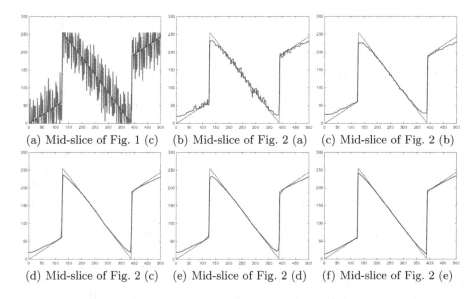

(a) Mid-slice of Fig. 1 (c)   (b) Mid-slice of Fig. 2 (a)   (c) Mid-slice of Fig. 2 (b)

(d) Mid-slice of Fig. 2 (c)   (e) Mid-slice of Fig. 2 (d)   (f) Mid-slice of Fig. 2 (e)

Fig. 3. A middle slice for detailed comparisons.

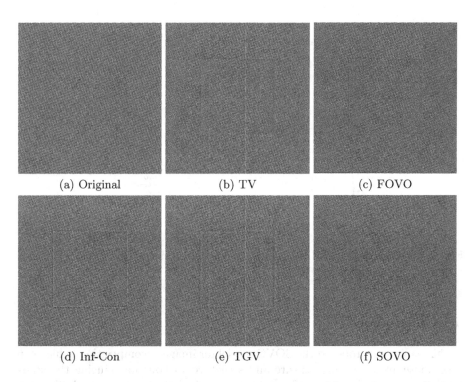

(a) Original   (b) TV   (c) FOVO

(d) Inf-Con   (e) TGV   (f) SOVO

Fig. 4. Decomposed textural components of image Fig. 1 (c) by different methods.

that TV, Inf-convolution and TGV models simply use the L^2 norm to capture the oscillations while the FOVO and SOVO models apply L^1 norm of $|g|$. The latter is more capable of extracting oscillatory pattern from the textural signal.

In Fig. 3, the middle slices of the image Fig. 1 (b) are shown and its corresponding decomposition results are shown in Fig. 2. The conclusions are consistent with Fig. 2. It is more obvious that the proposed model outperforms the other fours in terms of capability of textural removal and smoothness and sharp edges/corners preservation.

To examine the quality of the decomposed structural parts in Fig. 2 as the number of iteration increases, we have plotted the evolution of SSIM values in Fig. 5. In the horizontal axis, instead of the number of iterations we put the absolute CPU time calculated by multiplying the number of iterations and CPU time per iteration. We also designed fast split Bregman algorithm for TV, Inf-convolution, and TGV models for comparison with our algorithm. The fixed point iteration method is applied to the FOVO model. For TV and FOVO models, their corresponding SSIM value increases at the beginning of iterations and then drops when the staircase effect appears. The SSIM values of Inf-convolution, TGV and SOVO models increase dramatically before 5 seconds and then remain stable. The split Bregman is also used for TV, Inf-convolution, TGV and SOVO models, and performed faster than the FOVO model.

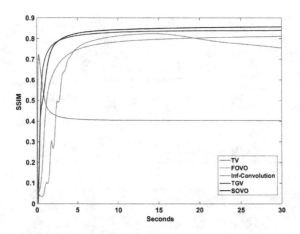

Fig. 5. Evolution of the SSIM index with absolute CPU time for the examples in Fig. 2.

6   Conclusion

In this paper, we proposed the SOVO model for image decomposition, which can decompose an image into texture and structure without introducing the staircase effect. The advantages of our model also include computational efficiency by using the split Bregman algorithm to avoid numerically solving high order

PDEs. We provided the implementation details of the split Bregman algorithm for solving the proposed model efficiently. Extensive experiments demonstrate that the proposed second order model performs better than the existing image decomposition models. Our future work will extend the proposed SOVO model to image denoising and decomposition [23] for real world applications such as medical imaging [24].

Acknowledgments. This work is supported by National Natural Science Foundation of China (No.61305045, No.61303079 and No.61170106), Qingdao science and technology development project (No. 13-1-4-190-jch).

References

1. Wang, G.D., Pan, Z.K., Dong, Q., Zhao, X.M., Duan, J.M.: Unsupervised texture segmentation using active contour model and oscillating information. J. Appl. Math. 2014

2. Duan, J.M., Pan, Z.K., Liu, W.Q., Tai, X.C.: Color texture image inpainting using the non local CTV model. J. Signal Inf. Process. **4**(03), 43 (2013)

3. Duan, J.M., Pan, Z.K., Liu, W.Q., Tai, X.C.: Fast algorithm for color texture image inpainting using the non-local CTV model. J. Global Optim. **62**(4), 853–876 (2015)

4. Vese, L.A., Osher, S.J.: Modeling textures with total variation minimization and oscillating patterns in image processing. J. Sci. Comput. **19**(1–3), 553–572 (2003)

5. Meyer, Y.: Oscillating patterns in image processing and nonlinear evolution equations: the fifteenth Dean Jacqueline B. Lewis memorial lectures. Am. Math. Soc. 22 (2001)

6. Rudin, L.I., Osher, S.J., Fatemi, E.: Nonlinear total variation based noise removal algorithms. Physica D: Nonlinear Phenom. **60**(1), 259–268 (1992)

7. Lysaker, M., Lundervold, A., Tai, X.C.: Noise removal using fourth-order partial differential equation with applications to medical magnetic resonance images in space and time. IEEE Trans. Image Process. **12**(12), 1579–1590 (2003)

8. Goldstein, T., Osher, S.: The split Bregman method for l1-regularized problems. SIAM J. Imaging Sci. **2**(2), 323–343 (2009)

9. Bergounioux, M., Piffet, L.: A second-order model for image denoising. Set-Valued Var. Anal. **18**(3–4), 277–306 (2010)

10. Bergounioux, M., Piffet, L.: A full second order variational model for multiscale texture analysis. Comput. Optim. Appl. **54**(2), 215–237 (2013)

11. Duan, J.M., Pan, Z.K., Ying, X.F., Wei, W.B., Wang, G.D.: Some fast projection methods based on Chan-Vese model for image segmentation. EURASIP J. Image Video Process. **2014**(1), 1–16 (2014)

12. Duan, J.M., Ding, Y.C., Pan, Z.K., Yang, J., Bai, L.: Second order mumford-shah model for image denoising. IEEE International Conference on Image Processing (2015) (In Press)

13. Bergounioux, M.: Second order variational models for image texture analysis. Adv. Imaging Electron Phys. **181**, 35–124 (2014)

14. Tai, X.C., Wu, C.L.: Augmented Lagrangian method, dual methods and split Bregman iteration for ROF model. In: Tai, X.C., Mørken, K., Lysaker, M., Lie, K.A. (eds.) Scale space and variational methods in computer vision. LNCS, vol. 5567, pp. 502–513. Springer, Heidelberg (2009)

15. Wu, C.L., Tai, X.C.: Augmented Lagrangian method, dual methods, and split Bregman iteration for ROF, vectorial tv, and high order models. SIAM J. Imaging Sci. **3**(3), 300–339 (2010)
16. Chan, T.F., Kang, S.H., Shen, J.H.: Euler's elastica and curvature-based inpainting. SIAM J. Appl. Math. **63**, 564–592 (2002)
17. Tai, X.C., Hahn, J., Chung, G.J.: A fast algorithm for Euler's elastica model using augmented lagrangian method. SIAM J. Imaging Sci. **4**(1), 313–344 (2011)
18. Zhu, W., Chan, T.: Image denoising using mean curvature of image surface. SIAM J. Imaging Sci. **5**(1), 1–32 (2012)
19. Hahn, J., Wu, C.L., Tai, X.C.: Augmented Lagrangian method for generalized tv-stokes model. J. Sci. Comput. **50**(2), 235–264 (2012)
20. Chambolle, A., Lions, P.L.: Image recovery via total variation minimization and related problems. Numer. Math. **76**(2), 167–188 (1997)
21. Bredies, K., Kunisch, K., Pock, T.: Total generalized variation. SIAM J. Imaging Sci. **3**(3), 492–526 (2010)
22. Wang, Z., Bovik, A.C., Sheikh, H.R., Simoncelli, E.P.: Image quality assessment: from error visibility to structural similarity. IEEE Trans. Image Process. **13**(4), 600–612 (2004)
23. Duan, J.M., Lu, W.Q., Tench, C., Gottlob, I., Proudlock, F., Bai, L.: Denoising optical coherence tomography using second order total generalized variation decomposition. J. Biomed. Signal Process. Control (2015) (In Press)
24. Duan, J.M., Tench, C., Gottlob, I., Proudlock, F., Bai, L.: Optical coherence tomography image segmentation. IEEE International Conference on Image Processing (2015) (In Press)

Sparse Coding for Symmetric Positive Definite Matrices with Application to Image Set Classification

Jieyi Ren[1] and Xiaojun Wu[2(✉)]

[1] School of Digital Media, Jiangnan University, Wuxi 214122, China
alvisland@gmail.com
[2] School of IOT Engineering, Jiangnan University, Wuxi 214122, China
xiaojun_wu_jnu@163.com

Abstract. Modelling videos or images with Symmetric Positive Definite (SPD) matrices and utilizing the intrinsic geometry of the Riemannian manifold has proven helpful for many computer vision tasks. Inspired by the significant success of sparse coding for vector data, recent researches show great interests in studying sparse coding for SPD matrices. However, the space of SPD matrices is a well-known Riemannian manifold so that existing sparse coding approaches for vector data cannot be directly extended. In this paper, we propose to use the Log-Euclidean Distance on the Riemannian manifold, which naturally derives a Riemannian kernel function to solve the sparse coding problem. The proposed method can be easily applied to image set classification by representing image sets with nonsingular covariance matrices. We compare our method with other sparse coding techniques for SPD matrices and demonstrate its benefits in image set classification on several standard datasets.

Keywords: Spared coding · Covariance matrices · Riemannian manifold · Image set classification

1 Introduction

Symmetric positive definite matrices, in the form of covariance descriptors, are becoming increasingly popular to address classification problems in the field of computer vision [1]. For example, covariance descriptors are used to represent image sets for their robustness to measurement variations in image set classification, which has broad applications like video surveillance, multiple views classification with images and long-term observations classification [2].

As covariance matrices, covariance descriptors are obtained by calculating the second-order statistics of feature vectors at a number of observation points. These feature vectors could be extracted form intensity value, coordinates, gradients and filtered intensity value of image pixels. As studied in several researches [3–5], the SPD matrices lie on non-linear manifolds. Hence, neglecting the intrinsic geometry of such manifolds and applying Euclidean geometry could result in poor performance and other undesirable effects. However, [6, 7] introduced that a Riemannian structure combined with proper Riemannian metrics is more suitable for analyzing SPD matrices.

© Springer International Publishing Switzerland 2015
X. He et al. (Eds.): IScIDE 2015, Part I, LNCS 9242, pp. 637–646, 2015.
DOI: 10.1007/978-3-319-23989-7_64

Sparsity is a very popular concept in signal processing [8] and it indicates that natural signals like images could be efficiently represented with only a few non-zero coefficients over a overcomplete dictionary of basis atoms [9]. Since sparse coding has gained outstanding achievement for vector data, it is worth looking forward to similar success when applying it to SPD matrices. Nevertheless, extending sparse coding to SPD matrices is nontrivial as computations on Riemannian manifolds (like similarities or distances) involve non-linear operators. As a consequence, the solution of sparse coefficients for SPD matrices turns into a complex optimization problem. Generally speaking, there are two types of strategies to address the problem of sparse coding for SPD matrices. First, introduce proper linear decomposition method for SPD matrices and use appropriate statistical measurements to definite the reconstruction error [10, 11]. Although this strategy is practicable, it neglects the Riemannian geometry. Second, map SPD matrices into a locally flat Euclidean space or use kernel function to embed SPD matrices into a Reproducing Kernel Hilbert Space (RKHS) [12, 13]. This strategy usually suffers significantly when the size of dictionary is large. In this paper, we propose an approach to perform sparse coding for SPD matrices by mapping points on the Riemannian manifold to a Euclidean space via the Log-Euclidean Distance (LED) [14]. In contrast to prior works, we use the intrinsic Riemannian metric to derive a simple and nonparametric Riemannian kernel function, which is computationally efficient and preserves the manifold geometry sufficiently. We compare the proposed method with other state-of-the-art sparse coding techniques for SPD matrices and apply it to image set classification. The experiments on standard datasets demonstrate the efficiency of our method.

2 Preliminaries

In this section, we review the properties of Riemannian manifold and LED metric, including the derived Riemannian kernel.

2.1 Riemannian Manifold and Metric

A SPD matrix is a symmetric matrix with the property that all its eigenvalues are positive. As mentioned above, the space of $d \times d$ SPD matrices (denoted by Sym_d^+) is not a Euclidean space but a Riemannian manifold. For a point X on the manifold, its tangent space T is consisted by all the tangent vectors at that point. Each point on the Riemannian manifold has a well-defined continuous collection of scalar products defined on its tangent space and is endowed with an associated Riemannian metric [15]. A Riemannian metric makes it possible to define various geometric notions on a Riemannian manifold, such as angles, distances, etc. As the manifold is curved, the distances specify the length of the shortest curve that connects the points, also known as geodesics.

There are two major operations defined on the Riemannian manifold: the exponential map $exp_X : T \to Sym_d^+$ and the logarithmic map $log_X : Sym_d^+ \to T$, where $X \in Sym_d^+$ and $log(X) = \exp^{-1}(X)$. The exponential operator map a tangent vector v on the tangent space (depend on the point X) to the Riemannian manifold and the logarithmic operator is just the opposite. LED is a popular intrinsic metric for Riemannian manifold and it results in classical Euclidean computations as:

$$\delta_{\text{LED}}(X, Y) = \|\log(X) - \log(Y)\|_F, \tag{1}$$

where $X, Y \in Sym_d^+$ and F denotes the matrix Frobenius form. For a SPD matrix X, its eigen-decomposition is given by $X = U \Sigma U^T$ and the logarithm is defined as

$$\log(X) = U \log(\Sigma) U^T, \tag{2}$$

where $\log(\Sigma)$ is the diagonal matrix of the eigenvalue logarithms.

2.2 LED Kernel

The LED metric can be considered as mapping points on the Riemannian to a Euclidean space through logarithmic map. A conceptual illustration of this mapping is shown in Fig. 1. The tangent space here T_I is at the point of the identity matrix I and it is spanned by symmetric matrices. By computing the inner product in the tangent space T_I, a Riemannian kernel function on the Riemannian manifold $\mathcal{M}$ can be derived:

$$k_{\text{LED}}(X, Y) = \text{tr}[\log(X) \cdot \log(Y)]. \tag{3}$$

From [2] we know that the LED kernel satisfies the conditions of Mercer's theorem. Unlike other traditional kernel functions in [13, 16], our kernel function is simple and nonparametric. The LED kernel allows us to apply kernel sparse coding method to SPD matrices by taking the kernel function in Eq. (3) as input.

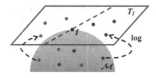

Fig. 1. Conceptual illustration of the LED metric

3 Sparse Coding

Sparse coding has been successfully applied to many computer vision tasks, e.g., image denoising, image decomposition and face recognition. In this section, we introduce our sparse coding method based on LED kernel.

3.1 Sparse Representation

Given a dictionary $\mathbb{D}$ with n atoms, $\mathbb{D} = [D_1, D_2, \ldots, D_n]$, $D_i \in \mathbb{R}^d$ denotes the i-th atom, $n > d$. With a sample $s \in \mathbb{R}^d$, sparse coding for vector data seeks to solve the optimization problem as:

$$\min_x \|s - \mathbb{D}x\|_2^2 + \mathrm{Sp}(x), \tag{4}$$

where $x \in \mathbb{R}^n$ is the sparse coefficient vector and $\mathrm{Sp}(x)$ is a sparsity inducing function. Since the original choose of $\mathrm{Sp}(x)$ (ℓ^0-minimization) is NP-hard, the most common form of Eq. (4) is solved via ℓ^1-minimization:

$$\min_x \|s - \mathbb{D}x\|_2^2 + \lambda \|x\|_1. \tag{5}$$

Sparse coding for SPD matrices could be generally considered as representing a sample point on the Riemannian manifold with a sparse "combination" of dictionary atoms (these atoms are also points on the same manifold). From [17] we know that the objective function of sparse coding on Riemannian manifold $\mathcal{M}$ can be formulated by adopting certain Riemannian metric $\|\cdot\|_{\mathcal{M}}$ as:←

$$\min_x \left\| S \ominus \tilde{\mathbb{D}} \otimes x \right\|_{\mathcal{M}}^2 + \lambda \|x\|_1. \tag{6}$$

where $\tilde{\mathbb{D}}\ominus$ is a Riemannian dictionary with n atoms and $S \in \mathcal{M}$ is a sample point. The operators $\ominus$ and $\otimes$ are subtraction and multiplication defined on the manifold. Instead of solving Eq. (6) directly, we rewrite it with a mapping function $\phi : \mathcal{M} \to \mathcal{H}$ as:

$$\min_x \left\| \phi(S) - \tilde{\mathbb{D}}_\phi x \right\|_2^2 + \lambda \|x\|_1. \tag{7}$$

To solve Eq. (7) efficiently, we use kernel sparse coding method in [17]. The original algorithm can be naturally extended to solve the kernel sparse coding problem by replacing inner products with kernel matrices, which are calculated from the kernel function in Eq. (3).

3.2 Classification Based on Sparse Representation

Given a new sample point $S \in Sym_d^+$, we first compute its sparse coefficient x over dictionary $\tilde{\mathbb{D}}\ominus$ with n atoms via Eq. (7). For all c classes in $\tilde{\mathbb{D}}\ominus$, let $\delta_i : \mathbb{R}^n \to \mathbb{R}^n$ be the characteristic function that selects the coefficients associated with the i-th class. Thus, $\delta_i(x)$ is a new sparse coefficient whose only nonzero entries are the entries in x that are associated with the i-th class. We then classify by using the Nearest-Neighbor (NN) classifier to the object class that minimizes the residual as:

$$r_i(S) = \left\| \phi(S) - \tilde{\mathbb{D}}_\phi \delta_i(x) \right\|^2 \qquad (8)$$

4 Experiments and Results

There are two categories of experiments in this section. First, we compare the performance of the proposed sparse coding method with previous methods on two classification tasks: face recognition and texture classification. In these two tasks, every sample in the training set is used to calculate the covariance descriptor as an atom in the dictionary. Second, the proposed method is applied to image set classification and its performance is evaluated with other popular image set classification methods and each probe image set is used to calculate the covariance descriptor as an atom in the dictionary. The classification approach of our method in all experiments is the same as stated in Sect. 3.2 and the original presettings of all comparison methods are strictly followed.

4.1 Experiments for Sparse Coding

In order to compare the classification performance of our method with the recent researches that focus on sparse coding for SPD matrices, the FERET dataset [18] and the Brodatz database [19] are adopted here for face recognition and texture classification respectively.

FERET: The FERET is a standard dataset used for facial recognition system evaluation. The 'b' subset of FERET dataset was selected, which includes 1400 face images from 200 persons, each person has 7 images. The face images were cropped and resized to 80 × 80. Samples from FERET dataset are shown in Fig. 2.

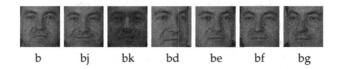

| b | bj | bk | bd | be | bf | bg |

Fig. 2. Face images in FERET dataset

Frontal face images with expression and illumination changes, named 'b', 'bj', 'bk', were used as training set. Non-frontal face images with different pose angle, named 'bd', 'be', 'bf', 'bg', were used as testing set. To calculate the covariance descriptor, the features consist of intensity value, x and y coordinates, filtered intensity values via Gabor filters with 5 orientations and 8 scales for each image are used and resulted in a 43 × 43 SPD matrices.

The recognition results of proposed method, Gabor feature based sparse representation (GSR) [20], Riemannian sparse representation (RSR) [16] and Log-Euclidean kernels sparse representation (LogEKSR) [13] are shown in Table 1. The proposed method outperforms GSR and RSR. The performance of LogEKSR with Gaussian kernel is a little better but our method is also comparable.

Table 1. Results on the FERET Dataset

	GSR	RSR	LogEKSR	Proposed
'bd'	0.77	0.795	**0.92**	0.895
'be'	0.93	0.965	0.99	**0.995**
'bf'	0.97	0.97	**1**	0.995
'bg'	0.79	0.86	**0.94**	0.915
Average	0.865	0.8975	**0.9625**	0.95

Brodatz Texture: Brodatz is a popular dataset used for texture classification and samples are shown in Fig. 3. There are 110 gray scale texture images and we use the first 10 of them. Each image was downscaled to 500×500 and then split into 100 sub images of 50×50. For each sub image, we use features of x and y coordinates, intensity value and gradients in x and y directions respectively. A 5×5 covariance descriptor was generated in a sub image. For each class, 10 sub images were randomly selected for training and other 90 were used for testing. Ten-fold cross validation experiments were conducted to obtain average results.

Fig. 3. Texture images in Brodatz dataset

The average classification results of proposed method, RSR and LogEKSR are shown in Table 2. We cannot find much difference in the classification rate between our method and LogEKSR, while the performance of RSR is a bit lower.

4.2 Experiments for Image Set Classification

Covariance descriptors are considered as a novel modelling tool for image set classification and it has shown promising results recently. We apply the proposed method to

Table 2. Results on the Brodatz texture

Method	Accuracy
RSR	0.836
LogEKSR	**0.903**
Proposed	0.887

image set classification and test its effectiveness on two standard datasets: ETH-80 [20] and Honda/UCSD [21]. Samples of these two datasets are shown in Fig. 4. There comparison methods include: Affine Hull based Image Set Distance (AHISD) [22], Convex Hull based Image Set Distance (CHISD) [22], Sparse Approximated Nearest Points (SANP) [23], Sparse Approximated Nearest Subspaces (SANS) [24] and Covariance Discriminative Learning (CDL) [2].

Fig. 4. Images in the ETH-80 and Honda/UCSD datasets

ETH-80: ETH-80 dataset contains images of 8 object categories including apples, pears, tomatoes, cows, dogs, horses, cups, and cars. Each category has 10 objects (e.g. different cars) and each object has 41 images of different views.

As in [24], we resized the images to 32×32 and treated each object as an individual image set. The covariance descriptor was generated from original image intensity values. For each category, only one image set was used for training and the rest 9 for testing. Ten-fold cross validation experiments were conducted to obtain average results and they are presented in Fig. 5. The proposed method achieved the highest classification rate at 0.781 and dramatically outperformed other methods, except for SANS.

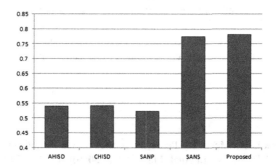

Fig. 5. Results on ETH-80 dataset

Honda/UCSD: Honda/UCSD is a standard video database for evaluating face tracking and recognition algorithms. It contains 59 video sequences of 20 different persons.

Each video contains approximately 300 to 500 frames covering large variations in head pose and facial expression.

A cascaded human face detector [25] was used to detect face in every frame of the videos automatically, and then each face images were cropped and converted to gray scale of 20 × 20. Histogram equalization was used to eliminate lighting effects. Each video generated one image set and the original image intensity values were used to calculate the covariance descriptor. We selected one image set as the gallery and the rest for probes for each person. The single gallery set was randomly divided into two non-overlapping subsets and each subset was used as a full image set. The final results in Tab. 3 were averaged over 20 trials (randomly selected gallery/probe image sets combinations). We see that the proposed method obtains the highest accuracy (Table 3).

Table 3. Results on the Honda/UCSD dataset.

Method	Accuracy
AHISD	0.878
CHISD	0.902
SANP	0.924
CDL	0.971
Proposed	**0.982**

5 Conclusion

In this paper, we proposed a sparse coding method for SPD matrices. In contrast to other approaches, we use a popular Riemannian metric, namely Log-Euclidean Distance to derive a Riemannian kernel and solve the sparse coding problem. Our experiments demonstrated that proposed method is a competitive scheme compared to other sparse coding methods for SPD matrices. Furthermore, we applied our method to image set classification and its superiority to state-of-the arts was confirmed.

Acknowledgments. This work was supported in part by the project of NSFC (No. 61373055) and the Research Project on Surveying and Mapping of Jiangsu Province (No. JSCHKY201109).

References

1. Tuzel, O., Porikli, F., Meer, P.: Pedestrian detection via classification on Riemannian manifolds. IEEE Trans. Pattern Anal. Mach. Intell. **30**(10), 1713–1727 (2008)
2. Wang, R., Guo, H., Davis, L.S., Dai, Q.: Covariance discriminative learning: a natural and efficient approach to image set classification. In: IEEE Conference on Computer Vision and Pattern Recognition, pp. 2496–2503. IEEE (2012)
3. Pennec, X., Fillard, P., Ayache, N.: A Riemannian framework for tensor computing. Int. J. Comput. Vis. **66**(1), 41–66 (2006)

4. Caseiro, R., Martins, P., Henriques, J.F., Batista, J.: A nonparametric Riemannian framework on tensor field with application to foreground segmentation. Pattern Recogn. **45**(11), 3997–4017 (2012)
5. Xie, Y., Vemuri, B.C., Ho, J.: Statistical analysis of tensor fields. In: Jiang, T., Navab, N., Pluim, J.P.W., Viergever, M.A. (eds.) MICCAI 2010, Part I. LNCS, vol. 6361, pp. 682–689. Springer, Heidelberg (2010)
6. Pennec, X.: Intrinsic statistics on Riemannian manifolds: Basic tools for geometric measurements. J. Math. Imaging Vis. **25**(1), 127–154 (2006)
7. Jayasumana, S., Hartley, R., Salzmann, M., Li, H., Harandi, M.: Kernel methods on the Riemannian manifold of symmetric positive definite matrices. In: IEEE Conference on Computer Vision and Pattern Recognition, pp. 73–80. IEEE (2013)
8. Olshausen, B.A.: Emergence of simple-cell receptive field properties by learning a sparse code for natural images. Nature **381**(6583), 607–609 (1996)
9. Wright, J., Yang, A.Y., Ganesh, A., Sastry, S.S., Ma, Y.: Robust face recognition via sparse representation. IEEE Trans. Pattern Anal. Mach. Intell. **31**(2), 210–227 (2009)
10. Sivalingam, R., Boley, D., Morellas, V., Papanikolopoulos, N.: Tensor sparse coding for region covariances. In: Daniilidis, K., Maragos, P., Paragios, N. (eds.) ECCV 2010, Part IV. LNCS, vol. 6314, pp. 722–735. Springer, Heidelberg (2010)
11. Sra, S., Cherian, A.: Generalized dictionary learning for symmetric positive definite matrices with application to nearest neighbor retrieval. In: Gunopulos, D., Hofmann, T., Malerba, D., Vazirgiannis, M. (eds.) ECML PKDD 2011, Part III. LNCS, vol. 6913, pp. 318–332. Springer, Heidelberg (2011)
12. Guo, K., Ishwar, P., Konrad, J.: Action recognition using sparse representation on covariance manifolds of optical flow. In: IEEE Conference on Advanced Video and Signal Based Surveillance, pp. 188–195. IEEE (2010)
13. Li, P., Wang, Q., Zuo, W., Zhang, L.: Log-Euclidean Kernels for sparse representation and dictionary learning. In: IEEE International Conference on Computer Vision, pp. 1601–1608. IEEE (2013)
14. Arsigny, V., Fillard, P., Pennec, X., Ayache, N.: Geometric means in a novel vector space structure on symmetric positive-definite matrices. SIAM J. Matrix Anal. Appl. **29**, 328–347 (2007)
15. Bhatia, R.: Positive Definite Matrices. Princeton University Press, Princeton (2007)
16. Harandi, M.T., Sanderson, C., Hartley, R., Lovell, B.C.: Sparse coding and dictionary learning for symmetric positive definite matrices: a Kernel approach. In: Fitzgibbon, A., Lazebnik, S., Perona, P., Sato, Y., Schmid, C. (eds.) ECCV 2012, Part II. LNCS, vol. 7573, pp. 216–229. Springer, Heidelberg (2012)
17. Li, Y., Ngom, A.: Sparse Representation Approaches for the Classification of High-Dimensional Biological Data. In: IEEE International Conference on Bioinformatics and Biomedicine, pp. 306–311. IEEE (2012)
18. Phillips, P.J., Moon, H., Rizvi, S.A., Rauss, P.J.: The feret evaluation methodology for face-recognition algorithms. IEEE Trans. Pattern Anal. Mach. Intell. **22**(10), 1090–1104 (2000)
19. Randen, T., Husoy, J.H.: Filtering for texture classification: a comparative study. IEEE Trans. Pattern Anal. Mach. Intell. **21**(4), 291–310 (1999)
20. Leibe, B., Schiele, B.: Analyzing appearance and contour based methods for object categorization. In: IEEE Conference on Computer Vision and Pattern Recognition, pp. II-409-15. IEEE (2003)
21. Lee, K.C., Ho, J., Yang, M.H., Kriegman, D.: Video-based face recognition using probabilistic appearance manifolds. In: IEEE Conference on Computer Vision and Pattern Recognition, pp. 313–320. IEEE (2003)

22. Cevikalp, H., Triggs, B.: Face recognition based on image sets. In: IEEE Conference on Computer Vision and Pattern Recognition, pp. 2567–2573. IEEE (2010)
23. Hu, Y., Mian, A.S., Owens, R.: Sparse approximated nearest points for image set classification. In: IEEE Conference on Computer Vision and Pattern Recognition, pp. 121–128. IEEE (2011)
24. Chen, S., Sanderson, C., Harandi, M.T., Lovell, B.C.: Improved image set classification via joint sparse approximated nearest subspaces. In: IEEE Conference on Computer Vision and Pattern Recognition, pp. 452–459. IEEE (2013)
25. Viola, P., Jones, M.J.: Robust real-time face detection. Int. J. Comput. Vis. **57**, 137–154 (2004)

Super-Resolution of Video Using Deformable Patches

Yu Zhu[1]([⊠]), Yanning Zhang[1], and Jinqiu Sun[2]

[1] School of Computer Science, Northwestern Polytechnical University, Xi'an, China
zhuyu1986@mail.nwpu.edu.cn
[2] School of Astronomy, Northwestern Polytechnical University, Xi'an, China

Abstract. We introduce the external examples to address the video super-resolution problem. Instead of using sub-pixel complementary information or self-similar examples, we propose the concept that the high frequency video details could be estimated from the external examples effectively. We prepare the high resolution dictionary by randomly selecting patches from the external high resolution images. For each low resolution patch in current frame, we select its neighboring similar patches within current and adjacent frames as the adaptive neighborhood. Their corresponding HR patches in the dictionary are chosen as the deformation candidates. Then we deform them to fit the LR patch. These patches are weighted combined and the final high resolution image is reconstructed by imposing the global constraints. Experiments show that our method outperforms the state-of-art methods.

1 Introduction

Video *super-resolution* (SR) estimates *high-resolution* (HR) frames from *low-resolution* (LR) sequences. There remain plenty of low-resolution videos made by old low-resolution devices on the various video websites, *e.g.* Youtube. This leads to great demands of super-resolving LR videos so that they can be viewed on the high-definition devices, *e.g.* high definition television (HDTV) and retina display for New iPad.

Traditional video super-resolution methods attempt to perform sub-pixel alignment between image frames, and thus the complementary information can be fused [1–5]. But sub-pixel alignment is easily affected by blur and noise. Moreover, even if we have good alignment there are limits to the enhanced resolution (up-sampling factor no more than 2 [6]). Moreover, accurate sub-pixel motion estimation acts as a challenging problem in video super-resolution especially for the arbitrary motion in reality.

Another promising way to enhance high frequency details is to use external dataset examples [7–14]. For single image, example based methods successfully enhance high frequency detail by estimating, or hallucinating them. The HR details are represented as either the best examples [7,15] or linear combination of the patches [8,9,13,14]. But at present few video super-resolution methods leverage this important information. Faced with the difficulty of motion estimation,

X. He et al. (Eds.): IScIDE 2015, Part I, LNCS 9242, pp. 647–656, 2015.
DOI: 10.1007/978-3-319-23989-7_65

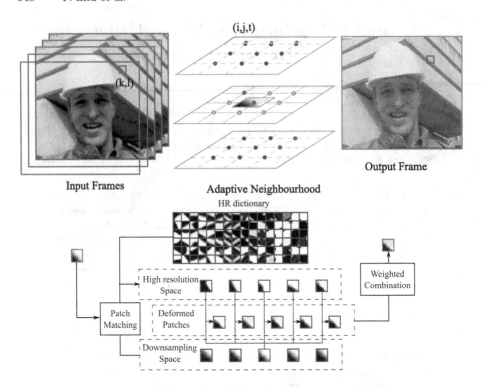

Fig. 1. Flow chart of our method. For a single patch centred at (k, l) in the current LR frame, we consider its neighborhood within the current frame (orange dots) and adjacent frames (blue dots). Similar patches to the centre patch are selected (red dots) to form the adaptive neighborhood $N(k, l)$. Then we find their corresponding HR patches in the HR dictionary within the high resolution space. And we deform them to fit the patch centered at (k, l) in current frame. Finally we combine them and impose global constraints to reconstruct the HR frame (Color figure online).

some researchers use local window or patches, adapting Kernel Regression [16–18] or non-local means [19] to avoid the explicit motion estimation. This resembles the local version of video super-resolution. But they still follow the search-fuse-deblur strategy, using patches from the video sequence itself instead of exploiting external HR patterns. Recently [15] has shown that single image super-resolution performance can be improved by deformable patches. In this paper, we extend this deformable example based method from single image to video.

This paper proposes a novel framework for video super-resolution. The main contribution is that we introduce example patches from the external dataset to improve the high frequency details estimation. The proposed method requires a HR dictionary, which is random selected from external HR dataset. Figure 1 shows the main idea of the proposed method. Our method is implemented in 3 following steps.

(1) Search the most similar neighboring patches of each patch in current frame. Then we obtain the corresponding HR patches by matching the neighboring patches to the HR dictionary. (Section 2)
(2) Deform the selected HR patches to fit the LR patches to perform local registration in the high resolution space. (Section 3).
(3) Combine the deformed patches with weighted coefficients to estimate the HR patch. Then impose global constraints to generate the HR frame. (Section 4).

2   Example-Based Video Super-Resolution

In this section, we elaborate the deformable example patches model for video super-resolution. As summarized above, the external dataset with rich HR details is more helpful than the sub-pixel alignment information, especially when the scale factor is large or the video is much contaminated by blur and noise.

Elad $et\ al.$ [1] proposed a classical super-resolution model. In this model, Y_t denotes t-th frame which is assumed to be generated from the HR scene $\boldsymbol{X}$:

$$Y_t = \boldsymbol{SHM_t X} + \boldsymbol{n} \tag{1}$$

where $\boldsymbol{S}$ is the downsampling operator, $\boldsymbol{H}$ the blurring factor, $\boldsymbol{M}$ is the motion vector and $\boldsymbol{n}$ the noise term. Since motion estimation is always too challenging [16,19,20], deformable patches are employed for local patch registration in high resolution space as shown in Fig. 1. Thus the motion vector M_t is ignored in Eq. (1).

In this case, example based methods create a joint space spanned by many HR/LR pairs extracted from the external dataset [8,9,13,15,17]. Given the external HR images, we get HR dictionary $\boldsymbol{D_h}$ and its corresponding downsampled version $\boldsymbol{SHD_h}$. We call the spanned spaces high resolution space and the downsampling space respectively.

In the downsampling space, the current observed LR frame $\boldsymbol{Y}$ and the downsampled dictionary $\boldsymbol{SHD_h}$ are given. Two issues are addressed.

First, we select the top K most similar patches from the neighborhood of center patch position (k, l) from both the current frame $\boldsymbol{Y}$ and the adjacent frames.

$$N(k,l) = \underset{\substack{N(k,l) \\ \#(N(k,l))=K}}{\operatorname{argmin}} \sum_{(i,j,t)\in N(k,l)} ||\boldsymbol{R}_{k,l}\boldsymbol{FY} - \boldsymbol{R}_{i,j}\boldsymbol{FY_t}||^2 \tag{2}$$

where $\boldsymbol{R}_{k,l}$ is an operator that extracts a patch with fixed size $p \times p$ centring at position (k, l). $\boldsymbol{F}$ is a high-pass filter that contains: $f_1 = [-1, 0, 1]$, $f_2 = f_1^\top$, $f_3 = [1, 0, -2, 0, 1]$ and $f_4 = f_3^\top$. Thus $N(k, l)$ links the adjacent frames, providing more candidates for HR details estimation. So we choose the patches with similar appearances. But they may differ and correspond to different HR dictionary elements. Thus the complementary information can be found in the high resolution space.

The second issue is that by minimizing energy function Eq. (3), we find the deformation HR candidates for each neighboring patch in $N(k, l)$ *i.e.* the indicator $\boldsymbol{\delta}$.

$$E_l = \frac{1}{2} \sum_{(k,l) \in \Omega} \sum_{(i,j,t) \in N(k,l)} ||R_{k,l} FY - \delta_{k,l,i,j,t} FSHD_h)||^2 \tag{3}$$

where Ω is the indices set for each frame. $\boldsymbol{\delta}$ indicates which element should be chosen from D_h for the LR patch located by (k, l, i, j, t).

In the high resolution space, we deform the HR candidates and perform linear combination to reconstruct the HR patch located in (k, l):

$$E_h = \frac{1}{2} \sum_{(k,l) \in \Omega} ||R_{k,l} X - \sum_{(i,j,t) \in N(k,l)} \omega_{k,l,i,j,t} \phi(\delta_{k,l,i,j,t} D_h)||^2 \tag{4}$$

where ϕ is the deformation operator that deforms the chosen HR dictionary element to fit the input. We give ϕ in Sect. 3. Note that Eqs. (3) and (4) share the same $\boldsymbol{\delta}$, linking the LR/HR space. Previous methods for single image use weighted combination [8,12,21] or sparse linear combination [9,14]. Here we use single deformable patch indicated by $\boldsymbol{\delta}$. $\omega_{k,l}$ is the weights when combining the deformed HR patch in the valid neighborhood $N(k, l)$.

3   Deformable Example Patches

In this section, we use deformable patches proposed by Zhu *et al.* [15] to model the deformation operator in Eq. (4). Following the deformation idea, we deform the candidates in the high resolution dictionary to fit the input low resolution patches. Thus the challenging motion estimation problem is avoided. Using Eqs. (3) and (4) we choose the basic candidate patch $B_h = \boldsymbol{\delta} D_h$, the deformed HR patch as follow:

$$\phi(B_h) = \alpha \phi_l(B_h) + \beta \tag{5}$$

Here local warp $\phi_l(B_h)$ and the intensity transformation by contrast α and mean value β are considered. They are modelled separately.

The local warp function $\phi_l(B_h)$ is modelled along the horizontal direction u and vertical direction v separately. Under the assumption of slow deformation filed, ϕ_l has the following form via first order Taylor expansion:

$$\phi_l(B_h) = B_h(x + u, y + v)$$
$$\approx B_h + B_{hx} \circ u + B_{hy} \circ v \tag{6}$$

where $\circ$ denotes point-wise multiplication. B_{hx} and B_{hy} are the derivatives of B_h along the x and y dimensions respectively. Note that all the patches and u, v are their vectorized version here and later.

Taking the degradation Eq. (1) into account, The energy function can be modeled as the error term plus prior term:

$$E(u, v) = ||SH\phi_l(B_h) - P_l||^2 + \lambda(||\nabla u||_2^2 + ||\nabla v||_2^2) \tag{7}$$

P_l is the LR patch located in k, l in the reference frame. Note that P_l and B_h are both normalized beforehand. The error term follows the basic constraint that the deformed and degraded HR patch should be consistent with the input LR patch. Based on the premise of Taylor expansion, the first order smoothness deformation prior is considered to regularize the deformation filed. λ is a balance parameter. The detailed solution on u and v can be found in [15].

Another issue is the estimation of the contrast α and mean value β. Thus the locally deformed patch can be scale to recover appropriate intensity. By minimizing the error between the degraded version of HR patch and input LR patch, α and β can be obtained as:

$$(\alpha, \beta) = \underset{\alpha, \beta}{\operatorname{argmin}} ||P_l - \alpha S H \phi_l(B_h) - \beta||^2 \tag{8}$$

The least square estimation is easily got as $[\alpha, \beta]^\top = (A^\top A) A^\top P_l$, with $A = [SH\phi_l(B_h) \ 1]$. 1 denotes the all-1 column vector with the same dimension of the degraded version of $\phi_l(B_h)$. Then, via Eq. (5) each deformed HR patch in $N(k, l)$ is estimated.

4 Frame Reconstruction via Patches

This section describes how to reconstruct the final HR image frames using these HR deformed patches within $N(k, l)$.

For each deformed HR patch $B_{k,l,i,j,t}^d = \phi(\delta_{k,l,i,j,t} D_h)$ within $N(k, l)$, we take a weighted combination pixel by pixel, then the s-th pixel of the reconstructed patch has the form:

$$P_{k,l}^s = \sum_{(i,j,t) \in N(k,l)} \omega_{k,l,i,j,t}^s B_{k,l,i,j,t}^{ds} \tag{9}$$

where $s = \{1, ..., p \times p\}$ indexes each pixels within the patch. Thus different weights are assigned for each pixel. The weights have the form $\omega_{k,l,i,j,t}^s = \frac{1}{Z} \exp(-(B_{k,l,i,j,t}^{ds} - \mu_s)^2 / 2\sigma_s^2)$ with Z the normalization factor. μ_s and σ_s^2 are the mean value and variance of all the K pixels in s-th position of each deformed patch.

Equation (4) models the relationship between the image and extracted patches. Let $\partial E_h / \partial X = 0$, we get the following HR solution:

$$X = \left[\sum_{(k,l) \in \Omega} R_{k,l}^\top R_{k,l} \right]^{-1} \left[\sum_{(k,l) \in \Omega} R_{k,l}^\top P_{k,l} \right] \tag{10}$$

where $R_{k,l}^\top$ denotes the inverse operation of $R_{k,l}$ that maps the patch to the position (k, l) in the reconstructed frame. $\sum_{(k,l) \in \Omega} R_{k,l}^\top R_{k,l}$ counts how many patches cover the same area. It depends on the patch size and overlap size. Obviously, Eq. 10 average all the overlapped patches.

Since all of the above models work on patch to provide only local constraint. The reconstructed HR image X is not necessarily satisfy the degradation model Eq. (1), so we impose the global constraint via back projection method [22] as the work [14, 15, 23] does.

5   Experimental Results

In this section, we evaluate the performance of our model on two real video with general motions: Foreman and Miss America. The competitors are three classical video super-resolution methods including Generalized Non-local means (GNLM [19]), 3D Iterative Steering Kernel Regression (3DISKR [16]) and Non-local Kernel Regression (NLKR [17]). We also give the result of the Deformable Patch for Single image Super-resolution (DPSSR [15]) on single frame, suggesting that extending the method to video is sensible.

5.1   Dictionary and Experiments Setting

It is common to prepare the dictionary by randomly selecting an enormous number of patches as in [7, 9, 13–15]. In this case, edge and corner structures are more invariant to the scale change than the dense texture details, thus of less ambiguity. In view of this, we use a logo set consisting of 14 logo images to make the dictionary cover more edge patterns. Because the logo set contains many simple structure, the patches extracted from the data is fairly redundant. We evaluate every two patches by normalized correlation (NC). For two vectorized patches a and b, $NC(a, b) = (a - \bar{a})^\top (b - \bar{b})/\|a\|\|b\|$. If $NC > 0.8$, we consider them to be the same patch. Using this NC threshold, we can shrink the dictionary size from the original 100000 to 1980 in our experiments.

In the experiments, the patch size is 7×7 and the regularization constant is $\lambda = 0.01$. For each frame, we consider 2 frames before and after current one, and for each center patch, we choose $K = 5$ candidates. The overlap is set to 6 (which is maximum). The zooming factor is 3 in the experiments. We use PSNR for image quality evaluation. High PSNR stands for good performance.

Our deformable patches based video super-resolution method is slow on CPU but very parallelizable on GPU. We implement our GPU code on Tesla C1060 to speed it up.

5.2   Video Super-Resolution Results

In this section we evaluate the performance on the testing video sequence Foreman and Miss America. The competitors are GNLM [19], 3DISKR [16], NLKR [17] and the single image super-resolution method [15]. Figure 2 gives the PSNR value frame by frame, showing that the proposed method performs generally better than the other methods. The example based methods (proposed method and [15]) achieve good performance, indicating that the external data set really improves the HR details estimation. Even using the example based method on

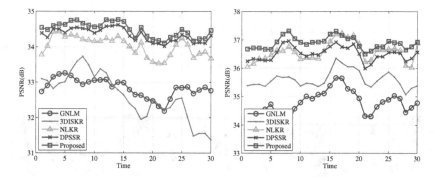

Fig. 2. The PSNR(dB) plots for video super-resolution. Left: Foreman and right: Miss America. The proposed method out performs other methods at almost all the frames

a single frame, comparable results can be achieved with the state-of-art method [17]. When taking the sequence information into account (*i.e.* the method we propose) the performance become much better, showing the proposed method acts as a good bridge before the example based single super-resolution and video super-resolution.

Fig. 3. Video super-resolution for the Foreman sequence (frame 8, zoom factor 3, PSNR in brackets). Left to right: GNLM (PSNR 33.22), 3D-ISKR (PSNR 33.30), NLKR (PSNR34.31), the proposed method (PSNR 34.76).

Figure 3 gives visual quality of the Foreman video. We show the result on 8th frame. From the figure, GNLM [19] generates severe block artifacts if few similar patches can be found, even at the edge areas there are little artifacts. 3DISKR [16] cannot preserve the straight structures well due to the non-robustness of its spatial-temporal kernel. NLKR [17] performs better than both these methods and has less artifact. But still it does not work well on the ambiguous edges with fine scale details, for example the boundary between the teeth and lips. Our method can produce sharper and more appropriate edges than the competitors, which can be seen both from the straight edges and the boundary near the mouth and the teeth.

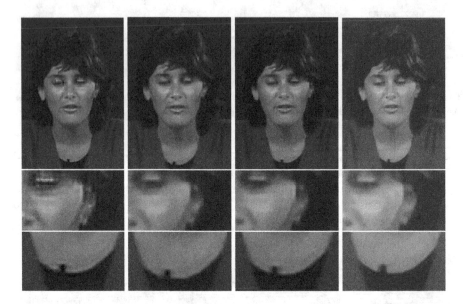

Fig. 4. Video super-resolution for the Miss America sequence (frame 8, zoom factor 3). Left to right: GNLM(PSNR 34.76), 3D-ISKR(PSNR 35.53), NLKR (PSNR 36.76) and the proposed (PSNR 37.32).

Figure 4 shows the comparison on the Miss America video sequence. Again the GNLM [19] generates block effects and ringing around the eyes and ears, even for the straight edges. NLKR [17] achieves similar result to 3DISKR [16], with still some blurring near the edges. By contrast, our method recovers the key structures well including the ear contour, the shape of the microphone, and the edge of the clothes.

We outperform the competitors since they only extract patches from the sequence itself. That is to say, their patches come from a set of downsampled, blurred patches with noise. If there are not enough similar patches nearby, then it is hard to avoid the artifacts as in the GNLM case. But the external example provides a way to hallucinate the finer details using the dataset of known high resolution images. That is the reason that we introduce the external dataset to video super resolution and the results support our decision.

6 Conclusion

In this paper, we introduce example patches from the external dataset to improve the high frequency details estimation in video super-resolution. Traditional sub-pixel complementary information is not sufficient to enhance the LR details accurately especially when the scale factor is large. Therefore we use external HR examples to supply more details. Deformable patches are also employed to make the selected HR dictionary more expressive. The experimental results show that our method outperforms the state-of-art methods. In future work, we will consider the impact of the patterns in the HR dictionary. The dictionaries with different patterns such as animals and flowers are also valuable to explore.

Acknowlegement. The authors would like to thank Prof. Alan Yuille from UCLA for the paper revision. The work is supported by Chinese Scholarship Council and grants NSF of China (61231016, 61301193, 61303123, 61301192), Natural Science Basis research Palan in Shaanxi Province of China (No. 2013JQ8032).

References

1. Elad, M., Feuer, A.: Restoration of a single superresolution image from several blurred, noisy, and undersampled measured images. IEEE TIP **6**(12), 1646–1658 (1997)
2. Elad, M., Feuer, A.: Superresolution restoration of an image sequence: adaptive filtering approach. IEEE TIP **8**(3), 387–395 (1999)
3. Farsiu, S., Robinson, M.D., Elad, M., Milanfar, P.: Fast and robust multiframe super resolution. IEEE TIP **13**(10), 1327–1344 (2004)
4. Park, S.C., Park, M.K., Kang, M.G.: Super-resolution image reconstruction: a technical overview. IEEE Signal Process. Mag. **20**(21–36), 1646–1658 (2003)
5. Liu, C., Sun, D.: A Bayesian approach to adaptive video super resolution. In: CVPR, pp. 209–216 (2011)
6. Lin, Z., Shum, H.Y.: On the fundamental limits of reconstruction-based super-resolution algorithms. In: CVPR, vol. 1, I-1171–I-1176 (2001)
7. Freeman, W., Jones, T., Pasztor, E.: Example-based super-resolution. Comput. Graph. Appl. **22**(2), 56–65 (2002)
8. Chang, H., Yeung, D.Y., Xiong, Y.: Super-resolution through neighbor embedding. In: CVPR, pp. 275–282 (2004)
9. Yang, J., Wright, J., Huang, T.S., Ma, Y.: Image super-resolution via sparse representation. IEEE TIP **19**(11), 2861–2873 (2010)
10. Gao, X., Zhang, K., Tao, D., Li, X.: Image super-resolution with sparse neighbor embedding. IEEE Trans. Image Process. **21**(7), 3194–3205 (2012)
11. Yang, J., Wang, Z., Lin, Z., Cohen, S., Huang, T.: Coupled dictionary training for image super-resolution. IEEE TIP **21**(8), 3467–3478 (2012)
12. Zhang, K., Gao, X., Li, X., Tao, D.: Partially supervised neighbor embedding for example-based image super-resolution. IEEE J. Sel. Top. Signal Process. **5**(2), 230–239 (2011)
13. Yang, J., Lin, Z., Cohen, S.: Fast image super-resolution based on in-place example regression. In: CVPR, pp. 1059–1066 (2013)

14. He, L., Qi, H., Zaretzki, R.: Beta process joint dictionary learning for coupled feature spaces with application to single image super-resolution. In: CVPR, pp. 345–352, June 2013
15. Zhu, Y., Zhang, Y., Yuille., A.L.: Single image super-resolution using deformable patches. In: CVPR, June 2014
16. Takeda, H., Milanfar, P., Protter, M., Elad, M.: Super-resolution without explicit subpixel motion estimation. IEEE TIP **18**(9), 1958–1975 (2009)
17. Zhang, H., Yang, J., Zhang, Y., Huang, T.: Image and video restorations via non-local kernel regression. IEEE Trans. Cybern. **43**(3), 1035–1046 (2013)
18. Zhang, K., Gao, X., Tao, D., Li, X.: Single image super-resolution with non-local means and steering kernel regression. IEEE Trans. Image Process. **21**(11), 4544–4556 (2012)
19. Protter, M., Elad, M., Takeda, H., Milanfar, P.: Generalizing the nonlocal-means to super-resolution reconstruction. IEEE TIP **18**(1), 36–51 (2009)
20. Zhang, H., Yang, J., Zhang, Y., Huang, T.S.: Non-local kernel regression for image and video restoration. In: ECCV, pp. 566–579 (2010)
21. Chan, T.M., Zhang, J., Pu, J., Huang, H.: Neighbor embedding based super-resolution algorithm through edge detection and feature selection. PR Lett. **30**(5), 494–502 (2009)
22. Capel, D.P.: Image Mosaicing and Super-resolution. PhD thesis, University of Oxford (2001)
23. Yang, J., Wright, J., Huang, T.S., Ma, Y.: Image super-resolution as sparse representation of raw image patches. In: CVPR (2008)

Author Index

Printed in the United States
By Bookmasters